The Bench Grafter's Handbook

Principles & Practice

The Bench Grafter's Handbook

Principles & Practice

by

Brian E. Humphrey

CRC Press
Taylor & Francis Group
Boca Raton London New York

CRC Press is an imprint of the
Taylor & Francis Group, an **informa** business

CRC Press
Taylor & Francis Group
6000 Broken Sound Parkway NW, Suite 300
Boca Raton, FL 33487-2742

© 2019 by Taylor & Francis Group, LLC
CRC Press is an imprint of Taylor & Francis Group, an Informa business

No claim to original U.S. Government works

Printed on acid-free paper

International Standard Book Number-13: 978-1-138-04622-1 (Paperback)
International Standard Book Number-13: 978-0-367-22484-4 (Hardback)

Library of Congress Cataloging-in-Publication Data

Names: Humphrey, Brian E., author.
Title: The bench grafter's handbook : principles & practice / author: Brian E. Humphrey.
Description: Boca Raton, FL : CRC Press, Taylor & Francis Group, 2019. | Includes bibliographical references.
Identifiers: LCCN 2019004176| ISBN 9780367224844 (hardback : alk. paper) | ISBN 9781138046221 (pbk. : alk. paper) | ISBN 9781351694933 (epub) | ISBN 9781351694926 (mobi/kindle)
Subjects: LCSH: Grafting. | Plant propagation.
Classification: LCC SB123.65 .H86 2019 | DDC 631.5/3--dc23
LC record available at https://lccn.loc.gov/2019004176

**Visit the Taylor & Francis Web site at
http://www.taylorandfrancis.com**

**and the CRC Press Web site at
http://www.crcpress.com**

You see sweet maid, we marry a gentler scion to the wildest stock and make conceive a bark of baser kind by bud of nobler race: this is an art which does not mend nature, change it rather, but the art itself is nature.

W. Shakespeare. A Winter's Tale. 1623. Act 4 scene 4. Polixenes (King of Bohemia) to Perdita (daughter of the King of Sicily).

Contents

PART TWO The Graft Components – Rootstock and Scion

PART THREE   Achieving the Union

PART FOUR   *Facilities: Structures and Equipment*

PART SIX   Role of Grafting in Conservation

PART SEVEN   Genera Specific Requirements

PART EIGHT   Genera Grafting Guide Tables

APPENDICES

Preface

In recent years, the number of published papers from practising grafters has declined. This may reflect changes in commercial pressures toward simpler, cheaper methods to meet public demand for competitive pricing, at the expense of product range and sophistication. The effect of this is likely to result in less time to acquire and communicate the combination of craftsmanship and knowledge required for success. There seems a worrying possibility that some of these skills will be lost, doubtless to be re-investigated and re-learnt in the years ahead. Consequently, now seems the appropriate time to consolidate and record existing information, hence production of *The Bench Grafter's Handbook – Principles and Practice*.

The importance of grafting to the world economy is significant. Grafted plants occupy thousands of hectares of land in many areas. They are found in private gardens, botanic gardens, arboreta, amenity plantings, orchards, vineyards, plantations of fruit and nut crops and, of course, plant nurseries, where most originate. Grafted seed mother plants can even form the basis of some of the world's forestry plantations.

Historians tell us that the ancient art of grafting may date back four millennia. It most probably evolved as a result of man observing natural grafts, then trying to copy the process. These early attempts would almost certainly have followed methods now known as approach grafting, where during development of the graft union, the two partners remain linked to their mother plants. Subsequently, the advantages of detaching a portion (the scion) of the chosen plant from its origin and grafting it to a recipient plant (the rootstock) encouraged investigation and eventual success in this strategy. Shakespeare's 'bark of baser kind' could then indeed be transformed by 'bud of nobler race'.

In the early stages, results were unlikely to have been good, being hampered by a lack of suitable tools, materials, knowledge of plant anatomy and satisfactory rootstock/scion combinations. Eventually persistent trial and error must have brought about sufficient improvement to establish the viability of the technique. The routes taken to achieve this cover centuries of development and are full of interesting stages, from technically sound and logical, to weird, even mystical. The fascinating history of grafting has been investigated in detail in some other publications*.

Many of the problems limiting success in the past have been resolved while a number remain to this day. To date, grafters have been heavily reliant on the experience and knowledge of their predecessors for the information currently available, and to an extent science has played a smaller part.

Expanding knowledge through experience, technology and science has resulted in improved tools, equipment and facilities, enabling grafting procedures to be upgraded. Some activities have moved away from the open field into areas providing a controlled environment for the graft and more comfortable working conditions for the grafter. It is these latter improvements which have resulted in procedures known as bench grafting.

Because of the advantages provided by greater control, bench grafting has become a major method of producing woody plants. The range of species involved is much larger than in the field, the potential for improved environmental facilities, further developments in mechanisation, possibly involving automation, make it likely that in the future more and more plants will be bench rather than field grafted. Already, grafting of non-woody vegetable plants is highly mechanised and automated. However, at present, and for the foreseeable future, grafting procedures for many woody plants rely heavily on the traditional horticultural abilities, skill and craftsmanship of the individual.

Over past decades, information from numerous sources has significantly increased knowledge of the topic. In the early 1950s through to the early 1990s, grafting procedures and methods were reported upon and discussed with a degree of enthusiasm and detail which have not be matched

---

* Garner (2013) pp.42–44, MacDonald (2014) pp.12–21 and Mudge, Janick, Scofield et al. (2009).

before or since. The impetus for this was very much due to the formation in the US in 1951 of the Plant Propagators Society, now the International Plant Propagators' Society (IPPS). In tandem with this was increased involvement by government agencies, in Holland at the Boskoop Trial Station, in the UK at several centres including East Malling Research Station and doubtless similar initiatives in the US. These were mandated to investigate aspects of nursery production which included, especially in Boskoop, investigations into aspects of bench grafting. In the UK and US, work was and is still focused more on cuttings and aspects of field grafting such as budding. Due to budgetary constraints most of this work in the UK has ceased and the Boskoop Trial ground no longer appears to carry out any of the sort of work described in previous reports.

My working life of over 50 years started in 1954 as a lad in a woody plant nursery where I was fortunate enough to have as a foreman a very good knifesman, who learnt his art at the famous nursery of Waterers, in Bagshot, near London. In the early 1960s, an initial two-year session in one of Hillier Nurseries' propagating units specialising in bench grafting further fostered my interest in the whole procedure.

At that time Hilliers justifiably claimed to commercially produce the world's largest assortment of woody plants for temperate climates. This endeavour was in part achieved by bench grafting an exceptionally comprehensive range of species, probably not equalled by any other institution before or since. Subsequently, my promotion led to responsibilities for the administration and operation of all aspects of the nursery's production. One of my duties was to produce schedules covering bench grafting of well over 1000 species of woody plants. I was also charged with developing new propagation facilities, including those for bench grafting. In the mid-1980s, I joined as head of production, Notcutts Nurseries, at that time with over 500 acres, one of the largest UK nurseries. This afforded me the opportunity to widen my experience in all aspects of nursery work, including bench grafting.

During my years in employment, I was fortunate in being able to take advantage of the UK's Ministry of Agriculture research and experimental work on behalf of the Hardy Ornamental Nursery Stock Industry. Sitting on the industry's R&D committees in various capacities for over 20 years gave an unrivalled opportunity to keep up with latest developments regarding propagation procedures in the UK and abroad. My involvement since the early 1960s with the American-based International Plant Propagators Society gave valuable insights into grafting procedures. It also provided the opportunity for visits to nurseries in North America supplementing those made to Holland, Germany, France, Belgium, Italy and South Africa, all extending my fund of knowledge and experience.

After more than 40 years in the industry, I decided it would be good to spend some years working for myself. With the help and support of my wife, who is also a horticulturist, we set up our own wholesale nursery specialising in bench grafting. For some years I personally grafted annually in excess of 35,000 plants in a range of species and forms. Since retirement I have become particularly interested in grafting endangered species in cultivation and from the wild, as well as to explore further the rootstock/scion combinations of a number of species which have so far proved difficult or impossible due to problems caused by incompatibility.

I have been fortunate to enjoy a career which has afforded opportunities to experience the practical, theoretical, managerial and financial aspects of many forms of hardy ornamental plant propagation, within which bench grafting has always played an important role and been of particular interest.

The Bench Grafter's Handbook has been written with the intention of providing information based not only on my experience but also on the experience of others. I have been anxious to include material on the scientific principles involved in successfully grafting woody plants. Without this basic knowledge, good results are less easily achieved, less easily interpreted and the path ahead for future progress is less clear.

The book is intended for use by a variety of individuals and organisations. These will include nurserymen producing plants for profit, botanic gardens, horticultural societies and public and private arboreta concerned with maintaining and enlarging plant collections. Those delivering horticultural education and training should also derive benefit from contents of the book; young people anxious to pursue a career in horticulture and undergo training to this end will find information which supplements their studies. While it is not expected that academics at the cutting edge of plant physiology and anatomy will obtain facts and information with which they are not already familiar, unsolved questions and problems are posed, which may hopefully serve as a basis for much needed future investigation. Plant conservation professionals ought to consider the use of grafting methods because they can often play a significant role. Amateur horticulturists who have adopted grafting as a propagation technique will find helpful and relevant information, some specifically aimed at their needs.

Brian Humphrey

Acknowledgements

The bibliography bears witness to the significant part played by the authors of the papers presented at the International Plant Propagators Society (IPPS) Proceedings, from 1951 onwards. Without this treasure house of knowledge, the author of any book on woody plant propagation would find the availability of information considerably reduced. The fund of knowledge and experience recorded by a host of highly skilled plant propagators willing to uphold the motto of the society, "To seek and share", has become an irreplaceable asset to any student of plant propagation and has proved an invaluable aid in writing this book.

I have to thank the late Robert Garner, author of *The Grafter's Handbook*, for his interest in my career at an early stage. I well remember him coming to give Kew students one of his electrifying lectures in the late 1950s. I seem to recall we kept him so long with questions he nearly missed the last train from London to his home. Subsequently, he visited the Hillier propagation unit, led by Peter Dummer, where I was then working, and we were able to take advantage of his fund of knowledge. His book has been the mainstay of grafter's information for decades.

I have also been fortunate to have had as a friend the late Bruce Macdonald, the author of a major book on woody plant production, *Practical Woody Plant Propagation for Nursery Growers*. His book stands as a main contributor to our knowledge on the topic and the section on field grafting and bench grafting still serves as a reliable guide.

Aside from these, I have a large number of individuals to thank for their support, assistance and encouragement, always freely given, some over the many years it has taken to produce this book.

Nicholas Dunn, the third-generation nurseryman of F. P. Matthews, Tenbury Wells, Worcestershire, UK, has been a loyal and supportive help to me since the very early days. He has always given encouragement when inevitable setbacks and problems have occurred, and when requested always provided invaluable technical expertise. I am enormously grateful to him.

Topics with a strongly scientific base have presented a particular challenge and I must record my grateful thanks for assistance in this sphere, most particularly to Professor Donald Pigott, previously Director of Cambridge University Botanic Garden and Dr. Keith Loach, past Head of Nurserystock Research at the Glasshouse Crops Research Institute Littlehampton Sussex, UK. My grateful thanks also to Dr. Richard Harrison-Murray of RHM Science, previously of the East Malling Research Station; Dr. Ross Cameron of Sheffield University; Dr. Peter Gasson of Royal Botanic Gardens, Kew; Dr. David Hanke of Cambridge University; Dr. AJ Harris of Oberlin College, Ohio; Dr. Hugh Morris previously of Ulm University, Germany; and Associate Professor Tom Ranney of North Carolina State University. My thanks are also due to Professor Geoffrey Dixon of Green Gene International, UK for help on technical matters and also his valuable practical advice and assistance on acquiring a publisher for the book. For help with an associated topic, Conservation, I am grateful to Dr. Paul Smith of Botanic Gardens International, Kew and Dr. Chuck Cannon of the Centre for Tree Science, Morton Arboretum, Illinois. In this connection and for other assistance related to graft compatibility tests, I must express my grateful thanks to Tony Kirkham Head of Arboretum Royal Botanic Gardens, Kew.

For assistance with the chapter on Grapevine production, my thanks are due to Chris Foss of Plumpton College, Sussex, UK.

Individuals within the nursery industry in the United Kingdom, mainland Europe and the US have been kind enough to spend time responding to requests for information despite the demands of running a nursery or nursery department. Sometimes this included my visiting their establishments. My very grateful thanks are due to the following nurserymen and nurserywomen who have provided loyal and supportive assistance often on a repeated basis over months or years.

In America: W. 'Bill' Barnes – Barnes Horticultural Services, Pennsylvania; Tim Brotzman – Brotzman's Nursery Inc., Ohio; Talon Buchholz – Buchholz and Buchholz Nursery, Oregon;

T. Burchell – Burchell Nursery Inc., California; Bobby Green – Green Nurseries, Alabama; Bill Hendricks – Klyn Nurseries Inc., Ohio; Mark Krautmann – Heritage Nurseries, Oregon; Stacia Lynde – Carlton Nurseries, Oregon; Guy Meacham – J. Frank Schmidt and Son, Oregon; and Brian Upchurch – Highland Creek Nursery, North Carolina.

In Holland: Dick van der Maat – Boskoop; and Ruud van der Wurf – Boskoop. My special thanks to Klaas Verboom – Boskoop, for many valuable suggestions and help with Dutch/English translations.

In Belgium: Carlos Verhelste – Jabbeke, for extending to me his unrivalled knowledge on hot-pipe systems and construction.

In the United Kingdom: Jude Bennison – Bennison Peonies, Lincolnshire; Steen Berg – Yorkshire Plants, Yorkshire; Ivan Dickings – Notcutts Nurseries, Suffolk; Douglas Harris – Penwood Nursery, Hampshire; David Harwood – Sandy Lane Nursery, Suffolk; Chris Lane – Witch Hazel Nursery, Kent; David Millais – Millias Nursery, Surrey; John Richards – John Richards Nurseries, Hereford; Chris Sanders – Bridgemere Nurseries, Cheshire; Derek Spicer – Kilworth Conifers, Leicestershire; Robert Vernon – Bluebell Nursery, Derbyshire; Richard Ward – Golden Grove Nursery, Lincolnshire; Lee Woodcock – Palmstead Nurseries, Kent; Tom Wood – Oakover Nursery, Kent; and last but by no means least, Andrew Wright – F.P. Matthews, Worcestershire.

Most of the photographs are my own, but to the following I must record my thanks and appreciation of their efforts on my behalf for those pictures recorded in the relevant captions: Steen Berg, Tim Brotzman (US), Nicholas Dunn, John Dyter, Dr. Todd Einhorn (US), Maurice Foster, Paul Janssen (Netherlands), Guy Meacham (US), Hans Wahler (Germany) and Xiangyin Wen (Guangdong, China). My special thanks are due to Jim Mountain of The Walled Garden plant centre and nursery in Suffolk, UK for the set of pictures of grafting, knife-work and tying-in techniques. The enormous trouble and care he took to secure these pictures are reflected in the final result.

Apart from fellow nurserymen, many others have provided much valued and appreciated advice in details of plant taxonomy, provision of plant material and technical assistance on various matters from specialist crops to grafting knives, suggestions on layout, text, and health and safety matters; to these I extend my grateful thanks: John Anderson, John Attwood, Jim Ballington, Lawrence Banks, Robert Bloom, Dr. Allen Coombes, Susan and John Cornwell, John Dyter, Maurice Foster, Carol Gurney, Peter Gregory, Andrew Hewson, Dr. Brian Howard, Tom Hudson, David Jewell, Roy Lancaster, Andrew Luke, Dr. Paul Murphy, Lady Diana Rowland, Keith Rushforth, Tony Schilling and Herr S. Schwille.

Randy Brehm, a senior editor with my publisher Taylor and Francis, has provided loyal support from the outset. My sincere thanks go to her for supporting my request for her company to publish the book and her always efficient, helpful response for any queries which inevitably occurred during its preparation. Laura Piedrahita, on the editorial team of the same company, has provided great help in the difficult process of formulating the book for printing and my sincere thanks go to her for her patience and efforts in dealing with my many shortfalls in achieving this objective.

A special debt of gratitude must go to many past colleagues and members of the Hillier team of propagators – Graham Adcock, Peter Moore, Alan Postill and David Roberts. Sadly, heads of the propagation departments, Peter Dummer, Sylvester 'Vic' Palovski and Arthur Prior are no longer with us. As a visiting student from Kew Gardens, my early meetings with Vic Palovski at Hilliers were a revelation and did much to foster my enthusiasm for bench grafting. His deputy at the time, Peter Dummer, subsequently became my first boss. At the time, I considered myself a reasonably competent knifesman. I soon realised this was not the case. As an 'apprentice knifesman' I could not have had a better tutor and mentor than Peter Dummer, with unlimited enthusiasm for the skill and craftsmanship of grafting procedures, unrivalled skill in use of the grafting knife and interest in the art of bench grafting.

As always, family support is a vital ingredient in the production of a technical book of this type. I must thank my daughters Susan and Dora for their constant encouragement and interested support, laced with irreverent humour, throughout the long process.

My son Peter has been a major and crucial contributor to the production of this publication. His hours of work, suggestions on layout, structure and content and invaluable help from his expert knowledge on computer issues have been an indispensable aid. I feel sure that without his input the book would either never have reached completion or be delayed by many years. I am enormously grateful to him for his constant support, interest and assistance at all stages during the whole project.

Since the days we met as students at Kew Gardens, my wife Julie has given unwavering support and help. Her practical assistance, including the tedious process of entering the references, checking the figures, captions and text for inconsistences, has been an invaluable aid throughout the long process of writing the book.

My heartfelt thanks and dedication of this book go to her.

Author

Brian Humphrey commenced work in local nurseries in 1954 followed by two years National Service in the Royal Air Force before taking up a studentship at the Royal Botanic Gardens, Kew. A two-year studentship at Writtle Agricultural College (now Writtle University College) followed, and provided the opportunity to gain the Royal Horticultural Society's National Diploma (now Master of Horticulture) with Honours.

Joining Hillier Nurseries in 1963 as a propagator, he became a member of the Plant Propagator's Society based in the US in 1964 and aided in the formation of the first overseas chapter of the Society, the GB and I Region, in 1968–69. He was the inaugural President of the region in 1969 and was awarded the International Award of Honour by that Society in 1993.

Promotion at Hilliers lead to the position of Production Director and the opportunity to develop various aspects of the nursery, which, at that time, grew commercially the widest range of temperate woody plants in the world. Joining Notcutt's Nurseries in 1986 provided opportunities for further experience. Prior to retirement, over a decade was spent running his own wholesale nursery specialising in grafted plants.

Involvement and subsequent chairmanship of the UK ornamental nursery industry research and development initiatives with the UK Ministry of Agriculture led to the award of an O.B.E. (Order of the British Empire) in 1987. In 2013, his work in the nurserystock industry and in various RHS projects was recognised by the award of the VMH (Victoria Medal of Honour) by the Royal Horticultural Society.

He now lives in retirement in coastal Suffolk, eastern England and together with his wife maintains their garden with nearly seven acres of woodland and meadows of woody plants.

Introduction

ABOUT THIS BOOK

This book is about the fusion of skills and knowledge involved in grafting temperate woody plants. To provide a path through these diverse but closely linked topics, it is divided into eight parts. These are intended to provide a progression through all aspects of the skill, art, science and practicalities of grafting.

Part One – deals with the background knowledge, mechanics and craftsmanship of grafting. Essential information on the anatomy of plant stems and roots are coupled with detailed descriptions of grafting methods. An attempt is made to clarify strategies used and to install some order in the confused naming and categorisation of grafting methods. Grafting knives are discussed, as is the joining together and sealing of graft components. Work station and associated equipment are covered in following chapters.

Part Two – the vital components of grafts, i.e. rootstock and scion, are investigated. Specifications and aspects of provision and production for each are described.

Part Three – describes in detail underlying woody plant anatomy and physiology which promote the formation of a successful graft union. Incompatibility is also discussed, as well as the limitations it imposes and how it may best be predicted, avoided and overcome. After the grafts have been made, they must be subjected to the correct environment to achieve success. This vital aspect of control and amendment of the environment through horticultural procedures is described in detail.

Part Four – reviews and describes choice of major structures, propagation facilities and equipment for grafts and grafting. Future developments regarding artificial growing environments are considered and an assessment is made of their potential for replacing some conventional grafting systems.

Part Five – brings together topics investigated previously. These are used to describe grafting systems, methods and techniques to achieve best results for various plant groups.

Part Six – discusses the actual and potential role of grafting in conservation of endangered woody plants, both in the wild and under cultivation. Not only with regard to the benefits of its use in the conservation effort but also to the practitioners.

Part Seven – gives detailed accounts of specific requirements for genera which can be considered of major importance, or that have complex procedures which require more explanation than that available in Part Eight.

Part Eight – provides a checklist with details of genera which are, have been or could be grafted, with a broad outline of the procedures involved. In total over 200 genera encompassing over 2000 of their species, varieties, and cultivars are covered.

APPENDICES

A number of appendices with text, diagrams and figures supplement information in the main body of the book and provide detailed guidance on various aspects of topics previously discussed.

WHY GRAFT?

The first, obvious question is why is grafting necessary? The essential objective of propagating woody plants may be achieved using a number of horticultural procedures, many more easily accomplished and less costly than grafting.

One important aspect of grafting is its importance as a method of propagating plants which have proved difficult or near impossible to root from cuttings but where 'trueness', i.e. the retention of identical qualities to the parent plant, is essential to maintain desirable features such as control of size, quality of flower or fruit etc. These demands rule out the inherent variability of plants raised by seed.

Alternatives to grafting exist; the non-intensive layering or stooling technique is one but has a number of limitations and may be impractical in many situations. Micropropagation is also able to provide identical clones of the parent and has gained a significant following, especially in the scientific world. It is, however, most suited to producing large numbers of a narrow range of subjects, oil palms being an extreme example. Using this technique, woody plants new to the system often require many months of work establishing pathogen-free initiation material, and further time to investigate and devise protocols for proliferation, development and rooting. Grafting is more practical when a wide range of species and cultivars is required, sometimes as quickly as possible, often in numbers of dozens or hundreds, even thousands, rather than tens of thousands.

Grafting can provide a means of acquiring plants normally produced from cuttings which for various reasons are not available at the correct time. Scion wood may be collected when dormant (even old and senile) throughout a six- to seven-month period from late summer dormancy to just before bud break. Dormant scion wood may be stored for months, permitting easy transportation regardless of distance. Once established, grafted mother plants can subsequently become the source of cuttings procedures for those species which may be reliably rooted. In contrast, cuttings wood for most species needs to be active; timing for collection is often important, and relatively inflexible over a limited time period. Storage for this type of material can only be considered for comparatively short periods; consequently, transportation over long distances is often impractical. Apart from its use in horticulture, these qualities mean that grafting can have wide applications in the sphere of plant conservation.

As grafts are composed of two components, rootstock and scion, one of the most important features of grafting is that each may be exploited to enhance the characteristics of the other and, consequently, the resultant 'composite' plant. For a number of very large and important crops worldwide, particularly fruits and nuts, this has led to investigations into growth control, uniformity of cropping, pest resistance, disease resistance and a number of other highly desirable characteristics. These can be used to improve the performance and productivity of the grafted plant in a range of situations via one or both components. This unique attribute of grafts means they have a particularly important place in methods of plant propagation for crops where specific performance characteristics are required.

BENCH GRAFTING – DEFINITION

Bench grafting is a major category within grafting procedures and is the term used for any grafting not carried out in the field, instead taking place under cover, usually at a bench, often with the grafter seated. The root system (rootstock) is not planted and may be bare-root but, for many species, pot or plug grown rootstocks are preferred. Crucially, after completion, the graft is always placed in a controlled environment.

COMPARISONS BETWEEN FIELD AND BENCH GRAFTING

Field grafts are subjected to an environment which is dependent upon prevailing weather conditions. In some seasons, adverse environment will result in poor takes and can always limit the range of species successfully grafted. For bench grafts, use of additional heat, control of humidity, temperature and light have hugely extended the range of species which are successful. More recent developments, such as cold storage, have provided opportunities for new systems and extended the time period for grafting to the entire year.

The advantage of bench over field grafting to the personnel involved is considerable. There are few nursery operations more calculated to cause chilled hands and feet, combined with back pain, than field grafting in late winter. In the summer, ergonomics is the main issue; many field budders are only able to complete each working day with the liberal use of pain killers to combat back pain. In contrast the bench grafter can and should be comfortably accommodated in a well-lit area, impervious to weather conditions and maintained at suitable working temperatures.

RANGE OF SPECIES

In a cool, temperate climate such as the UK the difficulties of obtaining acceptable results from field grafting have an overriding influence on choice of target species. In such climatic conditions, possibly as few as twelve families of Angiosperms (hardwoods), containing some thirty genera of woody plants, may be considered for grafting or budding. Even within these not all species are suitable, but fortunately some of the commercially important ones do well (see Appendix A – List of Families and Genera Field Grafted for a full list). In northern areas no genera within the Gymnosperms (conifers) are commonly grafted in open conditions.

The control of environmental conditions possible in bench grafting means that the range can be hugely extended; any genera within the hardwoods or conifers can be considered. A list of candidates from species in the temperate zone could contain over 70 families embracing almost ten times as many genera as field grafting. This would range from those that are never or only rarely grafted to those that are usually propagated only by grafting (see Appendix B – List of Families Bench Grafted and with Potential to Graft, for a full list).

SIGNIFICANCE AND ECONOMICS OF BENCH GRAFTING

Grafting has an important role in the production of clonal forms of ornamental trees where a straight, strong stem is required. Grafted plants are also essential for much top fruit and nut crop production to obtain control of tree size, form, uniformity, cropping characteristics and sometimes tolerance of adverse growing conditions/soil pathogens, etc. The grafting of these crops often takes place in the field. The advantages of better results, greater convenience, wider range of target species, increased technical developments and superior working conditions have resulted in significant movement away from the field and towards more advanced bench grafting procedures.

Because it enables the use of rootstocks resistant to root diseases to which the selected scion varieties are susceptible, bench grafting is essential to meet requirements for the vast numbers of Grapevines and other major plantation crops such as citrus, stone fruit, nut crops and kiwi fruit (*Actinidia*).

Due to low output and high labour input, grafting is an expensive procedure compared with most alternatives. From a commercial perspective this disadvantage is offset by higher value of the final product. Competition is normally less severe because competitors do not have the skills necessary to provide a challenge. Nurseries adapted to supply high value grafted material are often more profitable and secure than those involved with large quantities of low value plants.

Part One

Bench Grafting in Practice

1 Grafting in Nature and in the Hands of the Grafter

Grafting is based entirely on natural processes. Common examples of natural grafts are seen in the interwoven stems of Ivy (*Hedera helix*) or interlacing branches of a beech hedge (*Fagus sylvatica*), but given appropriate conditions probably all dicotyledonous woody plants can unite. When in close contact and subjected to increasing pressure as growth precedes, the bark between the branches becomes squeezed and thinned to a point where they are able to bond together to form a natural union (Figure 1.1). Equally important is abrasion of adjoining surfaces, while the bark is still young and thin, caused by pressure and movement due to wind or other forces. To ensure a successful union there is a further crucial requirement: the participants must have close genetic linkage, that is, be compatible. In the wild this will invariably be of the same species; interspecific grafts are theoretically possible but rarely, if ever, occur while intergeneric grafts almost certainly never found.

Natural grafting of roots is very common because they are often in close contact, and soil structure frequently dictates that pressure is applied as incremental growth squeezes the partners together (Figure 1.2). Stability of groups of trees such as beech (*Fagus sylvatica*) in very exposed, windswept situations is often dependent upon root grafting, which can provide massive interlocked anchorage.

A pattern of natural graft development is emerging. These grafts are formed from genetically similar partners that are compatible, growing in close proximity, held fast together under pressure and exposed to mutual damage at the point of contact. Grafters mimic these conditions by cutting the surface of each component, placing the cut faces together and holding them tightly in place with a tie or other means.

Natural grafts are inevitably of the 'approach' type, as the components are still attached to their parent roots, whereas man-made grafts predominantly use detached scions. To ensure the ultimate plant produced is as shapely as possible, the shoot to be attached (scion) to the rooted portion below (rootstock) is aligned along the same axis as closely as possible. If, by mistake, the scion is grafted upside down, a union may be formed but the plant spends the rest of its life trying to re-orientate, resulting in stunted, deformed growth.

The most significant difference between natural and man-made grafts is in the components. In nature, almost without exception, only the same species unite. Under cultivation inter-specific combinations are normal. Grafts between genera can be successful; examples are seen with *Cytisus* scion on *Laburnum* rootstock, Pear (*Pyrus communis*) on Quince (*Cydonia oblonga*) and an entirely unexpected combination of the rare and beautiful *Melliodendron xylocarpum* succeeding on rootstocks of *Pterostyrax hispida* (Figure 1.3). In these examples the combinations involve genera from the same family Papilionaceae, Rosaceae and Styracaceae, respectively. Reports of successful grafts between woody plants from different families have not been found and it seems almost certain that they could not be successful.

VASCULAR CONTINUITY

Woody hardwoods and conifers have a continuous vascular system stretching from root tips to shoot tips. This comprises an inner core of xylem tissue (wood) separated from an outer core of phloem by a thin layer of cells capable of active division, known correctly as the vascular cambium, invariably shortened to the term cambium. It is this that is responsible for the formation of the woody plant's vascular system. The xylem cylinder contains vessels that carry water and dissolved mineral

FIGURE 1.1 Natural graft of a main branch and lateral branch on *Magnolia* 'Elizabeth'.

FIGURE 1.2 *Cercidiphyllum japonicum* with exposed surface roots. Natural grafts are visible as thickened and interlocked crossing roots.

FIGURE 1.3 *Melliodendron xylocarpum* 'Tregye', a rich pink flowered selection. On the left, two seasons after grafting to *Halesia carolina;* on the right to *Pterostyrax hispida*. Both grafts were made on the same day. *Halesia* subsequently failed to produce a compatible union. *Pterostyrax* has successfully united and in some seasons produced scion extension growth of 2 metres or more. After 8 years, no sucker growth is present on grafts of various selections which include white and pale pink flowered forms. It would appear the combination is entirely compatible. A good example of an unlikely but valuable inter-generic graft.

nutrients from the roots. The outer cylinder of phloem has vessels responsible for transporting carbohydrate-rich fluids (elaborated sap) from the leaves. Included in these transportation systems are substances vital for growth and development, the plant auxins (hormones) and cytokinins.

Grafting using detached scions interrupts the continuity of this system, as the two components act separately, in which situation the scion has a limited period to survive. The essential feature of successfully united grafts is re-establishment of the vascular system between the rootstock and scion, once again providing vascular continuity and ensuring survival of the scion.

Re-establishment can occur when a number of requirements are met:

- The respective components are compatible.
- The cambium of each component is well aligned, in contact or in close proximity with the other, and held together under pressure.
- Correct conditions are provided for the formation of callus. This is a mass of simple undifferentiated parenchyma cells, which have potential to develop and infill any voids between the cut faces of the rootstock and scion after they are placed together.
- Respective cambia of rootstock and scion are subsequently able to become linked within the callus to form a completed graft union.

The task of the grafter is to make cuts and placement of rootstock and scion accurately enough to guarantee good cambial alignment, thus providing the best chance of success. At the same time, sufficient pressure must be applied to the interface of the components to promote essential cell and tissue differentiation. This is normally achieved by tying the components tightly together. Finally, suitable environmental conditions must be provided to maintain the scion and promote callus development and subsequent union formation.

The structured vascular system of woody plants described above relates to those formerly known as dicotyledons, because of the diffuse vascular structure of monocotyledon stems and roots, cambial matching is almost impossible and results in much less success with this group. A few genera, mostly in woody Liliaceae, e.g. *Agave, Cordyline, Dracaena* and *Yucca,* enter into a growth phase where the vascular elements become arranged in a more orderly fashion, not dissimilar to that of dicotyledons, and, at this stage, grafting for some may be feasible.

2 Grafting Strategies: Categorisation of Grafting Methods

GRAFTING STRATEGIES

Before beginning a graft there is a need to decide between two main strategies: apical grafting and side grafting. Both utilize scions detached from the mother plant and grafted onto separate rootstocks, a process known as detached scion grafting. The choice of strategy influences grafting methods, rootstock pre-preparation, post-graft treatment and associated activities. These will be discussed here, in subsequent chapters and also as related to specific requirements for different genera, covered in Parts Seven and Eight.

A further strategy is to replace or, more often, supplement manual grafting methods by the use of machines or tools, to be discussed at the end of this chapter, as well as in Chapter 5 and Appendix K: Grafting Machines and Grafting Tools.

Somewhat different from other systems is a much less used, separate strategy with scions still attached to the mother plant, known as approach grafting; it is not considered at this stage, but described in Part Five: Chapter 18: Grafting Systems.

APICAL GRAFTING

In apical grafts the scion replaces the top of the rootstock stem, which is cut down to a chosen height, a process known in grafter's parlance as 'heading-back' or 'heading-down'. Grafts may be close to ground level, where they are said to be bottom-worked, or sometimes higher, where they are known as top-worked. Grafting methods most often produced manually are called by the terms 'splice', 'apical short' or 'long tongue veneer' and 'wedge'. Those produced by machine or grafting tool include splice and variants of inlay such as the 'V' notch and omega.

Apart from side grafting in the field using a single bud (budding), apical grafts constitute the most widely used strategy and huge numbers of apical bench grafts of plantation crops such as grapevines, fruit and nut trees are produced annually, many by grafting machines or grafting tools. To these can be added very large numbers of ornamental species, more commonly produced by manual methods.

SIDE GRAFTING

Side grafting involves totally or partially retaining the top portion of the rootstock above the graft. Scions are placed on the side of the rootstock stem at bottom- or top-worked height. The most commonly used side grafts are side veneers or side wedge.

Side grafts are normally used for more demanding deciduous species, and often for hardwood evergreens and evergreen conifers, most of which respond badly to removal of top growth of the rootstock. Side grafts are of particular importance for summer grafting when total removal of the leafy top can have adverse effects on the rootstock and grafting success.

Many think retention of top growth on leafy and leafless (deciduous) stems may help reduce and control sap flow from the root system. Correctly known as root pressure and to the grafter as bleeding, this can have harmful effects on performance, sometimes resulting in failure. The presence of shoots and leaves above the graft has other advantages in maintaining carbohydrate and protein levels in the tissues, promoting callus development and maintaining health of the rootstock.

APICAL AND SIDE GRAFTING STRATEGIES COMPARED

SPEED

Side grafts are slower to produce because:

- They use taller, less convenient rootstocks.
- They have more demands on tying-in than apical grafts because of retention of rootstock tops. This is particularly so where height of the rootstock prevents the grafter from passing his arm over and around the graft. Here a slower technique must be used (see Chapter 6).
- They are sealed only by brush application, in contrast to many leafless apical grafts where a faster dipping technique is possible.
- They will subsequently need heading-back by removing the rootstock portion above the graft union, where there is always the risk of the wrong shoot being cut off, extinguishing all efforts by a single ill-judged snip!

SPACE REQUIREMENTS

Because the top portion of the rootstock is retained, side grafts usually occupy more space than equivalent apical grafts, estimated to be between 10% and 25%, dependent upon the species and level of rootstock trimming.

GROWTH AND SURVIVAL

Apical grafting, with removal of all competitive shoots, is claimed to produce optimum growth, but in some situations side grafting may have advantages. Experience with summer side grafts of species, such as *Cornus florida* and *C. kousa* cultivars, as well as *Cercis canadensis*, 'Forest Pansy' and most Magnolias, have shown improved over-wintering survival compared with apical grafted equivalents. Scions that are poor, senile or diseased have much higher chances of producing a successful union when using a side rather than apical graft. The beneficial effects of light trimming and allowing 'sucker' development for promoting the growth of field budded 'T' and chip buds have been reported[*], and similar influences can occur in bench grafting.

The term 'sucker' requires further explanation. Grafters invariably use this term for any shoot which arises below the graft union. To be botanically exact, the term basal shoot should be applied to this type of growth and sucker only used for shoots that originate from the root system. However, the term sucker is in wide use by grafters and nurserymen, and for these reasons, despite being botanically inaccurate, it is maintained in this publication as the term for all shoots arising below the union.

CONVENIENCE

With side grafts it is often possible to rest materials used for moisture retention, shading, etc. directly onto the tops of the remaining portion of the rootstocks (Figure 2.1). Apical grafts must have a supporting structure above.

[*] Légaré (2007).

FIGURE 2.1   United side grafts of *Acer palmatum* cultivars conveniently covered by resting a thermal fleece on top of the rootstocks as part of the weaning process.

To start the shaping and training process of subsequent growth from the scion, side grafts may be conveniently tied to the stem of the rootstock before staking later after heading-back (Figure 2.2). There are numerous examples of this in the figures throughout the book. Apical grafts require timely staking and tying-in to produce a straight main stem after grafting, often not convenient in the spring rush of work.

To provide an effective way of maintaining humidity levels in an enclosed environment during the summer, unsealed side grafts can have their leafy tops misted with water (see Chapter 13: Grafting Environment).

More than one side graft may be placed on the same rootstock to become a multi-graft (Figure 2.3). This may be used to conserve rootstocks, quickly bulk up numbers, enhance quality and speed up development to saleable size. The method is further explored in the chapter on grafting systems (Chapter 18).

With failed apical grafts, re-using rootstocks is rarely possible but failed side grafts may often be re-used. There are opportunities to place the scion sufficiently high to permit re-grafting beneath the original should the first fail. The re-graft may be in the same season or, if a further failure occurs, it is often possible to hold the rootstock over for re-use the following year.

Heading-Back: A Specific Requirement of Side Grafts

Removal of the rootstock stem above the successful graft is necessary to complete the process of side grafting. For most deciduous species and some evergreens this normally involves heading-back summer grafts during the following winter/early spring, and winter grafts during the subsequent growing season.

The traditional method was removal just above the graft using grafting knives or disposable blade knives, followed by sealing the cut surface. Modern practice is to leave a modest (5–7mm) amount of wood above the union and use sharp secateurs for the operation. Sealing is not necessary as the natural process of wood compartmentalisation is able to seal off the cut stem (see Figure 11.4: Response to Wounding).

FIGURE 2.2 *Betula maximowicziana* showing the use of the side grafted rootstock stem above the graft to tie-in the scion and commence training a straight leading shoot.

FIGURE 2.3 Multi-graft of *Sorbus hedlundii* to *S. intermedia* rootstock. The grafting method used is the long tongue side veneer. Despite early union formation, this combination subsequently proved incompatible. Multi-grafting resulted in fewer lost rootstocks than would have occurred with single scion grafts.

Some species sensitive to root pressure problems, e.g. *Acer* and *Betula*, respond badly to heading-back at an early stage before extension growth has sufficiently developed. If done prematurely, the result is to promote bleeding and cause a significant reduction in graft survival. Species that fall into this category will be identified in Parts Seven and Eight.

Taking away a portion of wood in heading-back results in immediate overall reduction of stored carbohydrates and proteins available to the scion. Additionally, the loss of leaf bearing branch structure reduces the ability to produce carbohydrates through photosynthesis. Complete removal of all growth above the union represents a significant loss of foliage and it may be best to reduce rootstock top growth in stages. For very weak or dwarf types three or even four reductions may be required. When weak or senile scion wood has to be used, delaying heading-back for a season or more until scion health and vigour is restored may be the only hope of achieving success.

To allow the scion to build strength and substance for weaker types, or where old or poor scion material is used, gradual removal of rootstock shoots, leaves and/or buds over a prolonged period may be necessary. A technique of inhibiting but not eliminating side grafted rootstock growth above the point of union, called 'crippling'* by some, can be useful in aiding establishment of weak scion material. After the union is complete, snapping or partially cutting through the rootstock above the graft, then bending it over at this point allows on-going photosynthetic function but reduces apical dominance. If space is limited, it may only be feasible to control rootstock growth by repeated partial heading-back.

For evergreens, such as *Rhododendron*, continual removal of rootstock extension buds is effective in limiting rootstock competition while retaining leaves and photosynthetic potential (Figure 2.4). Once the scion variety has thoroughly established and regained viability, the rootstock is headed-back to the union in the normal way.

By using these techniques, side grafting offers opportunities for maintaining the health of the rootstock while the scion has time to rebuild.

GRAFTING: THE FIVE BASIC METHODS

Since grafting methods have developed, categorisation and naming of grafting methods has been disordered and inconsistent, unfortunately there seems to be no standardised system in general use. As a result, methods described can become muddled and the accurate description of important procedures may be flawed or misleading.

Over 30 methods of grafting using detached scions are described in the literature[†]. More are added by the use of specific methods for grafting tools and machines. A systematic approach towards categorisation of grafts linked to methods and names seems long overdue.

Methods can be categorised into five basics. These may be subjected to extra variations often influenced by grafting strategy, for example apical veneers versus side veneers. Occasionally, categories may be combined as in the complicated Gap graft (Pfarrer Dee's graft)[‡], where inlay, wedge and rind work together, but the fundamentals remain the same.

The five basic categories proposed are:

Splice Grafts

A transverse cut made along the stem of the rootstock, normally at an acute angle, but can be more obtuse to produce a flatter cut. A matching cut is made on the scion to provide vertical alignment of scion and rootstock when placed (spliced) together (Figure 2.5). This graft is sometimes referred to as a whip graft. An additional cut on the face of each slice permits the components to be held

---

* Hartmann, Kester, Davies Jr. et al (2011) p.457.
[†] Garner (2013) pp.126–217, 236–280 and Sziklai (1967).
[‡] Garner (2013) pp.191–192.

FIGURE 2.4   *Rhododendron magnificum* grafted to *R. calophytum* rootstock one season after grafting. Scion viability was maintained by allowing a significant portion of the rootstock to remain. To prevent competition, axillary extension buds were removed. Two terminal buds have been recently taken out, visible to the left of the scion; one lateral bud (lower left of centre) was missed, has developed further and will need removal.

together by means of an interlocking 'tongue', to produce what then becomes known as a whip and tongue graft, treated here as a variant of the splice graft and fully described later.

Veneer Grafts

Shallow cuts made almost parallel to the plane of the rootstock. The scion is prepared to fit into this rather in the fashion of the cabinet-maker's veneer (Figure 2.5). Some important variations exist and are defined by the length of the 'tongue', whether short or long, and, for side grafts only, whether or not the 'tongue' is replaced by a downward sloping cut. Some authorities* also describe variants of grafts closely linked to the tissues of the rind as veneer grafts; however, these would seem to fall within the category of rind rather than veneer and are an example of the confusion which exists.

Wedge Grafts

After heading-back the rootstock, cuts or splits are made into the apical end with no wood being removed, the suitably fashioned basal wedge of the scion being pushed into the cut to complete the graft (Figure 2.5). When the rootstock is cut at its diameter, as opposed to off-centre, the name 'cleft graft' is sometimes incorrectly used, causing some confusion. Cuts may be reversed, i.e. the base of the scion is split, and the apex of the rootstock cut to a wedge (as with some root grafts); the graft is then known as an inverted wedge. If the rootstock top is retained with side placement of

---

* Garner (2013) pp.164–169.

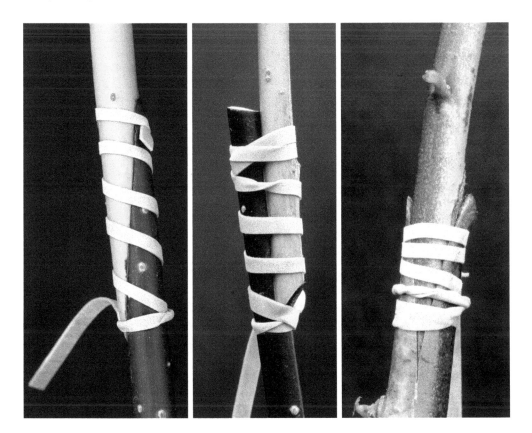

FIGURE 2.5   Three important types of graft for bench grafting. In these examples all are apical grafts. LEFT, splice graft. CENTRE, short tongue apical veneer graft. RIGHT, wedge graft.

the graft, it is here called the side wedge (the term 'oblique wedge' is sometimes used). This latter graft is sometimes confused with the long tongue side veneer graft, which may incorrectly be called by the same name. The differences between the two will be explained when grafting methods are described in detail.

INLAY GRAFTS

This apical graft is formed from two converging cuts made on the headed-down face of the rootstock or basal cut of the scion. In most methods these cuts are made at an appropriate angle from the edge of the face to meet at the centre, producing a wedge of wood which is then removed. The scion or rootstock is cut into a matching wedge, which is inserted to provide a firm fit (Figure 2.6). Adding to the confusion, if scion and rootstock are equal in diameter, and the cut-out wedge extends across the diameter of the headed-down face, the term cleft graft is sometimes used. Alternatively, cuts may be made to remove wood from the scion to produce a saddle shaped profile; the rootstock is then cut to a fitting wedge to form the saddle graft, a variant of inlay.

With modifications, inlay grafts are an important feature of many tool or machine-made grafts, to be discussed later.

RIND GRAFTS

The outer tissues (rind) of the rootstock or scion are prised away from the wood to provide a 'pocket', which accommodates the other portion of the graft, scion or rootstock, as appropriate. This graft

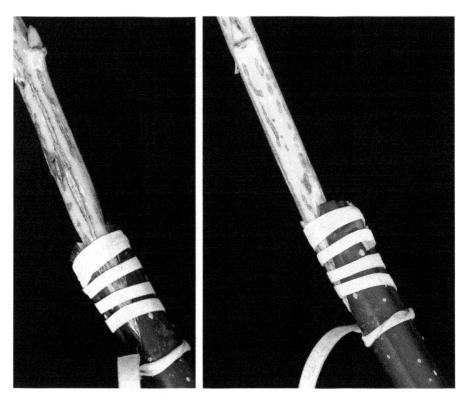

FIGURE 2.6   Inlay graft. LEFT, a longitudinal view. RIGHT, a side view, showing one of the two 'church windows' (almost dead centre of picture), present on each side of the scion; these are necessary to achieve good cambial matching.

may be placed at the headed-down portion of the rootstock for an apical rind graft (Figure 2.7) or on the side of the stem, as for 'T' budding. This latter should not be confused with chip budding, which is in effect a single bud short tongue side veneer graft.

THE THREE CATEGORIES OF IMPORTANCE TO BENCH GRAFTERS

For bench grafting performed manually, only three categories, splice, veneer, wedge and their variants, are commonly required* (Figure 2.5). In practice it is best to keep the choices to a minimum, enabling grafters to hone their skills through constant repetition and practice.

The inlay graft (often called cleft) is frequently cited as best for *Daphne*, but the long tongue veneer is technically superior and achieved more quickly. The saddle graft, a type of inverted inlay graft, often recommended for *Rhododendron*, is better replaced by the long tongue veneer, which is quicker to execute and most agree produces superior results, especially if used as a side graft. Advocates of rind grafting have failed to demonstrate its superiority over other methods. Comparing results achieved using chip budding (essentially a short tongue side veneer graft) with those from T budding (essentially a side rind graft), lower success rates from rind grafting can be expected. A further disadvantage of this method is that, to enable the rind to be lifted away from the wood, there must be natural or induced sap flow. As to be discussed later, rising sap or root pressure (bleeding) can often adversely affect results.

* Bailey (1920) p.137.

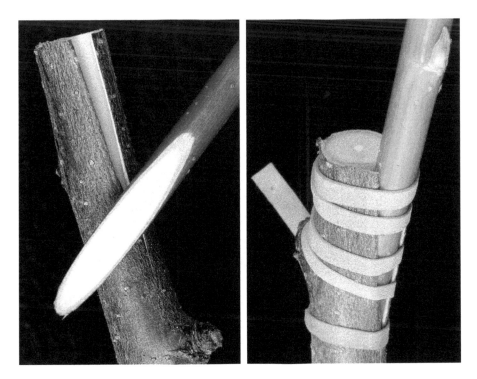

FIGURE 2.7 Apical rind graft. LEFT, showing components prepared ready for assembly. RIGHT, graft assembled and tied-in. Note the gap between lifted rind and central wood core. This must in-fill with callus cells to form a callus bridge before graft union formation is possible; because of the rather large void a fairly lengthy process. The graft requires careful sealing with grafting wax or use of non-porous enclosing ties to prevent these delicate tissues desiccating.

APICAL VENEER, SPLICE AND WEDGE GRAFTS COMPARED

A significant advantage of veneer grafts over splice and wedge grafts is that the varying diameters of the scion and rootstock can be precisely accommodated. Given reasonable matching, this is of little importance to the easily grafted species, but for more demanding types, the skilled grafter using a veneer graft is able to make an exact match by adjusting widths and lengths of cuts on rootstock and scion.

For those species particularly susceptible to problems caused by root pressure (bleeding) such as *Acer* and *Betula*, the splice graft has an important advantage. Left unsealed (un-waxed) excess sap is able to flow away from the cut surfaces and the graft drains itself. Humidity must be maintained to prevent excessive drying of the cut surfaces, but conditions for formation of healthy callus are achieved. In contrast, leaving veneer grafts unsealed, while helpful, may not ensure success because due to the configuration of the cut surfaces; sap may be 'dammed up' at the base of the scion resulting in infection of callus tissue, leading to, at best, infection of basal portions or, at worst, loss of the graft (Figure 2.8).

Despite the technical advantages of the veneer graft, opinions differ on its practicality compared with the splice. The main criticism, in many circumstances true, is that it is slower. When relatively large diameter or very hard wood is being used, the knife-work involved with the splice allows the grafter to exert more pressure on the knife blade and force it more quickly through hard material. For softer or thinner wooded species, the difference in speed is less significant, and for difficult species the better results achieved with veneer grafts may lead to less output but greater productivity.

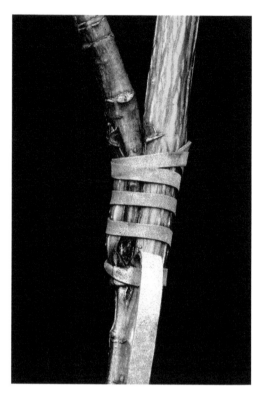

FIGURE 2.8   In this graft, sap has been 'dammed-up', causing dying at base of the scion. Sap, now dried up and showing as a white streak, has flowed from the base of the graft down the stem of the rootstock and over portion of the rubber tie. Graft union development continued above the dead portion and the graft was eventually successful. The scion is *Acer sikkimense* within the Macrantha section of the genus, grafted to *A. capillipes* rootstock.

The wedge graft undoubtedly has claims for attention because it is particularly well-suited to partial mechanisation and independent preparation of the components for separate assembly. The rootstock cuts may be made by an appropriate grafting tool or sharp secateurs. These features can be exploited to conserve demands on skilled workers and increase productivity of grafting teams. Normally the scion is prepared by hand, as tool-produced scions are sometimes characterised by ragged cuts compared with those by hand (Figure 2.9), but for easily grafted species a grafting tool may be used (see Appendix K).

Wedge grafts have the advantage over splice of more cambial contact, making them similar to the veneers, but suffer the same disadvantage of poor sap flow from the base.

MACHINE AND TOOL-MADE GRAFTS

Machine and tool-made grafts frequently rely on inlay or variants of inlay. Some grafting machines or grafting tools produce a 'V' shaped notch or 'saddle' with matching cuts on the other component so that the two fit tightly when pushed together (Figure 2.10). This type of graft requires tying-in to hold the two components in position. Another popular pattern for the most easily grafted species, such as Grapevines, produces matching omega shaped cuts which lock-in tightly when fitted and do not require tying-in (Figure 2.10). Unfortunately grafting machines and tools produce some level of serration and damage to the wood, particularly on the exit side of the cut, for many species this has an adverse effect on takes and consequently they are most suited to the more easily grafted types (see also Chapter 5 and Appendix K).

FIGURE 2.9 LEFT, apical wedge graft scion, prepared using a grafting tool. The rather rough, serrated edge sometimes produced is clearly visible and can reduce takes for the more demanding species. RIGHT, apical wedge graft scion prepared by hand, note in contrast the smooth finish to cuts on the scion. In both examples, one side of the rootstock has been removed for demonstration purposes.

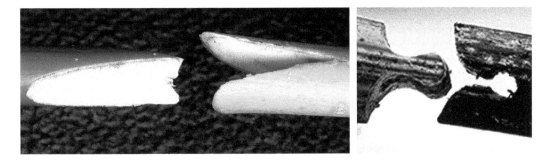

FIGURE 2.10 Inlay grafts. LEFT, 'V' notch or 'saddle' inlay graft produced by a grafting tool. This pattern will require tying-in to hold the components together, some call this a 'V' cleft graft. Better would be the term 'V' inlay graft. RIGHT, an omega inlay graft produced by a grafting tool the two components are slid into each other and lock tightly with no requirement for tying-in. The damage to the wood (serrations) often caused by this pattern means it is not suitable for more demanding species (photo Omega Star H+L Wahler Weinstadt–Schnait Germany).

3 Making the Graft

THE BASICS

Understanding Stem and Root Cambia

To achieve a viable graft union, it is essential that the cambia of rootstock and scion are able to unite, allowing the development of a linked, continuous vascular system. Re-establishment of essential vascular continuity is greatly aided by close alignment (good matching) of the respective cambia of scion and rootstock. If achieved accurately, it significantly improves the opportunity for graft union formation, and it is therefore essential for grafters to be able to recognise the cambium within the cut stem and root.

Cambium lies between the outer tissues of the stem and the inner woody core. Outer tissues are known in the UK as rind (hence rind grafts), a term used throughout this publication. The name bark is used for these tissues in the USA and universally by plant anatomists. When growing actively, rind can be lifted away from the centrally located wood to expose cambial tissue (cambium); this is moist and slippery to the touch, sometimes tinged green in the stem and populated by cells which function to develop new vascular tissues. Readily identified by experienced grafters, beginners may require guidance. Examples of different stem cambial position are shown in Figure 3.1.

Root cambium is also located between outer tissues and the central wood core. It usually lacks any green colouration and in younger roots is normally surrounded by a significantly thicker rind than in the stem; consequently, it is sited more deeply within the cross section, as can be seen in Figure 3.2.

Thickness of rind varies at different areas of the plant, changes with age and development and also to an extent between species. Cambium is located on the inner surface of the rind, and when making cuts, allowance must be made for differing rind thickness to obtain good cambial alignment.

Importance of Good Knifesmanship

Regardless of species, or quality of material, good knifesmanship requires the ability to produce accurate cuts to ensure good cambial alignment and contact. 'Well matched' or 'good matching' are terms used by grafters for such grafts. Other considerations are correct height placement, effective tying-in and appropriate sealing. Commercially, speed is very important, but with rare or endangered species the main objective is to achieve a positive result. An essential element of good knifesmanship is a well-designed and extremely sharp knife, to be fully discussed in Chapter 5.

Preparation of Components Prior to Grafting

Before grafting commences, the rootstock and scion will most likely require preparation to enable making the necessary grafting cuts. This involves removal of any side shoots obstructing access to the main stem and grafting position, trimming to height, etc. These preparations are described under the appropriate headings in Chapter 9 (Rootstocks), Chapter 10 (Scions) and particular requirements in Parts Seven and Eight – Genera Specific Requirements.

Holding the Knife – Support and Control

Most people naturally hold the knife correctly, using a 'three, one, thumb' configuration. The knife handle is held firmly in the palm using three fingers; the thumb is often placed at the heel or blade hinge point, the first finger positioned more or less opposite (Figure 3.3). The thumb provides grip but also imparts balance, aiding production of accurate, well-judged cuts. Changes in position occur

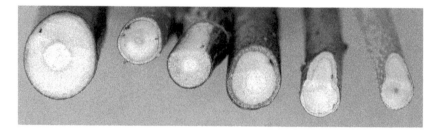

FIGURE 3.1   Cross section of 1 yr. woody stems. Showing position of the cambium (marked by a black dot) located just under the rind or bark. Left to right, *Acer* x *lobelii*, *Betula utilis*, *Liriodendron tulipifera*, *Magnolia sargentiana* var. *robusta*, *Quercus ellipsoidalis* and *Cornus kousa*. As will be discussed in Chapter 10, the pith content of some examples might be considered too high to constitute ideal scion wood.

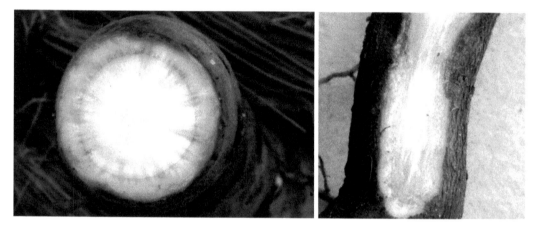

FIGURE 3.2   Root cambium. LEFT, cut end, transverse section, of Paeony root showing position of cambium (inner brownish ring surrounding central white wood). RIGHT, a longitudinal section of *Acer platanoides* root. Note: in both examples a thick 'rind', typical of root structure.

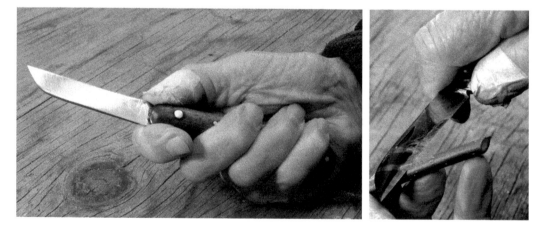

FIGURE 3.3   Holding the knife. LEFT, at rest, in this case a Tina 600A model for right-handed grafter. Showing the position of fingers and thumb. The thumb of the knife hand is protected by adhesive plaster. The flat face of the single-angled knife blade is facing the viewer. RIGHT, the knife in use, preparing a veneer graft scion. The flat face (no longer visible) aiding the production of a flat, smooth cut.

according to the type of cut; powerful cuts in thick wood require a firmer grip, the first finger moved closely to the other three, and the thumb held further back into the handle to increase the hold.

Delicate veneer grafts are performed with the first finger resting against the spine at the hinge point or beyond and the thumb resting on the tip of the heel (Figure 3.3). Individuals vary in precise placement of hand, fingers and thumb. Practice and experience will influence the favourite configuration.

Most grafters find making cuts on the rootstock is easier than on the scion because the bench, or pot, can be used to steady and 'line up' the knife hand. This highlights an important aspect of knifesmanship, which is to make use of any opportunity to rest some portion of graft components on a firm, steady surface, often using this support as a fulcrum for performing cuts. Techniques for achieving this will be covered in the detailed descriptions of grafting methods to follow.

SPLICE GRAFTS

The splice, also called a whip graft, is always an apical graft. It is achieved by cutting away at a suitable angle a portion of the stem at the apex of the rootstock and base of the scion and subsequently fixing or splicing the components together by tying-in or other means.

SPLICE GRAFTING METHODS

This graft is one of the most popular methods, having the advantages of simplicity and speed. Unless the diameter of rootstock and scion at the point of grafting are identical, geometry dictates that perfect alignment of each respective cambium is difficult and requires care and judgement. This may involve the need to adjust the final level of one component with respect to the other to produce what is known in grafter's parlance as a 'church window', to be described below. Despite this, good knifesmanship ensures that the match is sufficient for success. Preparing splice grafts can be done in either a standing or sitting position – the technique is basically the same. The sequence of cuts made is shown in Figure 3.4.

For heavy material, it is important that arms, bent at the elbows, are kept well up in the mid chest position and approximately in line with the direction of cut. This aids production of accurate, straight cuts. The knife blade is pulled through the stem of each component to produce a matching

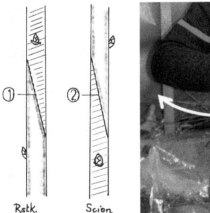

FIGURE 3.4   LEFT, the sequence of cuts to prepare a splice graft, assuming the rootstock (Rstk) is prepared first. The hatched-in area falls away and is discarded. RIGHT, a good stance for preparing splice grafts in a standing position when working on moderately heavy wood. Both elbows are kept approximately in line with the cutting direction of the slice. Arrows mark direction of arm movement and cut (John Richards Nurseries).

sloping slice. To achieve this, both arms are pulled away from each other in a controlled movement, the elbow of the hand holding the scion should be in an equivalent position to the elbow of the knife hand on the other side of the chest (Figure 3.4).

With lighter wood the cut may be achieved by applying most or all the pulling force on the knife hand (Figure 3.5).

The thumb of the knife hand is moved from the handle, placed more or less opposite the blade edge on the other side of the stem and positioned in a manner which aids control of the cutting edge of the blade. The wood is therefore held between thumb and blade; very gentle pressure is exerted by the thumb to steady movements and aid a controlled cut. The stem is held firmly at an angle of 30 to 50 degrees to the floor against the ball of the thumb by the fingers of the holding hand, and positioned so that when the cut commences the heel of the knife blade is close to the index finger; the entire length of the blade is then drawn through the wood (Figure 3.4 & 3.5).

At commencement of the cut, the knife blade should engage scion or rootstock with the cutting edge pointing upwards and *across* the plane at an angle between 50 and 60 degrees. Cutting angle of the knife *along* the plane of the stem determines the length of the splice. To remove a slice of appropriate length, a good aid to fixing the angle required is to imagine a line running across the shoot, commencing where the blade enters the wood and where it leaves*. As a rough guide, the best length of slice is five to six times the diameter of the stem; for most wood this will mean between 30 and 45mm long.

In addition to influencing the length of the splice, the cutting angle must also take into account thickness of wood and rind. Those with a thicker rind (normally rootstock) must be overlapped on the thinner to ensure good matching of inner edges and cambia. An extreme example of this is shown by the root graft in Figure 3.6. In this case, the difference in thickness between stem rind and root rind was so considerable that, to provide better alignment, a veneer graft (short tongue apical veneer) was substituted for splice.

For light or soft wood, the hands are best kept close to the body with the elbow of the knife hand pointing outwards more or less in line with the proposed slice. The forearm of the hand holding the

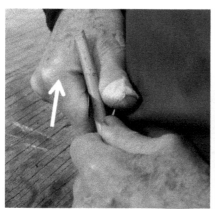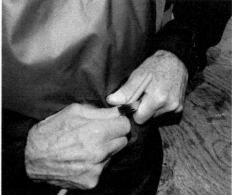

FIGURE 3.5   LEFT, a right-handed grafter preparing a splice graft working with relatively soft wood. The angle of the knife blade along the plane of the stem is clearly visible. In this example, the length of splice will be between 35 and 40mm. Arrow marks direction and angle of cut. RIGHT, the correct stance for a left-handed grafter with the same type of wood. The arm holding the stem (rootstock or scion) is best held closely against the lower chest/waist; the arm holding the knife is pulling the blade through, more or less in line with the cutting angle. In both examples the stem is shown held between the knife blade and the thumb of the knife hand to provide a controlled cut.

---

* MacDonald (2014) pp.133–137.

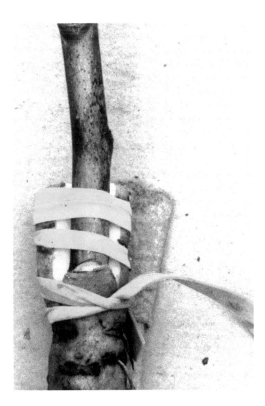

FIGURE 3.6 *Paeonia* root graft. Cambial alignment. The rind (phloem, cortex and periderm) of the root-stock, showing as white/pale grey tissue, significantly overlapping the scion. This emphasises the deeper placement of the cambium of the root, compared with the stem. To ensure best cambial alignment (matching), a short tongue veneer graft rather than splice is used.

rootstock or scion can be tucked against the lower chest/waist and most cutting action made by the knife hand (Figure 3.5). For all types of wood, arm movements should be kept as controlled and short as possible to avoid any 'flailing' action. At completion of the cut, the thumb of the knife hand should rest adjacent to or against the flat face of the blade close to the cutting edge. Apart from facilitating a controlled cut, this technique greatly reduces the possibility of producing an undesirable 'tail' (a sliver of wood left to extend beyond the slice) which, if present, must be removed by repeating the cut.

When the grafter is standing, the scion or rootstock is held by the non-knife hand, pointing upwards, with the apex of the rootstock or base of the scion pointing across the body towards the knife hand. Bare-root rootstocks are held by the base of the stem or the top of the root system; pot grown rootstocks are held by the pot in a similar position (Figure 3.7).

When working with light thin-wooded material, splice grafts may be produced while seated. Potted rootstocks are held by the pot while those bare-rooted at the base of the stem. Both are held across the front of the grafter's body at about 20–30 degrees to the horizontal (shallower than when standing) with the apex slightly angled towards the grafter; the scion is held firmly at a similar angle. For both components the knife is drawn through the stem as described.

An alternative for potted or bare-root rootstocks is to hold the pot or roots more or less vertically on the bench top to provide stem orientation similar to that in the field. The blade, held at more or less 90 degrees to the plane of the rootstock stem, is pulled upwards through the wood with the thumb of the knife hand placed behind the wood as previously described. This method, for obvious reasons, is favoured by those experienced in grafting in the field.

Seated grafters working with thicker and/or harder material using potted rootstocks may prefer to hold the rootstock using this vertical holding technique. Using the method when seated means that

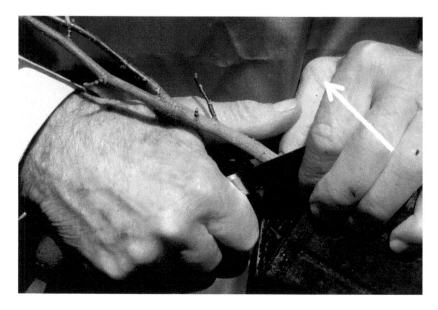

FIGURE 3.7   Holding a pot grown rootstock, showing configuration of hands and knife for preparing a splice cut. Arrow marks direction of the cut.

normal bench height is then too high and a second smaller bench, with possibilities for height adjustment, may be fitted at a lower level. Combined with a swivel chair, this procedure can be used with minimum disruption. A second bench is appropriate when using methods alternating between splice and veneer throughout the working day. For details of bench design to incorporate this feature see Chapter 7: The Grafting Bench as well as Appendix M: Grafting Bench Design and Construction.

Grafting success is always enhanced by good alignment of cambia of rootstock and scion (good matching). When scion and rootstock diameters and rind thickness of splice grafts are equal, matching slices may be made with relative ease, giving more or less perfect alignment of the respective cambia when the components are tied together to form the splice (Figure 3.8).

In practice the often thinner scion is cut through from side to side. To achieve matching scion/ rootstock cambia when diameters are unequal, the cut on the thicker rootstock must be made at a shallower angle to match the scion width. Unless further adjustments are made, this can still potentially result in a poor match with the broader cut at the apex of the rootstock. To correct this disparity the scion has to be raised to a point where cut widths on rootstock and scion are equal, resulting in a portion of the scion cut face being visible; this is known by grafters as a 'church window' (Figure 3.9). Unequal components with considerable disparity in size can result in an overly large church window. To avoid this, further refinements are possible. The rootstock cut may be slightly scooped at the base and the remainder of the cut made almost parallel to the plane of the stem, resulting in a configuration not dissimilar to a veneer graft. The scion is cut to shape so that it fits neatly against the rootstock (Figure 3.10). On rare occasions the scion will be larger and thicker than the rootstock and the descriptions above apply in reverse. However, a better choice in this situation may be an inverted apical wedge graft.

Every effort should be made to ensure cambial matching on each side of the cut; this should be the case for all grafts. Matching on one side only always reduces the chances of success and increases the chance of subsequent breakage.

Alternative Splice Grafting Method

An alternative method of making splice grafts particularly suited to those in a seated position is to pull the cutting edge of the knife towards the grafter. Graft components are held almost horizontally,

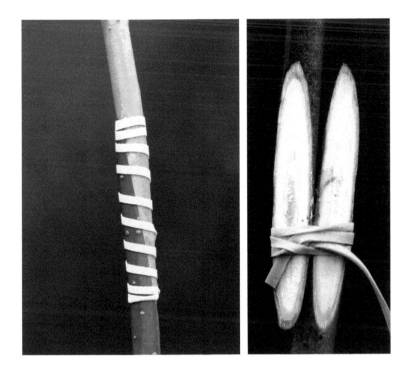

FIGURE 3.8   LEFT, a splice graft with equal diameter components tied-in. RIGHT, the same type of graft with equal diameter components; the cut surfaces are exposed showing the possibility of very good matching (cambial alignment).

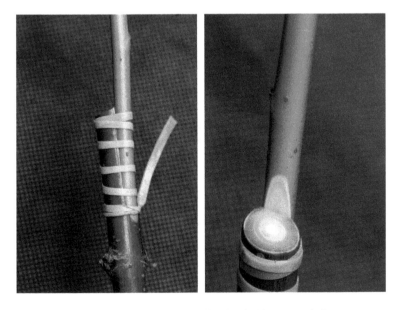

FIGURE 3.9   Cambial alignment. LEFT, a splice graft with tied-in unequal diameter rootstock and scion. RIGHT, the splice graft on the left showing its 'church window'. As can be seen the cambia of rootstock and scion are well aligned.

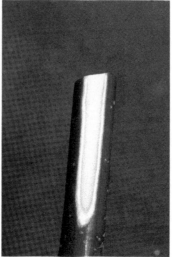
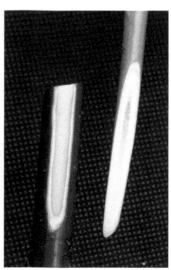

FIGURE 3.10   Modifications to the splice graft for substantially unequal diameter components. LEFT, show-ing the slice on rootstock slightly concaved (scooped). CENTRE, the slice cut on the rootstock can be seen, showing more or less parallel sides. RIGHT, the exposed cut surfaces showing good matching of rootstock and scion. A 'church window' is still required to ensure good cambial alignment.

the apical end pointing towards the chest for the rootstock and away for the scion; the 'holding hand' for rootstock or scion may be rested on the bench to provide extra support (Figure 3.11).

Before commencing the cut, the knife blade is placed almost at right angles to the plane of the wood in a virtually horizontal position and the thumb of the knife-hand is placed beneath the com-ponent to provide a conventional blade-scion-thumb configuration. With practice and constant use, this technique may be better suited to grafters in a sitting position than that previously described. It may not be so suitable when working with really heavy or thick wooded material because less 'knife power' can be generated. Grafters working with rootstocks in large pots sometimes favour this

FIGURE 3.11   Configuration of hands and fingers for the alternative method of preparing a splice graft, with cuts towards the grafter. Arrow marks the direction of cut.

method because, when the pot is laid on its side and suitably supported at the correct angle, the root-stock stem can be presented to the grafter at a convenient height and position for making splice cuts.

WHIP AND TONGUE GRAFT

In field grafting, the addition of matching cuts to rootstock and scion to produce an interlocking 'tongue' converts a basic splice into a whip and tongue. In the field the purpose of the tongue is to hold the two components together to enable the grafter to move ahead, leaving the scion attached ready for tying-in and sealing by an assistant. When the bench grafter ties-in his own grafts, preparing the tongue is unnecessary; time is better used tying the splice together and moving on more quickly to the next graft. The tongue reduces close fitting and some alignment of the cut surfaces which, for difficult species, may delay or reduce formation of a good union. If root pressure (bleeding) is a possibility, the interlocking tongues will prevent free sap flow from the cut surfaces. There is no evidence to support the view that whip and tongue grafts, when well established, make a stronger union than splice grafts.

In some circumstances for bare-root bench grafts of the more easily grafted species, use of whip and tongue grafts can be recommended. As with grafting in the field after preparation and assembly, the grafter is able to leave tying and sealing to assistants; consequently, skilled individuals can produce extremely high outputs (up to 300 grafts per hour are claimed). Well-executed whip and tongue grafts are technically superior and likely to provide better takes than the alternative high output methods of machine-made inlay or wedge grafts.

Whip and Tongue Grafting Methods

Preparation of the splice (whip) is exactly as described previously, for unequal diameter components a church window will still be required. Interlocking tongues are produced by making shallow cuts into and at right angles across the slices (Figure 3.12). This involves rotating each completed slice so that the cut surface faces the grafter. The scion will need to be inverted so that the apical end points downwards and towards the grafter.

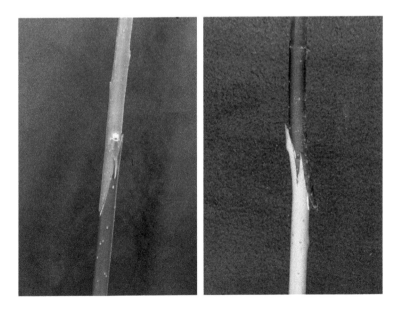

FIGURE 3.12   LEFT, a whip and tongue graft before tying-in. RIGHT, some distortion occurs as a result of the large tongue. The cut surfaces will pull inwards to provide good matching after tying-in. For some species (e.g. *Vitis*) tying-in is unnecessary as there is sufficient contact to allow the union to form satisfactorily.

The relevant component (rootstock or scion) is held at 60–70 degrees to the floor or, when sitting, 30–40 degrees to the bench and the hands holding knife and wood are brought together with thumbs holding knife and wood in close proximity (Figure 3.13).Using a knife technique similar to that for veneer grafts, a cut is made downwards into and at right angles across the slice to produce a flap called by grafters a 'tongue'. An angle of 7–10 degrees is required and up to one-quarter of the overall length of the slice.

As a guide, the tongue should commence about one-third from the top of the rootstock and, dependent upon length of tongue, about two-thirds to half-way along the scion. Positioning each tongue is critical because it is important that when the components are pushed together both are tightly interlocked and cambia properly aligned. Beginners may find it helpful to leave the knife in place at completion of the rootstock cut and to place the scion, properly aligned in the tying-in position, against the assembly. The cutting edge of the knife shows the correct point for commencing the cut on the scion (Figure 3.14). A small movement is made to produce a slight mark at this point. The knife is then removed and the scion inverted. Using the marked point, the appropriate tongue cut commences here and is made as previously described. Using this simple technique should ensure that when the components are slid together and pushed fully home the tongues interlock to provide perfect alignment of the stems and cambia. After practice and repetition, fixing the position of the cutting edge to provide good alignment becomes automatic. To aid fitment, the grafter may give a slight twist to the knife at the tip of the cut on the tongue, lifting the edge slightly but avoiding any possibility of causing a split.

After completion of the graft handling procedures, tying-in, sealing, bundling, etc. may involve assistants and some level of harsh handling; in this situation larger tongues provide more positive fixing of the components (Figure 3.12). An extra-large tongue may also be employed with Grapevine grafting, where the graft is left untied and held in position by tension of the interlocking tongues, an indication of how easily this species calluses and forms a union.

VENEER GRAFTS

The complexity of veneer grafts is such that an overview of the different methods is first given. The knifesmanship techniques required are then described in detail in the section below – Veneer Grafting Methods.

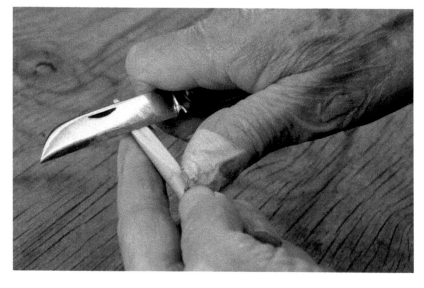

FIGURE 3.13   Configuration of the hands, knife and graft component to produce the cut forming the tongue of the whip and tongue graft. Note the similarity of positioning of the hands to the side veneer graft.

FIGURE 3.14   Whip & Tongue, the cut on rootstock (left) has been completed. Cutting edge of knife indicates correct position to begin the cut on the scion slice to produce the tongue. In this case the knife has been inserted deeply to create a large tongue for demonstration purposes. This method may be used for *Vitis* when tying-in is not required.

Veneer grafts are particularly suited to side grafting but can also be used as an apical graft. Arguably they are more difficult to execute than most other methods. They are appropriate for more demanding species because of their accuracy, the amount of cambial contact available and the possibility of achieving near perfect cambial alignment. Scions for these grafts are normally of conventional size but extra-large or single bud scions may be used for particular situations.

Veneer grafting does not adapt well to the technique of cutting scions and rootstocks separately for subsequent assembly by assistants; however, it is occasionally achieved by some grafting teams*.

There are three variants on the basic theme: Short Tongue, Long Tongue and, for side grafts only, Sloping Side Veneer.

APICAL VENEER GRAFTS

There are two key variations: Short Tongue and Long Tongue (Figure 3.15).

Short Tongue Apical Veneer (Apical Veneer)

In this graft a shallow, narrow triangular portion of wood ending in a basal flap or tongue is removed from the side of the rootstock; into this is placed the scion, cut accurately to fit (Figure 3.15).

When well-executed, this graft has consistently better cambial matching than the splice and can replace it for difficult species.

Long Tongue Apical Veneer (Modified Apical Veneer)

Also known as the Modified Apical Veneer, the name Long Tongue Apical Veneer, which has been previously used, is cumbersome, but provides a more descriptive title and therefore used throughout this publication.

---

* Hall (1977).

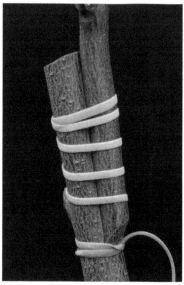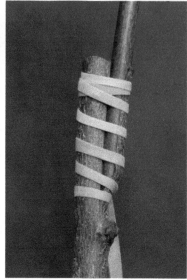

FIGURE 3.15   Apical veneer grafts. LEFT, a short tongue apical veneer. RIGHT, a long tongue apical veneer. The basal cut to produce the tongue of the short tongue veneer has been cut sufficiently accurately to be scarcely visible.

Similar to the previous graft but with the cut on the rootstock made in such a way that a long tongue or flap of wood is retained and the scion cut to fit beneath (Figure 3.15). Because the cambium under the tongue is not visible, there are special demands on the grafter to achieve a good match; therefore, producing this graft and its side graft variant requires more skill than most other methods.

Although no wood is removed, which should place it within the wedge grafts, the configuration of cuts, excellent cambial alignment and absence of a church window, are all characteristics of veneers; consequently. it is placed within this category.

There are advantages in using this graft; cambial contact is greatly enhanced by the additional amount of cambium in the long tongue (Figure 3.16) and presence of one or more axillary buds (stimulant buds) on the tongue further increases callus production.

Examination of hundreds of grafts has shown that callus development occurs initially on the tongue, often by many days, only on rare occasions initial development will be seen elsewhere. This effect can be so extreme that the graft fails to unite on its inner surface and becomes attached solely by the tongue (for further discussion see Chapter 11 and Figure 11.11).

When using old rootstocks with comparatively thick rinds or if there is considerable disparity in the diameter of rootstock and scion, use of the long tongue veneer to match a significantly thinner scion is not advisable. This is because, to achieve cambial alignment, insufficient live tissue remains on the tongue which, as a result, dies and is often followed by loss of the graft. With material of this type, a short tongue veneer is the only sensible alternative.

As an apical graft, the long tongue veneer is restricted to a few species such as *Rhododendron*, which derive benefit from additional cambial contact. In this case, the graft replaces the slower and less effective saddle graft.

SIDE VENEER GRAFTS

Side Veneer Grafts are divided into three key variants: Short Tongue Side Veneer, Long Tongue Side Veneer and Sloping Side Veneer.

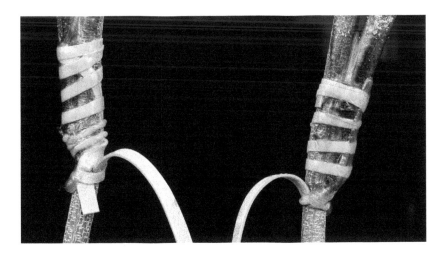

FIGURE 3.16   *Cercis canadensis* 'Forest Pansy'. Short tongue side veneer (left) and long tongue side veneer (right). Both grafts showing good callus development, but significantly more callus is able to develop on the long tongue. Note that no callus is evident along the inner edge of either graft interface. Callus development was visible in that position two or three weeks later. Compare the potential callus area of the long tongue veneer with that of the splice, which has no quick callusing tongue and half the contact area. This emphasises the advantage of this method over the splice. Notice the generous amount of wax at the divergence point between rootstock and scion on the long tongue side graft. This ensures no water can enter the interface between rootstock and scion at this vulnerable point. Rootstocks are *C. canadensis* 1+1P 6–8mm.

Short Tongue Side Veneer (Side Veneer)

Essentially the same as the apical short tongue veneer but with the top of the rootstock left wholly or partially intact (Figure 3.17). Less cambial contact is possible because of the absence of a long tongue but cambial matching is an easier procedure because the position of the cambium is fully visible and accurate cutting to achieve a near perfect match is facilitated. As explained previously, the method does have an important role when there is significant disparity between the diameter of rootstock and scion.

Long Tongue Side Veneer (Modified Side Veneer)

Personal experience and that of others* indicate that where it is possible to adopt this graft, 'takes' (successful grafts) out-perform other methods. As previously noted, it has the disadvantage of being among the more difficult grafts to perform (Figure 3.17).

After the graft union has formed, the rootstock can be maintained in good health until heading-back is appropriate. For scions which are weak, senile or from difficult species, this can be delayed by one or several seasons, allowing time for them to rejuvenate growth and develop into self-supporting plants.

A further important feature is the strengthening effect of the long tongue after tying-in, permitting the use of light, thin stemmed rootstocks, which, if the short tongue method is used, would be vulnerable to snapping off during tying-in and handling.

Side Sloping Veneer

Devised to lessen the effects of root pressure (bleeding), this is essentially the same as the short tongue side veneer, but with the cut at the base sloped downwards to provide a drip guide for any excess sap (Figure 3.18). It provides many of the good features of other veneer grafts regarding lining up of the cambia of rootstock and scion. To achieve any sap drainage, the graft must remain

---

* Wells (1955), Stoner (1974), Meacham (1995) and Leiss (1986).

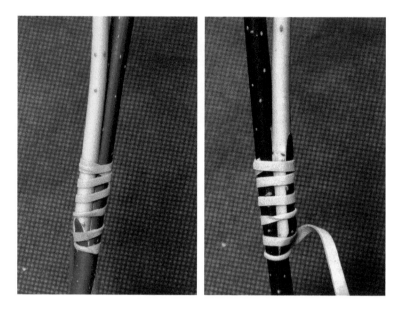

FIGURE 3.17 LEFT, a short tongue side veneer, showing on the scion a short steep angled basal cut and long, shallow angled cut to interface with matching cuts in the rootstock. RIGHT, a long tongue side veneer. The scion is showing two parallel cuts to interface with the rootstock stem, which has been cut to produce a long tongue. When tied-in this follows the contours of the scion.

FIGURE 3.18 Side sloping veneer.

FIGURE 3.19 Standing stance for preparing a veneer graft scion using light or medium girth wood (John Richards Nurseries).

un-sealed and is therefore only suitable for use within an enclosed environment. For badly bleeding rootstocks it may be the only hope of success. Some would argue that the splice graft meets many of these criteria but as an apical graft it lacks the additional moderating effect on root pressure of the rootstock stem above.

Veneer Grafting Methods

There are basic differences in movement between making splice and veneer grafts. For veneer grafts the knife and graft components are held lower, at low chest or stomach height, the elbows are held into the body and the cut is achieved by forearm, wrist and hand action (Figure 3.19). The only exceptions to this are for heavy or hard wooded material, where the cut surface of the scion providing the interface with the rootstock, and the basal steep angled cut may require splice knifesmanship techniques, capable of generating more 'knife power' (Figure 3.20).

Sequence of Cuts for Veneer Grafts

The diagrams below (Figure 3.21) illustrate the sequence of cuts for the preparation of short and long tongue veneers assuming the rootstock is first prepared.

The Rootstock – Short and Long Tongue Veneer

When making rootstock cuts for short tongue veneers potted rootstocks are best with a portion of the pot rested on the bench, held to the front, often against the grafter's stomach (an apron is advisable when executing this graft), with the stem apex pointing away at right angles to the shoulders (Figure 3.22). They are held leaning at an angle of 20–30 degrees to the bench for grafts placed at the base, or 50–70 degrees for those placed higher up. Short tongue veneers start at a point which will become the graft base with an angled cut of approximately 25–40 degrees, made into the stem to approximately one-third of its diameter. This cut is intended to match the profile of the cut to be made at the base of the scion. The knife is then withdrawn and a second cut is begun at a point which will be the top

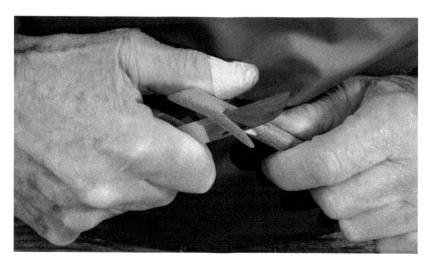

FIGURE 3.20 A grafting technique very similar to preparing the splice is required for preparing veneer scions from heavy and hard material. However, note the shallower knife blade angle to the plane of the stem to provide the shallower, narrower cuts required to match with cuts on the rootstock.

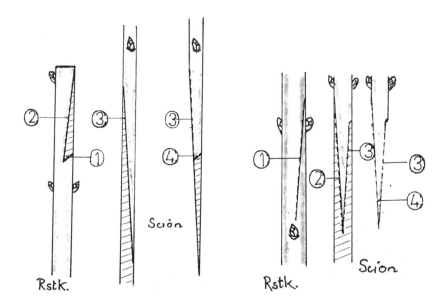

FIGURE 3.21 LEFT, a short tongue apical veneer, showing the sequence of cuts assuming rootstock is prepared first (Rstk = rootstock). RIGHT, a long tongue side veneer, showing the sequence of cuts assuming rootstock is prepared first. The hatched-in areas fall away and are discarded. For side veneers, the same sequence as for respective apical veneers but, as shown, leave in place the top of the rootstock above the graft.

of the veneer, dependent upon stem girth this is usually 25–45mm above the basal cut. For apical veneers it is important that this second cut should start about 2–4mm below the top of the rootstock (see Figure 3.23). Cuts started too high so that they slice into the stem at the top of the rootstock result in the need for an unwanted 'church window' and less well-fitting graft (Figure 3.24).

The cutting angle for the second cut is usually between 1.5–5.0 degrees but, in the early stages of practice, can be decided by using the imaginary line technique described for splice grafts. To provide support and control this cut is often best performed, by resting the thumb holding the knife

FIGURE 3.22 Configuration for producing cuts on the rootstock for short or long tongue apical or side veneers. For short tongue veneers, the basal cut is first made into the stem. Note the thumb of the knife hand can be rested on the pot to provide a fulcrum for producing an accurate, controlled cut. For apical grafts, use the same configuration except the hand holding the rootstock needs adjustment (normally holds the pot or bare-root root system) and the cut must commence just below the apex of the headed back rootstock; also see Figure 3.11. Arrow represents the arc of the cut.

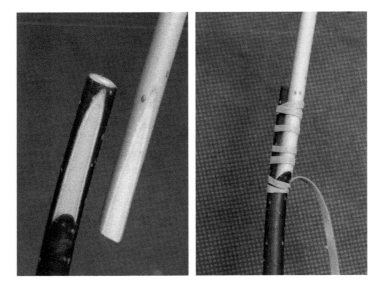

FIGURE 3.23 Apical short tongue veneer graft. LEFT, showing correct cuts on rootstock and scion to provide good matching. RIGHT, assembled and tied.

on the top of the pot, as shown in Figure 3.22. This gives stability, provides a fulcrum and allows controlled movement of the cutting edge of the blade towards the grafter. The cut commences in the toe half of the blade and finishes toward to heel end. To achieve this, the toe of the blade moves in an arc, around the fulcrum of thumb and pot (see arrow in Figure 3.22).

Once the long second cut reaches the basal cut, the resulting loose sliver of wood is flicked out to expose the cut surfaces. When the scion is prepared, the components are then placed together and tied-in (Figure 3.23).

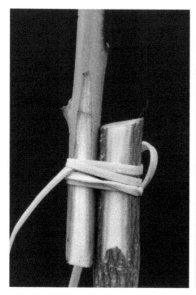

FIGURE 3.24   An incorrect graft for an apical short tongue veneer. LEFT, the rootstock cut is started too high, producing the need for an unwanted 'church window'. On the right, the graft is assembled and tied and showing the 'church window'.

Veneer grafts produced on bare-root rootstocks may be held in the hand or, preferably, the roots rested on the bench and the stem held at the same angle as potted rootstocks. In other respects, the knife technique is similar but as no pot is available to steady the knife hand it is important that the elbows are held well into the sides to aid stability.

When the graft is made at some height from soil level, as for top-worked standards, half standards or Patio Plants, a different technique is necessary. The knife hand is supported by the thumb of the knife hand, often combined with the thumb of the hand holding the stem, which acts as a pivot point for executing the rootstock cut(s) (Figure 3.25).

When preparing a long tongue veneer, positioning the rootstock is as described for the short tongue veneer. Only one cut on the rootstock is required, again using the pot as a fulcrum. This is a made at a shallow angle to produce very similar dimensions to those previously suggested but retaining the resulting flap (Figure 3.26). An additional but not essential refinement for side grafts only is to slightly scoop the cut at its apex. This produces a thicker stem section at the distal end of the tongue, providing a tongue profile more resistant to damage and desiccation than the very thin section produced by a flat entry into the stem. When completed, the knife blade is then re-positioned to just above the original cut and re-engaged with the stem to smooth and flatten the surface (Figure 3.26). This procedure involves very little movement of the knife hand and quickly becomes an automatic activity in making the graft.

The Scion – Short and Long Tongue Veneer

When heavy wood is being prepared, the technique for preparing the main interface scion cut has been described previously (see Figure 3.20) and is similar to that for making a splice cut.

When using normal or light wood, the scion is held at right angles to the body with the base pointing slightly upwards towards the chest. Then, using wrist and forearm, the scion is cut to match those on the rootstock. The first, main interface cut is achieved by pulling the blade through the stem, at an angle between 1.5 and 5.0 degrees, *along the plane* of the scion (Figure 3.27).

Two techniques are available for normal or light wood. The blade may be pulled into the wood using only wrist and forearm, as just described and illustrated on the left in Figure 3.27.

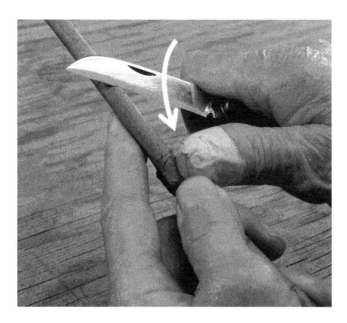

FIGURE 3.25   Configuration for a side veneer graft placed higher on the rootstock stem, possibly for top-worked or Patio plants. Note positioning of knife hand with supporting hand fingers and thumb. The thumb of the knife hand is used as a fulcrum to provide a controlled, accurate cut. Note the flat angle of the knife blade to the rootstock for producing in this case the tongue of the long tongue side veneer. Arrow represents the arc of the cut. For short tongue veneers, a basal cut at a much steeper angle is first made into the stem.

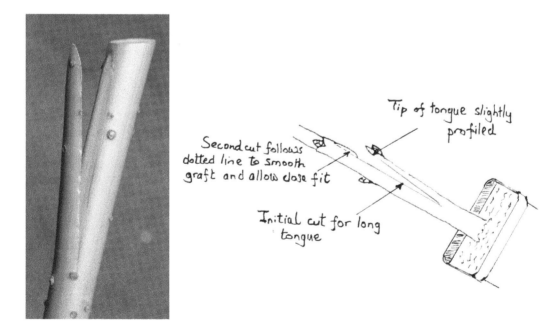

FIGURE 3.26   LEFT, the long tongue apical veneer, a single cut commencing 2–4mm below the top headed back surface of the rootstock. RIGHT, an additional but inessential refinement to the long tongue side veneer graft, see text.

FIGURE 3.27   LEFT, preparation of the main interface veneer scion cut without use of thumb. RIGHT, using the thumb with protection. Note supporting fingers beneath the scion providing a firm base for a flat cut and position of first finger resting on the spine of the knife blade.

The advantage of this is that very flat cuts are produced; however, occasionally the knife 'runs out', leaving an angle that is too shallow and the cut must be repeated.

The second method (Figure 3.27, right) involves placing the thumb, with suitable protection, such as a 'finger protector' or tape, on the base of the scion. Using some wrist action, coupled with hand clenching, the blade is pulled through the wood against the thumb. This technique provides a more positive cut, generally achieved in one pass and therefore often quicker; unfortunately, there is a greater possibility of producing a cut with a scooped profile, and to avoid this care is required.

After completion of this phase, the scion is turned over and reduced to a length matching that on the rootstock, normally between 25 and 45mm, by making on the base a sloping cut at an angle of approx. 20–35 degrees (Figure 3.28).

Long tongue veneers require the same main scion/rootstock interface cut as for short tongue. The scion is then turned over and a second fractionally shallower and shorter slice is made parallel to the first. The parallel cuts are made easier if, while being held between thumb and first finger, the scion is supported underneath by the second and third fingers (Figure 3.29). Achieving parallel cuts is difficult and largely achieved by practice. Feeling the lower cut surface with fingers supporting the scion gives a guide to its orientation and correct angle of the knife blade. Because a flat, smooth surface is less necessary with this cut, the technique described is used on light, medium or heavy wood and the 'splice' method described for heavy wood is not required.

The upper cut is best if slightly scooped at the start. This provides good matching with the flap on the rootstock and enables the continuing cut to provide more or less parallel positioning of cambium on each side of the stem, making subsequent alignment an easier procedure. The flexibility of the relatively thin tongue of the rootstock follows contours of the profile. The graft is completed by cutting the scion base to a length matching the tongue at an angle of 20–35 degrees as for the short tongue veneer.

Cambial Alignment for Veneer Grafts

Short tongue veneer cuts made on rootstock and scion are easily visible and good accuracy in achieving matching is readily accomplished. Long tongue veneers are produced in a similar way but suffer from the disadvantage of having the position of the cambium hidden by the tongue. To ensure cambial matching, it is necessary to ascertain the width and length of cuts on the scion are matching those already made on the rootstock. Also, it may be necessary to determine the relative thickness of component stems and rinds and use this to decide how much, if any, overlap is required.

FIGURE 3.28 Short tongue veneer scion cut to length with an (20–35 deg.) angled cut. Main rootstock/scion interface cut is underneath. The same configuration is used to cut long tongue veneer scions to length.

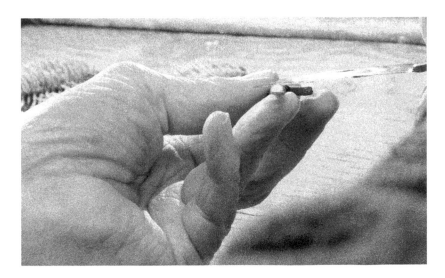

FIGURE 3.29 Long tongue side veneer scion, second cut. Scion is held by thumb and first finger. Flat interface cut beneath is supported by second and/or third fingers, which also act as a guide for making the second cut above and parallel to the first. The scion is finally cut to length using a steep sloping cut as shown for the short tongue veneer in Figure 3.23 and 3.28.

Only continual practice can achieve these requirements. For professional grafters with sufficient practice, it eventually becomes an automatic procedure. However, it is recommended that only when thoroughly confident in producing short tongue veneers should the long tongue be attempted. For amateur grafters, normally involved in short runs, the skill is hard to acquire, and using only short tongue veneers may be a better choice.

WEDGE GRAFTS

Although suitable for pot grown rootstocks, apical wedge grafts are excellent for bare-root grafting. They are well-suited to a teamwork approach; rootstocks and scions are prepared by hand or grafting tool, assembled and tied-in by assistants. These methods, coupled with systems such as hot-pipe, cold callusing or sub-cold callusing, may challenge the dominance of field grafting and budding for the more easily grafted species, traditionally produced in this way.

Wedge grafts may also be placed on the side of the stem, allowing retention of the top; some authorities use the name 'oblique wedge graft' for this*, but here it is called the side wedge graft to place it within the side grafted strategy.

Apical Wedge Graft

Closely related to the apical long tongue veneer, it differs because the rootstock is cut or split either down the centre or offset to provide access for a wedge-shaped scion. The graft requires less skill than veneer grafts, and geometry determines that a perfect match is not possible because cuts on each side of the scion mimic a double splice. This method can be used for species which are relatively easily grafted. Occasionally, cuts may be reversed: the scion cut as for the rootstock, the rootstock as for the scion (inverted wedge).

The rootstock cuts and the wedge-shaped scions can be produced mechanically; though the scion is best prepared by hand if ragged cuts are to be avoided (see Figure 2.9, Chapter 2: Machine and Tool-made Grafts). For 'easy' species (otherwise normally field grafted), a good strategy is wedge grafting using partial or fully mechanised methods, followed by placing the completed grafts into a controlled environment, such as a hot-pipe or a cold store for sub-cold callusing. See Appendix K: Grafting Machines and Grafting Tools for illustrations and description of mechanised grafting systems for wedge grafting procedures.

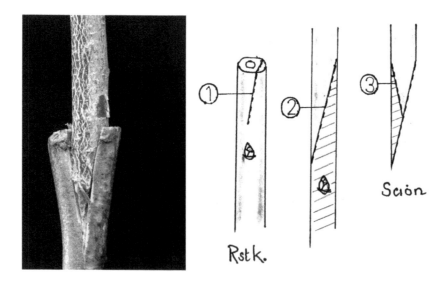

FIGURE 3.30   Apical wedge graft. LEFT, in this example the rootstock cut is made close to the diameter, almost parallel to the plane of the stem. RIGHT, the sequence of cuts is shown diagrammatically. Notice in the diagram that the rootstock cut is shown made off-centre, assuming that the scion diameter was less than the rootstock. The hatched area is discarded.

---

* Garner (2013) pp.189–190.

Rootstock Cuts for Wedge Grafts

Assuming the rootstock is prepared first, the sequence of cuts is shown in Figure 3.30. Positioning the rootstock cut is dependent upon scion stem diameter. Width of cut is greatest at the diameter (Figure 3.30), but to allow for smaller diameter scions it is important that adjustments of width are made by altering the starting position towards the edge (Figure 3.31). Depth of cut takes into account the length of the wedge, but is normally in the 20–35mm range. It is best to cut generously, as cut surfaces below the grafted area will quickly unite when tied together and on un-waxed grafts can provide an outlet in the early stage for excess sap exudation.

Making the cuts in stout rootstocks requires a heavy, sharp knife and some strength to pull the blade through the stem. Knife control is best achieved when the rootstock is held at an angle of 30–50 degrees to the floor or bench, the elbows are held well into the sides and the blade is pulled towards the grafter who is best wearing an apron. A slicing-pulling action commencing at the heel of the blade, slowly moving towards the toe, provides effective cutting. Generally, it is best to place the thumb of the knife hand on the stem and use this as a fulcrum for performing a controlled cut. Some prefer to rest the rootstock root system or pot on the grafting bench. Light grade or soft wood is always dealt with more easily.

Performing rootstock cuts on heavy wood is possibly more satisfactorily and safely achieved using grafting tools or machines. For small scale operations, well-sharpened double-cut or by-pass secateurs, particularly the former, may prove suitable (see Chapter 5 and Appendix K for further details).

Scion Cuts for Wedge Grafts

The objective is to produce an equal wedge-shaped profile at the base of the scion (Figure 3.31). The scion is held using the same stance and basic technique as for the splice graft. A slanting cut is made to remove wood from the scion, which is then turned over and the same procedure repeated on the other side. The angle of cuts to the plane of the wood can be similar to or slightly less than those for splice grafts. The two opposite angles reduce splice length and provide a wedge with an overall angle in the 8–17 degrees range (the wedge graft in Figure 3.31 has an overall angle of 9–12 degrees).

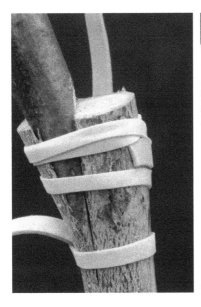 

FIGURE 3.31   Apical wedge: LEFT, wedge graft in position viewed from side. Note: the cut on rootstock is offset from the diameter to match the diameter of the scion providing good cambial alignment on each side. RIGHT, a hand-cut scion viewed face and edge-on.

Angles which are over-large produce a wide wedge, difficult to fit into the rootstock; too narrow give a flimsy scion, vulnerable to cut end damage and die-back. Species with light wood are often prepared while the grafter is seated; for hard, thick wood, standing may be preferred.

CAMBIAL ALIGNMENT FOR WEDGE GRAFTS

There is significantly more cambial area potentially available for contact compared with the splice; therefore, chances of graft union are good, even with rather poor alignment. Good cambial matching can only be achieved by leaving a church window on each side of the graft.

SIDE WEDGE GRAFT

This method, rather than the long tongue veneer, is sometimes used when 'multi-grafting' top-worked stems. It is especially relevant for more easily grafted types such as top-worked Flowering Cherries and Weeping Salix (Figure 3.32). The method used is similar to the apical wedge. Side cuts at an angle of approximately 15–25 degrees to the vertical are made at appropriate height into the rootstock. Wedge-shaped cuts on the scion are placed into these cuts. Normally, two scions are set opposite or slightly offset from each other and eventually provide a balanced head of branches.

FIGURE 3.32 Side wedge grafts on top-worked, weeping ornamental cherries. Note: cuts on the rootstock were made well off-centre to allow for differences in stem diameter. Placement of the two scions on the grafts at the left-hand centre and centre show the first stage in producing a well-balanced head (John Richards Nursery).

4 Fine Tuning Grafting: Positioning; Sequences; Training; Rates of Work

POSITIONING GRAFTS

An important aspect of grafting is positioning rootstock and scion cuts relative to each other and the root collar or soil surface. Consideration must also be given to avoid placement difficulties and, when appropriate, to include a stimulant bud.

Height above Root Collar or Soil Level

Graft placement above root collar or soil level is a matter of choice or, for commercial growers, specifications laid down by industry standards. The majority of species are grafted close to soil level, often 25–60mm above, when the graft is known as 'bottom-worked'. To accommodate features such as pendulous growth habit, top-working at a specific height to produce a half standard on 1.2 metres stem or a standard on 2.0 metres stem may be required. Other heights can be chosen, an example is the so called 'Patio plant'; this can be a tree, shrub or conifer grafted at a height calculated to enhance its aesthetic appeal, often between 45 and 80cm.

Ornamentals grown for their bark effect are normally bottom-worked as low as possible, but a few are deliberately top-worked with flowering types to take advantage of their attractive stem and provide a mix with floral interest. An example is the practice of top-working Japanese Cherries onto stems of *Prunus serrula*, with its appealing polished bark. The name Sheraton Cherry has been coined for these trees.

A number of species, including top fruit and nut trees, are grafted so that the rootstock can influence factors that may include control of vigour, cropping capacity and disease resistance. Because scion rooting results in loss of this beneficial rootstock influence, it is essential that graft placement is high enough to prevent any possibility of adventitious root production from the scion making contact with the soil; a minimum 150mm has been suggested*. Some apple varieties grown for top fruit may be grafted higher at 450–500mm, so that branches and subsequent fruits are produced at a convenient height for harvesting; the term 'knip boom' is sometimes used for this type of tree.

In contrast, a minority of species are grafted as low as possible, at or below the hypocotyl (junction of root and stem), the objective being to subsequently plant the graft union beneath the soil to encourage scion rooting. The ultimate example of this is root grafting, or inverted root grafting when the scion is grafted to the root (see Chapter 18: Grafting Systems).

Rootstock and Scion Placement – Stimulant Bud – Grafting Methods

Some recommend making rootstock cuts between nodes because the smooth area without obstacles (nodes and buds) simplifies production of accurate, flat cuts. However, for long tongue veneers, the presence of developing buds, 'stimulant buds', on the tongue of the veneer acts as a powerful

---

* Garner (2013) p.221.

43

stimulus for the production of callus tissue. Including a stimulant bud increases the difficulty of knife-work, disrupting blade movement by the contours and differential strength of the wood. Good knife control is essential to avoid the blade slipping badly out of position but once mastered the extra effort is rewarded by better takes.

A stimulant bud can also, with advantage, be in close proximity to the small flap at the base of the short tongue veneer graft rootstock. Here, challenges are concerned with careful positioning of the basal cut. Splice grafts can benefit from a stimulant bud, which should be located close to the tip of the splice on the rootstock and, if possible, on the proximal end of the splice cut on the scion (see Chapter 11: Stimulant Buds for Further Discussion). However, stimulant buds do complicate tying-in procedures and need careful consideration when used with hot-pipe systems as the heating channel will encourage unwanted growth extension of any buds close to the graft union.

Positioning cuts on rootstock and scion needs thought. For side grafts, cuts on the rootstock must be placed to avoid barriers caused by the stem above, or it can be impossible to fit the components together at the tying-in stage. Using curves on the rootstock, or rootstock and scion, can provide the necessary clearance (Figures 4.1 & 4.2). Cuts on the rootstock require careful placement; for example, choosing an outside curve can result in the knife running through the wood and slicing off the tongue (Figure 4.2). Alternatively, the inside of a steep curve can make the cuts difficult and placing components together impossible. Veneer scions are cut on the outside of any curve, making flat cuts easier and reducing problems of entanglement with stems above.

Bends and curves in scion and rootstock that can be 'matched' require identification. To achieve good alignment when placed together, these cuts must be made as mirror images of each

FIGURE 4.1   Placement of side graft on the rootstock is important. LEFT, shows *Viburnum* 'Eskimo' grafted to *V. lantana*; notice how the graft is placed in the inner curve of the lower portion of the rootstock with a curve away from the scion just above. The scion cut is placed to take advantage of a slight curvature away from the rootstock. The result is clear space between the two components allowing easy placement and tying-in. Note the presence of a stimulant bud at the base of the tongue of the graft and the large amount of callus produced in consequence. RIGHT, *Rhododendron* 'Sandringham' long tongue veneer side graft to seedling *R. macabeanum* rootstock. Notice that advantage is taken of the natural curvature of the rootstock to ensure the cut made into it does not 'run out' and slice off the tongue. The scion is cut on an outside curve positioned so that it fits neatly against the rootstock, especially important for a scion with such large leaves.

FIGURE 4.2 LEFT, *Styrax serrulatus* grafted to *S. japonica*, left graft well-placed on inner face of curve of the rootstock. Right graft poorly placed on the outer face of a curve; without good knife control there is a strong likelihood of slicing through the tongue of this side veneer graft. For this graft, the inner face of the curve would not allow easy scion placement because of the presence of a lateral branch above. Note: in both grafts, good callus formation showing as a white line under the green wax. RIGHT, *Pinus bungeana* multi-grafted to *P. wallichiana*: both lower and upper scions have been cut on the outer face of the curve to allow good clearance of the rootstock shoots and leaves above.

other (Figure 4.3). Placement of apical grafts is less demanding as there is no danger of the scion becoming entangled in the rootstock stem above. It is normally possible to cut through appropriate faces of a curve on rootstock and scion to produce a suitably shaped profile.

It may be necessary to change the grafting method for thin scion wood grafted onto a thick rootstock. To achieve good matching, the long tongue side veneer may be substituted by a short tongue, to avoid the need for an unduly thin tongue with insufficient live tissue to ensure its survival. Difficulties of finding scion material of sufficient girth may require use of older wood at the base. A significant difference in wood diameter between rootstock and scion with rare or endangered species may occur because there are no opportunities for choice. A better match is often possible by raising the grafting position upwards into younger, thinner rootstock growth (Figure 4.3). This improves the chances of success, as, in addition to better matching, high grafts improve takes compared with those placed close to root level (see Chapter 11).

For placement in hot-pipe systems (see Chapters 16 and 18), side grafts require special consideration. To ensure good heat insulation it is essential that, when assembled, scion and rootstock lie in a reasonably flat plane, tightly 'sandwiched' between the plastic foam edging. To achieve this, careful choice of graft placement is necessary to match the profile of rootstock and scion. Sometimes it is impossible to place side grafts into slotted systems. In contrast, provided they are well placed, they normally fit well between solid cover plastic foam insulating layers (Figure 4.4).

PRIORITY – SCION OR ROOTSTOCK?

Many favour preparation of the rootstock first, but choice is influenced by a number of factors. For bare-root grafts, particularly with a large differential in scion and rootstock diameter, there is a case

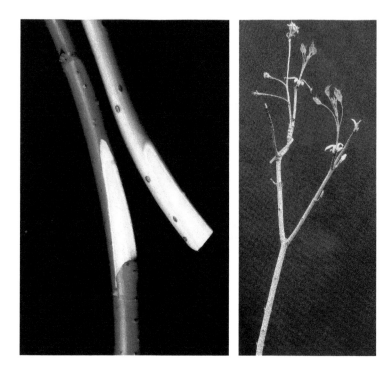

FIGURE 4.3   LEFT, short tongue side veneer cut on curved rootstock, mirror image cuts made on scion to provide a good match. RIGHT, scion of *Acer tenellum*, a very thin-wooded species grafted onto upper young shoot
of *Acer platanoides* rootstock. Note: rootstock cuts on the inner side of the curve and scion cuts on the outer.

FIGURE 4.4   *Magnolia* side grafts positioned in a hot-pipe system using plastic foam edging to insulate the
heating chamber. Grafts are placed carefully on the rootstock to provide a reasonably flat plane for close fitting
of the foam of the base and covering plate, not yet in position. Wooden block at centre of photograph is used
to help guide positioning of the cover. The thermostat probe is just visible between two top positioned grafts.

for following field grafting procedures and cutting the scion first. Most field grafters who have 'evolved' into bench grafting will automatically use this method. It has the advantage of giving a perfect blueprint of the dimensions of the cut surface of the scion and facilitates accurate judgement of matching cuts to be made on the rootstock. The disadvantage of laying the prepared scion down on the bench while making rootstock cuts may be avoided by holding single stem scions between the fingers (usually third and fourth) of the 'knife hand'. Practice is required to achieve this skill. Scions such as branched, multi-stemmed or evergreens are normally too large and are laid down while preparing the rootstock.

When using potted rootstocks, especially if veneer or side grafted, greater output is normally achieved by preparing the rootstock first.

SEQUENCE OF ACTIONS FOR PERFORMING THE GRAFT

Understandably, views differ on the best sequence of actions to produce the graft. If a grafting team is involved, a uniform approach aids efficient work. Consensus should be sought before arriving at a 'standard' procedure. Once agreed upon, the procedure should be built into training practices and everyone should be encouraged to follow it.

When preparing the rootstock first and using potted or bare-root rootstocks, a suggested sequence might be as follows:

Right-handed grafter apical or side grafting:

1. Pick up rootstock with left hand.
2. Pick up knife with right hand.
3. Trim rootstock with knife further, if required.
4. Make graft cut(s) on rootstock with right hand and return to bench with left hand – retain knife in right hand.
5. Pick up scion with right hand – transfer to left hand – trim further if necessary.
6. Make cut with right hand on scion to match rootstock (after completion the knife may be retained between the fingers, usually second and third).
7. Transfer prepared scion – left to right hand.
8. Pick up prepared rootstock with left hand.
9. Fit graft components with right hand and hold in place with left hand.
10. Pick up graft tie with right hand.
11. Tie-in graft with right hand.
12. Apply sealant (brush or dip) with right hand – while holding graft with left hand.
13. Place graft tied and sealed in handling tray with left hand.

Repeat sequence for next graft(s).

Left-handed grafter: sequence as above with opposite hands. Note: if possible, the materials and equipment should be arranged on the grafting table to facilitate left-handed grafters.

Suggested layout of grafting materials on the bench will be influenced by the sequence followed and is described and discussed in Chapter 4. In Appendix M: Grafting Bench Design and Construction, the grafting bench is illustrated and described.

TRAINING AND PRACTICE

Training is essential and best given on a person-to-person basis, with opportunities for frequent follow-up checks and advice. With good tuition, given time, commitment and suitable practice material, most individuals can achieve good results. Essentials are dogged determination and time to repeatedly practice the movements again, again, again and again! A few individuals are unable to acquire the necessary co-ordination and manual control, and their abilities must be channelled elsewhere. Once trained and in the grafting team, beginners should be re-appraised at sensible intervals and not expected, until ready, to attempt grafting the more demanding species.

An important issue is the choice and provision of an appropriate knife. Trainees frequently present themselves with a totally unsuitable type, incapable of producing and maintaining the cutting edge required to ensure success. Reasons for this are lack of information on available models and an understandable lack of financial commitment. Good quality grafting knives are not cheap, but the alternatives will not meet expectations. A comprehensive review of grafting knives is given in Chapter 5. At this stage, if financial considerations are important, a recommended choice for trainees is the fixed blade type, which can provide top quality blades and handle design at a substantially lower cost than folding blade alternatives. The Tina 685 model fulfils these requirements and is fully described in Chapter 5.

Genera such as *Aesculus* and *Alnus* with soft, rounded wood are good subjects for early practice*. Use of *Cornus alba* Sibirica, with red twigs, and *Cornus sericea (stolonifera)* Flaviramea, with yellow twigs, for the scion and rootstock provides visual impact[†]. These combine the advantage of being soft-wooded and fully compatible. Completed grafts can be wrapped in moist paper towelling, placed in polythene bags and stored in a warm chamber; an airing cupboard is suitable. At a temperature of ±24°C, callus formation is rapid, and in 2–3 weeks the union is complete. Subsequently the rootstocks may be rooted.

Cut shoots are commonly used for training practice but they have disadvantages. Whilst suitable for scions, the rootstock has roots attached either as a bare-root system or in a rootstock pot. It is probably not essential to simulate the bare-root rootstock, but a pot grown rootstock is significantly different from a detached shoot. Use of established, potted rootstocks for practice is commonplace but expensive, often resulting in insufficient material available for unlimited use. A better solution is to simulate potted rootstocks using model pots, called here practice holders, equipped with a gripper mechanism to hold a suitable shoot (Figure 4.5). These may be easily constructed at modest cost and have long life, allowing re-use for many years. Various models can be made to match pot sizes in use and give the authentic feel and handling characteristics of potted rootstocks (see Appendix C

 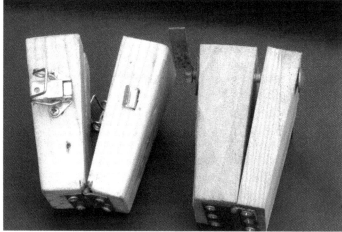

FIGURE 4.5   LEFT, a practice holder, in this case a copy of 9 × 9 × 12.5cm pot, with simulated untrimmed dormant deciduous rootstock in position. RIGHT, two patterns of holders using Protex clip (left) or long bolt (right). See Appendix C for details of construction.

---

* Macdonald (1986) p.187.
[†] Meyer Jr. (1977).

for details of fabrication). If several are available, it is possible to provide trainees with the chance to build speed and meet requirements for 'real life' situations.

Shoots of all types can be held in the holder, providing an opportunity for gaining experience with broadleaf and conifer species (Figure 4.6). It has proved ideal for demonstrating and practising skills involved with many grafting methods and reduces loss of valuable rootstocks.

Once the trainee has reached reasonable competence, it is important to provide genuine rootstocks and scions for grafting practice. These should subsequently be taken through normal procedures of tying, sealing and housing, under conditions that permit graft union development. Successful achievement of a take is a confidence booster for trainees who have worked hard to acquire the necessary skills.

RATES OF WORK

Grafting outputs are often a source of intense interest, debate and, occasionally, confrontation. Various figures for outputs are given in Appendix D: Outputs for Bench Grafting. An upper limit for veneer, side veneer and long tongue veneer is in the region of 100 grafts per hour, but 40–50, especially for more demanding species, can be considered acceptable. Incentive payments would normally increase output, but quality standards may suffer. Due to the extra time needed to produce the tongue, whip and tongue grafts are achievable in times similar to veneer. Many would argue that splice grafts can be produced more quickly and rates of 100–150 grafts per hour are frequent. Some growers would claim higher outputs than those above, but the problem of standardising procedures and work patterns make comparisons difficult. Undoubtedly, a few highly skilled individuals specialising in a particular method are able to achieve exceptionally high numbers of grafts per hour. In the absence of persons with this level of skill, grafting machines and tools, when used for suitable species, can sometimes substantially increase output.

FIGURE 4.6 Practice holders used to simulate potted rootstocks. LEFT, *Pinus sylvestris*; CENTRE-LEFT, *Chamaecyparis lawsoniana*; CENTRE-RIGHT, *Acer palmatum*; RIGHT, *Rhododendron* 'Cunningham's White'. Simulations of this type allow trainees to also gain experience in preparing the rootstock prior to the grafting procedure.

5 Grafting Knives: Grafting Machines and Tools

KNIFE SELECTION

A selection of knives is available for a varied range of requirements, from the most demanding to those for short term use. Knife design falls into three general types: folding blade, fixed blade and disposable blade.

FOLDING BLADE KNIVES

Despite the extra cost and weight of folding blade knives, the range of models on offer and their convenience for storage and safety make them by far the most popular. They are eminently suitable for more demanding grafting methods, but in some situations their cost may not be justified, and the cheaper fixed blade and disposable types are sufficient.

There are far fewer manufacturers of folding blade grafting knives now than in the past; one German make, Tina, dominates the scene. Other makers are to be found in Switzerland, France, USA, Italy and Japan. For models and types see Appendix E: Knife Types Other than Tina.

Before discussing the most suitable blade pattern, requirements for design and construction of a good grafting knife need consideration.

Design and Construction

General balance should be good, giving a comfortable feel with a point of balance well into the handle. Wooden handles are preferable to avoid any feeling of stickiness by adjusting to body temperature. Plastic or composition handles weigh less and, in certain situations, such as the wish to retain the knife in the hand while tying-in, a light small knife has advantages over heavier, more bulky models.

Best handles for balance and durability combine the use of suitable hardwood, such as walnut or rosewood, with good shaping, comfortably fitting into the contours of the hand. Rivets normally attach the handle to the metal frame; these should be of non-corrosive metal, ideally brass, well attached and sufficient in number to ensure strong, permanent fixing. Folding mechanisms should be positive in action, closing the blade down in a safe, controlled manner. The closed blade should fit snugly in position, the point fully enclosed within the handle.

Importantly, the blade should be set deeply into the handle to ensure it is well supported against sideways forces during use. Failing this, the blade will loosen within its mount and develop sideways movement. In worst cases, the blade eventually flops within the handle, becoming unserviceable. Before purchase it is recommended to carry out a visual inspection together with testing by applying sideways force to the blade; any knife showing slight movement should be rejected. Even expensive knives purchased from the best manufacturers have a low percentage with this fault.

Blade Steels

The knife blade is made from alloy steel formulated for this use. Some alloys are available off the shelf but many of the best knives are made from alloys using specific formulae. The most important characteristic is the ability to be sharpened to a very fine edge that is retained for a substantial period, ideally a working day. Blades with high stainless steel content resist corrosion but cannot form and hold the extremely sharp edge required in a grafting knife.

Knife blade steels need to be hardened through heat treatment, which directly affects their ability to hold an edge and resist wear. Other properties include flexibility, combined with sufficient strength to match forces generated when cutting through tough wood. Some degree of corrosion resistance is required together with the capability to be ground and formed into a working blade. Many of the highest quality knives are crafted individually by hand, using combinations of traditional skills and modern manufacturing processes.

Blade Design and Fabrication

Design and fabrication of the blade is crucial. Straight edged blades, known as 'Sheep's Foot', greatly facilitate sharpening and should always be chosen for bench grafting (Figure 5.1).

Curved blades have a place in field grafting where, due to the presentation angle of the blade, less bending is required, because to facilitate making cuts upward force can be applied more easily. Some bench grafters who have evolved their technique from field grafting still favour curved blade knives. When used for bench grafting, the extra sharp edge produced on a well-sharpened straight blade will obviate the need for any curve.

The majority of fixed and folding blade grafting knives have a thick edge called the spine, which gives strength and rigidity to the blade. The blade is then ground down by the manufacturer from the spine to the cutting edge to produce the inclined angle or primary bevel.

The spine width is reduced from heel to toe to improve balance and provide a sharper point. Finally, the manufacturer grinds a cutting edge, which is normally adjusted or removed by the user in the initial sharpening procedure.

The best grafting knives have the inclined angle (primary bevel) and cutting edge (secondary bevel) on one side only, the other being flat; the knife is then known as a single-angled knife. It is generally agreed that this design aids the production of flat cuts. Single-angled knives are made for right- or left-handed individuals. To clarify this, when the knife for the right-handed grafter is held vertically (blade uppermost) in the right hand, the cutting edge faces the shoulder with the flat edge facing across the body to the left and vice versa for the left-handed individual. It is important to be aware of this when purchasing grafting knives of this type. It is not possible to execute well-made, accurate cuts using the wrong orientation.

Most general-purpose knives have a double-sided inclined angle, and although not ideal, these can be used for grafting. The advantage of these is that they are suitable for use by right- or left-handed grafters.

Blade grind types producing the inclined angle, edge angle, and single- or double-sided blade profile will influence sharpening procedures to be discussed below and illustrated in Appendix F.

Folding Blade Models

The German company Tina offers the most important and comprehensive range of knives for grafters. Comfortable well-balanced handles of walnut are coupled with blades that sharpen to razor

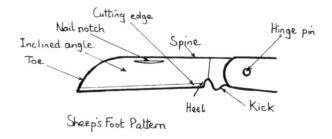

FIGURE 5.1   Sheep's foot blade of a right-handed, single-angled knife with parts named.

sharpness. Skilled sharpening will produce an edge, which is usually retained for the working day. After normal use, polishing the knife on a leather strop, very fine stone or sharpening machine, quickly transforms the blade back to the working edge. With correct re-sharpening procedures, the knife will remain serviceable for many years and 20-year old models (600A) are still in use.

Two models, the Tina 605 and 600A, are ideal grafting knives. The choice between them is mainly based on personal preferences and requirements, but the narrower blade of the 605 makes perfectly flat cuts more difficult.

The Tina 605 is a single-angle, smallish knife; weight 57gm; handle length 100mm; blade length 60mm; blade width 12.5mm; better suited for species with thin wood, ideal for most bench grafted subjects. These features make it possibly the most popular knife used for bench grafting. Its lightness and size make it ideal for use by women grafters (Figure 5.2).

The Tina 600A is a single-angle, comparatively large knife; weight 110gm; handle length 120mm; blade length 77mm; blade width 15mm; it can be used for heavy, thick wood but also fragile, delicate grafts. The broad blade facilitates production of flat, well-matched cuts, which promote fast development of cambial connection between rootstock and scion. Because of its size and weight, it is not well-suited to holding between the fingers while tying-in the graft, so for this reason not favoured by some grafters. Superior sharpening tools will give this knife an unrivalled cutting edge.

FIGURE 5.2   Tina 600A knives showing the inclined angle and cutting edge of a single-angled, left-handed grafting knife on the right and flat face on the left. The knife in the centre is the smaller Tina 605.

Although substantially more expensive than the 605, the larger blade gives a longer serviceable life and the economics are open to debate (Figure 5.2).

When making a choice, the requirements described for a quality folding blade knife also apply to the fixed blade variant. Known in the USA as 'Stationary Knives', they have advantages over conventional folding blades. Simpler construction results in reduced cost and nearly 50% less weight for an equivalent size. The absence of a folding mechanism means the blade is held very positively and is much less inclined to become loose in its mounting. Disadvantages are less choice of blade sizes and patterns, and a need to protect the blade from damage and avoid accidental injury. Unfortunately, these knives are rarely sold with a durable blade cover, essential when the knife is not in use. Various simple covers held in position with elastic straps, or similar, can be devised to overcome this deficiency. Figure 5.3 shows the use of a leather glove finger held in place with an elastic strap located within a small notch cut into the wooden handle end. This provides a cheap, simple and efficient method of protection, easily fabricated 'in house'.

The single-angled Tina 685 provides an excellent choice when quality and economy is important, such as for those at the training stage. Details are as follows: weight 30gm; handle length 105mm; blade length 65mm; width 12.5mm. This has a sheep's foot design blade with inclined and cutting edge angles on one side, flat on the other and is very comparable to the 605 folding model at a significantly reduced price. Left- and right-handed models are both available* (Figure 5.3).

Blade care, sharpening and general care of the fixed blade knife is as for folding blade models.

DISPOSABLE BLADE KNIVES

These knives have rather thin handles and the exposed blade is short, making execution of some grafts difficult. Further disadvantages are that the blade is double-edged, so flat cuts are not as easy to achieve. Working with hard wooded species quickly blunts the cutting edge.

Normally the choice is between two designs: an elongated blade and a double-ended blade:

1. The elongated blade has a short section, the cutting portion, protruding from the handle at one end. When blunt, the next un-used section is exposed by a thumb-operated sliding mechanism and the old portion is snapped off at a preformed indentation. This procedure is repeated as each section becomes blunt. Eventually the blade is used up and needs replacing (Figure 5.4).

FIGURE 5.3   Tina 685 fixed blade knife. LEFT, a right-handed model is shown. RIGHT, a simple, easily removed cover, made from a leather glove finger, attached by an elastic strap, held in position by a notch cut into the handle.

* S.Schwille from Tina Messerfabrick (pers. comm.).

2. A double-ended blade held within a well-designed, comfortable metal handle. When blunt, the blade is removed and reversed to expose a new edge. Eventually replacing the double blade is necessary (Figure 5.4).

This design provides a more comfortable, better balanced knife than the elongated blade design but suffers from many of the disadvantages noted previously.

Economical and not requiring sharpening, disposable blade knives have a role for use in certain, less-demanding procedures. Although they do not meet many of the criteria for good knife design and construction, they are suitable for soft wooded (herbaceous) species such as Gypsophila, or where only simple, short cuts are required, such as for the scion of wedge grafts.

KNIFE SHARPENING

Blade design and the cross-section profile of the blade (grind type) influence the appropriate sharpening procedure for any knife. Economically priced knives are normally of uniform thickness with a cutting edge (edge angle) to make cuts; better quality knives have a profiled blade with an inclined angle and cutting edge Best knives for grafting have one side flattened, leaving the inclined angle and cutting edge on one side only (see Appendix F for details of grind types).

HAND SHARPENING – EQUIPMENT AND MATERIALS

The new knife will require sharpening as it is supplied with a factory-finished cutting edge, which is often too steep. Knife sharpening is a crucial operation. Properly carried out it can produce a razor-like finish, which significantly aids the speed and accuracy of making grafts.

Most grafting knives are sharpened by hand, normally using flat sharpening stones lubricated with oil or water. Producing the final cutting edge is achieved by using stones in sequence according to their abrasive qualities, commencing with coarse grade (<1000 grit) and finishing with a fine grade (6000–10,000 grit) or ultra-fine grade (30,000+ grit). Because of the cost of ultra-fine stones, most final edge finishing and polishing is achieved with leather strops dressed with suitable abrasives.

For full descriptions of types of sharpening stones, i.e. natural, synthetic, etc. (see Appendix G: Types of Sharpening Stones).

FIGURE 5.4 Two patterns of disposable blade knife. Top, a double-ended blade pattern. Bottom, a snap-off blade pattern. Notice the relatively short length of blade protruding from the handle of each type of knife.

Choice of Cutting Edge Angle

For single-sided (single angled) knives, with one flat face, the recommended cutting edge angle is 20 degrees but the best knives can maintain 17.5 degrees. When working with thick and hard wooded species such as *Quercus*, *Fagus*, as well as some of the Rosaceae, and using mainly splice grafts, the choice of 20–22 degrees may be wise. With thinner wooded species such as Japanese maple, *Cornus* and *Hamamelis*, an angle of 17.5 degrees provides extreme accuracy, speed and minimum effort to perform the cuts. Double-angled knives unavoidably produce a wider cutting edge because each side contributes to the angle; therefore, angles less than 25 degrees are not normally possible without loss of edge strength.

Stone Choice and Maintenance of Sharpening Stones

A flat stone is essential for straight blades and although rarely possible the stone should ideally be wider than the length of the blade. More complicated movements are required if a narrow stone is used, reducing the possibility of grinding a perfectly flat, accurately angled edge. Stones distorted and showing loss of flatness through wear must be repaired or replaced. Flat diamond stones or wheels, known as lapping plates, may be used to true-up worn sharpening stones before re-use.

Periodically, it will be necessary to clean the stone using paraffin for oil stones, water for wet stones. Exceptions are stones used without lubrication, which must be kept scrupulously clean. Stones should be covered when not in use as dirt can damage the blade's cutting edge and reduce sharpening efficiency. Use only light oils for lubricating oil stones; thick oil is not suitable as it causes the blade to float over rather than bite into the stone. For ultra-fine grit stones, it is advisable to thin the oil further with paraffin. Wet stones require water for lubrication, best applied by a small hand syringe. All lubricants should be applied evenly over the surface.

Procedures

It is important to fix the stone securely in position to prevent movement during sharpening, most easily achieved by mounting it in a Mounting Box, or position and fix it at right angles to a bench stop. Usual placement of hands and fingers during sharpening is shown in Figure 5.5. Right-handed individuals hold the knife handle in the right hand, then steady and fix the blade with the fingers of the left. Positions are reversed for those left-handed.

FIGURE 5.5 Sharpening a grafting knife on a sharpening stone. Configuration of fingers and hands for right-handed grafter is shown. Note that the knife blade needs to be placed at an angle to the plane of the stone to ensure the entire length is in contact.

As the blade is moved along the stone, skill and practice is required to make the small adjustments to configuration necessary to maintain the chosen edge angle. Opinions differ on whether the knife should maintain contact or be raised on the return movement. Removal is likely to disrupt delicate adjustments and continuous blade contact is recommended. Less pressure should be applied on the return movement.

If the stone is sufficiently wide, the blade is placed at right angles across the width, held at the cutting edge angle and pushed away from the operator along the length of the stone.

The process of pushing away and returning with less pressure is then repeated until a flat even edge at the chosen angle is produced. It is important that the angle of blade to stone is maintained throughout the hand sharpening procedure. This is not easy; time well spent for beginners is to practice using a wooden blade (a wooden ruler is suitable) and block covered with fine grade abrasive paper, to simulate knife and stone. This will provide the operator with a feel for appropriate speed and the best method of holding the blade at the correct angle.

For stones that are narrower than blade length, one of two options can be taken. The blade is placed on the stone at an angle so that the entire length of the blade comes into contact with the stone (Figure 5.5). The alternative method is to use a circular motion (Figure 5.6) so that the full length of the blade is sharpened.

Double-sided blades require the procedures described above for each side, increasing the time taken and demands of sharpening.

Techniques for Curved (Hooked) Blades

Curved-edged (hooked) blades present special difficulties and require a number of adjustments to the procedures described above. Grinding an accurate cutting edge angle is more difficult and, due to unavoidable variations along the length of the blade, is likely to finish with an increase of 2.0–2.5 degrees over that recommended. A maximum angle of 22 degrees should be the aim; double-angled blades may finish with a higher cutting edge angle of 25 degrees or more.

Maintaining the smooth curve produced during manufacture requires care and skill, the tendency is to gradually grind away the curve producing a straighter outline. This is an almost inevitable result if a flat, broad sharpening stone is used. Curved bladed knives require rounded stones, with the profile of the curve maintained. Sharpening may be aided by using narrower sharpening stones; the sharper the curve, the narrower the stone. The stone is placed flat on the bench, as for straight-bladed knives, and sharpening achieved using the previously described circular sharpening

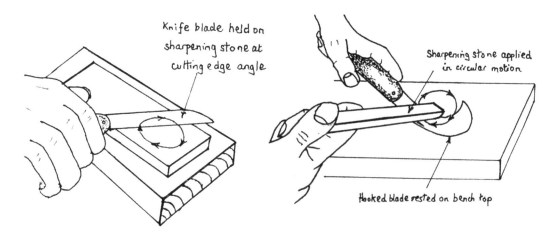

FIGURE 5.6 LEFT, a circular sharpening technique used for a straight-bladed knife which is too long for width of sharpening stone. RIGHT, one technique for curved (hook) blade using a narrow sharpening stone applied to the blade.

technique (Figure 5.6). As sharpening proceeds, some adjustments of the knife angle across the stone will be required to make allowance for the curve. An alternative method, particularly suitable for steeply curved blades, is to rest the blade on a bench and apply a narrow curved stone to the edge rather like using a file (Figure 5.6). Sharpen using circular movements while holding the stone as accurately as possible at 20–22 degrees to the blade.

STAGES IN SHARPENING – BURR REMOVAL – HONING

After grinding-in to the correct edge angle using a sharpening stone of 500–1000 grit grade, a tattered, serrated and loose edge, known as a burr is formed. This is visible with a ×10 lens and can be detected by touch. When viewed under the lens it is seen as a tissue-thin serrated projection of metal along the sharpened edge. It should have developed equally and evenly along the length of the blade and must be removed to allow final sharpening.

Once the burr has formed the first stage is completed. Subsequent phases are to remove the burr and smooth and polish the faces. This is divided into two or more steps using a stone of 3000–5000 grit (medium grade) followed by a fine (honing) grade finishing stone of 6000–10,000 grit. The same procedures and techniques described above are used for these finer grade stones. Use of honing stones may be replaced by honing procedures using leather.

Honing

Honing refers to the final stage of sharpening and involves the use of a finishing stone, leather, or other material to remove any burr on the cutting edge and produce a polished finish to the blade. Stropping is fully described below and used to describe one method of honing, using leather or leather substitutes, applied to the blade in a particular way.

Honing should continue using movements as before until the burr is barely discernible. Having completed the angled cutting edge side, turn the blade over and press the flat side against the stone, honing this until the burr is no longer detectable by sight or feel. The double-angled blade requires honing at the correct cutting angle on each side until the same result is achieved. The sharpest edge can only be produced by finally honing on 10,000 grit or ultra-fine 30,000+ grit stone. For substantially less cost, equally sharp knives can be produced using leather-faced, bench or paddle strops, dressed with sharpening pastes or sprays.

Curved blades also require honing but there may be difficulty in finding suitable fine grit rounded honing stones. To produce correct curves, it may be necessary to adjust the profile of the stone with diamond lapping plates or wheels. Wooden backing boards for leather bench or paddle strops can be shaped to the desired profile, the leather fixed to this using impact glue. For single-angled curved blades the flat side is honed on a flat hone as for straight blades. The aim should be to finish with highly polished surfaces on each side.

Stropping

In practice stropping is a popular and much used method of removing any burr and polishing the blade to sharpness. The stropping technique is to hold the knife against a dressed, leather surface known as a strop which is suitably supported by a bench or paddle backing (see Appendix H). Dependent upon configuration, the knife is held against the strop, at the cutting edge angle or flat, as appropriate. It is pushed down the strop with the cutting edge facing the operator, the spine facing away (Figure 5.7). On the return stroke it is turned over on its spine, pulled towards the operator, while being held against the strop with the cutting edge facing away. In both movements remember that the cutting edge angle must be polished against the strop only by *pushing or pulling away*; pushing forward will result in damage to strop and edge. To enhance sharpness, ensure that stropping continues until both the inclined angle and flat side are smooth and polished to a mirror-like finish.

Using hand stropping techniques over time inevitably produces a convex cutting edge (see Appendix H). This is bound to reduce sharpness; however, a highly polished, reasonably angled,

FIGURE 5.7   Paddle strop being used to hone a right-handed grafter's knife. The flat face of the knife is held flat against the leather, which has been dressed with Tormek honing paste, and is pushed away from the operator. For right-handed grafters at the end of travel the knife is turned over on its spine (rotated clockwise) so that the cutting edge faces away, and while being held at the cutting edge angle, it is pulled towards the operator and handle end of the strop. At the end of travel towards the grafter, the knife is rotated anti-clockwise on its spine and pushed away, held flat on the flat face. This activity is repeated and continues until the knife is fully honed. Arrows mark only direction of travel and not rotation. For left-handed grafters, procedure is the same except rotations at the end of each strop (still only on the spine) are reversed.

convex cutting edge provides satisfactory results for a long period. To restore original sharpness, it may eventually be necessary to re-grind the cutting edge to the desired angle, as for new knives.

See Appendix H for a description of types and materials for hones, strops and corrosion inhibitors.

Cutting Edge Retention

To retain the cutting edge on a properly sharpened knife, regular honing or stropping is required. Usually once or twice a day is sufficient, more often with extra heavy/hard material. The aim is to retain the cutting edge angle and polish the surfaces. Unless the knife is abused by chipping, cutting into gritty, dirty material or corrosion due to neglect, there is only rarely a need to resort to coarse, medium or fine grit stones.

KEY POINTS FOR EFFECTIVE HAND SHARPENING

The skills of knife sharpening by hand are under-rated and appear to be not well understood. They require practice and experience. The keys to success are:

- Maintain the correct cutting edge angle as accurately as possible.
- Sharpen until an even burr is produced along the length of the blade.
- Hone and polish edges to remove the burr.
- Continue polishing until a smooth mirror-like finish is produced on all surfaces.

Of these, maintenance of the correct angle for the cutting edge is the most difficult. Almost inevitably, the cutting edge is reduced first to a multi-bevelled edge and subsequently, due to honing or stropping, a convex edge (see Appendix F for full description). A narrow angled, highly polished, convex edge provides a very sharp knife, allowing most grafters to complete fast and accurate work. A flat or concave cutting edge will further improve the ease, speed and accuracy of their performance.

MACHINE AND MACHINE-AIDED KNIFE SHARPENING

There are alternatives to the hand sharpening procedures described above. However, there appears to be no fully automated sharpening machine for use by grafters. The aids and machines described below require some level of manual input. Good equipment properly used enables accurate, flat or concave cutting edges, to be consistently produced.

Sharpening Aids

These fall into one of two categories:

1. Sharpening guides. The knife blade is placed in a mobile holding device (guide), which can be adjusted to set the desired angle. Blade and guide are pushed across the sharpening stone surface simulating hand sharpening procedures. As the angle is mechanically fixed, maintenance of consistent, accurate grinding and honing is possible.
2. Rod-guides. The knife is held in a fixed position by clamping into a jig. A sharpening assembly, rather like a small metal file, makes contact with the blade at the desired angle by use of a second jig, which is adjusted to give the required sharpening angle. The edge is ground onto the blade using the same action as for a hand metal file.

For further description and discussion of sharpening aids see Appendix I: Sharpening Aids.

Sharpening Machines

These are characterised by the use of a power source, normally an electric motor, which drives an abrasive wheel or belt. To achieve the sharpest, most durable edges, certain design features are necessary. A basic requirement is a slow revolving speed to avoid overheating the blade steel, causing loss of hardness. Most workshop grindstones revolve too quickly and are unsuitable. As previously discussed, ultra-fine grit finishes are required and must be available.

The best sharpening machines are capable of producing the sharpest possible cutting edge. They can also retain this in near perfect conditions for many grafting seasons (Figure 5.8). Hand sharpening is unable to rival their accuracy and speed, and for medium- and large-scale grafting programmes the cost of such machines is warranted. For further description and discussion of sharpening machines and aids see Appendix J: Knife Sharpening Machines.

TESTING FOR SHARPNESS

Several tests have been suggested for ascertaining sharpness of the blade. One recommendation is the ability to cut paper. This does not provide a good guide if the blade is applied in a sawing motion, even a well-sharpened meat carving knife will have this effect. The blade must only be lightly pressed against the paper edge. The standard method amongst male grafters is to see if the fore-arm can be shaved with and against the direction of the hair; if this is possible it is a good indication of sharpness. Lady grafters must use the paper cutting test or ask their male colleagues to assist.

KNIFE MAINTENANCE

At the end of each working day, after thorough cleaning on a sterilant pad, the knife should be stropped/honed. This ensures removal of accumulated plant sap and any dirt and restores the

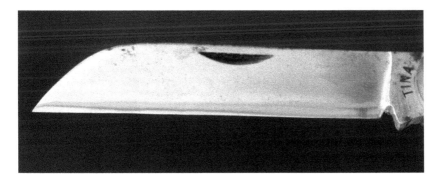

FIGURE 5.8   Tina 600A grafting knife for a right-handed grafter, with a sheep's foot pattern blade. This is sharpened using a Tormek Super Grind sharpening machine. Note the even, parallel cutting edge which has a concave profile at a precise angle of 18 degrees, also the polished surface of the inclined angle face. The upper edge is the 'spine', with the point known as the 'toe' on the left. To the right is the 'heel' and below this the projecting portion (below 'Tina') is the 'kick'.

polished surface. The blade should then be smeared with a light oil to prevent any possibility of corrosion, which can rapidly develop on the best steels. At regular intervals the hinge point should be sparingly oiled, excessive lubrication can result in undesirable oil seepage into the handle.

For lengthy periods of storage, the knives should be housed in conditions of low humidity, not left in the propagation house (see Appendix H: Honing – Strops, Honing Machines and Corrosion Prevention).

After much use, the blade will become narrower as use and sharpening wears it away. Eventually the blade of folding knives will not fully close into the handle. The unprotected point (Figure 5.9) may cause injury. This must be corrected to ensure the blade fits properly by removal of a portion of the 'kick', a procedure requiring care as the blade can be easily damaged. The recommended hand file must be used only when the blade edge is properly protected. A convenient way is to cover the edge with several layers of newspaper held in position with insulating tape or 'Sellotape'. With care the 'kick' can then be filed back until the blade is able to be fully housed within the handle. If available, a light weight mini-grinder can replace the file, but care is needed or too much 'kick' is removed and the blade becomes inaccessible within the handle.

GRAFTING MACHINES AND GRAFTING TOOLS

Achieving and maintaining output from grafting machines or tools is very dependent upon grading and selection of plant material for suitability and uniformity. Species most suited are those that may be easily grafted and are capable of producing straight stems, Grapevines being the perfect example. Suitable stems may be promoted by using appropriate pruning techniques requiring the establishment of stock mother trees, together with good cultivation practices and pruning systems (see Chapter 10: Scions).

Rootstocks should have uniformly straight stems and exhibit well-balanced, reasonably symmetrical root systems. For best results, rootstock and scion should have equal diameter at the grafting position. When possible, these criteria should be applied to all grafting material; if they cannot be achieved, results are often unsuccessful and grafting by hand is the only feasible alternative.

GRAFTING MACHINES

No machine has so far been produced that exhibits the level of flexibility, accuracy and precision of an experienced grafter working with a knife. However, with suitable material, the level of skill required is reduced and higher sustained output is possible compared with manual grafting methods.

FIGURE 5.9   Knife blade worn down showing a small amount of cutting edge at the 'toe' (point) end, consequently the 'kick' needs grinding back.

Apart from the possibilities of higher output the advantages of machine over hand grafting have been identified* as follows:

- A single day of training is usually sufficient to achieve competence. Significantly less than for grafting by hand.
- Injuries such as laceration, muscle strain and repetitive strain syndrome are virtually eliminated.
- Machines have the ability to use large/hard wooded material, reducing wastage because less is rejected as oversized.

After each component is cut to shape by the grafting machine, hand assembly is normally necessary. Once assembled, scion and rootstock may be held together by disposition of the cuts, as in Omega grafts; fixed by mechanical aids such as staples; or tied together by hand (see Appendix K: Grafting Machines and Grafting Tools and Alternative Graft Fixing Methods within Chapter 6).

Grafting machines have their place for certain species. Classic examples are Grapevines (*Vitus*) but also Cape Gooseberries (*Actinidia*) and to a lesser extent some of the Rosaceae. A significant factor is the number of grafts required; to set up an efficient machine grafting system, there has to be a need for many thousands of grafts of the same species. Some large companies in the USA, Europe and Australasia have developed machine-aided systems 'in house' to meet particular specifications. One for fruit and nut trees utilises compressed air as the power source and requires five teams of

---

* Patrick (1992).

employees to service the machine*. An assembly line is used to independently prepare scion, root-stock, attach the components, wax and store for further handling. This system has been developed to meet the requirement for 2.5 million plants per year.

GRAFTING TOOLS

Grafting tools may be distinguished from grafting machines because more manual input is required to operate them. Many rely on hand operated secateur-like devices, which cut variously shaped profiles subsequently fitted together by hand. When fitted with mechanical aids, such as pneumatic power, tools are converted to achieve substantially increased output and might be considered as grafting machines. Used manually, with their greater flexibility, grafting tools have fewer demands on choice of material and target numbers. As a result, it is possible for the grafter to use some degree of judgement regarding positioning of cuts and component parts. Flexibility is increased but output is reduced and there is need for higher levels of skill. Not surprisingly, when using grafting tools, competent hand grafters achieve significantly better results than inexperienced operators.

For further details, illustrations and descriptions of grafting machines and tools see Appendix K: Grafting Machines and Grafting Tools.

---

* Patrick (1992).

6 Ties – Tying and Sealing

BASIC PRINCIPLES

It is essential for graft union formation that the components are held together under pressure. This recreates conditions naturally occurring within the stem and root of living plants, where considerable internal compressive forces induce essential cell differentiation and structural development (see Chapter 11 for further discussion).

Pressure on the graft interface, and consequently on any developing tissue in between, may be achieved in various ways, usually by tying together (tying-in), but alternatives such as pegs or pins can be considered. Some machine-made and hand-made grafts are not tied as sufficient pressure is generated by the strength of the wood under tension.

The amount of pressure applied is the subject of debate. Some suggest that "over-zealous tying-in can strangle a graft union at birth", and that grafts can be tied too tightly*. Unless extra strong materials such as wire or strong nylon cord are used, the popular tying materials very rarely result in this problem. Occasionally, in the active growing period of late spring/early summer, cambial activity, often combined with root pressure, may allow the rind to be easily torn away from the wood and at this time extra care is required, particularly when tying-in long tongue veneer grafts.

Once union development is completed, ties which do not degrade must be removed to avoid excessive constriction and eventual damage.

TYING MATERIALS

The choice of tying materials is dependent upon various criteria, listed as follows[†]:

- Readily available at economic cost.
- Easily manipulated regardless of working temperature.
- Able to exert sufficient tension without breaking until union formation occurs.
- Break-down at an appropriate time after graft union formation, avoiding the need to remove the tie.
- Must be durable in storage.
- Must flow well if used with a tying aid or machine.

To this list could be added:

- Pre-cut to specific lengths and sizes.
- No danger of adhesion when picked up individually from bulk.
- Impermeable to water.
- Permeable to air.
- Does not react with sealants applied to the surface.
- Some authorities[‡] maintain that translucency is important. This allows light for photosynthesis, which takes place in chloroplasts developed in the callus of some species.

An added bonus but implying a different type of tie, would be:
- Capable of self-adhesion without the need for tying-off.

---

* Hyde (1996).
† Macdonald (1986) p.192.
‡ Tereschenko & Duarte (1998).

NON-ENCLOSING AND ENCLOSING TIES

Ties fall into two main types:

1. Non-enclosing ties, formed by a series of coils around the graft leaving gaps in between, which require sealing under many environmental conditions.
2. Enclosing ties are those which enclose the graft completely and, provided the material used is waterproof, obviate the need for further sealing to prevent water loss.

NON-ENCLOSING TIES

Non-enclosing ties constitute the traditional method of fixing scions to rootstocks. Early ties were plant based, e.g. raffia, bass and even reeds. Subsequently, string (fillis string) was used as a replacement for these unstable and inconsistent materials. Today rubber ties are the most popular of non-enclosing types. They are supplied as strips pre-cut to convenient and uniform sizes. Most used are 3.5mm wide by between 120 and 140mm long. Larger strips 6mm wide by 200mm long should be chosen for species with thick wood such as *Juglans*, *Quercus*, etc. Larger strips than these are available but rarely needed for bench grafting procedures. Those with thin and/or soft wood such as *Daphne* are best suited to a lighter tie, 2mm wide by 120mm long (Figure 6.1).

Rubber has several advantages in use*; its elasticity exerts pressure on the graft, it is extremely flexible and is easily tied using a half-hitch, tying-off knot. A convenient bunch of ties is placed in a holder at a strategic position on the bench. It should be possible to select individual ties from the bunch and it is very important that they separate easily when picked out. Those which stick together must be separated by the grafter, an additional operation, substantially slowing output and causing huge annoyance. Manufacturers dress ties with powder to reduce adhesion, but it is still recommended to rub them through vigorously before final use on the bench.

The elastic nature of rubber allows union development before constriction becomes a problem. Many rubber ties currently in use are light degradable and break down after several weeks avoiding

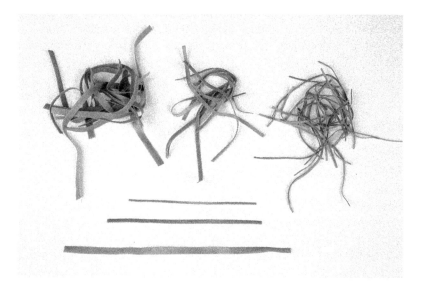

FIGURE 6.1   Non-enclosing rubber ties. Pre-cut to lengths, from the left, 200mm × 6mm, centre 140mm × 3.5mm and on the right 120mm × 2mm.

---

* Lagerstedt (1969).

the need for removal. To prevent premature degradation, these must be stored in dark, preferably cool conditions. Loss of bud grafts in the field due to premature break down of the tie highlights the importance of a tie retaining its strength long enough to allow slow forming unions to fully develop*. Early disintegration is more likely in the field due to the influence of sunlight; bench grafting ties in covered, mostly shaded structures are less affected[†]. Rubber ties which are buried do not break down, and may be left in position to allow girdling and promote scion rooting.

Tapes derived from PVC or polythene can be used as enclosing ties but also as non-enclosing if twisted or spaced when tying-in. When they are non-enclosed, sealing may be necessary to prevent drying-out. If hot wax is used as a seal, it must be applied with care as these ties may be softened by heat, resulting in loss of tension. Degradation rate is uncertain, and unwrapping is necessary after union formation. Consequently, an unsealed strip is left behind the graft, permitting the tie to be easily cut for removal (see Figure 6.10 in the Sealants and Waxes section).

Tying-In Procedures and Techniques for Non-Enclosing Ties

It is usual to commence wrapping coils around the graft starting at the top. Some grafters are adamant that tying-in should be done from the bottom up (normal in field grafting) but most would agree that good alignment and prevention of slippage between scion and rootstock is best achieved by wrapping from top downwards. Methods differ on how to hold the end of the tie while the scion and rootstock are aligned and pressed together. Trapping it with the first finger against the rootstock at a point just behind the top of the graft is usual. The cut face of the scion is held firmly against the equivalent face on the rootstock by the thumb; therefore, the graft components, with the tie, are pinched together. If a long tongue veneer is used, the tongue is also held within the 'pinch'. This method is referred to here as the 'single finger tie' (Figure 6.2).

An alternative tie-holding technique referred to as the 'double finger tie' is favoured by some, especially for splice grafts. The end of the tie is held by the second finger against the nail of the first which, with the thumb, is holding the graft components together. The tie is held pinched between the first and second fingers while tying-in proceeds (Figure 6.3). The advantage claimed for this is that more feel and control and subtle changes in position, angle and pressure are possible. This technique has significant advantages for ultra-thin, brittle or misshaped scions; very thin weak rootstocks; or

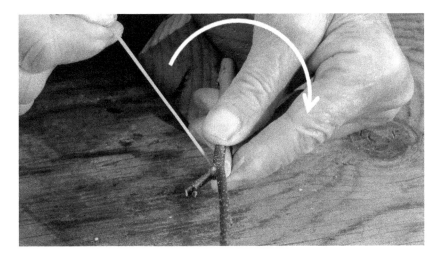

FIGURE 6.2 Tying-in, using 'the single finger tie' technique, showing configuration of hands and fingers for a right-handed grafter. The right hand 'winding hand' continues to provide coiling round rootstock and scion.

---

* Howard (1990).
[†] Alley (1965).

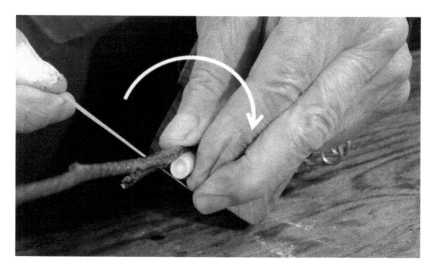

FIGURE 6.3   Tying-in using 'double finger tie' technique, showing configuration of hands and fingers for right-handed grafter. Note: the scion is held in position with the thumb.

side grafts with evergreen scions (especially conifers), which do not always sit snugly against the rootstock because of stiff leaves. Tying-in splice grafts is often more easily accomplished by this method, especially when using potted rootstocks, as it is often more convenient to hold the scion in place with the first finger rather than the thumb. For veneer grafts the scion is normally held in position by the thumb and the single finger tie is usually satisfactory.

For either method the tie is initially fixed in position with two overlapping coils to prevent slippage and trap the end. As coils are made, tension on the tie holds the rootstock and scion together. For particularly demanding or valuable grafts, it is a wise precaution at this stage to further check that the cambia are well aligned. Before tie coils are continued down the stem, it is possible to make tiny adjustments by slightly pushing the scion from one or other side to ensure alignment is as accurate as possible. Wrapping is then continued downwards to the base of the graft. For the 'double finger tie', the end of the tie can be held between the fingers, pulled down the back of the rootstock stem behind the graft and trapped under the coils as wrapping proceeds. This provides a neat and professional appearance (Figure 6.4).

Coiling is achieved by passing the winding arm over and around the graft in a circular motion as shown in Figures 6.3, 6.4 and 6.5. Some grafters favour retaining the knife in the hand during this procedure, claiming less interruption to work flow and increased output. Skill and practice are required to achieve this safely and helped by the use of smaller knives such as the Tina 605 or 685. For side grafts, retention of the knife when tying-in is possible but adds to the difficulties.

Potted rootstocks are often held on the bench while tying proceeds and may be tilted and slightly rotated in an opposite direction to synchronise with the coiling action. Bare-root grafts are frequently held more or less horizontally above the bench (Figure 6.5). Side grafts can often be treated in the same way.

Tall rootstocks which are side grafted present more of a challenge. Holding the graft (bare-root or potted) in an almost horizontal position can help, but despite this the tying arm cannot always be passed over the rootstock stem while still holding the tie and with the other end held in position against the graft. In this situation, the 'coiling-end' of the tie must be passed from one hand to the other. Initially, the hand used for coiling, the 'coiling-hand', is passed round the graft to fix the tie with two overlapping coils. Once completed, to aid control and speed, the 'receiving hand' is held in position by holding the rootstock just below the graft between the second, third and fourth finger (Figure 6.6). It then receives the tie, which is held between the first finger and thumb, from the 'coiling hand'. The 'coiling hand' revolves in the reverse direction beneath

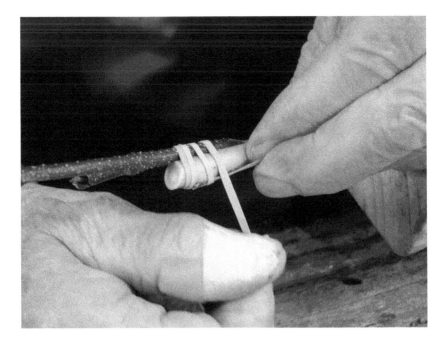

FIGURE 6.4 Initial two coils hold the tie in place. Holding the end of tie between double fingers provides a neat and practical finish to the tie. Demonstration by a right-handed grafter.

FIGURE 6.5 Hand and finger configuration for tying-in a splice graft on a bare-root rootstock.

the graft to gather the tie end from the 'receiving hand'. It then reverses direction and continues passing the coil around the graft until the tie is in a position to be caught again by the 'receiving hand'; this retains the tie and moves it round until the revolving hand can hold the tie and repeat the procedure.

This action continues until the number of coils required is completed and tied-off.

This process can slow tying to half or one-third of the usual speed, and thus it is advisable to maintain rootstocks at a convenient height during the growing season. For deciduous species side

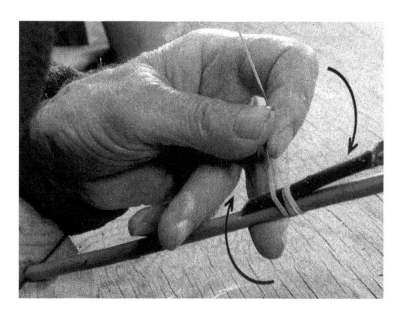

FIGURE 6.6 Configuration of hand and fingers for the 'receiving hand' (left hand) of a right-handed tier tying-in a side veneer grafted rootstock too tall for the arm to pass over. The hand is held in place by second, third and fourth fingers placed on either side of the rootstock stem. The hand revolves on these to pass the tie to the 'coiling hand' beneath the stem. The 'coiling hand' (right hand) continues the rotation and passes the tie to the 'receiving hand' in the position shown in the figure; this continues until the graft is tied-in. Tying-off is accomplished by trapping the last coil with the index finger of the hand holding the rootstock to achieve a loop, through which the end of the tie is passed, and a half-hitch knot completes the procedure.

grafted during the winter, cutting back the rootstock to a convenient height just prior to grafting to facilitate arm movement during tying-in does not appear to influence takes.

Distance between wrapping coils is adjusted to take account of features, e.g. buds, poorly fitting gaps, etc. along the length of the graft. Take extra care when tying the extreme tip of both splice and veneer grafts; damage is possible to this vulnerable portion by over-compression and abrasion. If possible, coils should not wrap over buds, especially stimulant buds. Short tongue veneer grafts should have a coil positioned at the base of the tongue to put pressure at this point, and the tip of the tongue should not be covered; this is illustrated in several figures throughout the book. Gaps between the interface of scion and rootstock due to imperfect knifesmanship can sometimes be closed by extra pressure on strategically placed coils.

As a guide, spacing of coils can be slightly less than, or up to three times greater than, the width of the tie. The more coils applied, the longer the time taken, and over a working day this can be significant. If the rootstock is likely to bleed, it is advantageous to leave a gap in the coils at the base of a long tongue veneer graft to allow excess sap to escape. Rubber ties may be wound round to produce overlapping coils which perform as enclosing ties; however, unless wide versions are used, this is a lengthy procedure and sealing may be a better option.

Some grafters maintain that ties must always be placed so that twists are avoided. This seems questionable as pressure applied by ties, flat or twisted, is unlikely to differentially influence union development or cause any damage to the stem. Flat ties are aesthetically more pleasing, but they require more care, attention and time.

Tying-Off

Non-adhesive ties require tying-off to fix them to the graft. A half-hitch knot is invariably used. Two methods are in use, both relying on the need to produce a loop on the last coil at the tying-off point.

Method 1: The end of the tie is tucked into an enlarged loop and then pulled to tighten and secure the knot. The loop is produced after coiling has been completed by passing the tie over the raised first finger of the hand holding the rootstock (Figure 6.7); the tie is then rotated round the stem once more until the loop is reached. The end of the tie is held between the thumb and first finger, against the nearest strand of the loop, and thumb and finger are then rolled against each other, dragging the tie between the strands. Once through, the end is caught and pulled to tighten and lock in position; practice is required to ensure quick, positive results. This is a flexible, versatile method which can be used to fix a tie, even where access to the loop is difficult. Using this method, polythene, PVC and most other plastic ties are more difficult to tie-off than rubber.

Method 2: Ideal for most bare-root grafts, it is arguably quicker but requires good access to the stem, which ideally should be straight and well above the root system at the point of making the tie. After completing the coils, the thumb and first finger are rested against the rootstock, fractionally below the graft, and a final coil is wound over both, creating a loop above. The tie is given another rotation until it rests between the thumb and first finger (Figure 6.7). It is then held firmly, 'nipped' between them, and pulled downwards through the loop and locked in position. This method is satisfactory for rubber or plastic ties.

Enclosing Ties

Enclosing ties impermeable to water are made possible by the development of modern synthetic materials. They either require tying-off at completion of the tying procedure, or are self-adhesive and held in place by stretching the material and pulling one coil firmly against the other.

Tying Procedures for Enclosing Ties

Non-adhesive, impermeable enclosing ties include those of polythene, PVC related plastics (Parafilm, etc.) and rubber tying strips 12mm wide or greater. They are normally cut to convenient lengths, as for rubber ties, and due to lack of adhesion the tying method is almost identical. Wrapping coils should not be twisted but overlapped to give a complete seal. The width of enclosing ties reduces the chance of damage to delicate tips of the graft, but care should be taken to reduce pressure when covering buds on scion or rootstock. Polythene tape 8–12mm wide is generally the most practicable material for bench grafts. Wider ties can be used for thick wooded species, and for bare-root grafts of strong growing tree species such as ornamental cherries, flowering Crabs and fruit trees. The graft union of apical grafts may be completely enclosed by the tie, largely avoiding the need for sealing, but the cut tip of the scion or any other exposed cuts must be sealed. Tying-off using the half-hitch knot is rather more difficult than for non-enclosing ties, and combined with the need for some sealing, causes many grafters to prefer the use of rubber tying strips combined with use of a sealant*.

FIGURE 6.7   LEFT, shows hand and finger configuration for tying-off method 1. RIGHT, hand and finger configuration for tying-off method 2.

---

* Byrne & Byrne (2004).

Self-Adhesive Enclosing Ties

A new generation of self-adhesive enclosing ties is now available. Buddy Tape® and Medelfilm® are best known and described as synthetic polymer films; both are supplied as tape in rolls. A patent application for a similar self-adhesive enclosing tie using a polyalkylene polymer was filed in the USA in the late 1990s*.

These ties avoid the need for final tying-off, are impermeable to water but permeable to air, have the ability to apply compression to the graft and self-degrade after some months. The material can be supplied un-perforated or with perforations, allowing easy separation along their length. Choice of perforated versions (at 50mm spacing) will produce ties which are too long for most bench grafts, but the tape breaks off easily and for most grafts two or three lengths can be produced from each section. These features have led to their increasing use. Several sizes are available, a number suited to field grafting and budding. For bench grafting, the favoured choice are the narrowest width tapes available, currently 18mm; they are non-adhesive until stretched, when their length may be increased by up to eight times.

Tying-off is eliminated, the coils sticking to each other where they are stretched, overlapped and finished by pressing together below the graft. An I.P.P.S. summer grafting discussion group reviewed a number of types of grafting ties, polythene tapes, rubber bands and Buddy Tape®. Several delegates commented favourably on Buddy Tape®, noting that it stretches like cling-film, does not contract and is waterproof and biodegradable. One grower voiced the opinion that it was "about twice as quick as the alternatives"[†]. Some grafters use self-adhesive tape for all bench grafts except certain conifer species where it is considered that non-enclosing ties, which allow excess sap to escape between the coils, are an advantage[‡]. A further benefit of using flexible materials such as Buddy Tape® is the avoidance of wax sealant cracking when grafts are subject to frequent handling such as can happen in field planting[§].

Tying using self-adhesive ties is most easily achieved if the graft can be rotated to wrap the coils. This is best suited to bare-root grafts when the completed graft can be lifted off the bench and rotated by spinning between thumb and fingers. Grafts to pot grown rootstocks are best dealt by the mechanical method described below. Ideal are apical grafting methods such as wedge and whip and tongue, which hold the scion and rootstock firmly together during handling. For this application, un-perforated tapes are preferable as unwanted breaks can cause problems. The tape can be mounted in a holder fixed to the bench allowing it to feed out in a controlled manner. After tying, grafts may be gathered in conveniently sized bundles and together dipped to seal the cut surface of the trimmed end. Used in this way, self-adhesive enclosing ties demonstrate their potential as an efficient, high-speed tying system.

Mechanised Tying-In

A system has been developed using sewing machine motors, fully controllable by a foot operated pedal, to partially mechanise tying with self-adhesive enclosing ties[¶]. Potted rootstocks are placed in a holder attached to the motor shaft and grafted while in position, using side veneer methods. Tying the graft is achieved by fixing the tape to the graft; the pot is then rotated while the tape is fed out by hand. Foot pedal control makes it possible to adjust speed and duration of turns to achieve even coverage. When completed, the tape is easily pinched off, simultaneously producing a good seal.

This method has the added advantage of allowing the knife to be more easily retained in the hand during the whole procedure. A rotating holder, such as described above, facilitates the use of

---

* Tereschenko & Duarte (1998).
[†] Hyde (1996).
[‡] G. Meacham (pers. comm.).
[§] N. Dunn (pers. comm.).
[¶] Meacham (2008).

self-adhesive ties for grafts on potted rootstocks and, it is claimed, speeds up tying-in and reduces the potential for long-term muscle strain caused by manual tying methods.

ALTERNATIVE GRAFT FIXING METHODS

Grafts are not always held together by wrapping with grafting ties. Machine-produced grafts designed to hold without ties rely on a combination of wood tension and carefully designed cuts which lock the components together. A commonly used pattern is the 'Omega', based on the Greek letter, Ω (see Figure 2.10 in Chapter 2).

When sealing is not required, other options are available. The tomato industry has developed a method of fixing tomato grafts using a miniature clothes peg. Some thin and soft wooded scions and rootstocks of woody species can be attached utilising the same technique. Examples are early-summer grafts of deciduous azaleas and *Daphne* (Figure 6.8). The Dutch Trials Station in Horst compared the rubber tie method with the use of clothes pegs for holding splice and long tongue veneer grafts of *Alnus*, *Betula* and *Corylus*. Average takes obtained for splice grafts was 50% for rubber ties and 47% for clothes pegs. For long tongue veneer grafts 60% were successful using rubber ties compared with 47% for those held by a clothes peg*.

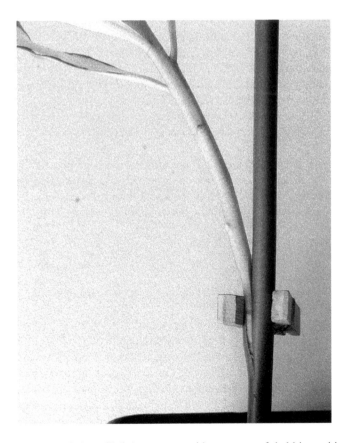

FIGURE 6.8 *Daphne odora* 'Geisha Girl', long tongue side veneer graft held in position with a miniature clothes peg. For easily grafted types such as *Daphne*, this is a quick means of fixing grafts and usually results in fairly good takes. This type of graft is placed in an enclosed environment and normally left unsealed.

---

* Kloosterhuis (1975).

An instant adhesive to replace ties and seals has been under development by a specialist adhesives company*. It comprises an adhesive gel which surrounds the graft and is claimed to provide enough support to enable formation of a successful union. It has been trialled with success on rose buds (presumably a 'T' bud which itself exerts some pressure); it is biodegradable and subsequently disappears. Whether the compressive strength is sufficient to ensure differentiation of cells and tissues in splice and veneer grafts has yet to be established.

SEALANTS AND WAXES

In supportive environments with sufficiently high humidity, non-enclosing ties do not require sealing to prevent moisture loss (see Chapter 11, Section: Moisture, Humidity, Oxygen for further discussion). To ensure graft survival in non-supportive environments, the graft and the cut end of scions must be sealed to prevent moisture loss from the cut surfaces. Enclosing ties require only exposed cut portions of the scion to be sealed as the interface is protected beneath the tie.

Graft sealants fall into two main groups, those which are fluid at ambient temperatures, known as cold sealants, and those which require heating to make them fluid, known as hot waxes. A further group, known as hand mastics, rarely used for bench grafting, require only slight warming with the hands to allow them to be pressed around the graft to achieve a seal.

The important requirements of graft sealants have been identified as:

- Not too high a melting point to cause injury when applied, the generally recommended upper limit is 75°C, but judgement should be used for thin barked species and scions having less well-ripened wood.
- Not too low a melting point to permit excessive softening or melting in high temperatures.
- Not too fluid to allow penetration between graft cuts[†].
- The ability when dry to produce a perfect, unbroken seal, without cracks or pin-holes.
- Protection of the graft from air-borne diseases and pests[‡].

Hot waxes provide an excellent means of sealing grafts. There is no penetration between cut surfaces because once the material contacts much cooler wood, almost instantaneous hardening from the inside commences. Cold sealants rely on evaporation and therefore harden from outside to inside, and while the sealant is fluid, penetration between graft cuts is a possibility, potentially causing loss of the graft.

Hot Sealants

Hot sealants invariably contain one or more materials referred to as wax, although often the main ingredient is resin.

Despite this the popular term 'waxing' is used for procedures involving the application of hot graft sealants. Formulae for mixing waxes are available[§]. Most comprise mixtures of resin, blended with other ingredients which include beeswax, paraffin wax, tallow, linseed oil, Venetian Red, Burgundy Pitch and sometimes inert solids (Fuller's Earth). The addition of inert solids and Burgundy Pitch is to ensure the wax remains solid in high ambient temperatures, making it suitable for use in hot climates. The disadvantage of these is that they require higher temperatures to flow well and quickly solidify into threads between wax-pot and graft. Venetian Red produces a red coloured wax, but other pigments are available. In the UK, a popular wax is coloured green (FPM Professional Grafting Wax), while on the continent a much used German wax is red (Rebwachs WF).

---

* Farrell (2011).
† Lagerstedt (1969).
‡ Macdonald (1986) p.189.
§ Garner (2013) pp.112–114 and Bailey (1920) pp.169–171.

A cheaper option is paraffin wax (candle wax), which is popular and used in significant amounts for application by dipping. Because its adhesive and anti-cracking properties are not as good as those previously described, it is suited to use with bare-root grafts held in boxes for callusing and subjected to minimal handling until graft union is complete.

Melting Hot Waxes

Hot waxes require a heat source to be melted and retained in a fluid state. Volume production warrants the use of purpose-built, thermostatically controlled heated containers of several litres capacity.

Some grafters cut costs by using a bucket placed in a water bath, kept hot by a heater such as a propane gas ring beneath. Small- and medium-scale producers find domestic slow cookers, incorporating a water bath and range of settings, convenient and effective (Figure 6.9).

To reduce risk of damage, grafts should be dipped and removed as quickly as possible. Convenient placement of small heaters and wax pots, as shown in the grafting bench layout, is required for individual grafters waxing their own grafts (Chapter 7). Centralised brush waxing or dipping is best served by large heaters placed strategically in the grafting shed.

COLD SEALANTS

Convenience makes cold sealants an attractive alternative to hot waxes. Penetration is prevented or eliminated by ensuring that the sealant has sufficient viscosity at the time of application. Water-based sealants must not be diluted to restore old samples or achieve more coverage. At present several types of sealant are available:

- Bituminous paints, normally as aqueous emulsions, can be dangerous if applied too wet. Penetration of bitumen emulsion, showing as stained tissue in and around the cambium when the grafts were separated, has caused loss of *Prunus* grafts*. Arbrex® (a bitumen

FIGURE 6.9   Domestic slow cooker used for melting grafting wax is suitable for small-, medium-sized grafting programmes (Penwood Nursery).

---

* Macdonald (1986) p.190.

product used by Arborists) resulted in complete graft failure when applied new and too liquid, but when older and more viscous it was satisfactory*. Bitumen products have been combined with rubber latex resulting in a flexible, waterproof, durable and safe material such as Ryset® sealant.

- Water-based acrylic sealants are popular with some, Plant Seal® being an example.
- Polyvinyl acetate paints now appear to be used mainly as an adhesive in building construction, but some formulations may still have a use as graft sealants.
- Cold resin-based waxes are less often used than their hot wax alternatives. The same basic ingredients as in hot waxes are used but with solvents such as methylated spirit added to liquefy the mix. Once painted on the graft, the solvent evaporates leaving the wax in a solid state. A polyvinyl ester formulation with kaolin and what is described as a 'proprietary aqueous mixture', from Farwell Products in western USA, is now widely used as a graft sealant. It is formulated as a viscous brush-able yellow material, called Farwells Grafting Compound®. As with many of the cold sealants, there have been reports of this product seeping between the graft surfaces. Accurate flat cuts, held tightly together by firm tying-in, are less vulnerable to this problem.

APPLICATION METHODS

Sealants are normally applied by brush or by dipping the entire scion and rootstock to just beyond the grafted area.

Brush Application

Brush application is used for side grafts, evergreen species and summer grafts. Painter's brushes 1–2cm wide are suitable, high-quality brushes are not necessary. Non-enclosing ties require sealing to ensure all cuts are covered, paying particular attention to the diverging gap between stock and scion of side grafts. When applying by brush, it is usual to leave an un-waxed strip at the back of the graft; this conserves wax and provides access to the tie if required to allow subsequent cutting and removal (Figure 6.10).

Enclosing ties on side grafts can be conveniently sealed using a smaller artist's brush for the small gap between the diverging rootstock and scion, plus any cut surfaces on the scion.

Dipping

Dipping into molten hot waxes provides a quick and efficient way of sealing grafts of deciduous species. Wax usage is substantially more than for brush application, but the method ensures that sealing is complete, unlike brush application where misses can occur. Use of potted rootstocks increases the possibility of compost and debris falling into the wax pot causing pollution, eventually making the wax unusable. This may be reduced if a simple removable pot cover is slid in position before dipping; replaceable versions can be fabricated from tough cardboard boxes by cutting out to a pattern; more durable tin plate covers are also suitable (Figure 6.10).

Use of cold sealants for dipping is not recommended because of the danger of penetration into the graft union.

---

* Howe (1976).

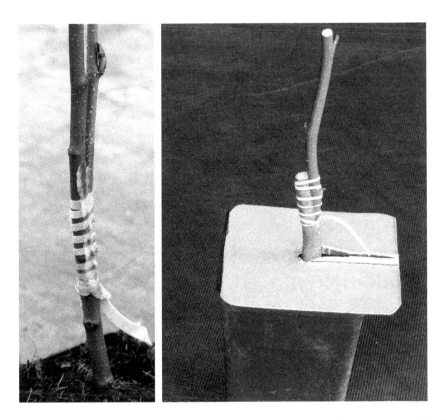

FIGURE 6.10   LEFT, *Betula* graft sealed with hot wax, applied by brush to leave an un-waxed strip at the back, conserving wax usage and providing access for tie removal for non-degrading ties. RIGHT, simple cardboard cover over the potted rootstock, held in place to prevent compost falling into the wax pot while inverting the graft for dipping.

7 Work Station: Equipment and Associated Ancillary and Safety Equipment

CENTRALISED, INTEGRAL OR MOBILE WORKSTATIONS

Whether the work station is centralised, integral or mobile, it will be influenced by the scale of the enterprise and grafting programme. For large enterprises, centralised purpose-built buildings may be appropriate. These are often amalgamated with other facilities such as office accommodation and potting/containerising areas. Grafting systems may also influence the choice. If both winter and summer grafting are in use, a significant facility will be needed for much of the year and a centralised location may be best. To provide good materials handling in these circumstances, consideration must be given to utilising palletised systems, possibly combined with roller conveyors, monorails or other materials handling equipment.

Work stations integrated into the growing area are often preferred; they are sited either within a greenhouse bay or part bay, across one or more bays, or in the access corridor. They have the advantage of closer proximity to graft standing-down positions, which facilitates efficient handling systems such as roller conveyors, gantries or monorails.

Although cost savings by workstation integration into grafting/growing areas are achievable, it is important that working conditions are good with no danger of over-heating or excessive humidity. In this respect, glasshouses may be more easily insulated against temperature extremes than polythene clad structures.

An opportunity for efficient handling and good working environments is provided by mobile work stations, designed for use in external areas or alternatively positioned in access corridors close to grafting polytunnels or glasshouse walk-in enclosures. A suggestion for a mobile grafting work station design and layout is to be found in Appendix L.

WORK STATION – LOGISTICS OF HANDLING GRAFTS AND ROOTSTOCKS

An area for assembly of rootstocks prior to movement into or within the work station will be required in most grafting operations (Figure 7.1). To prevent re-wetting, dried-off pot grown rootstocks must be protected from rainfall during transport and at the assembly point. Bare-root rootstocks must be prevented from drying during handling procedures by the use of polythene bags, or covers of polythene or damp materials.

Sealed, dormant deciduous grafts for cold callusing and hot-pipe systems may be transported by customary handling methods.

Unless the work station is in close proximity, grafted evergreen hardwoods, conifers, summer grafts and unsealed grafts will require protection from desiccation between completion of the graft and standing-down in the final position. This is achieved by a holding tent, adjoining or within the work station (Figure 7.1), and the use of enclosed transporters. For details of holding tent design and construction see Chapter 14 section Poly-Tent and Holding Tent. Movements in and out of the work station indicate a need to carry out a full critical path analysis of the operation. This may highlight design requirements and unexpected problems and hold-ups which need to be overcome.

FIGURE 7.1   Single person workstation for a grafter, integral within a polytunnel and viewed from the entrance. Box of rootstocks for movement into grafting position are assembled for easy access on the bench just outside the workstation (left-hand corner of picture). 'Holding tent' position and access point is visible at the far end. To the left on the floor, an electric heater and hand sprayer for humidifying the holding tent is just visible. Mid-bench cut-out (visible picture centre) marks the grafting position. Bank of fluorescent tubes above provides illumination. Fixed height bench top is placed on Boskoop trays. Rootstocks prepared for grafting are boxed and tilted forward to aid handling. Corner bench cut-out (foreground centre) used to hold rootstock stems for top-working. See Appendix L for suggested dimensions and further details.

WORK STATION EQUIPMENT

A well-equipped work station centred on the bench, combined with appropriate tools and procedures, allows grafting to take place safely and efficiently. It is important that furniture and equipment within the work station is moveable. This enables re-organisation to take account of changing work patterns and demands.

THE GRAFTING BENCH

A bench is required whether the grafter is seated or standing. It should enable close and easy access to rootstocks, scions, tools and ancillary equipment. Bench top design and layout follow a pattern commonly used in horticulture for other repetitive operations such as cuttings preparation (Figure 7.2) and is illustrated and described in Appendix M: Grafting Bench Design and Construction.

Plywood is an excellent material for work-tops and, if required, enables additional cut-outs to be incorporated to aid handling by holding rootstock stems for top-working, etc. A semi-circular cut-out from a central position on one side of the bench top provides better access to materials and improves efficiency but reduces flexibility.

FIGURE 7.2   Grafting bench showing layout for a *left-handed* grafter in a standing position, containing a central, semi-circular bench cut-out to provide better access for the grafter. On the left, a square cut-out for holding stems for top-working. Because it is only used by a single individual, fixed working height is achieved by placing the bench top on Boskoop trays which can also provide shelving for grafting accessories (ties, cleaning pads, etc.). Optional placement of a cabinet for coloured labels storage on the left, Boskoop tray containing prepared rootstocks tilted for easy access, heated wax point (slow cooker) with waxing brush on right, blue rubber ties held in shallow saucer, small hand brush, lightweight secateurs, cleaning pads (including narrow strip for cleaning rootstock stems), grafting knife and prepared (de-leaved) scions all positioned on left-hand side of central cut-out. The position of tools and equipment will need reversing for a right-handed grafter.

The top surface should be matte finished to avoid reflected glare from natural or artificial light; a non-absorbent surface aids cleanliness. Smooth plywood painted with several coats of matte paint or varnish is ideal but unpainted exterior grade or water-proof plywood is satisfactory.

To aid handling for seated grafters, some favour sloping the top slightly downward by 5–10 degrees towards the working position. A shallow upturn may be fixed to part or all of the lower edge, reducing the possibility of items, especially the knife, rolling or being knocked off. For grafters who prefer standing, the bench is best not sloped.

Dimensions are fixed on an individual basis but should be sized as generously as possible and will be influenced by whether or not items such as waxing pots, label holders, etc. are placed on the bench or elsewhere. For a single grafter a suitable bench of dimensions 1400–1800mm long by 900–1200mm wide with a central cut-out will accommodate a waxing pot, label holder and a reasonable reserve of rootstocks packed in one or two holding trays placed on the bench in front or to the side of the grafting position (Figure 7.2).

When splice grafting is carried out while seated, some grafters prefer to hold the potted rootstock vertically. This is best achieved on a lower bench and it may be appropriate to have a double bench system. This can comprise the full-sized bench to which is attached at one or other side (for right- or left-handed grafters) a smaller bench placed 200–250mm lower (see Appendix M).

HEIGHT ADJUSTMENT RELATED TO SEATING AND METHOD

Height adjustment of the bench, seat or both is important for comfort and optimum working position. Grafters who are standing also need height-adjustable benches to meet ergonomic requirements. Adjustment may vary according to the grafting method. If splice grafting to potted rootstocks only is employed, a bench height at least 100–200mm lower than that used for veneer grafts may be preferred (see Appendix M for Bench/Seat Design options).

Layout and Handling

Arrangement of graft components, tools, tying materials, etc. is influenced by the chosen grafting sequence discussed in Chapter 4, and is illustrated in Appendix M.

Potted rootstocks are best placed in shallow boxes, normally plastic trays; in Europe the Boskoop tray is widely used (Figure 7.2). The tray should be tilted towards the grafter at an angle of 20–30 degrees; this speeds-up grafting by reducing the distance necessary to reach forward to gather the next rootstock. To allow room for tilting the rootstock while making grafting cuts, it is often best to place the tray off-centre to the right or left as required. Bare-root rootstocks are best in a three-sided deep box with the open side facing the grafter. A deep Boskoop tray laid on its side with the upper side removed works well. Provided scion and rootstock diameters are well matched and graded; the rootstocks can be packed so that the roots point outwards facing the grafter and in rows above each other. For less well-graded samples, rootstocks are best placed in a small heap so that choice of matching diameters can be made by the grafter.

Grafting Procedures and Bench Design

Grafting procedures may influence bench design and arrangement. When more than one individual is involved, basic arrangement may be replicated with sensible modifications providing good access to rootstocks, scions, etc. A method such as apical wedge, using bare-root rootstocks, lends itself to separate preparation and tying-in of the components. If this procedure is followed, careful monitoring is necessary to keep the various activities in balance. This is facilitated when a team works around a single bench of suitable size and shape (Figure 7.3).

When potted rootstocks are used, each grafter is normally responsible for the complete range of tasks; where several grafters are involved, a linear arrangement of single or double grafting benches is best. The use of roller or power conveyors to supply rootstocks, scions, etc. to individual grafting benches improves efficiency and output. Sometimes sealing (waxing) is done centrally and the completed grafts are fed to a waxing point on a parallel conveyor system (see also Chapter 7: Work Station Design and Specification as well as Appendix M: Grafting Bench Design and Construction).

SEATING

Making veneer grafts and wedge grafts while seated is common practice. For splice grafts, particularly when using thick, heavy wood, grafters often remain standing or half-seated on high stools.

Many years and hours spent crouched over a grafting bench are not conducive to a healthy spine. It has been estimated that 50% of people in the industrialised world suffer from some form of back complaint, strongly related to poor seating design*. There is a case for standing up for all forms of grafting, or changing between standing and sitting on a daily or part-daily rota. Pressure on the discs of the spine caused by sitting rather than standing leads to 40% to 90% more stress on the back[†].

For details of seat design see Appendix N: Seat Design.

Seat and Bench Placement

The relative placement of bench and seat is important. Seating too high results in increased pressure on the underside of the legs while too low produces excessive weight on the 'seat bones'. The

---

* Cornell University Department of Design and Environmental Analysis (2015).
[†] Cornell University Department of Design and Environmental Analysis (2015).

FIGURE 7.3   A possible grafting team layout. Partially mechanised apical wedge grafts being prepared. An in-house fabricated tool is used to produce cuts on rootstock (lady on right of picture). Most of the team produce wedge-shaped cuts for the scions. Completed graft components are assembled separately. Supervisor on the left monitors progress and adjusts tasks of team members as required. Careful analysis of work procedures, layout and dispersal can significantly influence productivity in tasks of this type (Verhelste Nursery).

optimum provides good leg and arm position resulting in minimum stress on the spinal column and associated musculature. Over the long-term, careful consideration of these issues minimises lost time due to health problems. See Appendix N for further details.

LIGHTING

Detailed work involved in performing a graft demands good lighting. Of the light sources available, fluorescent light is currently the best choice due to a number of factors: reduced glare, good light spread and distribution (light is emitted from a long tube rather than a concentrated point), good light diffusion, low temperature output, relatively low ultra violet (UV) output, long life (×10 to ×20 longer than incandescent bulbs) and its ability to perform best at room temperature. Developments in LED lighting technology may result in a need to review provision of lighting in the future.

Older male grafters no longer furnished with a dense thatch of thick hair are advised to wear some form of head protection when working under artificial lighting; the level of UV emission could eventually risk causing skin damage, especially in close proximity to the light source.

For recommended light levels see Appendix O: Lighting Levels.

ANCILLARY EQUIPMENT AND ACTIVITIES – SAFETY EQUIPMENT

This may include all or some of the following:

- Grafting knives and knife sharpening equipment (discussed previously).
- Ties/graft fixing materials and sealants (discussed previously).
- Secateurs.
- Cleansing equipment and materials for bench and plants.
- Knife blade cleansing and sterilisation materials.
- Inspection equipment for grafts.

- Labelling and marking materials.
- Protective clothing.
- Production planning, computer use and recording.
- First aid equipment and information sheets.

SECATEURS

Four design types are available: 1) By-pass; 2) Anvil; 3) Double-cut; 4) 'Mini'-secateurs.

1. By-pass: This design is by far the most popular in use in the UK and consists of two curved blades, with the material to be cut being placed between them. When pressure is applied to the handles, the blades by-pass each other severing the material in the process. Models range from a heavy-duty pattern with a swivel handle for the thickest wood to light, small-duty types. It is important that secateurs are well maintained with sharp cutting blades. For most species this design is suitable for making cuts at the apex of the scion, and may also be used to head-back the rootstock to the point of grafting.
2. Anvil: One sharp, straight-edged blade closes down on a flat anvil blade when pressure is applied to the handles. Clean cuts can be produced but there is some bruising and crushing on the anvil side. This damage is slow to repair and potentially liable to infection. Damaged wood must be removed with a knife before completing the graft.
3. Double-cut: Two straight-edged blades are designed to meet each other edge-on when pressure is applied to the handles. With sharp blades, well set-up to meet precisely, cuts of exceptional quality are produced. In general horticultural practice, the double-cut is less used than the by-pass probably because to work well, more attention to blade sharpness and adjustment is required.
4. 'Mini'-secateurs: Essentially scissor-like tools. Because of their lightweight construction and small dimensions they are often better than secateurs for delicate operations such as leaf removal, scion side shoot trimming, rootstock cleaning up, etc. Sharpening these tools is difficult and many are designed to be discarded after some use.

BENCH AND PLANT MATERIAL CLEANLINESS

A small hand brush should be available for sweeping the bench.

If possible, initial cleaning of rootstocks by assistants should take place before arriving on the bench.

Rootstock stems may need brushing, even light polishing, to remove grit, loose bark, etc. A strip of rubber-backed carpet tile cut to convenient size is excellent for this purpose. Some large-scale operations, particularly involving heavy grade bare-root rootstocks, use a powered circulating brush. Grafters should always avoid foreign material becoming lodged anywhere on, or close to, the cut faces of the graft, where it can become a source of infection and risks damaging the cutting edge of the knife.

At the end of each working day a thorough clean-up is necessary and use of industrial vacuum cleaners and other equipment may be appropriate.

KNIFE BLADE CLEANING AND STERILISATION

To reduce the spread of some diseases, especially those from bacterial and fungal infections, it is wise to clean and disinfect the knife blade at intervals during the grafting session. This is best achieved by periodically wiping each side of the blade on a conveniently placed cleaning pad soaked with a suitable disinfectant. Those recommended include domestic sterilants based on sodium hypochlorite*, peroxyacetic acid, 50% alcohol (unspecified type)† and other alcohols (70% iso-propanol) including methylated spirit.

A pad can be soaked in the chosen material with a clean, damp, untreated pad alongside. Pads can be cut from a rubberised plastic-backed carpet tile, which provides a good, moisture retentive cleansing surface. The knife blade is wiped on the sterilant pad and, dependent upon the disinfectant used, afterwards on the untreated pad. The blade is wiped on these at intervals, every 30 minutes is sensible, or when there is a change in variety. If the grafter has suspicions regarding the health of the material being handled, treatment after each graft is appropriate. Blade cleaning quickly becomes almost automatic and part of the grafter's routine.

INSPECTING GRAFTS

An important piece of equipment is a good quality x10 lens. This enables close monitoring of callus and union development, giving indications of progress and suitability of environment (see Chapter 13: Grafting Environment – Monitoring Grafts for a full account).

LABELLING AND MARKING

Grafted plants should be accurately labelled. Choosing between many available methods is depen-dent upon several factors. Long commercial runs of bare-root grafts may be packed in labelled boxes or pallets. Grafts on potted rootstocks are usually stood down in position on beds or benches. Conventional layout is to label the first plant in the first row running from front to back, left to right, continuing until the first plant of the next variety, which receives the new label. It is wise to use a reversed label on the last plant of each variety; this reinforces the marking and has the added advan-tage of creating the same layout if plants are viewed from the opposite side of the bed.

For small batches, less than 50–100, this method is less satisfactory, as subsequent handling increases the likelihood of mix-ups and mislabelling. To avoid such problems, grafts may be labelled individually, a slow and tedious procedure (Figure 7.4). An alternative is using coloured 'stick-in' labels to establish the identity of each. This system is best with a written label attached to the first

FIGURE 7.4   Labelling. LEFT, use of coloured labels combined with loop-lock name label. RIGHT, use of loop-lock written labels for each graft.

---

* Garner (2013) p.111.
† McPhee (2007).

plant of the batch and a separate record kept of name and identifying colour (Figure 7.4). Where numbers of different species are being grafted, colours can be combined to provide more options. Coloured labels can also be used where more than one rootstock species is used to spread the risk of incompatible combinations or to trial different species for compatibility. For example, rootstock *Tilia cordata* could be designated with white labels while *T. platyphyllos* with red. Colours allocated to scion varieties grafted to these may then be added for identification while retaining that of the rootstock, a useful procedure in recording various combinations. Future technology may permit convenient use of barcodes as a means of labelling plants during propagation phases.

PROTECTIVE CLOTHING

A full-length apron is convenient to use and waterproof versions provide better protection from damp, bare-root or potted rootstocks.

Scion preparation requires thumb protection for some methods described in Chapter 3: How to Graft. Protection can be provided by sticking plaster or micropore adhesive tape; an alternative is a purpose-made thumb protector, usually of crepe rubber or something similar. It does make tying-off using a half-hitch knot more difficult and for this reason the plaster or micropore option is sometimes preferred.

If chemical treatment of scion or rootstock wood is used, it may be necessary to wear very lightweight gloves when handling the treated material and during grafting procedures.

As noted previously, there is often a need for head protection when grafting under artificial light.

PRODUCTION PLANNING, COMPUTERS AND RECORDING DATA

Essential to production planning is fixing target numbers for grafting output. From these, levels of staffing and facilities required can be calculated.

Because two distinct components, rootstock and scion, are required, there are more constraints on forward planning for grafting than for most other propagation systems. A full range of rootstock species established in suitable pots may not be available for purchase, often bare-root plants must be acquired for potting in-house. Unless potted rootstocks are purchased for immediate grafting, decisions on how many rootstocks of a given type are required for potting must usually be fixed 12 months ahead to allow time for establishment. Consequently, last minute decisions on numbers in response to market trends may be limited.

Grafting facilities as an aid to maintaining plant collections in botanic gardens, arboreta, etc. must carry a representative range of pot grown rootstocks for use should the need arise. How comprehensive will depend upon management strategies, resources available and range of species present in the collection (see also Chapter 19: Conservation).

Use of computers in planning and recording has become routine for many establishments. These systems permit comprehensive recording and monitoring of many aspects of grafting procedure. Input information as work progresses enables close and accurate monitoring of output, and may also be used to establish costings (Figure 7.5). Careful thought is required to avoid skills of the grafter being dissipated by time spent at the computer terminal; specialist advice should be sought regarding the use of computers.

Because data collection is so easy, the danger of much unwanted, unused data accumulation should be avoided. Basic requirements, to meet targets and subsequently record successful results, will include information identifying scion and rootstock, grafting date, grafting system (cold callus, hot-pipe, etc.), numbers of each species grafted as well as results (takes).

Additional data can be readily collected for future use, but records focussed on the topic being investigated are essential. Drawing conclusions from subtle changes without the benefit of properly replicated and analysed trials repeated over a number of seasons is always questionable, and probably accounts for some of the 'myths' within the world of grafting. Significant topics, such as the

FIGURE 7.5   Use of computer station for recording grafting output (Golden Grove Nurseries).

effect of various grafting methods, lengths of scion storage or comparisons between various root-stock species are examples of those most likely to provide reliable answers. It is wise to involve team members in decisions on which investigations should be chosen, providing the opportunity for all to take part and be involved in potential changes and improvements.

FIRST AID EQUIPMENT, HEALTH AND SAFETY AND RISK ASSESSMENTS

In many countries there is a legal requirement for provision of first aid equipment and materials at the work place. Additionally, some approved, related basic training on their use must be given to nominated staff. First aid kits in the UK are designed to help deal with trauma injuries. Typical contents include eye wash, adhesive bandages, band aids, sticking plasters, sterile dressings (gauze pads, non-adherent pads containing a non-stick Teflon layer), bandages (gauze roller bandages, adhesive elastic roller bandages), saline solution and antiseptic wipes and/or sprays. Bandages may be used for securing dressings; saline is used for cleaning wounds, possibly followed by antiseptic wipes or sprays.

The most frequent accident during grafting is a cut, normally to the hand, usually a finger on the hand holding rootstock or scion. Fortunately, experienced grafters rarely cut themselves, but the extremely sharp knives in use are capable of inflicting deep wounds. Most common accidents are caused by momentary loss of concentration when picking up objects or moving the hand across the bench. These result in relatively shallow cuts which are inconvenient, inhibit work output and need some attention, normally an adhesive bandage or sticking plaster.

Following a cut, bleeding should be stopped by applying pressure or raising the injured hand above the head. Clean the wound under a running tap or with saline wash, dry the area with suitable sterile material and apply a waterproof adhesive dressing such as plaster or micropore. Professional medical help is only required if:

- The knowledge and/or skill of the person dealing with the injury are challenged beyond safety levels.
- Bleeding cannot be stopped for any reason.
- Large loss of blood (an arterial cut, for example).

- The possibility of tendon injury (e.g. lack of function of finger movements) or nerve damage (loss of sensation near the wound).
- Large wounds (long or deep) which may need assessment for suturing (stitching) should be dealt with at the local A&E Department.

All persons working with soil and plants should receive routine injections against Tetanus; a booster may be given at the A&E Department following an injury.

After some hours of work, grafters, especially with advancing age, sometimes develop stiffening of tendons in one or more fingers. In addition to some discomfort this will reduce output. The condition may be worse in cold conditions and wearing warm fingerless gloves can help, as may use of a hot air blower directed at the affected hand. Various rub-in gels or creams often reduce symptoms.

In many countries, Health and Safety regulations carry a requirement to display posters and notices relating to employers and employees obligations towards health and safety in the work place. Employers may consider it appropriate to display further information on First Aid procedures; it is important that these are prepared after consultation with qualified authorities to ensure that they meet all legal requirements.

RISK ASSESSMENTS

Employment legislation is increasingly framed to ensure that a risk assessment is carried out for work-place situations. In the event of an accident, if a risk assessment has not demonstrably been carried out, a successful action for liability could be brought against the employer.

A comprehensive risk assessment is best prepared by a qualified individual or member of the work force who has received appropriate training. Risk assessments concerned with grafting will cover major features such as the need for only trained individuals to carry out knife-work associated with grafting, but are likely to include some of the following:

- Environment – lighting, temperature, humidity, radiation and noise.
- Furniture and equipment – seating, grafting bench, waxing and knife sharpening stations.
- Operator performance and working conditions – job design, hours of work (breaks, etc.).
- Worker participation – worker consultation, information for workers, safety equipment/ training.
- Publications have been produced listing the potentially toxic plant species*, and it may be wise to consult one or more of these to decide on a possible need to include details in the risk assessment.

---

* Leon (1994).

Part Two

The Graft Components – Rootstock and Scion

8 Rootstocks: Origin and Underlying Principles of Anatomy and Physiology

The rootstock is a crucial component of the grafted plant. Rootstocks can have significant influences upon scion growth, health and many other aspects of performance. The scion can also affect the rootstock, though normally less significantly. The influences of origin, anatomy and physiology are discussed in this chapter; categories, specifications and other important issues are discussed in Chapter 9.

ORIGIN

The origin of the rootstock can have influences on performance and contribute towards advantageous or disadvantageous characteristics. Rootstocks normally originate from one of the following:

- Seedling origin.
- Clonal origin.
- Root portion.
- Unrooted stem.

Anatomical and physiological features of these types differ and affect use and performance in grafting procedures.

SEEDLINGS

Seed raised plants constitute by far the most commonly used rootstocks for ornamental species, where the visual appearance of the scion is of major importance. This contrasts with many fruit and nut crops where rootstocks from vegetative propagation are selected for their influence on size, uniformity, cropping and suitability for a range of soils and soil conditions.

All seedlings exhibit varying degrees of juvenility, the pre-flowering stage in the life of the plant. This varies in length between species and between non-woody and woody plants, the latter normally retaining juvenility much longer than the former. There is a transition in development between juvenile and adult plants when characteristic differences are blurred.

Apart from an absence of flowering in the juvenile phase, some species show marked differences between juvenile and adult plants. The deeply lobed leaves of juvenile Ivy (*Hedera*) become entire when adult, and in many of the Rosaceae, stems can be thorny in the juvenile phase before gradually reverting to reduced or non-thorniness when adult. Another manifestation is the retention of dead leaves throughout much of the winter period by many deciduous species while in the juvenile phase, such as in Fagaceae.

As well as external differences, there are internal properties linked to the plant's physiology, such as significantly improved rooting capacity of cuttings taken from juvenile mother plants. A number of comparisons between grafts made on juvenile and more mature rootstocks have shown results in favour of juvenility. In some instances, a particularly difficult-to-graft species, Red Flowering

Gum (*Corymbia ficifolia*), has responded well to the use of juvenile seedling rootstocks*. Young seedlings of *Grevillea robusta* used as rootstocks for this difficult-to-graft genus gave 88–100% success compared with 62.5% for older seedlings[†]. Investigations into grafting Gardenia showed that juvenile rootstocks enhanced results compared with older alternatives[‡]. Some difficult to graft tropical trees, such as Clove (*Syzygium aromaticum*), have also exhibited the beneficial effects of juvenility[§]. Juvenile rootstocks of some conifers, e.g. *Picea*, reportedly show an improvement in takes of 5–10%[¶]. Juvenility may be particularly relevant for conifers since at the fully adult phase they have a significantly lower proportion of potential callus producing cells (parenchyma cells) than hardwood species**. The increasing use of plug-grown conifer seedlings, which exhibit juvenile characteristics, takes advantage of this trait.

Retention of juvenility is dependent upon a number of factors, of which age is the most important; generally, seedling rootstocks are grafted well before their juvenile characteristics have totally disappeared. Rootstocks produced by intensive techniques, such as plug production of seedlings under protection, exhibit juvenile characteristics for a longer period than those produced conventionally in open-ground seed beds. These sources all display more juvenility than rootstocks produced by vegetative means – cuttings, layering or division. But even vegetative sources can be rejuvenated to an extent by hard pruning and other cultural techniques. An extreme example of vegetative rejuvenation is seen in micropropagated material.

The serious disadvantage of seedling rootstocks is their genetic variability and consequently their indeterminate influence on the scion. Even with apparently suitable combinations, there are effects on growth rates, health, vigour and compatibility. This normally has little significant effect on ornamental characteristics but for some it causes problems of incompatibility in various stock/scion combinations. *Quercus* are an extreme example in which a proportion within any clone can fail, despite being grafted onto seedlings of the parent species.

Plants which exhibit apomixis (seed production without pollination) are an exception and produce seedlings of identical genetic composition. For woody species the range is limited, many within Rosaceae, and even here a small proportion may contain a percentage of seedlings showing normal variability.

CLONAL

Clonal rootstocks propagated vegetatively have the advantage of a fixed and constant genetic make-up. This enables establishment of their suitability as rootstocks in respect to compatibility and influences on the control of growth, cropping, resistance or tolerance to disease and adverse soil conditions. In general, rootstocks derived from a clonal source can and should exhibit good vigour and health; they are normally beyond the transitional phase of juvenility.

ROOTS OR ROOT SYSTEMS

Roots are generally responsive to grafting because the relatively large rind/cortex comprises a high percentage of parenchyma cells and ample supply of stored carbohydrates. The main difficulties are obtaining good subsequent root regeneration and identifying the position of the cambium to enable a good match with the scion.

A portion of a much larger root system may be used as a rootstock, or alternatively, for a limited range of species, the whole root system of a young plant severed just below the hypocotyl (the point

---

*   Ryan (1966).
†   Dupee & Clemens (1981).
‡   Mustard & Lynch (1977).
§   Garner (2013) p.42.
¶   Decker (1998).
**   Widmoyer (1962).

of transition from shoot to root). The former may be obtained from older, more mature roots; the latter may have juvenile features similar to those previously described. Whole seedling root systems without aerial portions are occasionally used to overcome the problem of suckering from genera, such as *Tilia*, which produce basal growths from the stem and not from the root*.

Use of rather mature material is not the disadvantage which might be expected as, apart from a thicker cortex often with more stored carbohydrates, roots retain juvenile characteristics for an extended period.

Because of the flow of auxins rootwards from the stem, the root (like the stem), does not continue to function effectively if polarity is reversed. However, roots can be grafted to the scion in an inverted position, a feature sometimes utilised as a 'nurse root', providing limited support to the scion until it is able to produce its own adventitious root system. Once this is accomplished, the original rootstock withers away (see Part Five: Root Grafting, Nurse Grafting as well as Parts Seven and Eight regarding Genera Specific Requirements for further discussion on root grafting).

Unrooted Stems

These are an integral part of cuttings-grafts when the stem must be induced to form roots at the same time as forming a graft union. They fall into one of two main groups:

Group (1): Unrooted stems, used as hardwood cuttings comprising mature, one-year old wood from species with a propensity to root easily; either having pre-formed root primordia (root initials) within the stem, e.g. *Salix*, *Populus*; or selected species and hybrids without pre-formed root primordia but, given appropriate conditions, are particularly easily rooted, prime examples being some *Vitis*. In both types, stem rooting is sufficiently reliable to permit graft union development and adventitious root production to take place more or less simultaneously.

This group has good reserves of stored carbohydrates in the stem. The pith/wood ratio (ripeness) is satisfactory; easily grafted species are involved and rooting is rapid. Grafting can take place before rooting with good takes expected. Those with pre-formed roots develop a root system once severed from the parent plant and placed in soil or growing media. Darkness and moisture at ambient soil temperatures stimulate the root primordia to extend into the surrounding substrate.

Mature stems without or with only few root initials (Grapevine rootstocks, etc.) require more treatment and manipulation to initiate adventitious roots. This normally involves application of growth substance (rooting hormone) combined with additional heat providing temperatures in the 22–27°C range.

Group (2): Unrooted leafy stems of semi-mature wood required to become rooted during the union forming process.

Adventitious roots must be induced to form on the unrooted stem in tandem with the formation of a viable graft union. Such material is relatively low in stored carbohydrates, and to maintain these at a level sufficient to produce adventitious roots, photosynthesis needs to continue throughout rooting and union formation. This requires a proportion of the rootstock leaves to be retained and the graft to be housed in an area with sufficient light levels, despite some essential shading to control temperature. The technique works well with genera such as *Rhododendron*, which are able to photosynthesise efficiently under low light conditions.

Although this material is in the semi-mature state and comparatively low in stored carbohydrates, it has the advantage of copious young tissue capable of active cell division and differentiation and is therefore able to quickly produce a graft union. Adventitious root formation will need to be stimulated by the use of synthetic rooting hormones (IBA, IAA, etc.). Because of the potential negative effect of these substances on graft takes, IBA is best chosen as it is very weak for all auxin responses except rooting. Additionally, it is relatively immobile, an important property when applied in fairly close proximity to the graft.

---

* Pigott (2012) pp.365–366.

SIZE OF ROOTSTOCK

In principle, the larger the rootstock, the more stored carbohydrates and proteins it contains, but inappropriate size can make grafting carpentry more difficult. The positive influence of a rootstock with a large active root system on takes of field-budded *Acer platanoides* 'Crimson King' has been demonstrated[*]. For bench grafts, the influence of root system size is less obvious. Using pot grown or bare-root rootstocks with comparatively restricted root systems, bench grafted *A. platanoides* 'Crimson King' would be considered a fairly easy subject, with high takes being usual. It is likely that greater control of the environment outweighs the beneficial effects of a large unrestricted root system. Despite this, it may be wise to use deep containers for rootstocks; a 9 × 9 × 12cm pot is a better choice than the usual 9 × 9 × 9cm. This pattern also has the added advantage of providing a larger volume of growing medium without loss of ground area.

ROOT PRESSURE (BLEEDING) AND ROOT ACTIVITY

Though opinions on this issue differ, root pressure or bleeding is often considered as having a strongly negative influence on graft takes. The arguments for and against are fully discussed in Chapter 11. At this stage, it can be identified as an emission of sap (cell sap) from the cut surface of the rootstock. This is prevalent in some species, especially when potted rootstocks in wet compost are combined with high root temperatures (bottom heat).

A number of methods to combat the effects of root pressure are discussed throughout the book. Of these the most important, drying-off, is described in detail in Chapter 9 Drying Off. It has been suggested that side grafting mitigates the effect of root pressure. For bottom-worked side grafts, personal experience indicates that this is questionable; certainly, length of wood above the graft, which might be considered an important factor, appears to have little, if any, influence.

However, tall rootstocks, allowing the scion to be grafted higher on the stem, show a significant reduction in bleeding, which seems to be correlated to the height of the graft above the root system.

Many authorities stress the importance of root extension and activity at the time of grafting[†]. However, it is important to realise that early in the season root and bud activity result in a drain on stored food reserves, particularly so with young, white, extending roots, but less with older, non-extending, brown roots[‡]. Loss of food reserves may be at the expense of those required to aid graft union formation.

Active root tips, with or without extension growth, are important in the synthesis of cytokinins, which are involved in the production of callus and cell differentiation. Examination in late winter of a wide range of species of pot grown rootstocks, including those dried-off, has shown the presence of active root tips before root extension has taken place (Figure 8.1). These rootstocks have been housed in unheated structures, and temperatures may often be only slightly above, occasionally below, open-air ambient. It is not certain whether or not these observations represent a reliable and species-wide pattern of root activity, personal experience indicates they are consistent. Possibly because of the presence of active root tips, demanding grafts are more successful on established potted rootstocks than bare-root alternatives. Rootstocks at the stage described above, root tips active but not extending, may well equate to the ideal state for apical grafting procedures identified by some authorities[§].

A significant advantage of bare-root rootstocks is that they rarely cause problems with root pressure. This is because many young extending roots and all water-absorbing root hairs are desiccated during handling procedures; they are able to re-form only after undisturbed conditions prevail, by which time graft union development is normally well advanced. It can be concluded that root

---

[*] Howard & Oakley (1997).
[†] Upchurch (2005), MacDonald (2014) p.109, Larson (2006), Meacham (2003), Bush (2003) and Wolff (1973).
[‡] Dixon (2016).
[§] Dorsman (1966).

 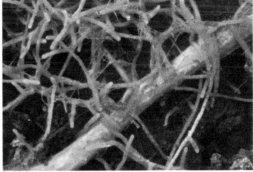

FIGURE 8.1 *Betula pendula* root system, removed from dried-off pot in early February. LEFT, showing no root extension but active root tips. RIGHT, a closer view of same.

extension is not a requirement for successful grafts involving bare-root rootstocks. Logic raises the question of how essential they are to rootstocks grown in pots.

Experience has raised further speculation concerning the role of root systems in graft union formation. In severe winters, root-balls of pot grown rootstocks in hot-pipe systems (to be described later) can become frozen. It is assumed that in this condition movement of bio-chemicals is restricted. Despite this, success rates of a range of species were satisfactory and included: *Crataegus chungtiensis, Malus sargentii, Sorbus henryi, Carpinus cordata, Styrax serrulata, Quercus* x *hispanica* 'Ambrozyana', *Wisteria sinensis* and *Wisteria floribunda* cultivars. Whether the root systems in this observation were continuously frozen throughout the grafting period is uncertain, but root temperature was low for a prolonged period. A similar experience has been reported for *Betula*, where, in a cold callusing system, the roots of pot grown rootstocks (*B. pendula*) were frozen for several days; despite this, takes were good and not adversely influenced*.

The assumption that root activity and linked cambial activity is necessary for success may in part be explained by the requirements of 'T' budding and rind grafting, where activity is essential to allow separation of rind from wood. It is possible that woody plants' response to wounding, involving the production of callus tissue, overrides other responses. For further discussion on root activity and root extension, see Chapter 11.

---

* C. Lane (pers. comm.).

9 Rootstocks: Categories; Acquisition; Specifications; Culture

Methods of propagating rootstocks by seed and vegetative methods are well covered by other publications and, apart from specific situations closely linked to the grafting procedure, for example cuttings for cuttings-grafts, or particularly demanding seed-raised species such as *Daphne*, they are not discussed in detail here. Any relevant information is given in Genera Specific Requirements – Parts Seven and Eight.

CATEGORIES OF ROOTSTOCKS

Dependent upon the chosen grafting system, rootstock categories fall into five types:

- Bare-root – after lifting most of the soil, it is removed by shaking manually or mechanically.
- Pot or container grown rootstocks – generally in small plastic pots, 9cm square or similar; plugs, cells, or larger sizes for special use, such as top-working.
- Rootballed rootstocks – comprise the root system lifted with some surrounding soil wrapped in a suitable material, normally hessian (burlap). These are not often used but appropriate for larger rootstocks in certain situations, e.g. top or frame working. Unwrapped root-balls are used sometimes for fibrous rooted species such as *Rhododendron*.
- Root system, without inclusion of any stem portion – a less often used choice of rootstock. Complete root systems are normally obtained after severing the seedling plant at the hypocotyl. Part root systems (piece root grafting) are obtained by utilising only portions of the root system. Aspects of root grafting are fully discussed in Chapter 18.
- Unrooted stems – a further category restricted to a few species but some of real significance, e.g. Grapevines, *Salix* and *Rhododendron*, can be grafted to unrooted stems. Roots and graft union formation are simultaneously induced after grafting. Such procedures constitute cuttings-grafts, discussed in Chapter 18.

ACQUISITION

BOUGHT-IN ROOTSTOCKS

Most grafters obtain rootstocks from external suppliers. The advantage of this will depend upon sourcing items of appropriate quality, graded to meet requirements.

Purchasing policy usually follows one of two routes: competitive pricing influencing choice of supplier; or a lower emphasis on price, and greater emphasis on close liaison to provide quality and optimum delivery times. Choice of easily grafted species may favour the first option, particularly if bare-root rootstocks are the main requirement. With this approach, it is sensible to have a growing season inspection to ascertain that quality standards and grading will meet expectations. With good grafting and growing on procedures, results from economically priced rootstocks are often indistinguishable from more costly options.

For more demanding species and certain grafting systems, better results are obtained using top-quality rootstocks delivered at the appropriate time. Pre-delivery contractual agreements with the supplier, coupled with good communication and liaison, should ensure a reliable supply.

IN-HOUSE PRODUCTION

This strategy enables total control of range and quality according to a specific time scale, and production of small numbers of rare species unavailable from commercial sources.

Some grafters have selected specific strains (clones) of rootstocks which they consider have superior features. These may include enhancement of growth rates; resistance to, or freedom from disease; better compatibility with chosen scion varieties; and, for vegetatively produced types, better rootability. To retain exclusivity, these selections are frequently produced in-house.

ROOTSTOCK SPECIFICATIONS – TERMS USED

Specification of rootstocks is based on a mix of age, cultural methods and dimensions. These need to be combined to provide meaningful information for their use and application in grafting*.

Standard specifications for forestry transplants grown in a seed bed may be applied to rootstocks. Under this system, a one-year old seedling is designated 1+0 and two years old is 2+0.

If the plant is raised vegetatively, for example from cuttings or layers, convention determines that the term then becomes 0+1 or 0+2, respectively.

If plants are transplanted after one season's growth, the seedling designation becomes 1+1, or if vegetatively raised, 0+1+1. Sometimes seedlings may be left in situ for two seasons then transplanted and grown-on for one further season becoming a 2+1 item.

Potted seedlings are labelled with a 'P', so that a one-year old seedling grown in a pot for a further year is 1+1P. A one year cutting raised plant, potted for one year, becomes 0+1P.

An alternative to transplanting is to carry out a procedure known as undercutting. This involves pulling a blade beneath the soil surface at a pre-determined depth to sever all roots beneath the blade. The resultant plant has close similarities to the transplant. Undercut plants are designated with a 'u', so that a two-year-old undercut rootstock would be 1u1. Unlike transplanting, undercutting can be carried out during the growing season and it is therefore possible to have a one-year old undercut rootstock designated ½ u ½. This is particularly relevant to species such as the Oaks or Walnuts that typically produce a single tap root which can be encouraged to branch using this technique.

Forestry grades described can be further refined by adding the diameter measurement of the seedling at soil level ('neck collar' diameter) to sizes of relevance to rootstock specifications. The neck collar diameter of a one-year old deciduous broadleaf seedling (1+0) normally falls in 4–6mm or 6–8mm range; a two-year old transplant (1+1), including conifer seedlings, produces sizes of 4–6mm, 5–7mm, 6–8mm, 8–10mm or, for vigorous types, 10–12mm or larger.

For example, a popular size for bench grafting is 1+1P 6–8mm. This is the specification for a one-year seedling, lifted, potted and grown-on for one year in the pot, achieving a neck collar diameter of 6–8mm.

Best specialist rootstock growers specify age, stem girth and height, so it is possible to have a complete description of the grade being offered and to choose the optimum size for a given purpose.

Recommendations covering rootstock specifications related to particular species and uses are given in Part Seven: Genera Specific Requirements and Part Eight: Genera Grafting Guide Tables.

---

* Wood (1996).

BARE-ROOT DECIDUOUS ROOTSTOCKS

SPECIFICATIONS TO MEET REQUIREMENTS

All rootstocks must appear healthy with no evidence of disease and have a well-developed, fibrous, evenly branched and balanced root system. Each plant must have a single stem of reasonable length and be free of large side shoots at the location of the graft.

1+0, 1+1, 1u1, and 2+0 seedlings are normally required in diameters of 6–8mm, 8–10mm, 10–12mm.

Vegetatively propagated rootstocks (cuttings or layers) for a range of species designated 0+1+1; 0+2, are normally required in diameters of 6–8mm, 8–10mm or, for top fruit, 8–10/10–12.

Many rootstocks for top fruit and nut crops, as well as a lesser number for ornamentals, are raised vegetatively. These are often from layers or hardwood cuttings and an increasing number from micropropagation. Such rootstocks are expensive to produce and have a high value compared with most of their seedling counterparts. The majority are destined to provide the basis of fruit or nut plantations because of their influence on growth, field performance and cropping. Consequently, the highest standards of grading are required in respect of trueness to type, dimensions, freedom from disease and good root development. In some instances, this will involve carefully monitored material originally derived from elite (virus free or virus tested) sources, often originating from government agencies.

TRANSPORT – RECEPTION AND HANDLING

Plants should be transported in covered vehicles and protected from direct sunlight. In cold weather, suitable protective material should be used to prevent frost damage to the root system.

Labelling has to be sufficient to avoid all possibility of mix-ups occurring during unloading procedures. Plants must be well-packed in moisture proof, co-extruded polythene bags (white external, black internal), as opposed to clear polythene, which can over-heat. Within the bags, the plants should be moist but not over-wet. Root and stem tissues have to be plump with no evidence of desiccation or shrivelling. There should be no physical damage beyond normal lifting procedures, and bundles must not have been tied so tightly as to cause stem damage. In the case of species with very easily damaged bark (e.g. Acers), rubber ties are recommended.

Delivered plants should be checked to ensure that they conform to the required specifications. Top trimming is then normally carried out to reduce bulk and limit moisture loss, a hedge cutter or shears may be appropriate for bare-root rootstocks. For large numbers, it is more efficient to pass bundled rootstocks through a horizontally mounted cutter bar. Subsequent procedures will depend upon whether the stocks are due for a period of storage or for immediate grafting.

PREPARATION FOR IMMEDIATE USE

Rootstocks to be used for apical grafts are normally headed back to between 50–100mm above the root system; this will be determined by rootstock and scion diameters and grafting method. Hot-pipe systems may need slightly more length to accommodate thickness of the channel insulation. If side shoots are present, the remaining stem portion is normally trimmed-up to provide at least 75mm of clear stem. For genera prone to suckering, such as *Syringa* and *Viburnum*, this procedure may include removal of any lateral buds. Rootstocks destined for side grafting are cut to leave sufficient stem above the planned grafting position, normally 150–200mm.

To ensure cut surfaces are free of material likely to damage the cutting edge of the grafting knife, rootstock stems should always be cleaned of any existing debris and adhering soil. Root trimming is usually required; some species will withstand severe root pruning, which will enable those for hot-pipe grafting to be closely spaced in the heating chamber. Tree species are best with minimal root pruning, leaving sufficient to support good extension growth.

To streamline and quicken grafting procedures, re-grading for collar diameter may be carried out at this stage. After preparation, rootstocks are packed for convenient handling by the grafters, as discussed in Chapter 9.

STORAGE (SHORT- OR LONG-TERM)

Bare-root rootstocks should preferably be grafted as soon as possible after reception, but short-term storage is often necessary to enable large quantities to be dealt with in rotation. Longer term storage may be required, particularly if scion wood is cold stored. This strategy extends the grafting season and is sometimes necessary to cope with very large grafting programmes.

After trimming plants may be re-packed within sealed (tightly tied) co-extruded bags and stored in sheds or barns for a period. Safe storage for several weeks is possible in window-less, non-insulated buildings with a concrete floor and ventilation on the north side. Ideally this should be sited in the shadow of evergreen screens.

In the absence of cold storage, for medium to long periods before use, laying-in is a popular procedure. Well-protected sites in shady areas with good access are ideal. After the bundles are trimmed and opened, the roots are spread in the trench as widely as possible to obtain good contact with the laying-in material. To facilitate this, light, sandy soil is excellent – if necessary fortified with additional sand and lighter materials such as composted bark. It is essential that laying-in beds are well-drained, but the plants must not be allowed to dry out. Unwanted attention from mice and voles during the winter months require control by routine baiting.

Prolonged storage is best in refrigerated cold stores, the plants first being removed from packing or packs opened before placement. High humidity conditions, through the use of humidifiers, or a 'Jacketed' cold store, must be maintained at temperatures in the −0.5°C to 2°C range. Above 2°C, Botrytis becomes more active. Below freezing point, temperatures can increasingly pose a risk of damage, minus 2°C being the safe minimum for most hardy temperate species. Botrytis (Grey Mould) causing extensive die-back and rotting of shoots, roots and foliage is evidence of lengthy storage in poor conditions.

BARE-ROOT EVERGREEN ROOTSTOCKS

REQUIREMENTS – SPECIFICATIONS

These include species such as *Cotoneaster, Ilex, Kalmia, Ligustrum, Rhododendron* and a range of conifers including: *Abies, Cedrus, Picea, Pinus* and *Taxus.* Most evergreen species will subsequently be potted for use the following season, when they constitute the 1+1P or 0+1+1P grade.

Choice of grade will affect their suitability and to achieve necessary stem girth, slow growing species, such as many of the conifers, will normally be 1+1 or, more often, 2+0 seedlings at the time of potting.

TRANSPORT – RECEPTION AND HANDLING

Transportation should follow the guidelines for bare-root rootstocks.

Evergreens closely packed in bundles for prolonged periods will result in foliage becoming diseased, rotting and rendering the plants unusable; it is important that bundles are opened, checked and handled as suggested for deciduous species.

Many evergreens are grafted using side rather than apical methods, and care is required when trimming the top. This is best done on an individual basis, facilitating good judgement of height. Removal of side branches to allow access for the graft may be carried out, but for plants destined for potting and growing on, this may be delayed until the plants are established. Root pruning may be required to facilitate potting up.

STORAGE

Options for short periods are to lay-in as described; for medium periods, stack upright in suitable boxes with moisture retentive material worked between the roots and place in a cool shed, barn or an unheated growing structure. Cold storage facilities as discussed previously are required for longer terms of up to 12 weeks; the plants should be loosely re-bundled and held in high humidity at 0.5°C to +2°C.

ROOT-BALLED ROOTSTOCKS

There are no specifications covering this category, but a wide range of species and sizes can be used. For example, large plants of *Acer palmatum*; a range of conifers, e.g. *Abies*, *Cedrus* and *Picea*, are well suited; even *Magnolia* has been suggested*. Root-balls may be wrapped conventionally in root-balling squares. Smaller plants of species with a network of fine roots such as *Rhododendron* are lifted with soil attached to the root system and further wrapping is unnecessary.

Management of root-balled grafts is as for those potted or plug grown, see below.

ROOTSTOCKS FOR TOP-WORKING

Height is the guiding factor in deciding appropriate grades of stems for top-working. A few species, such as *Salix*, may be sold to a specification which includes stem diameter. Over-sized stems are less of a concern for top-worked species because side veneer or wedge grafts are commonly used and can be adjusted to account for variable stem diameters.

ROOTSTOCKS FOR CUTTINGS-GRAFTS

A limited range of species, sometimes in very large numbers, are grafted to unrooted stems. The principles of this system have been discussed in Chapter 8 and requirements for specific genera will be covered in Parts Seven and Eight.

One group comprises those species, e.g. *Salix*, which produce root initials within the stem. Stems for this purpose are produced by planting generously spaced stock plants in rich moist soil. Pruning involves annual cutting back to ground level to produce the type of shoot required. The resultant strong, straight, one-year old stems, at least 2.5 metres long, form the foundation for top-worked species such as the various dwarf or weeping forms of Willow.

Grapevines are propagated by cuttings-grafts, and provision of rootstock stems for these mirrors requirements for scion wood, to be described for climbers in Chapter 10. Conventional cuttings used as unrooted rootstocks for Rhododendrons are standard procedure for some grafters. Rootstock varieties are selected from a small range of easily rooted types. Of importance is the establishment of sufficient numbers of mother plants to support the cuttings material which may be required for a cuttings-graft programme.

Of less importance, but a favoured method for some grafters, is use of the cuttings-graft technique for Junipers. Here the rootstock is normally *J.* x *pfitzeriana* 'Hetzii'; further details are given in subsequent relevant parts of the book. The guidelines for managing scion mother plants for evergreens and conifers, set out in Chapter 10, also applies to those supplying material for cuttings-grafts.

---

* Macdonald (1986) p.527.

POTTED ROOTSTOCKS – SOURCED FROM OTHER PRODUCERS

REQUIREMENTS – SPECIFICATIONS

Rootstocks may be purchased as 'finished' pot grown items. The advantages are easy adjustment of numbers allowing response to increased or reduced demand as well as substantial labour saving. Disadvantages include: quality standards may not reach in-house production, overall cost may be higher, no control of potting compost and chosen pot sizes may not reflect best practice. Most potted rootstocks are supplied in 9cm square pots, usually as 1+1P grade. Other sizes to meet specific requirements are rarely available.

TRANSPORT AND HANDLING

Transport and delivery follow the recommendations given for bare-root rootstocks. To prevent any risk of damage, the plants should be transported in single layers in an upright position. To conserve space, many suppliers opt for packing the plants horizontally above each other in layers within suitable containers such as cardboard boxes or wooden crates. Alternative methods, such as vertical packing in co-extruded bags used for bare-root plants, are sometimes favoured. When these methods are used, care is required to avoid damage while packing and unpacking, particularly important with species having thin, soft bark, such as many of the Maples. Scarring and bruising caused by mechanical damage creates extra work for the grafter who must avoid such areas.

RECEPTION FOR GRAFTING SHORTLY AFTER ARRIVAL

Sample pots should always be knocked out for inspection before further handling takes place. This provides opportunities to check for perennial weed roots, root disease and pests. By far, the most important pest is the larvae of Vine Weevil (*Otiorhynchus salcatus*). If any larvae are found, all pots must be knocked out to ascertain the severity of the infestation and decide on appropriate action, which may include insisting the supplier removes the infected material.

Just before grafting operations take place, handling is facilitated by packing into suitable units such as Boskoop trays. As these are packed, trimming can be conveniently carried out to expose the stem for graft placement, at the same time removing any debris and cleaning the stem.

Some weeks before grafting, many favour standing potted rootstocks, without additional water, in a growing structure maintained at 10°C or above. The purpose is to initiate root extension and reduce compost moisture content (see below – Drying-Off Pot Grown Rootstocks). This aspect has been further discussed previously and in Chapter 8 Root Pressure and Root Activity.

RECEPTION FOR USE AT A LATER DATE

After inspecting the plants for quality, health, and removal of any weeds, handling potted rootstocks for use at a later date involves standing the plants down in suitable areas for growing on, following guidelines to be discussed further in culture/growing conditions.

POTTED ROOTSTOCKS – PRODUCED IN-HOUSE

In-house production of potted rootstocks has several advantages over the alternative in potentially better quality, range of types available, easy accessibility and the possibility of growing to precise specifications with regard to shape size and container size. Raising seedlings for potting in-house can provide complete control of production but many prefer to buy-in seedlings for later potting.

SPECIFICATIONS

Specification of the bare-root grade to be potted must allow for subsequent growth in the pot. An increment in stem diameter of 2–4mm for the first growing season is normal, but this is influenced by species, choice of growing conditions, potting date and pot size. There is a tendency to purchase over-sized plants for potting up, resulting in over-large rootstocks requiring more space, trimming and an increased likelihood of producing stems over-sized for grafting. This can add to demands on knife skills and possibly the need to change to a less desirable grafting method, i.e. short tongue rather than long tongue veneer.

POT SIZES AND TYPES

Finished rootstock pot sizes are normally 9×9×9cm, 9×9×12.5cm or 1-litre deep pattern (Figure 9.1). There is an important range of woody species which produce a single, strong, unbranched tap root; the most extreme include *Carya*, *Juglans* and *Quercus*. These are best raised in large plugs subjected to air pruning, resulting in a more branched and balanced root system. Often the only available plants are raised in traditional open-air seed beds as 1+1 or 1+2 seedlings, requiring severe root pruning to fit into normal pots resulting in an adverse effect on growth and takes. This type of root system can be retained more or less intact by the use of extra deep, rigid plastic pots, or polythene polybags, up to 35cm deep (Figure 9.1), both of which provide significantly improved results and superior subsequent growth.

Larger pots may be required for Patio plants and top-worked rootstocks. A 2-litre-deep pattern provides a good compromise between the need to provide sufficient compost volume and reasonably close spacing.

Some grafters favour the use of rootstock pots which provide a so-called root controlling pattern, e.g. Air pots; these add to production costs and, unless strong evidence emerges to support their choice, seem unlikely to be widely adopted by commercial growers.

FIGURE 9.1   Pot sizes and patterns. Front row left to right – 9 × 9 × 9cm^2, 9 × 9 × 12.5cm and 1litre deep. Back row left to right – deep 'rose' pot, extra deep poly-bag pot and extra deep rigid plastic pot.

PLUG GROWN

Plugs, otherwise known as cells, are becoming an increasingly important method of producing seedling raised rootstocks. They comprise individual plastic cells moulded together into conveniently sized trays (Figure 9.2). Various modifications to this basic design are available and include those having short runs of cells which are hinged to open lengthwise, enabling easy extraction of the plant in its growing medium. An important version of this is known as the Rootrainer®.

Specification and Types

Cell size is measured in cubic capacity and for rootstocks chosen sizes range from 50cc–350cc. The major decision concerning size is related to whether it is planned to graft directly from the plug or pot-on into a larger container for further growth. Species which are plug grown to grafting stage respond best in relatively deep patterns, 110–150mm, at plant centres of 45–65mm and cell volumes between 150 and 350cc. Cell walls are profiled to discourage spiralling and encourage roots to grow downwards towards a large open drainage hole at the base. During cultivation a branched root structure is induced by supporting the trays above ground level; consequently, when the roots grow downwards, they reach open-air conditions at which point the growing tips are desiccated and killed (air pruned), lateral roots then develop, eventually developing a branched fibrous root system.

Requirements and Handling

Discussion with the supplier may be necessary to establish requirements for handling. Most plug producers remove the plants from their cells and deliver specific numbers in wrapped bundles.

This may be acceptable for sizes due for immediate grafting or moving-on into a larger plug or pot, but retention in the plug may be preferred for plants due to be held, pending grafting at a later date.

FIGURE 9.2   A 35 cell plug tray of direct sown, one-year *Magnolia kobus* seedlings. Cells are at 52mm centres 115mm deep and approximately 200cc in volume.

ROOTSTOCK CULTURE – BARE-ROOT

Bare-root, seed raised rootstocks are important for many ornamental species, and bare-root clonal rootstocks are usual for top fruit crops and some ornamentals. Increasing use of the hot-pipe system has resulted in more emphasis placed on bare-root as alternatives to pot grown.

By far the greatest numbers are grown in the field, either produced from seed, cuttings or layers.

Soil – Sowing – Seedling Density

Soil conditions are important; light organic soils are favoured and can normally be expected to produce the best fibrous root systems. Correct sowing density is essential to provide good quality one-year seedlings.

Herbicides for Rootstocks Growing in the Field

Weed control is an important issue in the culture of seedlings in the field and use of herbicides is common practice. Herbicide use on field-budded tree species has had variable effects on results; *Robinia* have reacted adversely* and others may be similarly affected. In other investigations, untreated rootstocks produced poorer results than those treated, presumably because of the effect of weed competition[†].

Negative effects of herbicides can generally be linked to poor practice which might include:

- Use of specific herbicides on species known to be susceptible.
- Using higher than the recommended dose.
- Short timespan between application and propagation procedure.
- Application to immature material.

Herbicides used sensibly are unlikely to have adverse effects on grafts[‡].

ROOTSTOCK CULTURE – POT GROWN

Establishment

It is generally accepted that well-established rootstocks produce better grafting results than those bare-rooted or recently potted. Consequently, most pot grown rootstocks are grown in the container for one growing season before grafting takes place.

Summer/early autumn grafting reduces the year-long timescale but is sufficient to allow full establishment. However, some allowance must be made for the reduced growing period, particularly during the early part of the summer, and to provide the desired stem girth a slightly larger grade rootstock (+1–2mm Ø) should be chosen for potting.

For economic reasons with easily grafted species, potting may be delayed until the autumn before winter grafting without seriously jeopardising results. For genera particularly prone to bleeding, such as *Acer* and *Tilia*, some use delayed potting as a means of reducing root pressure. If this is the main aim, potting-up just before or just after grafting has the most significant effect.

Genera having a branched fibrous root system, such as *Acer* and *Cornus*, can be retained in the rootstock pot for two or more years before grafting without prejudicing good results. Many rootstock stems for top-working grown in 2-litre pots or larger are held in the pot for more than one growing season.

---

* Howard & Oakley (1996).
† Atwood (1999).
‡ Ahrens (1979) and Briggs (1977).

POTTING COMPOSTS

Ideally, a purpose made, quick draining rootstock compost should be used. After significant drying-off, consideration must be given to the difficulty of re-wetting a predominately peaty mix. To address this problem, some grafters favour use of loam in the mix. Personal experience is that addition of loam is not ideal because a good quality sample is difficult to obtain and adds significantly to costs. Addition of medium grade sand to aid re-wetting is a better solution and can supplement the existing grit content (usually 5% for most mixes) to give an overall mineral content of 10% or 15%. More recently, the good hydration characteristic of wood fibre has been recognised and addition of 10% to the compost can increase its wettability when dry by 50%*.

Peat can be augmented by other major bulky organic materials; in the UK, composted pine bark is the most important of these. It has low water absorption compared with peat and consequently improves aeration and drainage. It also accepts water more readily than dry peat. For species requiring drying-off, the addition of pine bark is always beneficial. The amount incorporated varies, normally between 30% and 45%, higher amounts being restricted to species which are susceptible to root rots, e.g. some conifers, *Camellia*, *Daphne* and *Magnolia*. As with coir, bark has a de-nitrifying effect and supplementary nitrogen may be required.

Other forestry product materials, such as composted wood chips, wood fibre and sawdust, have promoted the introduction of various products with affinities to bark, but with higher water holding and nutrient exchange capacities. These may be blended with bark to produce peat-reduced or peat-free composts with characteristics well suited to pot culture. An additional advantage of these alternatives is that they can reduce weed problems, especially Liverwort and Moss.

When deciding upon levels of fertiliser to be added to the mix, it is important to be aware of requirements for various groups. Certain types (e.g. *Acer palmatum* and genera within Styracaceae) have a low nutrient requirement and excessive use of slow release fertiliser inevitably results in losses, especially when plants are under protection during the winter period. For these species, it is advisable to always use the lowest or lower than recommended rates. Any shortfall can always be mitigated by use of liquid feed or top dressing.

Use of slow-release fertilisers requires care because the rate of nutrient release is increased by bottom-heat, and if nutrient levels become too high, root scorch is possible. This is more likely when 8–9 month or 12–14 month formulations are used; safest is a 16–18 month product used at 10%–15% below recommended rates. Long-term formulations such as these are helpful in maintaining nutrient levels if rootstocks are retained for more than one season. At these lower rates, some short-term supplementary feed may be required during the growing season.

GROWING CONDITIONS

Views differ on ideal beds to provide optimum growing conditions. Good light levels promote compact balanced growth and a high root-to-shoot ratio; this benefits union formation but woodland species will require suitable shading. Use of capillary beds has the advantage that water supply to each pot is assured, unlike overhead irrigation systems, which for small containers depends upon very uniform coverage.

For convenience, plants are often set down pot-thick, but ideally, more spacing allows light to penetrate through the foliage. It is thought this has a beneficial effect on basal stem viability, the area where grafts are normally placed.

After potting-up and standing down, regular top trimming is important in managing growing conditions for potted rootstocks (Figure 9.3). It helps offset the effects of close spacing and induces even growth because dominant individuals which might overshadow others are controlled. Importantly the rootstocks need to be maintained at a suitable height to facilitate tying-in side

---

* Jackson (2018).

FIGURE 9.3 Trimming pot grown rootstocks, in this case *Liriodendron tulipifera* to be used for side grafting. A hand-held hedge trimmer is a suitable tool for small scale use.

grafts. Frequency and height of trimming will be influenced by the species; as a guide, a maximum of three trims per growing season is reasonable. There are indications that removal of portions of the rootstock just before field budding can have adverse effects on the take* and the same response may apply to summer grafts. Consequently, for summer grafting, it may be advisable to have as long a delay as possible between the last trim and making the graft.

Plug-grown rootstocks are either potted-on or grafted soon after receipt. It is not recommended to retain plug grown rootstocks in plugs once they reach optimum size; however, if this is unavoidable they should be placed in suitable trays (Boskoop trays or similar) to prevent rooting through. They may then be stood down on beds receiving overhead water, frequently monitored for nutritional levels, and regularly trimmed.

WEED CONTROL STRATEGIES

Weeds in blocks of small pots can represent a significant and costly problem, and weed control is of major importance. Comments on herbicide use on field grown rootstocks apply here, but the confined volumes of small pots make adverse effects more likely. Herbicides with very little or no root activity are ideal. In the absence of these, applications only of known, generally safe residuals can be considered.

Because no mechanical systems have yet been developed, cultural methods of weed control for potted rootstocks rely on good practice and hand weeding, supplemented by limited herbicide use.

Strategies for minimising weed problems are:

- Try to ensure that material coming on site from other sources is not harbouring weed problems.
- Total weed control in the surrounding area with a herbicide such as glyphosate.
- Where there is no chance of run-off reaching the growing area, total residual herbicides can replace or supplement glyphosate and will give at least six months' control.
- Empty standing-down beds should be cleaned-up before occupation with contact herbicides, followed by a suitable residual.

---

* Howard & Oakley (1997).

- Standing-down beds should be totally weed-free before being occupied.
- Some weed germination and moss and liverwort may be discouraged by capillary bed systems, which allow pot compost surface to be maintained in a fairly dry state.
- To ensure an initially weed-free surface before standing down, the surface of capillary beds can be covered with light-weight capillary cloth, woven fabric or perforated black polythene, replaced each season.
- If weeds are present or subsequently develop, timely hand weeding is vital to ensure they are removed to prevent flower and seed. If there is insufficient time to totally remove the weed, picking flowers or seed pods is better than no action.
- Use of mulches or pot covers has grown in significance as the range of available herbicides has reduced. An overall mulch of bark or wood chips is possible but less effective as small grade material must be used on small pots. Individual pot covers can only be considered for larger pots.

Pesticide Use

Emphasis on monitoring and control is important as pest and disease problems assume more significance in closely spaced plants, particularly those under protection. Vine weevil, Red Spider Mite (*Tetranychus urticae* and related species) and aphids can assume epidemic proportions if allowed to go unchecked and have an adverse effect on takes. It is recommended to seek specialist advice on current pesticide choice, the role of biological control methods and other strategies to limit problems caused by pests and diseases.

DRYING-OFF POT GROWN ROOTSTOCKS

Drying-off is a method used to combat the effects of root pressure, referred to by grafters as bleeding. The influence and significance of root pressure will be fully discussed in Chapter 11 and the varying requirements of different species in Parts Seven and Eight.

Root pressure is mainly concerned with pot grown, deciduous rootstocks grafted in the late winter/spring period but some evergreens including conifers can also exhibit symptoms. Certain species which are particularly susceptible, such as *Acer*, *Cornus* and *Tilia*, are liable to bleed during the summer/early autumn period and will require treatment as discussed below, though extreme drying-off at this period is not recommended. Bleeding can be alleviated in a number of ways. In warm callusing systems involving supportive conditions (poly-tents), by far the most effective is drying the rooting medium to a point where water absorption by the roots is significantly reduced.

Most recommendations for early spring grafting involve moving dormant rootstocks into warmer conditions and withholding water. Usually it is suggested this procedure is commenced 2–3 weeks or longer, before grafting is due. This has the double effect of inducing root activity and subsequently lowering the water content of the rootball.

Those not wishing to initiate early activity can achieve effective winter drying-off by placing potted rootstocks in an unheated covered area and withholding all water. For deciduous species, temperatures are kept as low as possible, but not below freezing, for long periods.

This procedure results in dormant rootstocks with no bud or root extension but with compost moisture levels low enough to inhibit root pressure. Personal experience is that when rootstocks in this condition are warm callused, graft union development proceeds rapidly and effectively.

An efficient cold drying-off system is to place rootstocks into convenient containers such as Boskoop trays and load these onto tiered pallets (Figure 9.4). The whole unit may then be transported to a suitable location such as a covered lean-to or shed, sited in a cool north facing position. Good access should be provided so that the pots can be regularly monitored for moisture levels and checks made to ensure no rodent damage is occurring. In these conditions, it normally takes several weeks for pots to dry-off. Timing of placing into storage is a key factor and best before heavy winter

FIGURE 9.4   Potted rootstocks packed into Boskoop trays loaded onto a three-tiered pallet moved from storage shed to a position close to the grafting work station. A few plants (*Betula*), top shelf right hand side, have been knocked out of their pots as they were still too moist and required to dry more rapidly.

rain soaks the pots. If the procedure is delayed, and pots are moved into drying positions in a very wet and leafless state, they may not have dried out sufficiently before grafting is due to start. To speed-up the process, a labour-intensive practice, such as removal of the pot (de-potting), must be used. In some milder areas where leaf fall is delayed, it may be necessary to use defoliants, enabling an early start to allow adequate drying-off.

Levels of drying are an important issue and will be influenced by choice of grafting system. Warm callusing under conditions of high humidity, especially if combined with bottom heat, will require rootstocks significantly dried-off. Judging the amount of drying is largely dependent upon experience, precise measurements are difficult to obtain and guidelines are generally not available. 50–60% of the moisture content of the potting mix has been suggested*; in practice it seems hard to convert this into meaningful guidelines. Some species including *Acer* and *Betula* may require drier conditions than others. Species such as *Juglans* scheduled for placement into warm callusing poly-tents can scarcely be over-dried, and the compost can appear 'bone dry' with no apparent moisture content.

The correct level can be ascertained only by a combination of appearance, weight and close inspection of sample plants.

When optimum dryness is reached, further drying must be prevented, either by grafting immediately or placing in conditions of high humidity. Polythene sheet may need to be placed over the top and down the sides to reduce further moisture loss.

Leafy rootstocks, particularly during the summer, are potentially damaged by over drying; experience and care is required to achieve correct moisture levels. Those which have been dried

---

* Macdonald (1986) p.529.

may require use of heavy shading (80%+) to hold them in a viable state until grafting takes place. After grafting, once placed into a 'safe' high humidity enclosed environment, dehydration is greatly reduced and further drying-out is largely prevented until graft union formation permits watering.

Most cold callused grafts are grafted in the 'just moist' stage; severe drying is not required. Grafts placed into a hot-pipe system are also much less demanding. With this system there can be a danger of over drying and some grafters make no attempt at pre-drying rootstocks.

The drying-off process can be difficult, demanding and time consuming; if incorrectly carried out, it can lead to a reduction rather than improvement in grafting success. However, most agree it is essential for some species when using warm callusing systems, especially when combined with bottom heat.

10 Scions – General Principles – Acquisition – Stock Mother Plants – Selection and Collection

SCIONS AND CUTTINGS COMPARED

The scion should not be equated with a cutting, where the aim is to induce new organs in the form of adventitious roots. For most species, this requires young, active stems containing tissues with plentiful parenchyma cells, capable of significant regeneration and differentiation.

In contrast, grafting builds on wound repair systems, active in all shoots which includes those described as well-ripened, often dormant, and normally well supplied with adequate amounts of stored food. For most grafting systems younger active material, often with low food reserves, is much less suitable and, because of its lack of survival capacity, is to be avoided.

PRINCIPLES FOR SELECTION OF SCION WOOD

Scion selection follows the basic principles; trueness to name, obvious health and vigour, avoidance of any 'off' types due to mutations, pest and disease infestation and physical damage. Most recommendations are to use well-balanced and ripened current years' wood. This is characterised by high levels of stored food, good potential to form callus tissue, normally well supplied with healthy buds. Such material is usually selected from the basal portions of the shoot, often discarding the apical portion. Length can vary but is frequently in the 100–150mm range with 3–5 healthy buds. Occasionally, when material is scarce or especially valuable, it may be appropriate to select dual or single bud scions (Figure 10.1).

Extra-large shoots, known as 'water sprouts', should be avoided. These frequently constitute soft or unripe wood with excessive amounts of pith and low amounts of stored carbohydrates (Figure 10.1). Their size makes it difficult to achieve well-matched grafts and the combination of adverse factors means they are likely to fail at an early stage. Alternatively, small, thin scions, often with high pith content, have low food reserves and poor survival potential. If these are the only choice, it is often better to include a thicker, older base of two-year wood*.

When selecting top fruit scions, extra vigilance may be required to maintain certain mutations influencing factors such as fruit colour, e.g. 'Queen Cox' and 'Galaxy', where reversion back to the original form is a possibility and not detectable until fruit is formed. Other scions which always need careful monitoring and selection include material from chimeras such as + *Laburnocytisus* 'Adamii', foliage mutations causing variegated colourways, upright growing (fastigiate) selections, weeping forms and dwarf mutants or selections of various genera. In every case, choices should be made which reflect best examples of the desired features and maintain the value of the original variety.

---

* Upchurch (2008).

FIGURE 10.1   Choice of scion material. LEFT, *Cercis canadensis* 'Forest Pansy' scion mother plant. Scions have been removed, leaving behind unsuitable material such as oversized 'water sprouts' and too thin shoots. RIGHT, *Davidia involucrata* 'Sonoma', shortage of material has prompted use of a single bud scion, side veneer grafted to seedling rootstock. Note: stimulant bud on the long tongue side veneer has extended.

Virus infections may cause growth and foliar distortion or chlorotic leaf blotches, but some may be virtually undetectable. Latent viruses are of this type; only material routinely monitored under laboratory conditions can ensure it is virus free.

DIFFERENCES BETWEEN 'RIPE' AND 'UNRIPE' SHOOTS

Ripeness or well-ripened requires further explanation. The horticultural terms ripe (firm) and unripe (soft) have a physiological and anatomical basis, but for practical purposes only anatomical features can be used to distinguish between the two. In a few species, pale-coloured current year's shoots interspersed between darker ones, can be an indication that they are unripe. In particular, for yellow or red stemmed types a paler hue possibly suffused with green, is often a sign of 'softness'. Firmness, or resistance of the shoot to bending, is a useful method which may be used to make comparisons between several shoots of the same age and type; those which are very pliant often indicate lack of ripeness and should be avoided.

A further indication of ripeness can be used as a confirmation guide to previous decisions. When harvesting scions, normally using secateurs, sample cross sections can be closely examined more precisely using a ×10 magnification hand-lens. Shoots which are judged to be unripe always have a larger area of pith than those which are ripe, i.e. they have high pith-to-wood ratio; often over 20–25% pith-to-wood.

Where possible, always choose scions with a low pith-to-wood ratio; less than 10% pith is recommended. More cross sections will reveal that at the base of the shoot the pith content is lowest, emphasising the importance of using ripened basal portions (Figure 10.2 & 10.3). To establish whether or not it is wise to use material cut from nearer the tip of the stem, additional sample sections can be viewed to make a judgement on relative ratios. For easily grafted species, such as many

FIGURE 10.2   Ratio of wood to pith. LEFT, *Cornus* x *rutgersensis* scion wood, on the left, small pith to wood ratio provides ripe, good scion material; middle, large pith-to-wood ratio, too soft for suitable scion; right, intermediate ratio, less than ideal, but possible. RIGHT, *Cercis canadensis* 'Forest Pansy' – good, ripe shoots. At the base of the left shoot a bud is visible. This can have a beneficial effect, potentially providing a larger area for callus production.

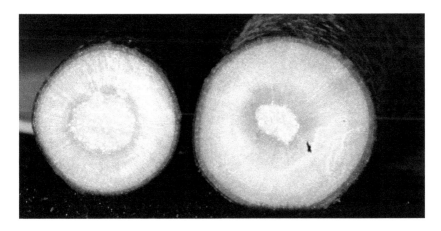

FIGURE 10.3   *Cercis canadensis* 'Forest Pansy'. Left, shows a section across upper part of stem which is too soft and unsuitable for scion wood (high pith-to-wood ratio, ±20%). Right, a basal section, low pith-to-wood ratio (3–4%) providing good scion wood for this demanding species.

in Rosaceae, much more latitude is possible and reasonably high pith-to-wood ratio can still permit good results. Growing in maritime climates a demanding type, such as *Cercis canadensis* 'Forest Pansy', pictured in Figure 10.3, normally requires only scion wood selected from the base of the shoot and showing low pith-to-wood ratio (<5%).

SCION ACQUISITION – NON-COMMERCIAL SOURCES

External sources provide scions from publicly or privately owned plant collections, or commercial suppliers.

PLANT COLLECTIONS

Many grafters have sources of scion wood external to the nursery. These can comprise collections in privately owned gardens, arboreta, larger institutions and botanic gardens.

Such arrangements can have substantial advantages for the recipient, saving the need for land set aside for stock mother plants, offering the possibility of redressing shortfalls in propagation material and providing a source of material for a wide range of species. The hugely diverse collection of species offered for sale by Hillier Nurseries during the 1960's to 1980's was heavily supported by material collected from a separately run extensive arboretum adjoining the nursery.

The disadvantages of such sources are relatively slow rates of collection and limitations on quality which may not match purpose grown mother plants. Shortfalls in quality may often be addressed by use of material with a two- or even three-year basal portion, which can be expected to produce acceptable results (Figure 10.4). For some, including *Acer palmatum* cvs, *Fagus*, *Liriodendron*, *Magnolia*, *Quercus* and others, use of scions with a basal portion of two-year wood may be preferred over one-year old material (see Parts Seven and Eight).

SCION ACQUISITION – COMMERCIAL SUPPLIERS

OTHER NURSERYMEN OR SPECIALIST SUPPLIERS

Acquisition is possible by purchase of scion wood from other nurserymen or specialist suppliers. Many can offer a comprehensive range covering the important commercial species, and some grafters rely on this source rather than their own. Quality of material coming from purpose grown stock mother plants is usually excellent. It is not normally possible to obtain rare or non-standard items, such as scions with a base of two-year-old wood. Scion wood is normally supplied on a 'ready to use' basis and the obvious advantages are convenience and a reduction in own labour costs.

OVERSEAS SOURCES

Southern hemisphere nursery producers in Australasia, notably New Zealand, have taken advantage of exceptionally good growing conditions to export certain high-value woody plant species into

FIGURE 10.4   *Magnolia* 'Sweetheart', scion from a garden collection, showing the use of two-year wood at the base. Experience has shown such material is suitable and provides very successful grafting results.

northern hemisphere nurseries. Transporting scion wood by the same route appears not to have been exploited but has potential for certain high value species such as *Acers*, *Cercis*, *Cornus* and *Magnolia*. Dormant material can be collected as early as late May/early June, but more usually July to August, and transported to nurseries in the northern hemisphere. Grafting at this time permits rapid graft union development and extension growth in the new home well before the onset of leaf fall and dormancy. Grafts of *Cornus controversa* 'Variegata', using scions from New Zealand, are sufficiently successful to indicate the potential of this strategy; *Magnolia* has also shown promise (Figure 10.5).

High Health Status Material

Clonal material of known provenance and health status is available for a limited range, mostly focused on fruit crops. This usually originates from government or state-sponsored agencies responsible for providing foundation plants tested and certified as free from pests, diseases and virus infections. Such material is made available to registered growers who subsequently supply to production companies distributing to fruit and ornamental tree nurseries.

In Holland, which has possibly the most significant scheme in Europe, annual cost of maintaining and developing tested material is in the region of 200,000 Euros (as of 2016). These costs are recouped by sale of material to growers in Holland and internationally. In the UK, the Plant Health Propagation Scheme (PHPS) provides a range of material to ensure a guarantee of health and trueness to type. Highest health status foundation material is distributed to designated growers. Plants fall into three grades, a 'top', Super Elite; a 'middle' grade, Elite; and finally, a Certified grade. These are distributed to fruit tree raisers where they have to be grown on approved sites, which involve varying levels of isolation dependent on grade of material. They are inspected regularly, including soil inspection and, subject to grade, strict quarantine regulations may be necessary. The various grades have a maximum stipulated life span before requiring replacement.

In the USA, schemes such as the Pacific Northwest (Oregon) are often State-based and mostly relate to fruit crops. The National Clean Plant Network (NCPN), involving 17 States, is concerned with Grapevine, fruit and nut trees and follows a path very similar to that described for Europe.

FIGURE 10.5 *Magnolia* 'Honey Tulip' single bud scion taken from plant imported from Southern Hemisphere (New Zealand), one month after grafting. Photographed in the UK in early August.

The Dutch scheme (Vermeerderingstuinen) covers the following crops and genera (2016); Apricot, Almond, Apple, Asian pear, Cherry, Quince, Medlar, Nectarine, Pear, Peach and Plum, all of which are available as virus tested items. The following list of ornamentals are also offered, those marked * are virus tested and the number of types available is shown in brackets: *Acer* (19), *Caragana* (3), *Corylus* (3), *Fraxinus* (8), *Gleditsia* (5), *Laburnum** (2), *Malus** (38), *Populus** (22), *Prunus** (56), *Pyrus** (6), *Robinia* (5), *Salix** (7), *Sophora* (now *Styphnolobium*) (3) and *Syringa** (28).

SCION ACQUISITION FROM INTERNAL SOURCES

Two main strategies for acquisition of scion wood on the nursery are either from growing on stock or dedicated stock mother plants. Using growing on stock as a source can result in substantial savings in land use because no stock bed areas are required; additionally, some labour saving is possible. Dedicated stock mother plants can ensure a guaranteed supply of high-quality scion material from plants selected for exhibiting best traits and trueness to type.

Final choice is likely to be influenced by the type of nursery. Those with a big acreage of field grown stock, especially trees grown to reasonable size, may be less dependent upon stock mother plants because of the reservoir of scion material available from prunings, etc. Smaller nurseries, especially container plant growers, are more likely to have a need for stock plants.

SCIONS FROM GROWING ON STOCK

For many species, scions gathered from growing plants, immediately or eventually destined for sale, are a convenient and satisfactory method of obtaining propagation material. There are advantages because for some species shoots can be removed with the dual purpose of shaping the plant and supplying scion wood. Material obtained is often good quality, sometimes with juvenile characteristics, good vigour and high health status.

There are disadvantages; a danger of errors in naming from other suppliers, caused by inadequate labelling and misidentification; young plants do not always produce ideal material, particularly if donor plants have been fed and watered heavily, when over-soft wood with low survival potential may lead to lower takes*; finally, there can be possible effects of atypical habit on young vigorous growth.

SCIONS FROM DEDICATED CONTAINER GROWN MOTHER PLANTS

Stock mother plants grown in containers provide a way of enhancing growth rates and a means of conveniently housing them in protected growing conditions continuously or for specific periods (Figure 10.6). Combining excellent growing media with the elevated temperature, controlled irrigation and sometimes extended day length can promote higher growth rates than are possible in open beds, and may be used to forward the availability of scion material to meet particular requirements. This can be useful to quickly bulk-up new plant introductions or rapidly respond to pre-placed contracts.

Containerised stock plants may be conveniently grown in protected structures, a necessary requirement for species liable to winter cold damage or those which respond best to high summer temperatures. Good examples of species responding well to container culture are found in: *Acacia* sp; *Albizia*; *Caragana* sp (not *arborescens*); x *Chitalpa*; *Citrus*; *Daphne bholua* cultivars (in this case large containers, 25+ litres, are required to avoid the problems of small pots) (Figure 10.6); *Diospyros kaki* cvs; *Edgeworthia*; *Erica* Cape Heath sp; *Eurya* sp; *Firmiana*; *Grevillea*; *Halimodendron*; *Lonicera hildebrandtiana*; *Magnolia* some evergreen sp (*M. grandiflora* cvs and the *Michelias*, etc);

---

* Ryan (1966).

FIGURE 10.6   LEFT, *Magnolia* 'Black Tulip' grown in 28-litre containers, resulted in accelerated scion wood production in the early years to meet forward contracts. RIGHT, *Daphne bholua* 'Jacqueline Postill' growing in 28-litre containers under protection for much of the year, housed in open air conditions during late June to mid-September.

Melliodendron; *Olea europea* cvs; *Pittosporum* some sp.; and finally, *Vitis* grown for Green Grafting scion wood.

SCIONS FROM STOCK MOTHER PLANT BEDS

Provision of stock mother plants to supply propagation material is regarded by many as worthwhile and, for some, essential. Advantages can be identified within two categories, strategic and technical:

Strategic Advantages
- If the range and quantity of material required is carefully matched to stock mother plant yield, in practice a procedure more easily proposed than achieved, availability of material will meet the demand. When successful, this can be a valuable business asset.
- Careful planning and foresight can lead to the elimination of shortfalls due to sudden popularity and demand from other commercial growers. This is particularly significant when, as is often the case, the shortage is universal across the industry.
- Material held in stock plant areas on the nursery site may well avoid the effects of imposition of embargoes (movement, etc.) due to sudden plant health issues. The dangers of spreading pests and diseases by transporting plant material from one region to another are now widely recognised, and becoming of increasing importance. It seems certain that further restrictions on movement will apply in the years ahead and self-sufficiency in the availability of propagation material will be of greater importance.

Technical Advantages
These may be listed as below:

- An opportunity to establish accurately identified species, varieties and cultivars. This may involve preliminary trials to establish best clones within a representative group. Unfortunately, variability is commonplace and an inferior form may be unwittingly chosen unless some preliminary effort is put into ensuring only superior selections are made. Once selected, only plants of this specific clone should be represented in the stock area. This will prove a positive aid to the production of uniform crops.

- Treatment of the mother plants through correct pruning, feeding, pest and disease control, etc. can produce best opportunities for evenly graded scion wood, a significant factor in production of uniform grafting output, takes and performance.
- Easier opportunities for manipulation of growth. This may have implications for scion type, i.e. stem girth, straightness, apical dominance, multi-branched, etc.
- Problems of plant health can normally be quickly identified and steps more easily taken to isolate and eliminate the problem.
- Record keeping for numerous characteristics is facilitated, including comparative success rates, growth rates and susceptibility to adverse factors such as wind and frost damage.
- Rapid collection of scion material; enabling the use of less skilled labour for some species is more easily achieved.
- As an aid to sales, stock blocks can provide a centralised area for training in plant identification and demonstrating desirable features.

Disadvantages

Disadvantages centre largely on costs, firstly of land occupied and, possibly more significant, costs of setting up and maintaining the area for the duration of its use. One estimate for costs of maintaining stock plants for cuttings production (ignoring setting-up and collection costs) was 1.5% of the selling price*. The lower productivity of scion wood compared with cuttings wood may raise this figure.

Some grafters hold a strong opinion that wood collected from the growing crop is often superior to that from stock plants. This conclusion may reflect the age of the plant material, which from stock beds of many years' establishment is likely to be less physiologically juvenile than a recently grafted equivalent. This highlights the need to maintain the quality of material obtained from stock beds. It is essential that appropriate pruning and all associated cultural practices are adequately sustained. Failure to do this will immediately result in a reduction in quality and question the whole process.

YIELD OF SCION MATERIAL FROM STOCK MOTHER PLANT BEDS

Yield of scion wood for a given quantity of plants/area is so dependent upon various disparate factors that it is difficult to provide accurate figures for planning purposes. Allowance must be made for time to bring the mother plants into production and also to factor in a replacement policy. Life span of the plants is variable but mostly influenced by species choice. Despite a heavy pruning regime, many of the non-bleeding types, which includes a significant number of genera, will have a life of 20 years or more. Those which bleed heavily such as *Betula*; those susceptible to Bacterial Canker (*Pseudomonas syringae pv. syringae*), such as the Cherries and Plums; those susceptible to Coral Spot (*Nectria galigena*), such as *Acer*, *Cercis* and *Robinia*, cannot be relied upon to provide healthy material beyond 15 years. However, a stock mother plant of *Cercis canadensis* 'Forest Pansy' growing here is still producing good quantities of scion material after 20+ years; although, Coral Spot is now beginning to attack some of the older branches.

Productivity is significantly influenced by the age of the mother plant; it will also depend on the plant group, trees versus shrubs for example, and on growing conditions. In favourable areas, such as the west coast of North America, many tree species become productive after only three to four years† while in the UK four to five years or longer must be allowed. In southern England, *Chamaecyparis lawsoniana* 'Stewartii' produced 15 useable shoots per plant when three years old, which had increased to 200 when five years old. Similarly, *Elaeagnus pungens* 'Maculata' had

---

* Anderson (1976).
† G. Meacham (pers. comm.).

increased from five to 45 shoots and *Rhododendron* 'Pink Pearl' from three to 25 over the same time period*.

For a wide range of ornamental tree species, 50 scions per tree after reaching full stature under a fairly hard pruning regime is a reasonable estimate, but productivity can vary significantly under different cultural conditions. Two hundred scions per tree have been suggested for the most productive species in favourable areas. Over the past decade the large, 20-year-old *Cercis* 'Forest Pansy', mentioned previously, has been able to produce annually 150–200 top-quality scions.

ESTABLISHMENT AND MAINTENANCE OF STOCK BEDS

SITE AND SOIL – INITIAL ASSESSMENTS

Stock bed areas may be in use for decades and correct initial choice of site and soil is crucial. Basic questions, such as choice of soil pH to suit particular groups of plants, must be addressed before detailed preparation takes place.

Most stock beds are situated in open areas, but some species respond best to overhead shade, indicating the need for a site in woodland areas[†] or provision of lath houses. Use of protected conditions within polytunnels or glasshouses is less often a necessity for scion wood production than it is for cuttings. Requirements for siting emphasise the importance of fencing against rabbits and deer, wind protection, good aspect, good drainage and, if possible, reasonably level land. However, if space is limited, sloping land, possibly too steep for efficient field cropping, can be usefully utilised.

Soil preparation is vital; key factors in making an initial assessment are listed below:

- Where possible, soil maps for the area should be obtained. This will give an identification of soil type and origin, which in turn provides guidance on other major features.
- Attempts should be made to discover the history of the site from previous owners or local records. Investigations may reveal aspects having influences on soil characteristics and possible problems such as carry-over of soil diseases.
- Initial land assessment should include digging core holes to ascertain soil conditions at depth. Excellent guides are available on practicalities of digging, core sampling and interpretation.
- Testing for soil characteristics is important and basic tests for pH and nutrient status should always be carried out.
- In the UK, some tests are mandatory for registered stock areas supplying super elite, elite or certified material. Tests to investigate soil-borne disease and pests are essential, particularly for Phytophthora and Verticillium Wilt, but might also include disease-causing nematodes (*Meliodegyne, Pratylenchus, Xiphinema*, etc.).

SOIL PREPARATION

Results of investigations listed above will provide guidance on actions to be taken to prepare the soil for planting. The following are likely to be included in pre-planting activities:

- On some sites a land drainage scheme will be required.
- Presence of aggressive perennial weeds requires effective treatment before planting.
- Sub-soiling required at depths relevant to the findings of the test holes.
- Addition of lime or flowers of sulphur may be required to adjust pH levels.

---

* Scott (1979).
[†] Vanderbilt (1960).

- A pre-land preparation cover crop is strongly recommended once basic operations are completed. Grasses have a very beneficial effect on soil structure and allow use of selective herbicides to control perennial dicotyledonous weeds while the cover crop is in place. The cover crop should remain in place as long as possible, ideally some years, before planting-up begins

PLANTING PROCEDURES AND DESIGN

Provision of irrigation for the site allows planting to take place more or less independently of season. Optimum timing for quick establishment and growth is during mid/late spring.

Planting design will depend upon overall policy. Some favour wide spacing, 3–4 metres, with crop cover strips (often grass) between the rows. This strategy is best for larger growing species such as trees and large shrubs (Figure 10.7). On closer spaced rows of 2 metres or less, cultivation or herbicides are used. Mechanisation using mini or vineyard tractors allows relatively close row spacing and provides a good compromise between labour saving and land use. Planting in closer spaced double rows with a wider alley between allows space to be conserved, whilst maintaining the efficiency of mechanisation.

Row layout can sometimes be altered and used to advantage, for example, by running rows east to west and planting shade-loving species between rows of taller plants. Such compromises can save space and resources. Protection may also be provided by planting tall evergreens on the windward side of shorter, wind susceptible types.

Good labelling is essential and may be further aided by alternately planting species with different foliage type or colour so that the differences are apparent. Grafting programmes are normally carried out within species groups, and time for scion collection is saved by planting blocks of related types in close proximity.

WEED CONTROL

Herbicides may have an adverse effect on graft success, and many are concerned about their use. Ground cover materials e.g. black polythene or ground cover fabric, can be laid along the row and

FIGURE 10.7 Stock mother plants on 3.65-metre-spaced rows, 1.8 metres apart in the row. Grass cover crop in centre of rows with herbicide strip beneath trees. On the left *Betula;* on the right, mixed Rosaceae trees. All on 0.9–1.2-metre-high stems with trained, closely spaced main branch system supporting scion producing shoots (Witch Hazel Nursery).

initial planting made through holes cut to provide access for the mother plants. Use of mulching materials such as bark, wood chips, etc. is a successful way of reducing weed problems but soil nutrient levels, especially nitrogen, may be affected.

As discussed in Chapter 9, some residual herbicides have not caused a measurable reduction in takes and their use is permissible. Careful application makes it possible to use contact herbicides with no possibility of adverse residual effects. This group of chemicals can be safely incorporated into a strategy of chemical and hand weeding.

Late in the season, some systemic herbicides, such as glyphosate, can be safely sprayed into the row just above ground level to eradicate or check weeds which have survived other weed control measures. It is essential that only totally dormant tree and shrub species grown on stems which have well-developed bark is treated in this way. Provided the weed is carrying foliage or green stems, this procedure is very effective.

PESTICIDE AND FUNGICIDE APPLICATION

Routine monitoring should be carried out for problems caused by pest and disease. These will follow recommendations for control measures concerning woody plant problems covered in other publications. Those growing material under PHPS regulations in the UK, and high health status schemes in Europe and the US, are subject to rigorous requirements for maintaining freedom from pests and diseases.

PRUNING AND TRAINING STOCK MOTHER PLANTS

Correctly pruned and trained mother plants are essential for a plentiful supply of top-quality scion wood. When a wide range of species is involved, this may require several different strategies and methods which are described below:

DECIDUOUS TREES

The first major decision is whether to grow a main stem or prune back to produce multi-stems more or less from ground level. Most opt for some form of main stem, the height of which is fixed to suit husbandry policy; a 0.6–1.2 metres high clear stem aids scion collection; it also provides access to the base of the plant facilitating weed control when applying herbicides or cultivating close to the stem (Figure 10.7).

Once a sufficient main stem height is reached, it should be headed back to encourage development of a side branch system. Training and further pruning is often required to ensure this comprises well-spaced, symmetrically arranged branches of equal vigour. They will form the long-term basis of scion producing shoots which develop from this basic branch structure. In the process, some thinning-out may be required and extra strong shoots, likely to unbalance the formative branch system, may be reduced in vigour by being pulled downwards by ties or weights.

The aim should be five to seven main lateral branches per stem. Over time these are allowed to develop and eventually pruned to form further lateral branches which produce the scion wood.

The method described produces a scion-bearing branch system arising from a closely grouped position on the main stem. This affords easy management and convenient collection, but it can produce a low percentage of shoots either too large or, because of some overcrowding, too small and not well ripened (Figure 10.8).

An alternative option for a number of species involves allowing the main stem to extend further and produce spaced lateral branches along its length. This more open habit produces shoots which are uniformly well ripened for a longer period than the method just described. Despite this, after some seasons, extra vigorous species, such as strong growing *Acer*s, *Fraxinus*, and *Tilia*, produce wood from higher branches which can become increasingly strong compared with those

FIGURE 10.8 *Betula* showing lateral scion-bearing shoots from a closely spaced main branch system pruned back after scion removal. Note: there is a quantity of extra heavy and thin shoots produced, in part as a result of over-crowding.

lower down. This problem can be reduced by careful, considered pruning, involving a harder regime on the upper shoots. Occasionally, early summer pruning is necessary with the object of allowing middle and lower shoots to develop more extensively and ripen in concert with higher ones.

The ideal but rather labour-intensive solution is to grow selected types on a modified fan or espalier system when vigour between branches can be more easily controlled. This is best achieved by appropriately spaced wires running down the row to which the main branches are tied. A further option is to tie main branches to each other in a near horizontal layout. With these training methods, scion material is produced from laterals which are allowed to develop in a controlled manner, along the length of the branches, not unlike pleached trees. Good management ensures very evenly graded, well-ripened scions season after season. Not all species respond to this treatment, and there will be some degree of trial and error required.

Heavy pruning to produce multi-stems commencing close to ground level is best suited to making stems for cuttings-grafts production of species, such as *Salix*. Some favour this approach also when large, heavy scion wood is required, as for "Long John" Grafts (see Chapter 18).

Deciduous Shrubs

There may be more emphasis placed on collection from saleable material for this group, but the advantages of dedicated stock mother plants apply.

A framework branch system is the basis for scion wood production but to retain health and vigour these may be subjected to a renewal policy. After a period of four to six years, the original branch is cut out and replaced by a younger substitute.

The majority of species respond well to an annual conventional pruning back by one-half to two-thirds, often the result of scion wood removal, supplemented by additional cutting out to promote and maintain shapeliness.

A few species require specific treatment, e.g. *Aralia*; these will be discussed in Parts Seven and Eight.

Removal of scion wood for summer grafting has a debilitating effect on the mother plant and may necessitate provision of additional numbers of stock plants to allow for pruning on a rotational basis.

For certain species use of large saleable specimen plants for scion material is an effective strategy, and often used as a summer scion source for certain species such as the *Acer palmatum* cultivars.

A few species require specific treatment for example *Aralia*, these will be discussed in Parts Seven and Eight.

Evergreen Trees and Shrubs

Evergreen shrubs include some important genera: *Rhododendron*, *Daphne* and *Camellia*. Good formative pruning in the early stages after planting, and for the following first few seasons, is necessary to build up a suitably extensive structure to provide ample material. Plants may need 'harvesting' on a rotational basis as removal of all suitable scion material from the same plant, season after season, results in significant loss of vigour and quality.

Magnolia grandiflora may require selective pruning to maintain sufficient material for substantial propagation programmes. This will normally take the form of ensuring a plentiful framework; lateral branches are developed by formative pruning in the early stages after planting. In many areas, cultivation under protection may be required, possibly using pot grown mother plants.

Conifers

Upright Formal Shaped Conifers

Scions from species grown for their formal shape, notably *Abies* (Silver Fir) and *Picea* (Spruces), are best obtained from managed stock plants with a central main stem and copious lateral branches forming sub-laterals, the terminal end of each providing a scion. To re-establish apical dominance, if terminal growths of this type are used, the subsequent growth of successful grafts will require staking for a period.

Some suggest that upright leading growths, achieved by decapitating the leading shoot after harvesting the scion wood of plants with six to eight tiers of branches is the only satisfactory way of obtaining suitable scion material*. This process is repeated annually, and the stock mother plant is eventually cut to ground level and discarded. Personal experience is that this treatment results in a fairly limited number of very strong shoots akin to 'water sprouts'. These are often not well ripened, over-long, too thick and result in scions difficult to match well with the rootstock.

Scions of terminal shoots from lateral branches carrying an adequate number of buds should produce extension growth with a balanced whorl of shoots at the base. Unfortunately, this is not always achieved, especially from scion wood collected from lower laterals. Not all buds grow out as expected and further growth and development extending over one or more additional seasons is often required to produce a shapely plant.

An alternative strategy, likely to speed up the process, is to use scion wood comprising a terminal shoot with a balanced whorl of branches at the base. Beneath this is included a basal portion of two-year wood, used to form the graft union. The method is discussed fully in Chapter 20: *Abies*, and Chapter 44: *Picea*.

---

* Sheat (1965) p.414.

Upright Informal Shaped Conifers

Best mother plants are those with a basic main branch structure, ideally comprising three or four stems of laterals and sub-laterals, which provide the required scion wood. Once developed, this structure is normally retained by annual collection of scions during the winter period. Even summer collection does not normally cause any disruption, but for weaker growing types some degree of rotation in the pattern of collection may be advisable.

Spreading and Dwarf Conifers

Spreading types normally produce a well-branched structure and the need is to keep pathways free of spreading branches. Dwarf types normally require inclusion of two- or three-year basal material to provide a scion of sufficient size. As a result, a larger number of stock plants must be available to allow scion collection on a two- or three-year rotation.

CLIMBERS

Large-scale grapevine production is best served by tall straining wire systems, possibly 3–4 metres high dependent upon vigour of the varieties selected.

These enable mother plant vines to be trained to provide fully ripened shoots with straight stems, suitable for grafting using manual or mechanical methods. High-level pruning platforms are required to facilitate constant tying-in and pruning away any unwanted shoots. Similar provision for many ornamental climbing species can be provided. Dependent upon vigour, height of wires can be adjusted between 1.8–2.4 metres.

Wisteria mother plants may be trained to poles 1.2–1.8 metres high and cut back to provide lateral branching. A basic frame-work branch system is built up and lateral branches which subsequently develop are used as scion wood. These are cut back to two or three buds each season and the prunings used as scion material (Figure 10.9).

SCIONS FOR WINTER GRAFTS

The term 'winter grafts' covers the period during which deciduous species are leafless and evergreens are non-extending and dormant. In temperate Northern Hemisphere climates, this can extend from early/mid-November to mid-April/early May.

DECIDUOUS HARDWOODS

Collection

In maritime northern temperate climates such as the UK, deciduous scion wood for winter/spring use should ideally be collected from mid-December to mid-January. A slightly earlier start is possible in continental northern areas. One-year old shoots are normally selected, leaving at least one bud still attached to the mother plant to produce shoots for next season's scion production.

If cold storage is not available, collection dates will need to reflect the sequence of natural bud break. Without cold storage, some constraints will be placed on grafting dates, as collection of early breaking material cannot be deferred indefinitely. Consequently, for some late breaking and demanding species (e.g. *Cercis canadensis* cvs), this could lead to late grafting past the optimum stage. To avoid this, cold storage facilities for storing scion wood are desirable if not essential.

Secateurs are invariably used to gather material which is cut from the mother plant, appropriately labelled and either placed loose into suitable receptacles or bundled and tied. Collection into individual bags for each variety is arguably quicker, subsequent handling is easier and desiccation is avoided. Bags of scions are transported to the reception area where storage and/or final preparation, trimming and possibly at that stage, bundling, can take place.

FIGURE 10.9 *Wisteria* scion mother plants trained to posts; before scion collection.

Suggestions that scions should not be collected in frosty weather have been made*. This does not seem to be supported by evidence and it is hard to see why, unless conditions are very severe, collection in temperatures at or around freezing point should constitute any danger to hardy temperate species, a view supported by some[†]. *Acer palmatum* has been collected in Pennsylvania commencing in mid- to late January in below-freezing temperatures. In this situation it is recommended that the wood should be allowed to thaw out slowly before use[‡].

Storage

Cool storage can take place in a cellar, suitable barn or shed. Scions of many species can be wrapped loosely in sealed, co-extruded polythene bags and placed in a cold position such as lying on an uninsulated concrete floor within a shed or barn. Birch scions have remained viable under these conditions for several weeks[§].

It is sometimes suggested to wrap moistened tissue paper, damp sphagnum moss or similar around the base of the scions, held in a polythene bag. This is not recommended as it can cause damage to the scion wood by water uptake (Figure 10.10). Best procedure is to pour some water in the bag, which is then upended and shaken out leaving only a modest number of water droplets. Once scion wood is inside, the bag must be very tightly sealed. Ideally, the scion wood should be surface dry before insertion into the bag.

Most recommendations for storage of dormant material involve refrigerators or cold stores. Low temperatures and protection from water loss are necessary. Various optimum storage temperatures

---

* MacDonald (2014) p.126 and Upchurch (2009) p.577.
[†] Bailey (1920) p.137.
[‡] Wolff (1973) p.339.
[§] Lane (1993).

FIGURE 10.10   LEFT, *Rhododendron* leaf over-trimmed and stored before grafting, causing substantial rotting. RIGHT, *Acer rubrum* cultivar scions packed with base of the scion surrounded by wet cotton wool. Note: stained, water-soaked wood at the base which must be removed when grafting takes place.

have been suggested, ranging from 0°C to 5°C. Temperatures close to freezing point will produce conditions for maximum lengths of storage. It is also essential that moisture levels in the scion tissues are maintained. This involves use of humidity-controlled stores or jacketed cold stores. Even in these conditions placing the scion wood in sealed polythene containers may be a safer option. Low temperatures will extend the period of safe storage. Hazel nut scions (*Corylus avellana* cvs) placed in sealed polythene bags can be kept in graftable condition for up to three months at 0°C. Temperatures below this are liable to cause damaged buds, and temperatures above shorten the storage time by periods corresponding to the increase in temperature. Above 2°C, Botrytis (Grey Mould) becomes increasingly active.

Trials carried out with Pecan (*Carya illinoinensis*) compared various procedures to sustain moisture content in scion wood for up to 70 days when stored at 2°C in a refrigerator. The scions were placed in sealed polythene bags with or without additional treatments. There was no advantage in sealing cut ends of the scion wood as is sometimes suggested, and the addition of water with an absorbent material also had no beneficial effect[*].

To provide prolonged storage for most species, a well-sealed slightly moistened polythene bag at a temperature of ±0.5°C is best. Cold stores with very accurate humidity control (difficult to achieve in practice), or jacketed stores, may permit scion storage without polythene wrapping, but this is generally a less safe option than sealed storage.

Preparation

Scions of dormant, deciduous species require little additional attention beyond selecting the best well-ripened portion, furnished with healthy vegetative buds. Any side shoots obstructing good access to the grafting area should be removed. The grafter or support staff may decide finished dimensions and cut to length accordingly.

EVERGREEN HARDWOODS SCIONS

Collection

Requirements and collecting procedures will closely follow those above. For hardwood evergreens, extra care must be taken to protect the detached scion wood from desiccation. Use of co-extruded polythene bags is important and adding moisture by light sprinkling with water is good practice. Bags of material should not be left in full sunlight on bright winter days. Some suggest collection should not take place in frosty weather; other experienced grafters disagree, and, in the absence of

---

[*] Nesbitt, Goff & Stein (2002).

frost-free weather in northern areas, invariably collect material during low ambient temperatures with no apparent disadvantage*.

Storage – Preparation

Scions should be placed in lightly moistened polythene bags closed sufficiently tightly to retain humidity.

When storing evergreen material, whole leaves only should be removed from the scions to reduce leaf area. To avoid leaf disease, reduction of individual leaves by cutting through the lamina should be delayed until just before grafting takes place. For most species, if not all, one cut across the lamina is all that is required, and several cuts to 'shape' the leaf should be avoided (Figure 10.10).

Leaf area reduction is carried out with three main objectives in mind: first, to expose the stem to provide access for graft placement; second, to reduce moisture loss through a reduction in leaf surface area; third, to reduce crowded overlapping foliage, likely to promote disease. For evergreen hardwoods, reduction of total leaf area by at least half is normal. This involves complete removal of selected leaves and/or reduction of individual leaves by half or more.

SCIONS FOR SUMMER GRAFTS

Scions in summer while active, i.e. growth extension still taking place, and dormant, i.e. growth extension ceased and terminal bud has formed, cover periods from May to late September/ mid-October.

SUMMER DORMANT DECIDUOUS AND EVERGREEN

Collection

Collection may be made approximately from mid-July to mid-October, when summer temperatures demand good protection from desiccation. Use of co-extruded polythene bags with some water droplets present is advisable, but covering bags with damp sacking during collection should not be required unless conditions are extreme. Early morning collection is an advantage, but the well-ripened shoots of summer dormant wood are much less vulnerable to damage than those still in active growth. De-leafing is normally delayed until the material is brought from the collection site to a more convenient area.

Storage

Summer dormant de-leaved scion wood may be stored under cool conditions at 4–5°C for several days before grafting. *Acer palmatum* cultivar scions with leaves removed held in a domestic refrigerator at 3–4°C were grafted successfully after four weeks storage. For summer dormant evergreens when leaves remain attached, two weeks is the longest safe storage period.

Preparation

The question of retention or removal of leaves from scions for summer side grafting is a contentious issue. It is suggested there is a need for a different strategy between side and apical grafts carried out in the summer.

For apical grafting, when the rootstock is headed-down during the summer, significant or total defoliation occurs and with it the ability to photosynthesise. To ensure some photosynthesis can take place, it is best to leave a proportion of foliage on the scion. Side grafts have the advantage of a leafy rootstock able to provide the support of photosynthates to the scion once the initial stages of graft union development are achieved and for these leaves may be removed. It is important to

---

* Thomsen (1978).

realise that de-leafing does not seriously jeopardise carbohydrate levels for ripened summer dormant scion wood.

Inconsistencies occur, for example, some recommend leaf removal on selected species only; one grafter reported removing leaves of *Acer palmatum*, but *Cercidiphyllum*, *Cornus* and *Hamamelis* had each leaf reduced by one-half to two-thirds. However, after three weeks, total defoliation was common[*] (this would equate to personal experience). Dutch grafters favour retention of *Acer palmatum* leaves but stress the importance of selecting only scions with totally healthy undamaged leaves, not easily achieved during normal collection and handling. Trials in Holland of late summer grafting of *Quercus coccinea* 'Splendens' compared retention or removal of leaves. Leaves retained produced a take of 81%, leaves removed 97%[†]. In North America, it is usual to remove scion leaves for many species.

Leaf removal significantly reduces problems of maintaining environmental support during summer side grafting procedures. Personal experience of a wide range of species is that it has a beneficial effect on grafting success. Scions carrying petioles with lamina removed are much less vulnerable to adverse conditions than leafy scions, and after four to ten days, petiole dehiscence occurs, resulting in further reduction of moisture loss. To produce an abscission layer and cause petiole dehiscence, the scion must be undergoing active cell division. This is a good indication that similar activity is in progress to produce essential callus for graft union formation. Petioles which shrivel and do not dehisce are an almost certain indication of graft failure.

Some consider leaf removal encourages production of unwanted new growth (Lammas shoots) in late summer. Lammas shoot production is significantly influenced by a number of circumstances: species/cultivar response, timing, rapid union formation and good subsequent environmental conditions. For many species, when grafted early and when other factors are favourable, it will occur whether or not leaves are retained on the scion.

Lammas growth may expose the young grafts to cold damage, especially if they are over-wintered outside, or in structures with no supplementary heating. Given reasonable protection, Lammas shoots are able to complete development and enter an abscission phase (up to mid-December in the UK) causing no harm, possibly a benefit, to the graft. They indicate that vascular connections between rootstock and scion are secure and such grafts overwinter as well as, or better than, those which had their scion leaves retained. Considering the range of species using summer bud grafting (budding) in the field and implicit in this is removal of the leaf, the opposition to leaf removal for the similar summer side grafted bench grafts would seem to be inconsistent.

If leaf removal is chosen, removal is by cutting through the petiole to remove the lamina but leaving at least half the petiole and associated bud still attached. Tearing-off leaves is not recommended as it is likely to cause damage to stems and buds.

SUMMER ACTIVE DECIDUOUS AND EVERGREEN

Collection

Only soft, semi-ripe wood will be available and collected during the period of active growth. Collection takes place in early summer when there is always danger of over-heating and desiccation. Handling procedures for actively growing scion wood are crucial if the material is to retain its viability. Co-extruded polythene bags containing some water droplets are required, and in very hot conditions, they should be contained within a wetted sack or similar. If possible, collection should take place in the early morning before the main working day commences. In the absence of cold storage facilities, only sufficient material for a day's grafting should be collected.

---

[*] Thompson (1988).
[†] van Elk (1978).

Storage

Cold facilities allow storage for two to three days at 5°C. Removal of very soft terminal growth should take place at the time of collection or shortly after. Once on the grafting bench, scions must still be kept in a turgid state which may involve some spraying over with a hand-held syringe. Once grafted, care must be taken with unsealed grafts to ensure the graft union is not drenched with water.

Preparation

Leafy scions of deciduous and evergreen species have a comparatively low content of stored carbohydrates, when collected in the early summer, while in active growth. These scions require leaf reduction but not total removal. This is normally accomplished by a combination of removal of a portion of the leaves present and, for species with large leaves, a reduction in the size of the lamina, the area of which is normally reduced by 50% to 75%.

CONIFER SCIONS

The selection of evergreen conifer scions has been discussed within the pruning and training of conifer stock plants. Collection procedures can follow those recommended for summer dormant leafy hardwoods.

HANDLING/PREPARATION FOR CONIFER SCIONS

Conifer scions comprise two major categories, deciduous or evergreen. Deciduous species, e.g. Gingko, have similar characteristics to dormant deciduous hardwoods. They are grafted when dormant and usual selection and handling of well-ripened wood cut to length, normally between 100–150mm, takes place.

Evergreen species are not grafted while in active growth but only during winter or summer dormancy. Procedures for handling both categories are influenced by three basic leaf types:

1. Those with scale-like leaves, e.g. *Chamaecyparis*, *Juniperus*, often borne on short, side branches will require some trimming to expose the graft area. Some shortening back of remaining side branches may be necessary to reduce potential moisture loss and avoid overcrowding.
2. Those with needle-like leaves comprising species such as *Pinus* sp., having comparatively long, well-spaced leaves held in small bundles in twos, threes, or fives. These must be removed from the stem at the grafting position, further removal by thinning out those above is normally required to reduce moisture loss and provide balanced, well-spaced foliage.
3. Those, such as *Abies* and *Picea*, with closely spaced needle-like leaves which often completely enclose the stem require a different leaf removal technique to expose the stem at the grafting position. This is achieved by holding unprepared scions more or less horizontally with the base pointing away from the body. Using a technique similar to that for making a veneer graft, the needles are shaved from the circumference with an extremely sharp knife (Figure 10.11). The objective is to expose a length of stem (40–50mm long) for graft placement. Care is required to avoid either cutting into the stem, which if damaged renders it unusable, or failing to remove enough of each leaf to allow good access. This is a demanding procedure and requires practice. Choice of scion wood with a two-year basal portion may still require shaving, but for some species, notably *Abies*, it may be possible to use a quicker and easier method of pulling rather than cutting to remove the needles.

FIGURE 10.11   Preparation of *Picea pungens* 'Oldenburg' scion, showing technique for shaving off the needles.

Storage

Storage procedures can follow those for deciduous scion wood. Careful closure of vapour-proof containers usually polythene bags or sacks are essential to prevent loss of moisture. Scions collected when completely dormant can be safely stored for some weeks at 0.5–1.0°C. Summer/autumn collected material is better maintained at temperatures of 1.5–2.0°C, when storage for up to one month is possible.

TRANSPORT OF SCION MATERIAL

Transport of scion material should follow the guidelines for storage described above but will include additional packing necessary to ensure safe movement from source to destination.

   For transport between sovereign nations or those outside national groups having special exchange arrangements such as the EU, there will invariably be a requirement for scion wood to comply with plant health regulations. These will prevent entry of certain species considered to pose a risk to the existing plant population of the importing country. The remaining 'allowable' items require the supplier to demonstrate that material has been inspected and certified as free of a range of specified pests and diseases. This normally involves a chargeable inspection by a qualified representative of the donor country, usually a government official, who has the authority to issue a phytosanitary certificate confirming the required 'freedom from infection' status.

   Packing for transport varies according to the distance and mode of travel. For local deliveries within the same area or country, simple packing involving vapour tight packages normally of polythene film, possibly wrapped in bubble film, placed in an outer cover, often of cardboard postal tube or cardboard box is usually sufficient. For more elaborate transport or longer distances use of a polystyrene tube to enclose the polythene inner, the whole surrounded by an outer cardboard box is able to provide protection from excessive temperature and mechanical damage. This type of package, often used for transporting liquid filled bottles, is very suitable for air delivery, frequently a necessary means of transporting scion material over long distances within an acceptable timescale.

POST GRAFTING SCION GROWTH – USE OF BUD GROWTH STIMULANTS

Scion buds of successful grafts which fail to extend but retain conditions of dormancy can cause delayed or reduced growth. Certain genera show this characteristic more than others; *Aralia elata* and *Rhododendron* are particularly prone and fail to respond despite procedures such as 'crippling' or removal of extension buds on the rootstock. Use of appropriate treatments has the potential to be a valuable aid for inducing bud break on recalcitrant grafts. To date, gibberellic acid (GA 4+7) used solely or in a mixture with benzyl adenine (BAP) appears the most promising (further discussion of specific examples is given in Parts Seven and Eight).

Part Three

Achieving the Union

11 Anatomy and Physiology of Graft Union Formation and Development

A full understanding of the basic principles of graft union formation is not essential to achieve good results. However, this knowledge is of real value when reviewing successes or failures and tackling new projects. The anatomical and physiological changes which take place within the graft to successfully achieve a union present a fascinating natural process and are well worth investigating.

It must be admitted this is not an easy topic; persistence is required to master the terminology and follow the complexities leading to an understanding of the whole activity. The limitations of this publication prevent a comprehensive discussion of all aspects of basic plant anatomy. Those interested are recommended to consult publications such as *Botany for Gardeners* by Brian Capon, or *RHS Botany for Gardeners*, which provide valuable background information to the topics discussed here.

BASIC WOODY PLANT ANATOMY

WOODY PLANT – PRIMARY STEM ANATOMY

In woody dicotyledonous species the simple non-woody primary growth of emerging seedlings and growing tips of shoots and roots continue developing with age to produce a more complex secondary growth structure. This eventually results in the woody stems, trunks and roots of hardwoods and conifers (Angiosperms and Gymnosperms).

Young non-woody stems have a centrally located tissue known as pith, composed of simple living parenchyma cells. These cells, a legacy of their ancient ancestral origins, the aquatic algae, form an integral part of various structures within the plant. They are very diverse in form and function, have a nucleus, and are notable for their ability to develop (differentiate) into other cell types. Added to this is their remarkable capacity to dedifferentiate, that is, to revert back into an actively dividing (meristematic) state. Of importance is that parenchyma cells are the 'building blocks' of a plant tissue known as callus, developed at the site of wounds, including those caused by grafting, and essential to the formation of a graft union.

Surrounding the pith is an enclosing cylinder of ground tissue* comprising mostly basic parenchyma cells but with the addition of two other closely related types with variously thickened walls, collenchyma and sclerenchyma. These provide some degree of structural strength to the young stem and root. At the surface are epidermal (skin) cells, which become impregnated with a fatty acid polymer known as cutin; this is not unlike polythene and acts as a protectant against desiccation of the delicate cells beneath.

Contained within the ground tissue are discrete strands of vascular tissue, arranged around the pith in a circular fashion stretching from apical buds to the hypocotyl (the transition point between stem and root). These are commonly known as vascular bundles, but more correctly as fascicules. Each vascular bundle comprises an inner tissue, known as xylem, composed largely of thick-walled vessels which eventually become non-living. This is separated from an outer layer of thinner walled

---

* Esau (1960) p.12.

135

phloem vessels by a narrow layer of tissue vitally important to grafters, the cambium, correctly called vascular cambium (Figure 11.1).

Below the hypocotyl, the vascular bundles merge to form in the root a centrally arranged xylem with no pith. As in the stem, on the periphery of this, are the vascular cambium and phloem. After secondary growth, apart from initially having a substantially thicker cortex layer, anatomical differences between stems and roots are minimal and need not be of concern to grafters.

WOODY PLANT – SECONDARY STEM ANATOMY

The early phase of secondary growth connects the vascular cambium from one vascular bundle to another by a 'bridging link' of dedifferentiated parenchyma cells originating from within the ground tissue (see Figure 11.1). This link, known as the interfascicular cambium, results in the development of a continuous cylinder of vascular cambium within stem and root.

The specialised meristems of the vascular cambium have a majority of cells, known as fusiform initials, with a specific structure of distinct, elongated prism shape. When the plant is growing actively, these cells repeatedly divide to become xylem (wood) on the inner side and phloem on the outer. Together these form a continuous cylinder or tube of tissue, stretching from root tip to shoot tip.

Spaced at intervals throughout the vascular cambium are small clusters of meristematic cells known as ray initials. These are less specialised than the fusiform initials, being rounded or only slightly elongated and consequently very similar to parenchyma cells. Ray initials are responsible for the production of secondary rays. These supplement, and eventually dominate, the primary rays formed by remnants of the original ground tissue after the vascular bundles are linked. When viewed in transverse section, primary and secondary rays radiate from the central pith like spokes of a wheel.

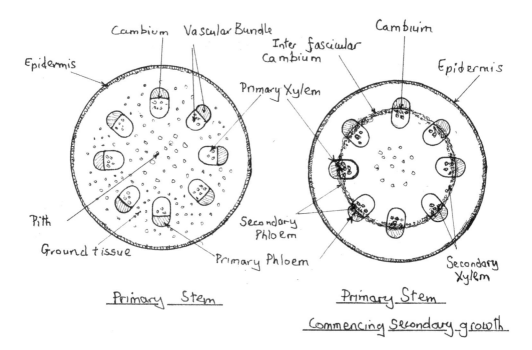

FIGURE 11.1   Diagram of transverse section of a woody plant stem. On the left, a primary stem showing vascular bundles (fascicules) embedded within pith and ground tissue surrounded by a surface layer of epidermis. On the right, showing the formation of interfascicular cambium within the simple parenchyma cells of the ground tissue. This links the vascular bundles, marking the commencement of secondary growth.

In structure, rays comprise groups of cells arranged in longitudinal bands of varying width and depth which may show up when logs are split. This is very evident in oak wood when they can be seen as pale patches, known to cabinet makers as figuring. Numerous secondary rays intersect the xylem and phloem; parenchyma cells within the ray form a vital living link between the two tissues. If the stem is cut or otherwise wounded, areas where the rays traverse the vascular cambium become major sites for the formation of callus. This is of real importance to grafters because callus is an essential ingredient of graft union development. It is eventually responsible for facilitating the connection between the cambium of rootstock and scion.

Once linkage of the vascular bundles and the resultant formation of the woody cylinder are completed, the plant has now entered into the secondary growth stage and exhibits the characteristics of woody species, represented in the drawing Figure 11.2.

Woody Plant Anatomy – Vascular System

In temperate climates xylem is produced each season during the growing period when, at this early stage, it is known as sapwood. Within hardwoods xylem cells differentiate to produce numerous elongated tubes known as vessels and tracheids; in conifers, only tracheids are developed. Vessels and tracheids fulfil the function of transporting water and dissolved mineral nutrients (sap) obtained from the roots to living plant tissues. Parenchyma cells (living cells) are also formed at randomly spaced intervals within the xylem of the sapwood. These are of importance because, together with those in the rays, they form a living link between the xylem and phloem and, in response to wounds caused by grafting, those close to the vascular cambium can dedifferentiate to form callus.

Sapwood is a storage area for carbohydrate (energy) reserves and provides significant mechanical support, especially in young shoots. Each season it is gradually pushed inwards by incremental layers of new xylem; in temperate regions this produces the characteristic pattern of annual rings. As a result of this inward movement, inner (oldest) portions of the sapwood eventually die and become

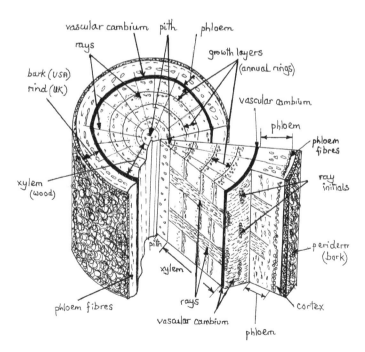

FIGURE 11.2   Diagrammatic representation of a woody stem. Transverse, longitudinal and tangential views of the stem are illustrated.

non-functional, when they are known as heartwood. This is the major constituent of trunks and branches; is the main repository for waste materials, and provides the principle mechanical support for woody plants.

Photosynthates (elaborated sap) are transported in the phloem from the leaves through sieve tubes in hardwoods and sieve cells in conifers. Photosynthates are the result of photosynthesis and are especially rich in carbohydrates which are stored within living cells, usually as insoluble starch. This is converted to soluble sugar when required to fuel growth and development. Synthesis of various plant growth substances (metabolites, auxins, cytokinins, etc.) controlling aspects of plant function occurs throughout tissues in leaves, stems and roots. Transportation of these is also through the vascular system.

Phloem contains a higher proportion of parenchyma cells than xylem. Those close to the vascular cambium, together with parenchyma cells in primary rays, secondary rays and xylem are responsible for production of much of the callus essential in graft union development.

Surrounding the phloem is the cortex, composed mainly of parenchyma cells often with chloroplasts containing chlorophyll, responsible for the green colouration of young stems. Cortex is important for providing storage of photosynthates and metabolites, especially carbohydrate reserves, and in some species can also be responsible for some callus production.

Incremental growth gradually pushes older phloem outwards from its origin in the vascular cambium. Cork cambium, known by plant anatomists as phellogen, subsequently arises as a continuous cylinder in the outer layers of the phloem or cortex. Phellogen is responsible for the development of the outer bark, otherwise known as periderm, external layers of which become heavily charged with waxy, waterproof suberin. As a result, these external tissues die and slough off, producing characteristic bark patterns on older stems and trunks.

All layers of tissue lying outside the vascular cambium, i.e. the phloem, cortex and periderm, are known as 'rind' by UK horticulturists, hence the term 'rind grafts'*. Rather unfortunately they are known as 'bark' by nurserymen in the USA. Plant anatomists also regard this layer as bark, dividing it into inner bark comprising the phloem and cortex, and outer bark or periderm, comprising the cork cambium (phellogen) and dead covering layer (bark) known as phellem or rhytidome.

GRAFT UNION FORMATION

UNION FORMATION – RESPONSE TO WOUNDING

Fundamental to the process of grafting is the reaction of woody plants to heal wounds. To protect portions of the plant that have been damaged, or subject to attacks by pathogens, new tissue development is stimulated in an attempt to isolate and trap any disruption or pathogen within a stem compartment. Compartmentalisation is sometimes seen at the top of the scion where, after uniting with the rootstock, the upper tissues are isolated and walled-off (Figure 11.3). For most species, when trimming scions to length, this response makes it safe to use sharp secateurs rather than insist on the cleaner cut of a knife. A similar approach can be taken when heading-back rootstocks of grafts which have been side grafted.

Wound healing is evident in all ages of woody plant stems; removal of large limbs from old tree trunks stimulates the development of wound-healing callus which eventually produces wound periderm to seal off vulnerable tissues. Use of young wood for rootstocks and scions is most desirable but if such wood is unavailable, use of somewhat older wood can still produce satisfactory results.

Wound response and production of callus at the site of wounds is modified in the grafting process by the addition of a further component, the scion, leading to the concept of 'wound healing in common', the basis of formation of a successful graft.

---

* Garner (2013) p.58, p.127 and Hartmann, Kester, Davies Jr. et al. (2014) p.436.

FIGURE 11.3   LEFT, a compartment ring developing on a single bud scion of *Sorbus hedlundii* showing as swollen area above the bud. RIGHT, a compartment ring developed in *Magnolia* 'Honey Tulip' showing walling off and dead tissue at the top of the scion.

STAGES IN THE FORMATION OF THE GRAFT UNION

The development of a graft union takes place in a series of stages; after the following account, these are illustrated by a diagrammatic representation of the process Figure 11.5 through 11.9.

First Stage – Necrotic Plate

This stage commences after cuts are made, when exposed surfaces die forming a brown or black necrotic plate composed of dead crushed cells and cell walls (Figure 11.4). The depth of this layer varies with species, length of exposure and sharpness of cutting tool. A blunt knife and prolonged exposure can cause deep layers, presenting a barrier to subsequent union formation. In dry air or low humidity, the dead layer is quickly permeated by suberin, a cork-like fatty substance which is virtually impervious to water and closely linked to the previously described cutin. The presence of suberin curtails moisture loss, seals off xylem and phloem and provides an early barrier to disease entry. It is important for grafters to realise that subsequent union formation can be adversely affected by excessive suberin formation (suberisation), caused by allowing cut surfaces to dry too much after grafting, by delaying sealing in a dry atmosphere too long, or sealing badly. Close fitting, well-matched grafts resist drying-out at this stage far longer (some hours) than those with large gaps between the components; differences between species may also be involved.

Timing for initiation of the necrotic plate stage depends upon species and temperature but is likely to be in the two to three day period. Some such as *Malus* are quick sealers; others such as *Juglans* are slow and this response can affect grafting success. After four weeks, some species, e.g. *Acer palmatum*, have failed to respond to wounding at temperatures at or below 4°C*. Others at this temperature, for example *Betula* sp., produce a necrotic plate and commence very slow cell division.

The necrotic plate is often not visible, but on unsealed grafts in a supportive environment, exposed tissue will develop the plate and subsequently stages of its breakdown and development of callus can be seen through a x10 lens. Breakdown of the necrotic plate occurs during callus formation; although, the dark staining on dead xylem tissues may be retained for a considerable time.

---

* Copini, Decuyper, den Ouden et al. (2014).

Second Stage

The second stage is the extension of living xylem and phloem cells from rootstock and scion into the necrotic zone* (Figure 11.4).

Third Stage

Copious cell division at this stage produces callus, a tissue largely composed of parenchyma cells. Contrary to the popular view that callus arises from vascular cambium, plant anatomists generally agree its origin to be from living cells within phloem and xylem, especially the rays[†]. The specialised cambial meristem cells (fusiform initials) are adapted to produce vessels, tracheids and sieve tubes only. The less specialised ray initials may well contribute to callus production.

Growth of callus cells involves rupture of the necrotic plate producing a pattern of black/brown stripes (strands) eventually becoming patches, before absorption into cells beneath (Figure 11.4).

Callus formation by division of meristematic parenchyma cells is an on-going process, infilling surfaces and voids between the graft interfaces. Eventually, callus cells from each component will become intermingled. Microscopic pores in the cell walls contain plasmodesmata, which are living tissue strands running from cell-to-cell allowing solutes to move between them. It may be at this stage that the foundations of any incompatibility start to develop. Although callus tissue is able to form within compatible and incompatible combinations, callus formation is no guarantee of compatibility.

At this stage, some movement of water and dissolved nutrients between rootstock and scion is possible. This is primarily achieved by movement from cell to cell but as a further aid, small discrete structures, (wound vascular elements) containing vessels or tracheids (water-conducting cells) often arise within the callus. The pith of some species, e.g. *Rhododendron*, is able to form meristematic cells which differentiate into vessel and tracheary elements, increasing transport and mechanical strength. Contrary to views often expressed, presence of pith at the graft interface need not be cause for concern, and cutting the scion to avoid exposing pith cannot be justified if it leads to poor cambial alignment. Reasons for disquiet over the presence of pith may be linked more to an association with unripe scion wood than the pith itself.

The processes just described helps retain scion health; however, without further union development, insufficient sap flow between the components means that grafts wilt and collapse after bud extension.

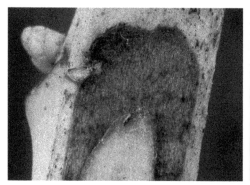

FIGURE 11.4   LEFT, *Rhododendron adenopodum* necrotic plate strands (showing as brown stripes on a pale green background) after three weeks at 10–12°C. RIGHT, *Rhododendron* 'Fantastica', showing necrotic plate breaking up into patches containing living cells after six weeks at 10–12°C.

---

* McCulley (1983).

† See for example: Kester (1965), Widmoyer (1962), McCulley (1983), Biggs (1992), Mudge (2013), Mustard & Lynch (1977), Begum, Nakaba, Oribe et al. (2007), Copes (1969), Starrett (2008) and Hartmann, Kester, Davies Jr. et al. (2014) p.444.

Given suitable conditions, callus production can begin after 5–10 days and may have developed sufficiently for stage four to commence at 14–21 days; this will often take longer, especially in many conifer species.

Callus production is so crucial to eventual success that factors influencing its formation and development will be investigated further at a later stage.

Fourth Stage

The success or failure of grafts to achieve a union is dependent upon the satisfactory completion of this stage. It involves development of a linkage between the vascular cambium of the rootstock and the scion.

Before a graft union is possible, a mass of callus cells forms what is known as a callus bridge[*], so called because, by infilling between the cut surfaces after the two components are placed together, these cells provide a 'bridge' between the rootstock and scion. Within this callus bridge, dedifferentiated (meristematic) cells, usually originating in the region of, but not from, the rootstock and scion cambia, form what in this publication is called the 'cambial strand'. In successful grafts, development of this strand continues until a linkage between the cambia of the rootstock and scion is formed. Once completed this connection re-establishes a cambial cylinder between the components; further growth and development provides vital vascular continuity, marking the beginning of a successful graft union. If no linkage is formed, vascular continuity is not achieved, and the graft fails.

Proximity of rootstock and scion cambia significantly affects the speed of connection and completion of the cambial strand[†]. Good knifesmanship resulting in close contact and alignment, i.e. 'good matching', ensures that voids between rootstock and scion are as few and small as possible. Because of the short distances involved, grafts which are well aligned and cut accurately allow portions of cambial linkage and graft union to form quickly, sometimes leading to development of a successful graft before all voids on the graft interface are filled. This can have significant benefits for the health and viability of the scion. Poor matching has the opposite effect, increasing chances of graft failure by delaying subsequent union formation, in some examples by as long as three months[‡].

If the cambial strand becomes elongated due to poor alignment, convolutions in its pattern occur. These can reach such proportions that subsequent union development is influenced, vascular tissue becomes distorted and movement of water and nutrients up and down the stem is impaired. Very extreme poor matching is not the cause of true incompatibility, but may lead to reduced growth, premature flowering, early fruiting, shorter life-span and increased sucker production, all symptoms reflecting those of incompatible grafts.

A further essential factor in graft union development is the need for pressure to be applied to the graft components. This recreates the substantial forces of natural compression present in the undamaged stem and root. Throughout the life cycle of the plant this is so significant that in woody plants it can cause innermost tissues within the xylem to become crushed and distorted by increments in growth[§]. Pressure makes possible differentiation of the simple parenchyma cells of the callus bridge and cambial strand into the organised connections between rootstock and scion, eventually leading to the formation of xylem, phloem and associated structures. Uncompressed callus cells at the edge of the cut surface, externally visible to the grafter, remain disorganised. After these tissues become permeated with suberin and change colour from white to orange or brown, they become part of the graft covering layer, known as the graft periderm.

Once continuity is re-established, the cambia of rootstock and scion renew their function of differentiating new xylem and phloem. Xylem formation is initially the most vital, ensuring flow of water, and preventing desiccation of the scion and any extending growth from scion buds; phloem

---

[*] Hartmann, Kester, Davies Jr. et al. (2014) p.444.
[†] Howard (1977).
[‡] Copes (1969).
[§] Esau (1960).

formation follows to complete full vascular function. Following completion of a successful graft union, the rootstock and scion tissues (xylem, phloem etc.) subsequently produced, totally retain their separate genetic identity. However, at the graft interface, these tissues become physically interlocked to provide a linkage as strong as in any other part of the stem.

Time-span for completion of vascular connection depends on the progress of previous stages and prevailing temperature, but often in the 17–30 day range, longer for many conifers.

DIAGRAMMATIC REPRESENTATION OF GRAFT UNION FORMATION

The following drawings (Figures 11.5–11.9) (adapted from Het Enten Van Boomkwekerijgewassen*) give an impression of stages in tissue development to achieve graft union between the interface of

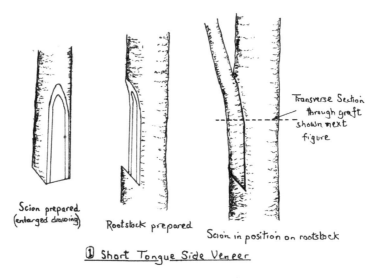

FIGURE 11.5   Short tongue side veneer graft. Enlarged view of transverse sections within the dotted lines is shown in Figures 11.6–11.9.

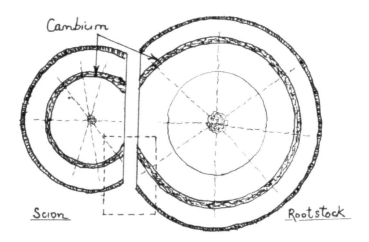

FIGURE 11.6   Transverse section drawing of the graft in Figure 11.5. The square marked with dotted lines is drawn in close-up in the figures below, showing diagrammatic representations of callus and early stages of union development between the cut faces of the graft union.

---

* Alkemade & van Elk (1989).

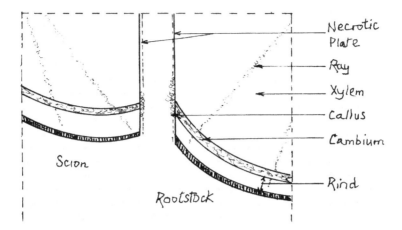

FIGURE 11.7   First to third stage – necrotic plate is broken up by the development of callus cells developing between the rootstock and scion cut surfaces, visible where the cut face exposes the cambium.

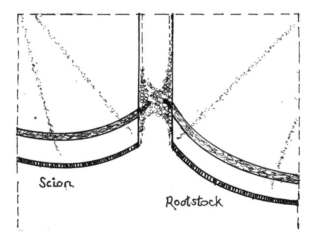

FIGURE 11.8   Third and fourth stages – callus development continues forming a callus bridge between rootstock and scion. Cambial strand commencing from scion and rootstock cambia.

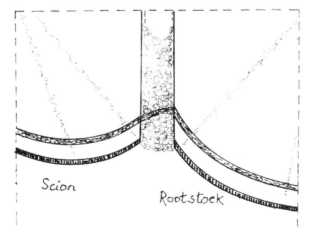

FIGURE 11.9   Callus bridge completed and cambial linkage between rootstock and scion is well advanced.

the rootstock and scion. Although a short tongue side veneer is depicted, development at the interface of the components follows a virtually identical process in all grafts.

FACTORS INFLUENCING CALLUS DEVELOPMENT

Callus development constitutes the vital third stage of graft union formation, and without it the graft is destined to fail. The following factors influence the process:

- Genetic influence.
- Rootstock/scion and grafting method.
- Temperature.
- Rootstock activity.
- Moisture, humidity, oxygen.
- Timing/seasonal patterns.
- Polarity.
- Stimulant buds.
- Root pressure.

GENETIC INFLUENCE

Speed of development and amount of callus produced can have a significant influence on the ease or difficulty of grafting a given species. In many environmental conditions, some species, for example in the Rosaceae, quickly produce copious callus and are therefore easily grafted. *Camellia*, other Theaceae, *Cercis* and many Styraceae are often minimal or slow callus producers, and represent a challenge to the grafter.

ROOTSTOCK – SCION – GRAFTING METHOD

In some species, most callus is produced in the rootstock, particularly when it is larger than the scion. The scion may produce most callus when rootstock and scion diameters are similar. In veneer grafts, callus development is almost invariably more advanced in the tongue than the rest of the graft, sometimes by several weeks. This is less noticeable in short tongue veneers but is nearly always present (Figure 11.10). In extreme cases, an effective union is produced only between the rootstock tongue and the outer face of the scion. The graft can fall away from the main rootstock being solely attached by the tongue (Figure 11.11). Such grafts often eventually develop normally and provide satisfactory specimens for planting but for those with a commercial interest, they are not saleable for several years.

TEMPERATURE

Between 5–32°C, the rate of callus formation increases differentially with rising temperature*. This rate may vary between species many responding best between 20–22°C. Those such as *Juglans*, Grapevines and Hickories (*Carya*) are best between 25–27°C. A number including *Betula, Fagus, Rhododendron* and most Rosaceae are still active at lower temperatures in the 8–12°C range, but if other factors such as freedom from disease and lack of excessive root pressure are satisfactory, they perform better in higher ranges (18–22°C). Temperatures above 28°C generally result in reduced callus formation. Above 32°C, callus formation for most species is inhibited; cells may be killed

---

* Hartmann & Kester (1959) pp.279–280.

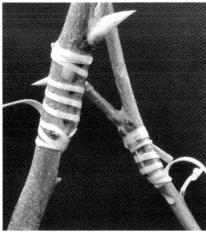

FIGURE 11.10   LEFT, *Acer palmatum*, short tongue side veneer, with a thin scion on a thick rootstock; note the callus on the tongue. RIGHT, *Carpinus fangiana*, on the left, graft callus is present on short tongue of side veneer; on right, graft callus on tongue of long tongue side veneer but no callus development evident yet on inner surfaces of either graft.

FIGURE 11.11   LEFT, *Acer shirasawanum* 'Autumn Moon' apical long tongue veneer. Showing only the tongue united. Note: dead top of the *Acer* rootstock developed after union formation, a result of compartmentalisation. RIGHT, long tongue side veneer of *Magnolia* 'Maxine Merrill' showing only the tongue united. The rootstock of *Magnolia* was a re-graft and the scar of a previous failed graft is visible.

above 40°C*. At temperatures below 6°C most temperate species require weeks to produce any significant callus.

In hot-pipe systems, once the graft union is warmed, callus quickly develops. However, because woody tissues have a high insulation value, the effect of heat is localised and retained inside the

---

* Hartmann & Kester (1959) p.280.

heating chamber. Outside this chamber tissues do not increase in temperature and remain dormant*; this is an important feature in the success of the hot-pipe system, to be discussed further.

ROOTSTOCK ACTIVITY, ROOT EXTENSION

The predominant view among practising grafters and some academic sources is that in the dormant season root activity, expressed by root extension, should be induced in potted rootstocks before grafting commences. This is presumably because it is considered to promote callus formation and subsequent development of the union. It is not suggested that the same should apply to bare-root rootstocks. For these, root extension would expose soft young roots to desiccation, and to attempt it would be unwise. For the highly successful hot-pipe system, root activity appears to be irrelevant to good results, and no suggestions have been made that it is required. Two genera, *Rhododendron* and *Juniperus*, specifically stated to require rootstock root extension are among the few recommended for cuttings-grafts techniques, when initially no roots are present[†].

These observations, coupled with the origins of callus and the condition of root tips discussed in Chapter 8: Root Pressure and Root Activity, must raise questions on the need for root extension before grafting. Personal experience across a wide range of species, using mostly potted rootstocks, has not demonstrated any significant advantage of rootstocks with active extending root systems over those not extending. These inconsistences suggest that there is a need for science-based investigative work into this potentially important aspect of graft physiology. Handling procedures, facilities and labour needed to induce root extension in dormant rootstocks can be significant, and practitioners should be certain that the extra inputs required are rewarded by better results.

MOISTURE, HUMIDITY, OXYGEN – EFFECTS OF SEALING

Parenchyma cells are thin-walled and highly susceptible to desiccation. If subjected to water stress, they react by becoming suberized, which inhibits further callus production. To prevent suberisation, humidity levels at or near 100% are considered essential. It has even been suggested that a film of water against the callusing surface produces still more callus on unsealed grafts of Apple[‡]. Experience among many grafters is that free water contact (drips) should be avoided as it provides conditions for infection, causing breakdown of delicate callus tissues and has many characteristics akin to root pressure problems.

There are several ways to ensure developing callus retains moisture: seal the graft union with grafting wax, use impermeable tape, bury the graft union in a moist but free draining substrate or substantially raise the humidity of the surrounding air. Wax sealants reduce the oxygen available to developing tissues which must then rely on diffusion from surrounding cells. Some species, such as Grapevines, are thought to be sensitive to this reduction and may be left un-sealed but maintained in conditions of high humidity. Personal experience is that a number of other species perform better without sealing, but it is not certain that this is linked to available oxygen. No detailed experiments for comparison of sealed versus un-sealed grafts have been found.

There are advantages and disadvantages to non-sealed or sealed grafts. A major disadvantage of non-sealing is the need to house grafts under conditions where high humidity must be maintained, requiring provision of relevant structures, equipment, monitoring and management. Advantages are that non-sealed grafts can lose excess sap through the cut surfaces (Figure 11.12), there is an easy opportunity to closely monitor graft development and more oxygen is available to the graft. Sealing

---

* Begum, Nakaba, Oribe et al. (2007) and Barnett & Miller (1994).
† Hartmann, Kester, Davies Jr. et al. (2014) p.454.
‡ Hartmann & Kester (1959) p. 281.

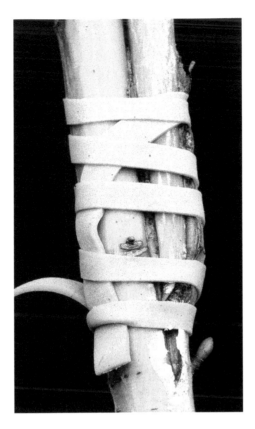

FIGURE 11.12   *Acer pectinatum* ssp. *pectinatum* on *A. rufinerve* rootstock, showing advantage of un-sealed graft allowing sap to escape and evaporate (showing as white stain). As a result of this, the graft survived, despite significant bleeding. For further discussion, see section: Root Pressure (Bleeding).

the graft has the disadvantage that if root pressure (bleeding) is a factor, sealing causes sap flow to become trapped within the seal, substantially increasing the possibility of graft failure.

Unsealed grafts require accurate knifesmanship to ensure good alignment of the cut surfaces and cambia of each component. Without such alignment, callus formed on exposed cut faces tends to form a graft periderm. This has the effect of seriously delaying or preventing the development of a callus bridge and cambial strand, resulting in a reduction in successful takes. Sealed grafts allow slightly more latitude in this respect. Therefore, unsealed grafts require more accurate knifesmanship than sealed ones.

Due to the deposition of suberin, callus produced at the exposed edge of the graft gradually changes colour from white to cream, eventually deep buff and brown, to form a graft periderm (Figure 11.13). At the pale, buff stage, humidity levels can be reduced without risk of damage to the developing callus tissues beneath. Extra ventilation is possible, enhancing suberin deposition and encouraging development of protective graft periderm, which has the effect of sealing-in moisture and reducing the danger of infection.

TIMING/SEASONAL PATTERNS/PRUNING

Callus formation is said to be greatest in the late winter and early spring period, possibly due to higher cambial activity*. Unless temperatures are too low, callus development will occur at any

---

* Kester (1965).

FIGURE 11.13   LEFT, *Pterostyrax psilophyllus* var *leveillei* graft. Callus is developing suberin (darker staining) and eventually becomes graft periderm. RIGHT, *Tilia dasystyla* subsp. *caucasica*, callus breaking through the wax sealant showing callus beneath, changing colour from white to orange/buff as suberisation occurs in the callus tissue. This eventually forms a graft periderm. At this stage, both grafts can withstand extra ventilation and complete hardening-off will be possible after a reasonable time period.

time as part of a response to wounding. Wound response seems to override any inbuilt seasonal patterns. It is stated that factors concerned with the effect of loss of plant mass caused by damage, i.e. pruning or browsing, instigate changes in biochemistry involving cytokinins and auxins, which are precursors to wound response[*]. This appears to conflict with observations made by Howard and Oakley[†] on the negative effects of rootstock shoot pruning before field budding, and the topic warrants further investigation.

POLARITY

Callus is expected to develop more extensively at the base of the shoot than the apex due to the movement of auxins rootwards. In the case of apical grafts, this works positively for the scion but negatively for the rootstock; thus, callus development should be more extensive in the scion. Seemingly the wound response is a more powerful stimulant for callus than polarity. An adequate supply of water and nutrients from the root system can often result in higher callus production in the rootstock than in the scion (Figure 11.14).

STIMULANT BUDS

Personal experience has shown that presence of a non-extending bud situated on the rootstock, and adjacent to the graft, has a positive influence on callus development. It is particularly effective when long tongue veneers are used, and ideally stimulant buds should be located at the end of the

---

[*] Ohse, Hammerbacher, Seele et al. (2016) pp.340–349.
[†] Howard & Oakley (1997).

 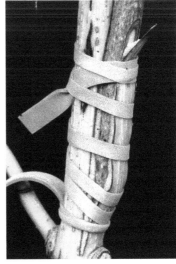

FIGURE 11.14   Polarity – sites of callus development. LEFT, callus development on a scion of *Juglans regia* 'Purpurea', showing on the left-hand side of picture callus tinged purple between the tie coils. RIGHT, *Acer longipes*, callus development in the rootstock but very little on the scion.

tongue. Here, in addition to stimulating callus formation, it helps maintain viability of the tongue (Figure 11.15 through 11.17).

Including a leaf with the stimulant bud can further improve results for some species, e.g. *Rhododendron, Camellia* and evergreen oak (Figure 11.18). Buds on the scion close to the graft area also have some influence in promoting the union but less so than those on the rootstock. Other work has shown that even the presence of leaf (needle) traces can stimulate callus development in *Picea**. Response to stimulant buds is possibly linked to the presence of additional vascular elements in stem tissues beneath the bud, which act as further sites for the generation of callus. Additionally, callus formation may be promoted by the presence of extra auxins as the bud breaks dormancy[†]. This may fuel cell division and extend targeted mobilisation of additional carbohydrates for bud development. All these supplement the normal wound healing response and may account for the reasons why stimulant buds can make the difference between success and failure with difficult-to-graft species or poor material.

Root Pressure (Bleeding)

Opinions differ on the effect of root pressure on callus development and subsequent results. Unfortunately, on the topic of bench grafting, no investigative work supported by replication and statistical analysis has been discovered. Some authorities state that grafts with moisture exudation (bleeding) around the union will not heal properly[‡] while others disagree.

Work on chip budding in the field demonstrated that bleeding had no adverse effect on success rates. In this situation it is concluded that copious sap flow is necessary to supply moisture and growth substances, particularly cytokinins, which may maintain chip bud viability, until a union is formed. Considerable bleeding of xylem sap lasting one to two weeks from chip budded rootstocks did not cause failure, and contrary to some suggestions, drying-off had no beneficial effect

---

* Barnett & Miller (1994).
[†] Hartmann, Kester, Davies Jr. et al. (2014) p.454.
[‡] Hartmann, Kester, Davies Jr. et al. (2014) pp.454–455.

FIGURE 11.15   Long tongue side veneer grafts. LEFT, stimulant bud at tip of tongue of *Acer pseudoplatanus* rootstock, grafted with a scion of *Acer* x 'Purple Haze'. RIGHT, *Rhododendron* 'Cunningham's White' rootstock after grafting with a scion of *R. davidii*. The stimulant bud is seen growing out at the tip of a tongue. Note the health of long tongue graft flap in both examples.

on results*. A number of grafters employing bench grafting methods agree with this conclusion†. Some recommend always watering the rootstock before grafting and note that rootstocks allowed to become dry adversely affect takes‡.

Divergence of opinion on this issue between practising grafters may be influenced by grafting procedures, especially the environments employed. Where temperatures have been artificially raised to higher than ambient, influences on the development and physiology of the graft are bound to occur and will affect root pressure response.

Many of the advocates of rootstocks which are wet or merely 'on the dry side' are using one of the following: cold callusing techniques involving low temperatures, grafts using bare-root rootstocks, potted rootstocks in a hot-pipe system where the root system is at low ambient temperature, or they are dealing mainly with species apparently not significantly affected by root pressure such as many in Rosaceae or sub-tropical or tropical species which may differ from plants in temperate regions.

The prevailing view amongst bench grafters is that for a number, if not all, of deciduous species and many evergreens, curtailing sap flow, usually achieved by reducing available moisture for uptake by roots, does improve results, often making the difference between success and failure. Results can be significant: nursery trials in the 1960's showed that *Juglans* grafted 'wet' produced a 7% success rate compared with 77% for 'dried-off', while *Betula* 'wet' gave 15%, and dry rootstocks

---

* Howard & Oakley (1997).
† S. Berg (pers. comm.) and D. Hatch (pers. comm.).
‡ McPhee (2007).

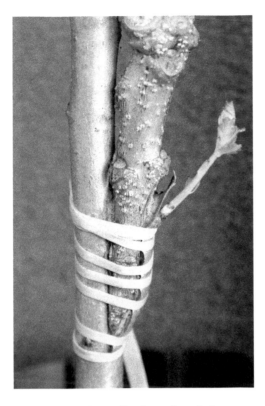

FIGURE 11.16   Influence of a stimulant bud on callus formation of a long tongue veneer of *Quercus canariensis*. Extension growth of the bud took place after some callus formation at the graft union.

FIGURE 11.17   LEFT, a long tongue veneer graft on *Acer palmatum* rootstock. The stimulant bud is growing out after significant callus development. RIGHT, close up of *Pterostyrax hispida* rootstock grafted with *Huodendron* sp. Showing a long tongue veneer graft with stimulant bud producing copious callus (note presence of harmless fungal hyphae).

provided 85% success. Other grafters have stressed the importance of using dried rootstocks for achieving good results*.

Personal observation of grafts showing symptoms of infection include white slimy mould formation and blackening of the callus, highlighting the deleterious effects of root pressure on callus and subsequent graft union development. On unsealed grafts, callus formation can be closely observed

* Hewson (2012), Brotzman (2012), Lagerstedt (1969), Wells (1986) and Humphrey (1978).

FIGURE 11.18 Stimulant bud (just commencing extension growth) with adjoining leaf on long tongue side veneer of *Rhododendron orbiculare* 'Sandling Park'. The rootstock is *R.* 'Cunningham's White'.

and appears significantly depressed at the point where sap accumulates. This may be in the basal region of the graft only, and upper edges can develop callus normally, sometimes allowing the graft to succeed (Figure 11.19).

Grafts completely inundated with sap are quickly and usually totally colonised by fungal and bacterial infections (Figure 11.20). Sealed grafts and those tied with enclosing ties are significantly more at risk. The fungi and bacteria involved are normally only weakly pathogenic, but bathed in nutrient juices and dead tissue from the necrotic plate, they quickly colonise faces of the graft, killing them by attacking the vulnerable parenchyma cells of developing callus. These infections are generally able to colonise only young soft cells and rarely penetrate far into mature tissues of rootstock and scion.

A number of factors influence the development of root pressure:

- Genetic variability.
- Stage of Growth – Timing.
- Root Temperature.
- Substrate Moisture.
- Side Graft, Height of Graft.

Genetic Variability

Copious bleeders are characterised by reduction or lack of tyloses in the main conducting vessels of the xylem. Tyloses are bladder-like cellular intrusions which act as microscopic balloons and expand in response to wounding to block the vessels. The number of tyloses present in xylem vessels varies with species; in some they are absent, e.g. *Acer*, *Cornus*, *Juglans regia*, and this may account

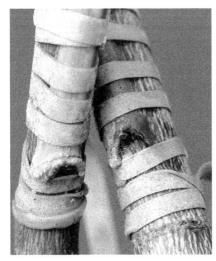

FIGURE 11.19 LEFT, *Acer palmatum* 'Oridono nishiki' showing influence of bleeding. Left-hand graft was dried sufficiently with no bleeding evident. Right-hand graft is showing some bleeding with a proportion of callus blackened and dying. Because grafts are unsealed, the right-hand graft will recover and develop a successful union. RIGHT, *Acer lobellii* showing the influence of bleeding causing death of base of the scion but healthy callus above. This graft formed a viable union and has now been planted in an amenity situation where it has so far reached 5 metres high.

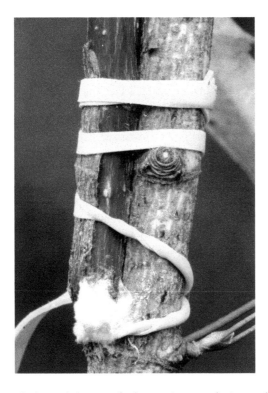

FIGURE 11.20 *Acer cappadocicum sinicum* grafted on to *A. cappadocicum* – bleeding causing extensive infection and probable loss. However, note some callus formation at top of the tongue.

for differences in the amount of bleeding between grafts in the same environment. Individual characteristics will be discussed within Parts Seven & Eight.

Stage of Growth – Timing

Rising temperatures in late winter/early spring result in increased sap flow. This continues for weeks, until a reduction or cessation of extension growth causes it to cease. Extension growth of the rootstock until late in the season can extend root pressure until mid-summer. After this time, excessive sap flow reduces and poses much less difficulty for rootstocks with normal irrigation and feeding regimes; hence the success of summer grafting. However, some including *Acer*, especially those in the Macrantha section, *Cornus* sp. and *Tilia* sp., often bleed into and beyond this period, and will still require varying levels of drying-off during the summer.

The amount of heat applied to the rootstock affects the initiation of growth activity and commencement of root pressure. Sap flow is invariably triggered in late winter/early spring by procedures involving some form of protection and additional heat.

Root Temperature

Temperature rise in the root system caused by bottom-heat has a significant effect on root pressure. Water uptake and flow increases with root temperature rising to an optimum beyond which it remains the same or reduces*. Many temperate woody plants reach maximum uptake when their roots reach temperatures in the 20–24°C range in line with transpiration increase. Most species which commence growth early in the season respond to temperatures in the 14–16°C region or below. This means that warm callusing involving bottom-heat will exacerbate bleeding, especially with genera already prone to this condition.

Conversely, the effect of low root temperatures is to reduce water uptake and bleeding. This may explain why drying-off is of less concern for grafters using systems with no bottom-heat, such as hot-pipe or cold callusing (Figure 11.21).

Substrate Moisture

Bleeding is reduced or stopped by restricting water available to the root system through drying the rooting substrate. Methods of achieving this and levels of drying have been discussed in Chapter 9: Drying-off.

Side Graft, Height of Graft

Suggested alternative procedures to drying-off are to use a side graft, or cut or drill holes in the rootstock beneath the site of the graft. Side grafts may reduce but do not prevent bleeding in rootstocks which are insufficiently dried. Given time, the unwieldy process of cutting or drilling the rootstock may affect sap flow but personal experience is that, unless grafting is delayed, reduction in sap flow following these methods is too late to prevent bleeding and infection. There is also the question of repairing the damage caused to the rootstock by invasive techniques such as cutting or drilling holes.

As discussed in Chapter 8, grafting higher on the rootstock stem is a successful strategy to reduce or overcome problems of root pressure (Figure 11.22)[†].

GRAFT HYBRIDS

The great majority of scientific investigation supports the view that despite close joining of tissues involving interlocking at the point of junction, the original graft components completely retain their genetic integrity and separate identity. Recently some exchange of genes in a limited range of

---

* Cooper (1973).
[†] Brotzman (2012).

FIGURE 11.21 Influence of root temperature on *Acers* in Macrantha section (*A.* x *conspicuum* 'Silver Vein'). Left, when combined with bottom heat, warm callusing encourages root pressure and scion activity. Note sap oozing from the graft union (orange slime covering blue tie) and evidence of extending buds of the *A. capillipes* rootstock. Right, hot-pipe with no bottom heat maintains rootstock dormancy. Lack of root pressure allows graft union development, showing as callus, just visible pushing through the wax seal as a white dot between the lowest tie coils, at the base of the long tongue.

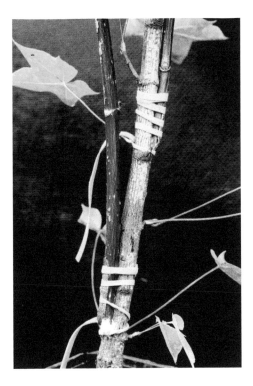

FIGURE 11.22 *Acer cappadocicum sinicum* grafted to *A. cappadocicum*, demonstrating the influence of the rootstock stem and side grafting on bleeding. The upper graft is successful but the lower graft, despite the stem above, is bleeding badly. This indicates that bleeding takes place despite a significant length of stem above the graft, but bleeding is significantly reduced when the graft is placed at higher level.

mostly non-woody species has been reported*. If this is the case, it will be a very rare occurrence and only possible under particular circumstances; furthermore, gene transfer seems to be restricted to an area close to the union.

So-called 'graft hybrids' have occurred following successful grafting procedures. These rarely occur at the callus formation stage and are due to layers of tissue from rootstock and scion becoming fused and producing a bud. When this bud grows out, it produces a shoot comprising varying tissue patterns known as a chimera, called by horticulturists a graft 'sport', examples being + *Laburnocytisus* 'Adami' and + *Crataegomespilus* 'Dardii'. Fusion can take several forms to produce woody plant chimeras, but often one layer of tissue is variously super-imposed on the other, when it is known as a periclinal chimera[†]. These are not true hybrids because the tissues, while fused, do not exchange genes and each layer of tissue retains its genetic identity.

ROOTSTOCK AND SCION INTERACTIONS

The influence of rootstock/scion combinations on ultimate size, disease resistance, etc. is well understood and for many fruit crops is covered in detail in other publications. It will be discussed further in Parts Seven & Eight: Genera Specific Requirements.

A valuable aspect of graft physiology is the ability to advance adult features. When scions from mature trees are used, resultant grafts exhibit adult growth characteristics much earlier than seedling

FIGURE 11.23   Influence of scion on rootstock growth. *Magnolia* 'Felix Jury', a large and vigorous hybrid, influencing growth of rootstock *Magnolia* deVos hybrid, variety 'Pinkie', a compact hybrid, which after 12 years shows similar girth and growth rate as the strong scion variety above. Rootstock is showing as greenish tinged wood just above soil level.

---

* Stegemann & Bock (2009) pp. 649–651.
† Crane & Lawrence (1952).

equivalents. Flower production from a grafted plant after two to five years is normal, compared with ten to fifteen for the equivalent seedling. This is seen in a wide range of species and makes grafted plants such as Wisteria and Magnolia first choice for amenity use when compared with the cheaper seedling equivalent. In selected seed mother trees of forest species, the same technique is used by foresters to reduce the period before seed production.

A converse mechanism to the above is also possible: an element of juvenility can be induced by grafting mature scion wood onto seedling rootstocks. If this procedure is repeated, using rejuvenated material from the original graft, further juvenility is encouraged. Repetitions of this procedure progress the condition in resultant plants. The technique is sometimes used to rejuvenate stock mother plant material, with the objective of improving rootability of resultant cuttings.

In some genera, rootstock growth may be influenced by the scion. In Magnolias, for some years after grafting, the stem girth of *Magnolia stellata* and *stellata* hybrid rootstocks have kept pace with a much stronger variety, *Magnolia* 'Felix Jury', and the graft union is barely discernible (Figure 11.23).

For a full discussion on inter-relationship of stock and scion, see Hartmann et al*.

---

* Hartmann, Kester, Davies Jr. et al. (2014) pp.469–493.

12 Compatibility

It seems appropriate to include graft compatibility in Part Three as it links closely to the overlying 'scientific' matters discussed previously. Investigation into DNA and RNA elements may be the route to eventually reveal the mysteries of this problem which for millennia has defied attempts to uncover its secrets.

The question of graft compatibility, the ability to form a permanent union, remains a primary cause for bias against the use of grafts. Concern is that the combination will fail at some point, either soon after grafting, when the grafter's efforts are wasted, or at some point later, when the problem becomes that of the recipient. Without microscopic investigation incompatibility in mature plants can only be recognised with certainty when the graft shows a clean break at the point of union (Figures 12.1 & 12.2). When this occurs, it indicates a failure of the components to rebuild a structurally sound and functional continuity of vascular tissues*.

RECOGNISING INCOMPATIBILITY

The first sign of likely failure due to incompatibility often occurs at the grafting stage or within the first two years of grafting, usually as a poor take despite good material, good technique and suitable environment. Grafts which do appear to succeed may grow poorly, possibly with unhealthy foliage colour and early profuse sucker production. Some incompatible grafts show swollen, ill-formed growth at the union (Figure 12.3). Often the scion subsequently dies and if tested by bending away from the rootstock normally parts with a clean break.

For those which survive the early years, there can be premature plentiful flower bud production, even when scions of non-flowering extension growth were originally selected. Grafts which survive into the second or third season may produce fruit. Occasionally, exceptionally strong growth can occur in the first year followed shortly after by the development of incompatibility symptoms. When the plant has reached sufficient size, incompatibility may show as a characteristic depression line at the point of union. The bark appears somewhat discontinuous in a narrow strip around the circumference, leaving the impression that it could be snapped off at this point (Figure 12.4).

A more concerning situation is when incompatibility is delayed, sometimes for decades, and a plant previously considered satisfactory in every respect suddenly fails and dies, or worse, breaks at the union and crashes to the ground, endangering safety and causing indiscriminate damage.

Stem breakage is frequently blamed on incompatibility, but does not always meet the criteria and may be caused by physical forces such as wind damage, pests or disease. Wrong conclusions for the cause of loss often result in a bias against grafted plants, founded on ignorance rather than facts.

Large differences in growth rates between stock and scion produce unsightly unions, which detract from aesthetic effect (Figure 12.5). This is not linked to incompatibility but emphasises the importance of selecting suitable graft combinations.

CAUSES OF INCOMPATIBILITY

"Four thousand years of practice and research in grafting woody plants have not provided significant answers to questions about the causes of graft incompatibility, nor allowed valid predictions to be made regarding potential incompatibilities between individual plants of most species of landscape

---

* Santamour Jr. (1988).

FIGURE 12.1 *Quercus rysophylla* grafted to *Q. palustris* rootstock. Graft failure after 8 years due to delayed incompatibility. Note: clean break of the union and some sucker growth just visible. The *Q. rysophylla/palustris* combination is usually successful, but the production of suckers over several seasons often indicates a level of incompatibility.

trees"*. These words of Frank Santamour sum up the present situation regarding compatibility, not only for landscape trees but other woody species.

Recent genetic work investigating the subject may eventually uncover the underlying causes. Whether the findings can be utilised to address the practical problems caused by incompatibility remains to be seen. The identification, through scientific methods, of mutually compatible types that may then be used as inter-stem mother plants, could have wide practical application and should encourage further investigation in this area (see below Future Progress – Strategies to Overcome Incompatibility and Double Working).

A few specific examples to account for incompatibility-like symptoms have been positively identified. Graft failure between *Juglans regia* and *Juglans nigra* was for some years considered to be due to incompatibility; now the role of Blackline disease (a strain of Cherry Leaf Roll virus) has been recognised and virus-free combinations are completely compatible (for a full explanation, see Juglans, Part Seven, Chapter 39). Another example is seen in fruiting Pear production when some scion varieties are grafted to quince rootstocks. A chemical (prunasin) produced by quince rootstocks can cause eventual loss. This is due to the production of cyanide caused by the breakdown

---

* Santamour Jr. (1988).

FIGURE 12.2 *Betula ermanii* 'Hakkoda Orange' on *B. pendula* rootstock. Graft failure after 5 years. Note smooth surfaces at breakage point, a clear indication of incompatibility. Note also a small segment showing as a white patch of torn fibrous tissue on the scion (upper right-hand side) and less obviously on rootstock. This appears to indicate some functioning linkage was established between the components and would explain the tree's survival and normal extension growth for a period of 5 years. In the centre of the graft, the imprint of the original rootstock/scion interface is still visible.

FIGURE 12.3 LEFT, *Quercus variabilis*, both grafts on *Q. cerris* rootstock – left is a compatible union, likely to thrive for many decades if not centuries. The right-hand graft shows chlorotic foliage colour and copious suckering, almost certainly to eventually fail. RIGHT, *Cercis chinensis* grafted to *C. siliquastrum*, an incompatible graft. Note copious sucker production and swollen uneven growth at the union. Other examples of *Cercis* using this combination have so far proved to be compatible and thrived here for 20 years. Both examples highlight the variability of seedling rootstocks.

FIGURE 12.4   *Malus trilobata*, which is not always compatible on all *Malus* rootstocks, showing characteristic ring of split bark at the point of union just above ground level. This may indicate eventual graft failure.

FIGURE 12.5   LEFT, *Fraxinus angustifolia* grafted to *F. excelsior* rootstock showing differences in growth rates. RIGHT, *Chamaecyparis obtusa* 'Aurea' grafted to a *C. lawsoniana* rootstock showing an unsightly union due to differences in growth rates and ultimate size. Both examples have long term compatibility.

of this chemical by the pear scion variety leading to damage and death of cambial cells and phloem tissues in the graft union and adjacent area. Eventually, over a period of time, sometimes as long as 20 years, the graft dies.

PREDICTING COMPATIBILITY

Records of past experience, built up over generations of trial and error, allow predictions concerning likely compatible combinations. This information may be found in the literature and especially in the proceedings of the International Plant Propagators' Society; additionally, as much data as possible is included on this vital aspect of grafting in Parts Seven and Eight of this book. The unique genetic characteristics of a seedling rootstock can always introduce a level of uncertainty into any combination previously recorded as successful, and because the majority of grafted ornamental woody plants rely on seed raised rootstocks, there is always the possibility of failure. For many genera this will be the rare exception while for some it can be a regular occurrence.

Where data from the past is unavailable, reasonable, but by no means totally infallible, guidelines for most if not all genera are found by linking taxonomic groups. Close botanical affinity is often a good starting point for deciding upon likely successful combinations. Closely related species frequently show mutual compatibility, often more likely if they share a common phylogenic origin. An outstanding example of this being the Magnolias where possibly all species and cultivars, including recently incorporated *Michelia*, *Mangletia*, *Manglietiastrum* and *Parakmeria*, are closely linked phylogenically and appear graft compatible. Although it has yet to be fully established, some awareness of phylogeny may prove a valuable aid to grafters seeking good rootstock/scion combinations.

Differing scientific sources, opinions, procedures and timescales can cause grafters to encounter problems in translating taxonomic and phylogenic information. Taxonomists seem to fall into two types: those who split the genus into small sets of taxonomic characters to create many groups; sometimes this is carried to a point where the genus itself is split into separate genera and those who lump taxonomic features to create broad groups with less recognition of separate characters and subsequent reduction in options. When making decisions on the most likely compatible combination, the 'lumping' approach, which may simplify identification and categorisation for taxonomists, is not always the best option for grafters.

Information now available on the World Wide Web (www.) can provide helpful guidelines. Where possible, this information is included in the Genera Specific Requirements in Parts Seven and Eight. Further information may be found by going on-line and entering into the search box – 'taxonomy of', or 'phylogeny of', followed by the name of the genera required. Often taxonomic and phylogenic investigations are linked on the same or closely related sites. In many cases, one or several comprehensive accounts are accessible. Using these guidelines to suggest suitable combinations for many genera can provide a useful aid to making decisions.

The whole subject is complex, with a strongly scientific basis. Negotiating the papers, understanding the terms and converting the data into useable form are difficult for those not versed in the topic. However, at present, when attempting previously unknown or unsuccessful combinations, it appears the only strategy likely to suggest comprehensive and reliable guidelines.

Despite the apparent links between taxonomy, phylogeny and graft compatibility, some genera show a percentage of failures even when cultivars of the same species as the seedling rootstock are grafted together. This must be due to seedling variation and is an indication that factors other than those discussed may be involved. To quote Robert Garner, "something more than kinship is required". To date, a total understanding of graft compatibility and therefore the possibility of making completely accurate predictions regarding compatible combinations remains an unsolved problem.

FUTURE PROGRESS – STRATEGIES TO OVERCOME INCOMPATIBILITY

Because of genetic diversity, using seedling rootstocks, even with a very close relationship, can result in individual combinations showing incompatibility symptoms. This has been demonstrated in both conifers, e.g. *Pseudotsuga** and hardwoods, notably including *Quercus*. At present, and possibly for a considerable time ahead, grafters will need to consider a broad choice of strategies for overcoming incompatibility.

Ideally the path ahead would be to identify and vegetatively propagate compatible rootstocks for the range of ornamental species, as has been done for many fruit and nut crops. This seems an unlikely prospect for the foreseeable future, mainly because of the vast number of species involved. This is also coupled with the difficulty of monitoring progress of plants which become dispersed

---

* Copes (1969).

singly or in small groups in various planting situations, where follow-up appraisal is difficult or impossible. Worse still, once planted out, unlike their fruit and nut counterparts, the plants have no measurable monetary value.

To provide consistent compatibility, ultimate size and form, and overcome soil-borne disease, attempts have been made to identify and vegetatively propagate rootstocks for some *Tilia* and *Acer platanoides* cultivars*. After an initial start in the UK, this has not been pursued due to financial constraints but work on Verticillium-resistant *A. platanoides* has been continued in Holland†.

A range of ornamental species are grafted using vegetatively propagated clonal rootstocks initially selected for fruit production. These are mainly in the Rosaceae family, examples being *Malus* grafted to Malling/Merton apple rootstocks, *Prunus* to Colt and Cob cherry rootstocks and *Prunus*/Plum to St. Julien A or Brompton plum rootstocks. Other genera utilising vegetatively produced rootstocks include *Rhododendron*, grafted to various selections including *Rhododendron* 'Cunningham's White' or *R.* INKARO®, and Weeping *Salix*, frequently top-worked onto stems of *Salix* x *smithiana* or *S. viminalis* selections.

DOUBLE-WORKING

This system currently has importance for particular problems in the production of commercial fruit, notably pears. Most pear varieties naturally grow into medium-to-large trees, unsuited for culture and harvesting in modern intensive orchards. Use of Quince rootstocks (*Cydonia* sp), having a strong dwarfing effect on growth, addresses this problem. Unfortunately, some important Pear varieties, e.g. 'Williams Bon Chretien', are incompatible on this rootstock. The solution is to use a 'bridging stem' (intermediate stem or inter-stem), identified as compatible with both the Quince rootstock and Pear scion variety (Figure 12.6).

Double-working has been applied to other fruit and nut crops (apples, walnuts, etc.), in most instances on the assumption that compatibility between the components already exists and therefore to provide combinations which combine desirable qualities of both rootstock and inter-stem (see Parts Seven and Eight for further examples).

To date, little investigation has taken place of double-working to overcome incompatibility in ornamental species. In some situations, it may have real potential for this purpose. Two different strategies apply.

Proven Inter-Stems

This category relates to inter-stems of material which has proved to be compatible with both graft components. To date, very few examples exist. Suggestions for potential inter-stems are made in Parts Seven and Eight, but apart from those for *Acer griseum* and *Tilia callidonta*, they have not been tested.

The procedure is to first graft the inter-stem to the rootstock and then immediately, or at the next following grafting period, graft the desired species to the inter-stem. Two apical grafts can be used but personal experience is that side grafts are preferable, especially for less than ideal material. Inter-stems can be very small, but it is probably best to use a reasonable 50–100mm long portion of wood.

An example of the technique is shown in Figure 12.6. The hybrid between *Acer griseum* and *A. pseudoplatanus*, *Acer* 'Purple Haze', has proven to be a reliable inter-stem between scions of *A. griseum* and the readily available rootstock, *A. pseudoplatanus*.

A further example is the choice of *Tilia chinensis* as an inter-stem for *Tilia callidonta*, shown in Figure 12.7. Previous attempts to graft *T. callidonta* to *T. cordata* were unsuccessful. The choice of *T. chinensis* was based on the close taxonomic relationship between the two species and the introduction of mutually compatible *T. chinensis* as an inter-stem provided a positive result.

---

* Howard (1995).
† Heimstra & van der Sluis (2005).

FIGURE 12.6 *Acer griseum* 'Fertility' (a fertile and therefore valuable cultivar of *A. griseum*) double-worked to an inter-stem of *Acer* 'Purple Haze' (showing extension growth and foliage), used between the scion variety *A. griseum* 'Fertility' and the rootstock *Acer pseudoplatanus*. The *A. griseum* buds are expanding and the combination has produced high takes of grafted plants of this clone of *A. griseum*. The long-term performance of this combination has yet to be confirmed. However, *A.* 'Purple Haze' seems totally compatible when grafted to *A. pseudoplatanus*, with no sucker formation after over 10 years.

The interaction between rootstock, inter-stem and scion variety is complex and not well understood*, made even more difficult to predict by use of seedling rootstocks. Figure 12.8 shows the effect of different *Tilia* rootstock/inter-stem combinations and their influence on the achievement or otherwise of compatibility.

Unproven Inter-Stems

Several genera with known severe compatibility problems present a particular challenge, *Quercus* being a prime example. The methods suggested all involve substantial time and effort and can only be justified when the scion variety is particularly endangered, valuable or important to commercial interests. One strategy is to assume that the successful grafts within any batch, sometimes as low as 5–10%, represent combinations where the seedling rootstock is compatible with the chosen scion variety. Only time can reveal whether a combination will later exhibit delayed incompatibility symptoms.

Having discovered a successful combination, the next assumption has to be that the compatible rootstock may be reliably grafted to seedling rootstocks of the same species. Experience shows that this is normally the case and the grafter is then in the position of having discovered a mutually bridging intermediate stem (inter-stem). A number of options are available.

---

* Garner (2013) p.284.

FIGURE 12.7   *Tilia callidonta* double-worked onto *T. cordata* rootstock using *T. chinensis* as the inter-stem. *T. chinensis* was grafted onto *T. cordata* the previous season and *T. callidonta* grafted onto the *T. chinensis* portion of successful grafts the following year. To complete the graft, the portion of *T. chinensis* (left side) above *T. callidonta* will need removal.

One method involves selecting a two- or three-year old successful graft. The scion is then removed together with a small section of rootstock. This now becomes a composite scion which may be re-grafted to a rootstock of the same species with every chance of a successful union. The remaining headed-off rootstock can be expected to produce suckers. These will provide material for scions, which may be re-grafted onto seedling rootstocks of the identical species, producing inter-stem mother plants for future use. If required, a range of inter-stem mother plants can be produced to meet requirements for particular combinations involving different species or cultivars.

This strategy, based on early decisions, cannot guarantee 100% success; subsequently delayed incompatibility between scion and inter-stem, or inter-stem and rootstock, is always possible. Examples of the use of mutually compatible inter-stems will be found in Parts Seven and Eight.

An alternative, more opportunistic method involves a decision based on very early scion growth. Several rootstocks are multi-grafted with the chosen variety and results will be variable, some failing to produce any takes, others a few, some excellent, followed by good extension growth (Figure 12.9). Decisions on good or poor short-term compatibility can be made at this early stage. As before, successful grafts may be detached as a composite scion. Re-grafting these to rootstocks

FIGURE 12.8 Two examples of *Tilia callidonta* double-worked using scion material from a mature flowering specimen. Grafted January 2015, photographed May 2017. Left showing a combination of *T. platyphyllos* rootstock with *T. chinensis* inter-stem. On the right is a combination of *T. chinensis* inter-stem but with *T. cordata* rootstock. On the left, observe sucker growth commencing from *T. platyphyllos* coupled with reduced scion growth and profuse flowering. On the right *T. callidonta* is growing well with good extension, no suckers produced and only one flower visible.

of the original species provides a graft with higher potential for a successful take and hopefully good compatibility.

Side grafts, leaving plenty of wood above the graft, afford an opportunity to remove rootstock material from above the successful graft and re-graft to suitable, closely related rootstocks to produce inter-stem mother plants. Early work with *Quercus* shows a potential use for the technique (see Figure 12.9).

OTHER STRATEGIES TO OVERCOME INCOMPATIBILITY

Hybrid Seedlings

Hybrid seedlings may be used as rootstocks to provide further opportunities for overcoming incompatibility. Many seedlings raised from botanic gardens, arboreta and private gardens show hybrid characteristics because of the proximity of other species within the same genus. Although a disadvantage in the context of maintaining true-to-type progeny, the possibility of using such 'off' types as rootstocks or inter-stems for closely related parent species and species cultivars should not be overlooked. It is likely that these will provide material with a better chance of compatibility than alternatives. An example is illustrated in Figure 12.10 using *Carpinus fangiana* seedlings from a botanic garden source; these were most probably a hybrid with *C. betulus* nearby. Acquisition of such seedlings is always likely to be on a 'one-off' basis and, as described previously, consideration should be given to taking scion wood from these for future use as inter-stem mother plants.

FIGURE 12.9 *Quercus*x *warei* 'Regal Prince', two rootstocks are shown grafted on the same day and subject to the same conditions. Scion growth on the left hand is excellent but on the right is poor, indicating likely incompatibility. When available, rootstock extension growth could be taken as scion wood from the left-hand rootstock to create inter-stem material. This may be either re-grafted to further rootstocks to form inter-stem mother plants or used on a limited one-off basis for a further batch of likely compatible rootstock–inter-stem– scion combinations.

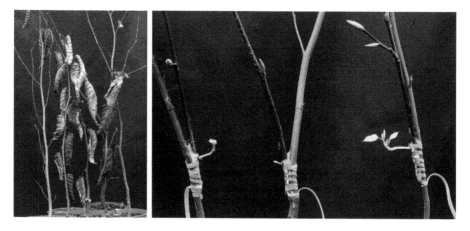

FIGURE 12.10 LEFT, *Carpinus fangiana* 'open pollinated' seedlings from a Botanic Garden to be used as rootstocks. RIGHT, the same rootstocks grafted with *C. fangiana* 'Wharton's Choice'. Note stimulant buds and callus formation on the long tongue of the veneer grafts, these were subsequently 100% successful.

Root Grafting

Root pieces taken from the original source plant and used as a rootstock is another strategy and has been suggested for particularly difficult genera such as *Quercus**. Success depends on availability of suitable material from the parent root system, which may not always be obtainable or of poor quality. Provided these difficulties can be overcome, grafts free from incompatibility problems are guaranteed (Figure 12.11).

Roots may also be taken from rootstocks which have demonstrated compatibility with the chosen scion variety. It can be assumed that compatible stem-grafted rootstocks will also produce compatible root grafts.

FIGURE 12.11 *Lithocarpus henryi* (Henry's Stone Oak) in the Sir Harold Hillier Gardens, Hampshire, England grafted to a root of the same species by Peter Dummer in 1962. An example of the grafter's extreme skill.

---

* Leiss (1988).

13 Graft Environment for Graft Union Development

This chapter discusses the principles and practice involved in providing the required environment during the formation of a graft union. There follows a discussion of graft treatments for influencing graft union development, treatments for suppressing graft union disease and cold storage of successful grafts to aid production scheduling and handling.

The major environmental factors for plant growth and graft development are considered to be light, temperature and water. In the earlier stages of graft union development, light is of importance almost entirely through its influence on temperature. At this stage, its effect is to increase the temperature of an enclosed or semi-enclosed environment. Temperature is also controlled by a number of factors, ventilation being significant among these. In addition to influencing temperature, ventilation has a profound effect on moisture levels in growing and propagation structures and consequently, at many stages in the grafting process, must be used with care and judgement. All these issues are essential to achieve good results; some might even argue they have more importance than the knifesmanship skills involved in making the graft. They are described and discussed here in detail.

Most grafts are reliant on components with ripened wood, well-charged with stored carbohydrate reserves and plant nutrients; the requirement of light for photosynthesis does not assume importance until graft union formation is complete and extension growth is well underway. It is important to recognise that the majority of deciduous species are grafted in the dormant leafless state when photosynthesis is hardly relevant. Using apical grafting methods during the summer/autumn period, the rootstock top and foliage are removed; however, results are normally better if some leaves are retained on the scion. This suggests that photosynthesis at some level may be an advantage for this type of graft.

Rootstocks of leafy side grafts and evergreen side grafts might be expected to require sufficient light for good levels of photosynthesis. In practice during the early stages of graft development, photosynthesis does not appear to be a limiting factor, and can be subordinated to the need to control excessive temperature by the use of heavy shading. It may perhaps appear that only grafts with soft, young, leafy scions, such as summer grafted *Daphne* or deciduous Azaleas, have a significant light requirement. However, because of a plentiful supply of active parenchyma cells in the young tissues, union development is rapid, and therefore heavy shading, which will seriously reduce photosynthesis, is permissible during the early stages while graft union formation quickly takes place.

Carbon dioxide and oxygen are critically involved in food manufacture and plant growth, but both are normally sufficiently available, and neither is likely to constitute a limiting factor for grafts.

Therefore, the environmental factors most influencing graft survival and development are water and temperature. The principles and practice of managing these will be investigated in the following pages.

WATER EXCHANGE AND LOSS BETWEEN THE PLANT AND THE ATMOSPHERE

As has been discussed in Chapter 11, loss of water from an unprotected graft interface will stop the development of tissues necessary for graft union formation. In addition, the scion, especially a leafy scion detached from its parent root system and unable to replenish lost water, is in a potentially lethal situation. To avoid failure, the grafter must take measures to prevent these events. Some knowledge of water relationships between the graft and atmospheric conditions surrounding it will provide an understanding of the factors involved.

Detailed knowledge is not essential. For those wishing to understand only the practical issues, the section below – Practical Methods of Preventing Water Loss in the Rootstock, Scion and Graft Union – will suffice. Enquiring further to establish a better insight into the principles will aid the grafter in dealing with the inevitable questions which occur when trying to obtain the best results from grafting procedures. A description of the main factors influencing water exchange in the plant follows.

Water exchange between the plant and surrounding atmosphere is shaped by the laws of physics and the structure and physiology of the plant. Water can exist in the air as water droplets – rain, hail, sleet or snow – but is mostly present as invisible water vapour. In certain weather conditions, particularly a sudden lowering of temperature, sufficient condensation occurs for minute water droplets to be visible as fog or mist. Because water vapour has volume it occupies space within a given volume of air, and consequently creates a pressure, known as vapour pressure (VP). In higher temperatures, if sufficient moisture is available, more water as water vapour can be held in the same volume, resulting in an increase in pressure. This creates a pressure differential between warm and cooler air. Provided no barrier exists between the two, to achieve equilibrium diffusion from higher pressure (higher water content) to lower takes place.

Plants absorb heat from sunlight during daytime and consequently become warmer than the air surrounding them. This raises water vapour pressure within plant tissues, increasing their potential for water loss to the cooler surrounding atmosphere. External protective coverings, cuticle, epidermis and periderm act as water-resistant barriers and ensure moisture is retained. In healthy plants, biological activity ensures controlled opening and closing of pores in this covering and regulates water vapour movement between the plant and the atmosphere. Grafting removes a portion of the protective barrier, potentially causing irreversible damage by allowing uncontrolled water loss from adjacent tissues.

For some types of graft, sealing the graft union, and any unprotected cut surfaces, is sufficient. Unsealed grafts, those which are leafy, freshly de-leaved or produce extension growth before graft union formation is complete, have the capacity to lose water in an uncontrolled way unless preventative action is taken. The atmosphere surrounding the graft and its potential influence on water loss is crucial. A major factor in preventing this loss is humidity.

Humidity – The Enclosed (Supportive) Environment

Humidity is the popular term used for atmospheric water vapour content. Relative humidity (RH) is a measure of the water content (vapour pressure or VP) present in the air at any given temperature divided by the maximum water content theoretically possible. This equation gives the relative humidity expressed as a percentage. When the air is totally saturated (at saturation vapour pressure or SVP) the relative humidity is 100%.

In practice this level of moisture is rarely maintained. Providing temperatures in plant tissues and surrounding air are roughly equal and relative humidity levels are in the region of 95–98%, any water loss from the graft due to diffusion to the surrounding atmosphere is very slow, and unlikely to be cause for concern. At most times of the year, grafts, particularly leafy grafts, will normally require shading to stabilise temperature rise within the tissues and the surrounding atmosphere.

Once achieved, very high relative humidity trapped with the plant material in an enclosed area then provides what is known to the plant propagator as an 'enclosed' or fully supportive environment, which is vital for leafy cuttings but also for grafting systems, such as warm callusing.

It is important to understand that if humidity is at a high percentage, increasing temperature without adding more water results in a lowering of levels. This is significant; every 10°C rise in temperature causes a drop in relative humidity by a factor of almost 2. For example: 100% RH at 15°C is a fully supportive environment. If no additional water is added, raising the temperature by 10°C to give 25°C causes RH to fall to a level in the region of 50%. This is a drying environment, potentially causing desiccation and death of unsealed and leafy grafts which have not yet become united. To allow enclosed/supportive conditions to be maintained, temperature control is therefore a crucial factor in managing environments for grafts.

Within growing/propagating structures, temperature of the atmosphere and plant tissues is controlled by shading and ventilation, used singly or in combination. Control of these to provide supportive humidity levels are an essential component of environmental management and will be discussed in detail throughout the chapter.

WATER – PRACTICAL METHODS OF PREVENTING WATER LOSS IN GRAFTS

Rootstocks

Potted rootstocks are kept moist by watering the compost; however, as discussed previously, rather than adding water, reducing water by drying-off may be of more significance for many grafting procedures. Bare-root rootstocks can have their moisture content maintained by laying into a suitably dampened substrate, wrapping in polythene or enclosing the roots in a dampened water-holding material, such as lightweight capillary matting (Figure 13.1).

Preventing Water Loss from Grafts using Dormant Leafless Scions

Other than at the union itself, this type of graft has a lower potential for water loss than others. The epidermis/bark covering the scion almost entirely prevents water loss, and only the graft interface and any exposed cut surfaces are able to lose significant amounts of water leading to graft failure. The graft union, along with any exposed cut surfaces due to trimming away side shoots from the scion, must be protected from water loss by one of the following methods:

- Sealing any cut surfaces of the scion and the graft union by dipping in melted wax after tying together (Figure 13.1).
- Applying sealant by brush to the graft union area and any additional exposed cut surfaces on the scion above the graft.
- Tying-in the graft union using non-porous ties. Any additional exposed cut surfaces on the scion above the graft must be sealed by brushing with wax or cold sealants.
- Burying the union together with the root system of bare-root grafts into a moistened medium. Unless the scion is also buried, as sometimes for example with Grapevines or Walnuts, any exposed cut faces of the scion beyond the union must be sealed, normally by brush.

FIGURE 13.1   LEFT, root system of bare-root grafts laid-in peat in Boskoop trays. Scion cuts and graft union are sealed by dipping into warm paraffin wax. RIGHT, grafts in shelved 'hot-pipe' structure, rootstock roots covered in damp lightweight capillary matting (partially drawn back to show bare-roots) (Yorkshire Plants). Polythene film (ideally milky) could be used for the same purpose.

- Burying the union when using pot grown rootstocks. This is now less popular because of the extra labour involved and the convenience of modern, non-porous, enclosing ties*.
- Enclosing grafts in a structure to provide an enclosed (supportive) environment and maintaining vapour pressure of the surrounding air at a point where moisture loss from the tissues is virtually eliminated. Here sealing is unnecessary as all cut surfaces and the union are protected.

Preventing Water Loss from Grafts using Leafy and De-Leaved Scions

- Enclosure in a fully supportive environment with relative humidity at or close to 100% to virtually eliminate moisture loss. In this publication, this type of system is called warm callusing – see Chapter 18.
- Use of dry fog, wet fog, mist or overhead irrigation surrounding the grafts, to maintain high levels of humidity sufficient to prevent water loss. In this publication this type of system is called Supported Warm Callusing – see Chapter 18.

High Humidity Combined with Sealing

Placing grafts in enclosed (warm callusing) or supported warm callusing environments should ensure adequate control of water loss. However, some grafters still prefer to seal grafts in enclosed conditions for the following reasons:

- Avoid damage to the graft if the supportive environment is interrupted for brief periods.
- Avoid potential damage caused by drips entering the interface of unsealed graft unions.
- Avoid free water entering the union interface from supported warm callusing systems using wet fog, mist or irrigation sprinklers.

TEMPERATURE – THE GRAFT ENVIRONMENT

Heat is a basic requirement for biological activity and, as has already been discussed in Chapter 11, plays a significant role in graft union development. In this section, the role of temperature in the context of environmental conditions is described and will be further considered in chapters on specific genera in Parts Seven and Eight.

Temperature of graft environments is influenced and controlled by combining shading and ventilation and/or by additional heat from an independent source. Less often, other techniques such as water in mist or fog form may be used to mitigate excessive temperature.

ADDITIONAL HEAT

From mid-autumn until late spring, and less certainly at other times, additional heat supplied by basal or space heating will be necessary in many grafting systems. Temperatures are invariably controlled automatically by thermostats set to provide optimum temperatures for graft union development.

When grafts are housed under enclosed conditions, the main requirement particularly for some hardwood evergreens and many conifers will be to avoid temperatures rising too high. This is normally likely during the summer, but due to the effect of bright sunny periods in late winter/early spring temperature rise can also be significant at this time. For example, in the UK throughout January and February, despite some shading, air temperatures within poly-tents set at 14°C were raised by 9°C to give 23°C during periods of bright sunshine. When forecasts indicate a period of low ambient temperatures coupled with clear sunny weather, it is a sensible strategy to reduce bottom heat settings by 2–5°C to compensate for a rise in daytime temperature. Recommendations

---

* Davis II (1982), Decker (1998), Hall (1977), Stoner (1974), Vermeulen (1983) and Wells (1955) pp.41–56.

made to set bottom heat at a given temperature, and maintain surrounding air within the enclosure several degrees lower, are impossible to achieve because of the heating effect of sunlight.

Shading

From the grafter's perspective, the major effect of light is to raise air temperature, resulting in a lowering of atmospheric humidity and increasing potential water loss. Shading is therefore essential to influence these outcomes.

Shading may be applied overall or to specific areas for particular requirements. In modern growing structures, a combination of shading at high level (normally at the eaves) and adequate ventilation reduces ambient temperature, and limits the need for more shading elsewhere (Figure 13.2).

Used in conjunction with enclosed environments, such as poly-tents, extra shade can be applied selectively to provide specific control of temperature and humidity.

Apart from its effect on temperature, light is essential for photosynthesis. However, as previously discussed, graft components are often in a dormant state and normally well supplied with stored carbohydrates; light requirements can be subordinated to the need to control temperature and maintain high humidity. Work in the USA on the effects of photosynthesis on Colorado Blue Spruce grafts, a species which might be expected to have a high light requirement, has shown it is not essential for graft union formation*.

Shading in High Humidity (Enclosed) Environments

Because ventilation can result in an unacceptable loss of humidity, shade is the only option for temperature control for most grafts during early stages of graft development. Insufficient shade in

FIGURE 13.2   High-level overall shading over roof trusses above poly-tents in a modern glasshouse. Here, the shade screen doubles as a thermal screen. Ventilation may be applied to lower ambient temperature. Note: normally in early February no extra shading material is required on individual poly-tent covers on the benches (John Richards Nurseries).

---

* Beeson & Proebsting (1998).

mid-summer causes temperatures to soon exceed 40°C, at which point callus development of many temperate species is inhibited or prevented.

Work at Glasshouse Crops Research Institute (Sussex, UK) investigated levels of shading to provide suitable conditions for rooting cuttings under polythene tents (poly-tents). The recommendations given for less light-dependent grafts are derived from those for cuttings. They are based on personal experience which has shown that higher shade levels can be used to more effectively control excessive temperatures.

In Southern England, assuming use of 40% shade screens, the proposed number of layers for grafts, and for comparison leafy cuttings, is shown in Table 13.1:

TABLE 13.1.

Suggested Shading Levels for Grafts Compared with Leafy Cuttings[a]

| Month | Grafts (Author's Suggestions) | | | Leafy Cuttings[b] | |
	Bright Day (No. of Layers)	(Shade Level on Bright Day)	Dull Day (No. of Layers)	Bright Day (No. of Layers)	Dull Day (No. of Layers)
January	0–1	(0%–40%)	0	0	0
February	1	(40%)	0–1	1	0
March	2	(64%)	1	2	0
April	3	(78%)	2	3	1
May	3–4	(78%–87%)	2	3	1
June	4–5	(87%–93%)	3	3	1
July	5	(93%)	4	3	1
August	5–4	(93%–87%)	4–3	3	1
September	4–3	(87%–78%)	2	2	0
October	2	(64%)	1	1	0
November	1	(40%)	1–0	0	0
December	0–1	(0%–40%)	0	0	0

[a] Where two figures are shown, the first figure represents shading levels for the first half of the month, the second for second half. When practicable, the highest levels of shading are best restricted to the brightest, warmest time of day, and reduced during evenings and early morning.

[b] Loach (1981).

During sunny periods in late June, July and early August, five layers of 40% shade screen will provide 93% shade. To avoid the possibility of too little photosynthesis to sustain the health of the grafts, it is best if this level of shade does not remain in place for over ±4 weeks on any one batch and, when possible on dull days, reduced during early morning and evening. Judgements must be made in this respect and rates of callus formation, graft union development, etc. should influence final decisions.

Although seeming excessive, this level of shading ensures that temperatures do not reach unacceptably high levels for prolonged periods. Insufficient shade can quickly cause extreme temperatures. On a bright, sunny day in July, in the relatively cool summers of the UK, an unventilated, unheated, walk-in grafting polytunnel with 40% shade reached a temperature of 51°C (124°F) by early afternoon. Fortunately the polytunnel was not being used for propagation purposes at the time. The high levels of shade suggested are not unsupported; work in the USA has demonstrated the significance of shade levels as high as 93% in promoting rooting in cuttings of *Quercus* and *Acer* species. Shade levels of 82% only improved results if a period of at least 10 days at 93% shade was included in the treatment*.

* Zaczek, Heuser Jr. & Steiner (1999).

Shade Level Calculations

Grafters wishing to make their own decisions on appropriate shading levels using more than one layer of shade screen may use the following calculations:

- One layer of 40% shade = 100% − 40% = 0.6 × 100 = 60% light transmission (L.T.) or 40% shade.
- Two layers of 40% shade gives 0.6 × 0.6 = 0.36 × 100 = 36% (L.T.) 100% − 36% = 64% shade.
- Three layers of 40% gives 0.6 × 0.6 × 0.6 = 0.216 × 100 = 22% (L.T.) 100% − 22% = 78% shade.
- Calculation for three layers of single screens of differing shade levels using the same method is, for example, using 40% + 20% + 50% shade levels: 0.6 × 0.8 × 0.5 = 0.24 × 100 = 24% (L.T.) 100% − 24% = 76% shade.

The shade levels calculated have been confirmed as reasonably accurate when checked by instrumentation*.

In practice, the number of shading layers required will be affected by other activities, such as shade provided by thermal fleece and other coverings over poly-tents, etc. Normally, thermal fleece of 15gm/m² grade has a shade level of ±12% so that three layers of fleece as an internal lining for a poly-tent will provide: 0.88 × 0.88 × 0.88 = 0.68 × 100 = 68% (L.T.) 100% − 68% = ±32% shade.

VENTILATION – TEMPERATURE AND HUMIDITY CONTROL

Two main ventilation regimes apply in grafting procedures and are used separately or jointly to provide optimum conditions for graft union development. In this publication the following terms are used:

- Macro-ventilation – ventilation of the whole structure housing grafts.
- Micro-ventilation – ventilation of enclosed environmental areas.

MACRO-VENTILATION

Ventilating the whole structure is commonly used for the following purposes:

- To affect overall temperature reduction.
- To provide lower temperatures for dormant deciduous grafts (cold callusing).
- To provide lower humidity for enclosed (warm callused) grafts when a change in surface callus colour from creamy white to buff and orange-brown becomes apparent. This indicates a suberisation stage has been reached where lower humidity is appropriate. To reduce ambient temperatures at this point, judicious macro-ventilation in conjunction with micro-ventilation is appropriate.
- To harden-off grafts which have united and have reached a stage where new extension growth is visible. Ventilation must be applied with care in the early stages of hardening-off as excessive loss of humidity can result in desiccation and collapse. Fully automated ventilation in many modern structures will need monitoring to ensure optimum conditions.

---

* Keith Loach (pers. comm.).

MICRO-VENTILATION

Micro-ventilation relates to the ventilation of enclosed environments (poly-tents, walk-in enclosures, etc.). As with macro-ventilation, it is used in conjunction with shading and the two must be thought of as a combined strategy of environmental management. Ventilating an enclosed environment normally results in a quick reduction of humidity levels and must always be given careful consideration.

A wide divergence of opinion exists on what is best. Some grafters favour placing grafts in an enclosed environment and leaving them unventilated until callusing and union formation is well advanced. Others monitor conditions frequently, at least daily, and some have recommended every three to four hours, applying ventilation if deemed necessary*. A sensible compromise is to limit ventilation for the first five to ten days and then monitor daily or more frequently to decide appropriate action. Once ventilation has been applied, follow-up checks are invariably necessary.

On all occasions, but particularly when ventilating for unsealed grafts, atmospheric conditions and levels of shading outside the enclosed environment must be taken into account. Ventilation will have an immediate and rapid drying effect on poly-tents or walk-in enclosures. Where differentials between enclosed and open environments are extreme, grafters need to be aware of the drying effect at the edges of the bed compared with those in the middle. To avoid uneven drying-off, there is a case for erecting low polythene curtains (150–200mm high) along the edges of the bed.

Open-air, Walk-in, Free-standing Grafting Polytunnels (see Chapter 14) require extra care because the obviously drier air from outside is immediately admitted and can have a severe drying effect. Often initial ventilation is best restricted to partially opening doors at each end and side ventilation not used until the grafts are showing some suberized callus.

Walk-in enclosures housed in glasshouses or polytunnels are more easily managed and can be constructed in lengths allowing for ventilation at end doorways. Use of overall automated shade screens above the enclosure, ideally coupled with targeted external shade on the structure itself, provide opportunities for creating optimum conditions.

SETTING-UP AND MANAGEMENT OF THE ENCLOSED ENVIRONMENT

A wide range of factors based on the principles previously discussed must be considered to aid decisions and provide guidelines for setting-up and managing an enclosed environment. Separating these aspects from details of practical design and construction, covered in Part Four: Facilities, is difficult, and some repetition and overlap is inevitable. In the interest of clarity, the two topics are dealt with separately.

BED DESIGN

Shade levels are the major means of reducing internal temperatures, but design and construction of the grafting enclosure can have significant influence. Covers with a low profile enclosing a large area (the most extreme example being contact polythene) will heat more rapidly and to higher levels than taller structures[†]. This response, known as EFAR (external to floor area ratio), indicates structures should be high-sided and have a square rather than a half-round profile, see Chapters 13 & 14.

Ideally, the grafting bed should be at ground level with no insulation beneath so that heat gained during daylight hours can be lost into the floor, which then acts as a 'store', returning heat to the atmosphere at night. Today, energy costs dictate that if bottom heat is installed, insulation beneath the heat source is essential. The resulting potential reduction of heat storage capacity caused by insulation may be mitigated by covering the insulation layer with a layer, 50–75mm, of medium grade sand or concrete, within which heating cables or other heat source is installed[‡].

---

* R. Thurlow (pers. comm.).
[†] Harrison-Murray (2002).
[‡] Harrison-Murray (2002).

The worst conditions for heat regulation are benches fitted with an insulated base, topped with a heating source of electric cables or hot water tubes, covered by a single layer of capillary matting. Due to weight constraints, for modern mobile bench systems, the capillary matting cover is usually the preferred option and a deep sand layer is not practicable.

Bed Layout – Condensation

When poly-tents or enclosures are positioned close to an outside wall without any heat source between, the temperature differential causes internal condensation, producing drips on grafts adjacent to the outer edge. To avoid this, structures should be at least 60cm. away from the outer wall. Ideally, a heat source and/or thermal screen should be installed between the enclosure and wall.

Very cold grafts placed into warm, humid, closed environments immediately become wet from condensate on the stems. This is undesirable for unsealed grafts, which cannot be safely dried-off by ventilation. A short warming-up period under a temporary cover to harmonise temperatures before standing down in their final propagation standing down area solves the problem.

Moisture Level and Temperature of Standing Down Substrate

The combination of moist propagation standing down substrate, potted rootstocks containing moist potting compost and bottom heat provides by far the major source of humidity within poly-tents and enclosures. Substrates are best 'dryish' rather than wet, excessive dampness can cause over-wet, 'drippy' conditions. Pure peat used as a standing down material should be moist, but not wet enough to permit water to be easily squeezed out using a firm grip. Mixing with other material(s) with a lower water holding capacity, such as bark or perlite, is recommended. Provided satisfactory water content can be maintained, coarse sand is the best choice for propagation standing down material because it drains freely and responds quickly to changes in applied moisture. Rootstocks appropriately dried-off contribute comparatively little to humidity levels within the enclosure.

In enclosed conditions, using bottom heat to raise the temperature of the substrate and air can significantly increase evaporation and humidity. The grafter is faced with the dilemma of choosing between optimum temperature for graft union formation and the danger of excessive wetness. Providing optimum conditions requires skilful manipulation of shading, ventilation and temperature, combined with a probable need for drying-off pot grown rootstocks and controlling substrate moisture.

Cover Materials

Polythene film has enabled grafters to enclose a range of structures, providing near-perfect sealing of water vapour and ensuring humidity levels are achieved and maintained. Once sealed, moisture within is trapped under the cover and contributes to the maintenance of humidity.

After setting-up the enclosed environment at the commencement of each grafting season, normal practice is to allow the enclosure to 'steam-up' and stabilise at a pre-set target temperature before placing grafts in position. More or less shade and additional water may be required to produce optimum conditions. Bear in mind that any breaks in the polythene covers (tears, poorly fitted joints and overlaps) can dramatically reduce humidity and must be repaired*.

Because of their lower humidity retention properties, when grafts are placed in traditional glass and wooden frame lights, the gap between grafts and cover is often made as small as possible to aid the maintenance of humidity. This is achieved by lying the grafts almost horizontally on their side, (at <30 degree angle) with the union facing upwards, enabling the glass cover to fit closely above; hence, the once often used term 'close case'. The system was popular for use with bare-root grafts

---

* Harrison-Murray (2002).

and also with difficult subjects, such as Blue Spruce cultivars grafted on to potted rootstocks, and is still in use in some areas today.

The superior vapour-proof property of polythene means that laying-down grafts placed under this material is unnecessary and rarely used. Despite this, some still favour the method for use with vulnerable soft scion material such as *Daphne* and deciduous Azaleas. The polythene may be laid directly onto the grafts, to provide so called 'contact polythene' (Figure 13.3) or supported on low hoops 50–75mm above. It is important to recognise that temperatures are likely to rise quickly in these conditions because the EFAR is very low.

Anti-Condensation Strategies

Condensation is formed when moisture-laden air touches the cooler surface of the polythene (or glass) cover and condenses into droplets. The discussion which follows particularly relates to unsealed grafts because of their vulnerability to condensation drips which form on the internal surface and fall on the graft. Drips running down stems into the graft interface can result in serious losses and account for the reluctance of many grafters to omit sealing. Sealed grafts, whether deciduous or leafy, apical or side, can also react badly to wet, drippy conditions with rots on foliage, buds and wood.

Formation of droplets is exacerbated by bottom heat, which increases evaporation from the moist substrate. Once formed, condensation can only be reduced by ventilation, but initially this has to be of short duration because of the danger of desiccation, and is consequently rather ineffective. A number of strategies mitigate the problem:

- Use of thermal fleece (crop cover fleece) as an internal lining between the supporting framework and the polythene cover is a very successful method of reducing damage from droplets. The fleece must be wide enough to completely cover top and sides of the support framework. For good results, more than one layer is required, three or more being most successful. These should be sewn or stapled together along the edges, with one or more

FIGURE 13.3   Floor level grafts covered with fleece-lined contact polythene. Note the eaves' level shading above, which, when combined with heat loss through the floor, mitigate high temperatures (Verhelste Nurseries).

sewn-in or stapled folds (tucks) running lengthwise and spaced round the profile. Without fixing together in this way, the moist layers stick together when ventilating, become 'rucked-up' and do not function so effectively.

- At the same time as reducing condensate drips, fleece linings provide an excellent moist layer over the enclosed grafting area, promoting consistent high humidity, (Figure 13.4). As the fleece is initially dry, time is required to achieve full humidification. Ideally, fleece linings should be replaced annually but sterilisation with a suitable treatment, e.g. Jet 5, peroxyacetic acid, may permit repeated use (Figure 13.5 & 13.6).

- During winter months the external surface of the poly-tent may also be covered with layers of fleece to provide extra insulation (Figure 13.6 & 13.7). These will have a shading effect and combined with those used internally, overall shading for the main structure is unlikely to be required (see shade levels previously recommended). As a result, sunlight is able to warm the unshaded structure making a contribution towards reduced heating costs.

- An anti-misting polythene film is available but has limited life, and personal experience is that drips can still occur unless 'gothic arch'–type profiles are used to support the cover so that run-off can take place (Figure 13.6).

- Use of space heating external to the enclosure rather than bottom heat within reduces condensation. However, even with external heating, the transparent cover normally remains cool relative to the dark, light-absorbing surfaces inside; consequently, some water

FIGURE 13.4   LEFT and RIGHT, broadleaf evergreen (*Rhododendron*) unsealed grafts in bottom heated, fleece-lined poly-tent. Note: dry foliage and no drips visible, despite cold air external to the tent and inside humidity over 95%. Three layers of fleece are used as lining.

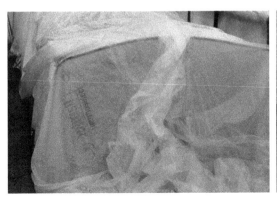

FIGURE 13.5   LEFT, grafting poly-tent, polythene cover pulled back showing internal three-layer fleece lining; RIGHT, a view of the fleece lining on inner surface of poly-tent cover.

FIGURE 13.6 LEFT, conifer grafts covered with polythene draped over the central wire along the moveable bench to produce a steep 'Gothic arch' profile. This allows condensate to run-off, preventing build-up of drips (Kilworth Conifers). RIGHT, poly-tent with external fleece covers drawn back showing only little condensate on inner surface of polythene cover despite warm internal temperatures (12–18°C.) but cold external temperatures (2–10°C).

FIGURE 13.7 Walk-in grafting polytunnel pictured during the winter period. This consists of a double skin inflated polythene exterior, lined with bubble polythene and two layers of fleece. Black, woven, light-grade ground cover fabric covers the 150mm-deep beds filled with sand. Within these is buried polythene hot-water pipes for base heating. Grafts requiring high humidity conditions in the summer period can be stood down on the open bed (not in use when photographed). On the right-hand bed, removable poly-tents are shown, the nearest with a three-layer fleece external cover and beneath a fleece-lined polythene cover pulled back. This contains warm callused deciduous species. Beyond is a further poly-tent containing *Rhododendron* grafts. Bottom-heated poly-tents rather than the entire tunnel are used for warm callusing during the winter when heating costs are a consideration. The tents are normally removed during summer grafting and the whole structure is used as an enclosed walk-in grafting polytunnel. A mobile gas heater set at 2°C for frost protection during the winter is visible at the far end. The hose-pipe on the nearest bed is fitted with a triple jet misting nozzle. This becomes a vital necessity when maintaining high humidity in the summer. The whole provides an economic, simple but effective facility.

condenses on it from the humid air trapped inside. Despite this, condensation is reduced, and some grafters adopt this procedure* (Figure 13.8). Daily ventilation, lifting and reversal of the covering polythene sheet, allow condensate to be shaken off, but some condensate inevitably sprinkles onto the grafts as the covers are removed.

MANAGEMENT OF THE ENVIRONMENT WITHIN WALK-IN STRUCTURES

For details of design and construction of walk-in structures, see Chapter 14 and Appendix Q.

High humidity environmental conditions can be scaled up from small structures, such as polytents, into much larger structures[†]. Walk-ins provide better access for monitoring development and opportunities for the introduction of improved labour saving and handling techniques. Two basic types are possible, either a walk-in enclosure built within an existing larger growing structure, or a free-standing, single-bay, walk-in grafting polytunnel.

Control and monitoring of humidity levels are much more easily achieved in walk-ins, and as a relatively large volume of moist air is involved, changes are less rapid. However, because of the relatively large air volume, there is a danger of too little humidity. This is best addressed by use of dry-fog equipment but to date few installations appear to exist; the obvious alternative is manual addition of water via hand held mist jets.

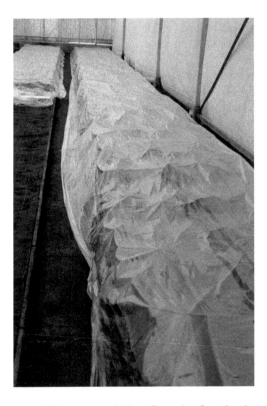

FIGURE 13.8 *Acer palmatum* cultivars: un-sealed grafts under floor level contact polythene, with only space heating (at 14°C). Contact polythene is occasionally lifted and condensate shaken off (mid-Feb.) (van der Maat Nursery).

---

* D. van der Maat (pers. comm.).
[†] Harrison-Murray (2002).

Damping-down the fleece lining of walk-in tunnels provides a useful reservoir of moisture, as does modest damping down the foliage of rootstocks of summer side-grafts, avoiding run-off onto the graft unions. To these activities should be added thorough wetting of floors and gable ends. Manual application must be accompanied by routine monitoring. Throughout hot, sunny periods, damping down in mid-/late morning, mid-day and again mid-afternoon, coupled with adequate heavy shading, may be required to maintain humidity at satisfactory levels.

Free-standing, walk-in grafting polytunnels should have shading applied externally for maximum cooling effect. To reduce levels of condensation, the use of inflated double-skinning, plus, if possible, internal bubble-polythene insulation, lined with one or two layers of horticultural fleece, is recommended (Figure 13.7).

MONITORING GRAFTS DURING UNION FORMATION

Duration and timing of ventilation, level of shade and temperature are influenced by many factors, a number relying on subjective judgements based on experience. To aid decision making, callus formation and development can be closely monitored by routine inspection, using a magnification (x10) hand lens. Observations are more easily carried out on unsealed grafts.

Slow and inadequate callus formation may indicate a need for higher temperatures and/or increased humidity. Premature browning or slight rotting of the callus may be corrected by extra ventilation. 'Bleeding' caused by root pressure or excessive condensate deposit can be mitigated for particularly valuable grafts by individually drying-off the graft union with absorbent tissue paper. Sometimes careful ventilation and reduced shading to produce lower humidity levels may improve the situation. With summer grafts, these strategies are more safely achieved when dealing with de-leaved rather than leafy scions.

Excessive, soft, 'fluffy' callus formation is a sign of too high humidity levels and/or temperature, too much shading or a combination of these. Failure to take appropriate corrective action can result in rotting and loss of grafts. Healthy callus tissue turning honey coloured indicates a stage where more ventilation can be safely applied, darker colours mean that graft periderm is forming and full weaning-off is soon possible.

Wax-sealed grafts are less easy to monitor until callus breaks through the seal, when the above observations apply. Sealed grafts are sometimes lost because problems cannot be detected sufficiently early to take remedial action.

Managing the environment through routine monitoring by an experienced grafter can result in substantially improved takes.

GRAFT TREATMENTS

INFLUENCING UNION DEVELOPMENT

Growth substances (rooting hormones), which have been highly successful in aiding production of adventitious roots in cuttings, have been investigated for their potential to improve results in grafts by promoting callus and union formation. In the Netherlands, applications to Birch grafts of fungicides alone and mixtures of fungicides with growth substances were compared. Without exception, the addition of a growth substance (indolyl butyric acid (IBA), indolyl acetic acid (IAA) or naphthalene acetic acid (NAA)) significantly depressed results[*]. Also, in the Netherlands, the growth substances (IBA, IAA and NAA) were applied as a scion dip to Pine grafts; all treatments depressed

---

* van Elk (1966).

takes[*]. Similar negative effects of growth substance treatment were found with treatments of IBA, IAA and NAA applied to *Robinia pseudoacacia* 'Frisia' grafts[†].

In the USA, the effects of IBA (Indole-3-butyric Acid) have been investigated on *Picea pungens* 'Hoopsii' using *Picea abies* rootstocks[‡]. This showed a 13% improvement in takes when the scions were treated with a 3-minute basal soak in 200 ppm IBA compared with untreated controls.

Little investigative work on the effect of applying cytokinins to promote graft union development has been recorded, despite their possible potential for success. Work has been carried out in Greece to test the effect of 6-benzylamino-purine (BA) applied to patch buds of *Juglans regia*[§]. Best results from using BA gave an increase in takes of 45% of treated versus untreated. The possible beneficial effect of these materials seems to warrant further investigation.

DISEASE SUPPRESSION

Several significant plant diseases, including *Fusarium, Rhizoctonia* and *Theilaviopsis*, attack a range of woody plants causing damaging infections and these, along with other pathogens, have potential to cause considerable losses of grafts.

Various fungicides may be applied as sprays or dusts to rootstocks or scions before collection, or during grafting procedure. Routine treatments applied before collection to scion mother trees and rootstocks of apple varieties gave a 70% take compared with untreated 48%[¶]. The beneficial effect of fungicide sprays on takes of *Acer platanoides* cultivars budded in the field has also been demonstrated[**].

Most convenient for treatment during the grafting process is dipping the cut surfaces of prepared scions, but other methods are used, such as spraying or dusting before or after assembling the graft. Use of fungicides on exposed graft tissue can be problematic. It is advisable to carry out trials before a full-scale launch into such treatments because, rather than promoting good results, they may reduce them by suppressing or damaging developing tissues. Thiophanate methyl is among the more successful and least phytotoxic; a mixture of this and difenoconazole may be used as a scion dip before attaching and tying them to the rootstock[††]. The addition of carbendazim to rootstock potting compost has further improved results.

International legislation is continually influencing use of pesticides; consequently, there will be a need for amendment and re-appraisal. Effective treatments will involve systemic, broad spectrum materials with low phytotoxicity. Bio-fungicides may eventually be most suited for use as graft treatments. See Appendix P: Effect of Fungicides on Graft Success.

COLD STORAGE POST-GRAFT UNION FORMATION

Use of cold storage for grafts after union formation aids production scheduling and can smooth workloads, enabling tasks such as potting or planting to be carried out with some control over timing. An efficient system for handling grafts in cold store is shown in Figure 13.9.

Cold callused or grafts from hot-pipe systems are most suited to cold storage because initial scion dormancy may be partially or wholly retained. Warm callused grafts, removed after union formation but before significant extension growth, can also be stored but always require careful monitoring. Temperatures at or below freezing point will cause irreversible damage to any young growth and immature tissues in the graft union. Grafters have reported problems with hot-pipe

---

[*] Joustra & Ruesink (1985).
[†] Joustra, Verhoeven & Ruesink (1985).
[‡] Kroin (1992).
[§] Pontikis, Papalexandris & Aristeridou (1985).
[¶] Mahlstede (1958).
[**] Howard & Oakley (1997).
[††] C. Verhelst (pers. comm.).

FIGURE 13.9   Successful grafts on pot grown rootstocks held in cold store and placed in carrying trays on Danish Trolleys and until required for potting-on or planting (John Richards Nursery).

grafts placed in cold store at temperatures which were initially too low, indicating that care and judgement is required even with what appears to be apparently dormant material.

Storage temperatures will depend on the condition of material and chosen species. At temperatures above 2°C the disease Grey Mould (*Botrytis*) is active and can cause damage to opening buds.

Routine sprays of fungicides appropriate for *Botrytis* control are helpful. Dormant grafts and those showing only slight bud activity can be safely stored at 1–2°C. Grafts showing bud activity, even short extension growth, should be stored at temperatures in the 4–5°C region and cannot be safely stored for periods of much over four weeks.

Successful summer grafts, once fully dormant, may be held in a cold store for prolonged periods. Under these conditions, over-wintering survival is often higher than for grafts housed in conventional growing structures.

Part Four

Facilities: Structures and Equipment

14 Protected Structures – High Humidity Enclosures – Shading Materials

The facilities described in this chapter will be required in various permutations to provide suitable conditions for most of the grafting systems to be described in Part Five. An exception is the system described as sub-cold callusing; for this and references to cold storage of rootstocks, scions and completed grafts made previously and in Parts Seven and Eight, specialised industrial cold storage equipment is required. These facilities require discussion with refrigeration engineers and contractors and considered to be outside the scope of this publication. For use in horticultural applications, the main considerations will be whether humidified or so-called jacketed cold stores are utilised. Final choice between the two will be dependent upon specialist recommendations and economic considerations.

Because of the specialised nature of the facilities required for hot-pipe systems they are dealt with separately in Chapter 16: Hot-Pipe Facilities – Design and Construction.

Traditional facilities such as cold frames have largely been replaced by greenhouses because of their advantages in labour efficiency, improvements in growing environments and energy conservation. The supply and construction of modern greenhouses by specialist companies is a significant industry. Innovation is constantly providing new opportunities for growers to obtain facilities of the highest standard to meet almost all requirements. Discussion with manufacturers on new or modifications to old structures is always worthwhile if not essential.

Materials normally used for cladding greenhouses are glass (glasshouses) or polythene (polytunnels); materials such as Acrylic/Polycarbonate dual panels or corrugated PVC panels are also used, particularly in the USA, less often in Northern Europe.

GLASSHOUSES

Glasshouse structures compared with polytunnels are characterised by better ventilation, slightly better light transmission and lower humidity. Glasshouse bays without additional equipment or modification can be used for cold callusing and growing on successful grafts. For grafts requiring supportive (enclosed) environments, additional equipment and/or internal structures are necessary. These may be fog or mist equipment, or enclosures such as poly-tents over ground level beds or benches. Walk-in, free-standing enclosures within areas of glasshouse or glasshouse bays are sometimes a preferred option and will be described later.

Multi-bay (multi-span) houses require integral lattice beam layouts to give uncluttered space for growing and propagation areas. Configurations can be arranged to provide a separate area for the workstation, where ventilation and heating can be separate from the growing area. This will allow more amenable working conditions for staff. Grafting bed/bench design can influence choices of post and lattice beam layout, mobile benching allows best utilisation of wide spaces. Some designs with overall heated flooring and moveable enclosures permit changes in layout to suit particular requirements (Figure 14.1).

Wide-span and multi-bay designs can be compartmentalised using static or moveable vertical screens within each bay or between the bays. Bays may be linked by concrete roadways (corridors) between or to one side, according to planning strategies, enabling different environmental zones

FIGURE 14.1   Overall floor heating in a multi-bay glasshouse. Poly-tent frames on a moveable base in the left-hand bay can be re-positioned throughout the floor area as required. The floor itself is constructed of no-fines concrete. The shading effect (40% shade) of manually applied screens above at eaves level is clearly seen in shadows cast on the floor (Hillier Nurseries).

to meet requirements for cold or warm callusing, growing on successful grafts and overwintering summer grafts*.

WALK-IN ENCLOSED STRUCTURES WITHIN A GLASSHOUSE OR POLYTUNNEL

In the absence of benching, polythene clad, walk-in enclosures can be built within glasshouse or polytunnel bays to replace poly-tents. Layout and dimensions can be adjusted to compliment glass-house bay size, but overall height is usually ±2 metres. Width of the enclosure normally takes account of existing heated ground level beds, handling systems and external structure bay width. Construction of an enclosure is described in HDC Report HNS 76, obtainable from the Agriculture and Horticulture Development Board (Kenilworth, Warwickshire, UK).

Lightweight metal or PVC pipes may be used as the supporting framework, as the enclosure is protected from open-air conditions. If joints at strategic points are push-fit only, not glued or welded, dismantling at the end of a grafting period is feasible and provides flexibility in use of space. Lining the structure with two layers of horticultural fleece provides an absorbent surface pre-venting drips and affords a reservoir of moisture, which contributes towards maintaining humidity levels, especially important when hand-held misting heads are used. To achieve the same objective more efficiently but at a higher capital cost, ultra-sonic dry-fogging heads can be installed within the enclosure.

The simplest method of providing ventilation is either to build enclosures in short lengths so that one or both ends can be opened to admit air, or clad the enclosure in such a way that the side clad-ding can be raised.

POLYTUNNELS

Polytunnels are significantly less costly to purchase and build than glasshouses and can provide appropriate environmental conditions for grafts. Adequate shading is vital, especially during the summer when temperatures can quickly become excessive.

Single span structures with a minimum width of 4.4 metres can be used for cold callusing of deciduous grafts. With minor modifications, such as covering access doors and ventilators with

---

* Clark (1979).

polythene covers to prevent moisture loss, they are suitable for easily grafted evergreen or semi-evergreen species such as *Cotoneaster.*

Polytunnels do not provide good environmental conditions for integral workstations, and for these it is worth considering either a centralised area or a purpose built bay, with features such as side ventilation, separated from others by vertical screens.

FREE-STANDING, WALK-IN GRAFTING POLYTUNNELS

Single span, walk-in grafting polytunnels, properly integrated into a planned layout, can provide very economical and effective structures for small- and medium-sized concerns, and in smaller sizes (shorter lengths) for committed amateur grafters. The structure, based on a 4.4-metre-wide polytunnel, may be erected and modified 'in house' using readily available and economical component parts. Heated ground-level grafting beds will be required for warm callusing. Heat loss during the winter period questions their suitability for economical warm callusing at this season. The smaller air volume of temporary poly-tents installed within the structure during the winter provides an alternative means of reducing heating costs (Figure 13.7). For summer grafting the poly-tents are removed and the entire house can be used as an enclosed, high humidity walk-in area (Figure 14.2). A portion of the tunnel could be used for mist propagation. Under these conditions, good results from cuttings can be expected with the added bonus of adding to humidity levels within the structure.

Use of an inflated double skin cover combined with an internal layer of bubble polythene significantly reduces heat loss, and raises internal temperatures in the region of 5°C above ambient. Further insulation is achieved by a recommended double fleece internal lining.

Ventilation is required; the installation of ventilating fans in the gable ends is a good choice, though these will require fully sealing-off when high humidity conditions are required. An alternative is the use of side ventilators, most easily constructed in the basal section of the tunnel. These comprise hinged and well-fitting wooden frames between vertical stanchions, clad with the same materials as those used on the side section, such as polythene lined with insulating bubble polythene or plywood.

See Appendix Q for details of design and construction of these structures.

FIGURE 14.2   Free standing, walk-in grafting polytunnels. Exterior shading has been manually installed over the second tunnel.

POLY-TENT

Poly-tents are the most usual structure used to provide enclosed conditions within larger structures.

Poly-tent cover supports (hoops) are placed over ground level beds or fixed over benches at approximately 1.2 metre centres. Design and fabrication of the hoops is normally carried out 'in-house'. A range of materials is suitable provided they are strong, flexible and can be formed into a suitable profile. Choices are usually from one of the following: flexible plastic pipe; heavy gauge (6 or 7 gauge) galvanised wire; 8–10mm diameter mild steel rod which is galvanised, painted or plastic-covered; small bore zinc-plated steel tube; or aluminium tubing. A suggested profile for bottom-worked grafts providing good EFAR and reasonable access down narrow paths between beds is shown in Figure 14.3. Dimensions must take into account available polythene sheet dimensions, bed widths and requirements for the grafting methods and systems in use. Top-worked grafts require taller purpose made tents.

Poly-tent hoops may be covered by so-called white propagation film, available in widths up to 3.6 metres. The shading effect of this can confuse shading levels, and it also has the disadvantage of not being UV-stabilised; as a thin 80-gauge material (20mu), it is particularly suitable as a contact polythene cover. A clear cloche film is available in widths from 1.85 metres to 3.7 metres with a thickness of 150 gauge (38mu). Unlike propagation film, it is UV-stabilised; additionally, coupled with greater thickness and strength, it is more durable and a better choice for covering poly-tent hoops.

Hoops placed over floor level beds may be held in position by pipes (tubes) or pre-formed holes cast into concrete floors. More flexibility is possible if moveable bases with locating points are available. A simple construction utilises 50mm galvanised angle iron base plates with locating pipes or spikes welded in position, to which the hoop may be attached (Figure 14.4).

Benches may be designed with fixing points to accept hoops or supports; alternatively, hoops may be fabricated to allow attachment to the bench (Figure 14.4 & 14.5).

HOLDING TENT

Holding tents, designed to replicate conditions in the poly-tent, are necessary for holding just completed, vulnerable leafy or un-sealed grafts, in a supportive environment, prior to transporting to the propagation standing down position. They may be constructed using poly-tent hoops clad with

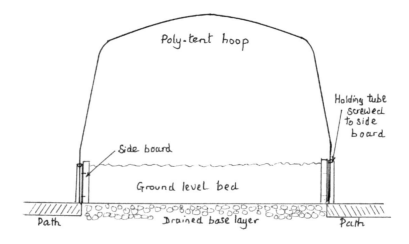

FIGURE 14.3   Poly-tent hoop sited on ground level bed. Suggested basic design; the dimensions can be decided to fit the layout. This design has good EFAR properties and permits closer standing-down towards the edge of the bed than a simple semi-circle pattern, and the slightly angled sides allow better access down paths between grafting beds.

FIGURE 14.4   LEFT, mobile, semi-circular, poly-tent hoops at floor level can be attached by a suitable metal rod, pipe or tube to a galvanised angle iron base plate making the whole assembly portable. RIGHT, a poly-tent hoop made for attachment to a bench. Fabricated in-house from galvanised 10mm Ø steel rod with angle iron base plate for fixing (by bolts) to the bench. The vertical sides of this type present some barrier to access down the path between benches.

FIGURE 14.5   Bench mounted, poly-tent support hoops in position before polythene cladding. Electrical heating mats are fitted beneath the black, woven, light-grade, ground cover fabric on the bench top. The white polythene sheet attached to the back wall provides some additional insulation against the glass sheeting exterior and a clean reflective surface when the poly-tent is in use. The covering polythene is normally 38mu, clear, cloche film (Sandy Lane Nursery).

polythene and placed on a bench adjoining the grafting position; a suitable design and layout is shown in Figure 14.6.

For convenient handling, it is important to provide easy access to the completed grafts often placed in carrying trays (Boskoop trays). Humid conditions must be maintained while the grafts await transport, most easily achieved by heavy shading and occasionally damping down the side walls and base. Top-worked grafts will require tall, bulky holding tents and, when possible, are best grafted in an area adjoining the propagation standing down position.

The need for holding tents can be eliminated by use of an integrated or mobile workstation, with necessary handling equipment (roller conveyors, etc.) appropriately positioned with access to the propagation standing down position (see Appendix L for an example).

SHADING – EXTERNAL

NATURAL SHADE

Use of natural shade can provide an economical and effective means of providing good conditions for grafts. The north side of a substantial Leyland Cypress screen provides a cool, shady site. Such plantings can be located alongside roadways providing access and a means of maintaining the screen with use of tractor-mounted hedge cutters (Figure 14.7).

SHADE HALL

Greatest heat reduction is achieved by external shading supported above the propagation structure. This is because of the combined effect of interrupted solar radiation and allowance for air movement between the surface of the structure and shading material. A significant disadvantage of supported

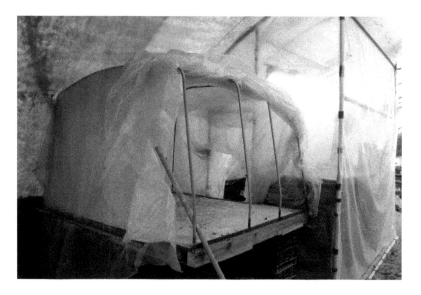

FIGURE 14.6   Holding tent adjoining the work station (fluorescent tube lighting visible). The tent is filled from the grafting bench via the access point located at centre of the adjoining polythene wall. Access to the completed grafts for the next handling stage is from the lifted side curtain. Both accesses can be closed by lowering the polythene curtains to maintain enclosed conditions within the holding tent. The tent itself is constructed from polythene-clad poly-tent hoops mounted onto a suitable basal framework. In this case, 50 × 50mm angle iron with fixing points welded at convenient spacings. The whole is built on a suitable bench. Movement from holding tent to standing down area is by multi-tier pallets. Dependent upon layout, roller conveyors or mono-rails could be considered as alternatives.

FIGURE 14.7   Natural living shade using x *Cuprocyparis leylandii* planted alongside a poly-glasshouse used to house grafts. Note: strong sunshine on right and in background, in contrast to the cool, shady area, which includes the structure. Photographed mid-June.

external systems is capital cost and vulnerability to snow load. The system, popularly known as a Shade Hall, provides the best possible environment for grafts placed in structures beneath, but it is rarely seen in the UK. See section on internal shading below for a discussion of shading screen types and characteristics.

ADDITIONAL EXTERNAL HOOPS

To give support to external shading materials, use of additional hoops placed externally to existing walk-in grafting polytunnels provides an alternative to Shade Halls. The next standard hoop size for 4.4 metres (14ft)-wide tunnels is 5.5 metres (18ft), and can be adapted for this purpose. These hoops must be spaced widely to permit access for re-cladding the underlying tunnel. Passing a rope along the length of the structure over the hoops, in the centre ridge position, allows the screen to be attached, pulled over before being spread down the sides, and fixed at the base with ties or clips.

DIRECT APPLICATION

Shade screens may be directly applied externally to polytunnels as the smooth unimpeded surface allows most types of shade material to be easily drawn over the outside in contact with the cladding. Although not as effective in cooling as the Shade Hall or external hoop supports, this provides an economical alternative.

For relatively small, i.e. <4.4 metres (14ft wide), structures most materials can be applied or withdrawn manually from ground level by a single person (Figure 14.8); hence, recommended shade levels for various light conditions can more easily be followed without the use of costly shade screen systems.

SHADING – INTERNAL/EXTERNAL

Most shading screens have been designed for internal use; however, external screens are offered by many manufacturers.

FIGURE 14.8   Walk-in, 4.4-metre-wide grafting polytunnel showing shade material (40% lockstitch pattern) positioned manually from ground level. Gable end is clad with exterior grade plywood. Gable end doors with clear polythene cladding will need separate shading. To increase shade levels, further layers can be easily applied by pulling over that existing.

Eaves-level internal screens provide a compromise between less temperature reduction but lower cost; most modern glasshouses and some polytunnels are designed to be equipped with this facility. Internally placed, enclosed grafting structures (poly-tents, etc.) at bench or floor level can be further shaded by laying shading materials directly on the surface of the covers.

Shading Screens

In modern growing structures, shading is frequently operated automatically. Different basic patterns are available, and the characteristics of these may be combined in various ways to influence environmental conditions. Adding reflective strips to the screen, especially when combined with areas of open weave, has the effect of providing shade, reducing ambient temperature beneath and conserving heat loss at night. A range of shade levels from 26% to 86% is available. Selective use of layers of screening material provides combinations to fulfil the recommended shade levels set out in Chapter 13. Specialist knowledge is important and should be sought when deciding upon the choice of options.

15 Ground Level Beds – Benches – Heating – Humidification

GROUND LEVEL BEDS WITHIN GREENHOUSES

Ground level beds facilitate the use of greenhouses with lower eaves and, consequently, overall height, an advantage if external shading systems are considered. Grafting houses built with structurally strong floors, such as concrete or no-fines concrete acting as ground level beds, can offer opportunities for a range of materials handling systems and maximum flexibility in use of ground area (Figure 15.1). They are also substantially cheaper to install than benching systems.

GROUND LEVEL BEDS FOR COLD OR WARM CALLUSING

The simplest beds for grafts which are cold callused at ambient temperatures within the structure are marked out areas on existing earth floors. In many cases, the beds are edged by pressure treated, 2400mm × 75–150mm wooden boards held in place by steel spikes driven into the ground. On poorly drained sites, land drains will be required and substrates, such as sand or gravel, placed to a depth of 75mm (at minimum) over existing soil. To aid maintenance beds are best covered with ground cover fabric or light capillary cloth.

Bare-root grafts may be laid-in in a suitable substrate within soil-based beds, but most grafters favour using trays or boxes, normally Boskoop Trays, for holding both bare-root and potted grafts (see Chapter 13, Figure 13.1).

Most floor level beds are constructed far more elaborately, either with designated areas, often edged beds, running the length of the house with concrete access paths between, or an overall floor covering of concrete or, less often, no-fines concrete. These layouts, commonly used for cuttings propagation as well as warm callused grafts, normally have the capacity to provide basal heat from a suitable heating system, to be discussed further (see Chapter 14: Glasshouse, Figure 14.1 and Chapter 15, Figure 15.1).

WARM CALLUSING – PROVISION OF HEAT FOR GROUND LEVEL BEDS

Provision of heat for grafts requiring supplementary heat (warm callused) is either by electricity or hot water. As capital costs are lower, the former is particularly suited to smaller installations. The disadvantage of electricity is that running costs are generally higher.

Electricity

Electrical heating cables of various types have been available for several decades and offer a reliable and safe form of heat source. The usual output is 10–12 watts per metre run, but confirmation with the supplier is recommended before calculating cable spacing and layout. Assuming good insulation within the bed, a loading of ±150 watts per square metre of bed area is recommended to maintain temperatures of 20–25°C for mist propagation beds in UK conditions. Grafting beds with less water applications (no mist) and lower input of 100–120 watts per square metre can be considered.

FIGURE 15.1 Ground level bed for warm callused grafts or cuttings propagation. The floor is constructed of no-fines concrete with integral bottom heating provided by polythene hot water pipes. Shown is a moveable poly-tent in course of assembly enclosing a mist line for cuttings propagation. Without mist, this structure would be equally suitable for grafts. The semi-circular, galvanised steel pipe hoops are pushed over metal support pegs welded into position on the galvanised angle base plate. This installation is being assembled along the length of the glasshouse bay; clear floor space provides the flexibility for other configurations (Hillier Nurseries).

Electrically heated mats or panels are a good alternative to cables, having the advantage of more even heat distribution. All mats, panels or cables will require controllers to accurately maintain temperature.

Hot Water

Some reliable systems have used plastic tubes of polyethylene, utilising standard compression fittings, and PVC pipes. Combined with glued PVC irrigation pipes, this can provide heating systems offering space and bottom heat functions. Significant cost savings are possible using this approach rather than a conventional, fully engineered heating system with steel pipe-work.

These materials require operating at temperatures not exceeding 38–40°C or softening, and eventual break-down of PVC and polyethylene pipe-work will occur. Boilers operating at 75–85°C require either a calorifier (heat exchanger) or mixing valve to provide the correct working temperature. Calorifiers are more costly than mixing valves but well refurbished, second-hand models can continue to give decades of reliable service. By utilising a reliable electronic thermostat to control water supply from the boiler, control of water temperature from the calorifier to plastic heating pipes is then a relatively simple procedure.

Plastic heating pipes made from PEX or BPEX, (a cross-linked polyethylene or polybutylene plastic), are able to withstand much higher temperatures than polyethylene and can be used as a replacement. At higher temperatures, manifolds and supply pipes will need to be fabricated using specialist expertise. Closely spaced small bore tubes fabricated from heat resistant synthetic rubbers such as EPDM can provide uniform overall heating within the bed. This enables use of relatively small amounts of covering materials, such as capillary mats, making them suitable for mobile and fixed glasshouse benching. Experience has shown that small pipes are eventually liable to become blocked, and it is recommended to have discussions with specialists on heating design and construction before installation.

Providing Bottom Heat and Space Heating

Large-scale systems relying on sophisticated temperature control and fabricated using latest technology are best designed and installed by specialists. A simple system which can be built with a minimum of engineering knowledge involves a well-lagged PVC mains pipe providing a reservoir of hot water at a maximum of 40°C from a heat source (Figure 15.2).

For small-to-medium scale operations, second-hand domestic oil-fired boilers can be obtained economically and are a sensible source of heat.

Bottom heat can be supplied from the hot water via flow and return manifolds, made from glued PVC fittings and compression connectors. These are then joined to polyethylene pipes buried in sand and spaced at 150 to 200mm centres (Figure 15.3). If space heating is required, this can be provided by 25 or 32mm PVC pipes mounted along the sides of the grafting house (Figure 15.3).

Grafting Beds

Traditionally, grafting beds comprise designated areas separated by pathways, dimensions adjusted to provide best space utilisation for poly-tents or other enclosures. Many designs involve raised beds surrounded by timber or concrete side-walls usually between 150 and 200mm high (Figure 15.3).

Good drainage beneath the bed is essential. Upon this is laid a suitable insulation material, such as 50mm thick polystyrene blocks wrapped in polythene sheet to prevent moisture ingress. Alternatives to polystyrene exist, and it is wise to seek specialist advice on the latest modern materials.

If heat is supplied from electrical heating cables or hot water pipes, sand is normally used to surround the heat source. This has the advantage of providing well-drained conditions and, providing it is kept reasonably moist, prevents cables from over-heating. Electric heating cables require a minimum 75mm depth from the surface. The depth of sand for hot water, using 20mm diameter pipes, should be a minimum of 100mm but should be adjusted to take account of pipe spacing. For wide spacing, say 200mm centre to centre, pipes can be covered more deeply, up to 150mm. As a general rule the deeper the sand the better because rapid changes in temperature, moisture and humidity are reduced.

FIGURE 15.2 Heating supply. LEFT, high-level, lagged, hot water PVC heating mains mounted on galvanised support posts. Supply from mains to walk-in grafting polytunnel is via a flexible, lagged pipe visible to the left. A circulation pump within the galvanised pump box pumps hot water into the grafting house on demand. RIGHT, pump box with access cover removed. Domestic central heating circulating pump is visible in bottom left. Valves and pipe-work are enclosed within extruded polystyrene insulation (blue) which is lining the steel sheet sides and supplemented by glass fibre insulation wool as required. The lagged heating main is mounted above and exterior to the box and is just visible, covered at that point with a domed steel roof for weather protection.

FIGURE 15.3 LEFT, a heating system comprising polythene bottom heating pipes installed in a deep capillary bed lined with black polythene, with 150mm timber side walls. Separate flow and return PVC manifolds fabricated in-house are shown connected to the 20mm in-bed polyethylene water pipes, carrying hot water at 40°C. Beneath the black polythene lining are polythene-film-wrapped, 50mm thick polystyrene sheets to provide insulation. Sand, normally filled flush to the top, has been partially excavated to expose the manifold. RIGHT, mounted on side walls for space heating, PVC rigid water pipes carrying hot water at 40°C.

Above the base insulation, construction may vary. There is much to be said for lining the bed with heavy gauge black polythene, in effect constructing a deep capillary bed (Figure 15.3). Filled with sand (medium-coarse) and supplied with water inlet and outlet points, as for capillary bed irrigation installations, water level and moisture content can be closely controlled. Within a poly-tent or contact-poly, this can influence moisture and humidity levels, supplementing or reducing that obtainable with shading and ventilation. As the water within the sand bed is warm, uniform heating is assured and hot-spots eliminated. General cleanliness is assisted if the sand surface is covered with light-grade ground cover fabric or capillary matting, which may be easily brushed over to remove plant debris, etc. and replaced at suitable intervals.

Electric heating mats or heating panels enable different construction. Because heat is distributed evenly over the area all that is required is insulation beneath and use of capillary matting overlying the heat source. Side walls may be much lower or omitted altogether and, consequently, construction costs are lower. Unfortunately, control of moisture levels and thermal storage is reduced in this type of construction. Consequently, they tend to be used on glasshouse benches, where weight limitations apply.

Bottom Heat throughout the Entire Floor

Heating the entire floor offers maximum flexibility in use of space and is a popular choice if mobile benching is not favoured. Usual means of heating is by hot water circulated within suitable pipework, generally polyethylene pipes supplied by PVC or steel header manifolds (Figure 15.4). In all cases, good insulation beneath is essential.

Most designs specify use of concrete as the preferred floor cover. Advantages are the possibility of very high bed to path ratios, efficient materials handling and quick, easy cleaning for pest and disease control. Disadvantages, apart from cost, are that the construction demands careful grading to fixed levels. Well-designed outfalls are required to avoid poor surface drainage and pools of water. Moisture fluctuations and drying-out can occur if water is not always available. This type of floor construction is best suited to systems employing mist or wet fog.

A rarely seen alternative to conventional concrete, well-suited to propagation areas including grafting, is the use of no-fines concrete. This is a rather specialised mix, best produced by an approved concrete supplier. Such floors have been laid in propagating houses which are still in use

FIGURE 15.4   No-fines concrete flooring. LEFT, showing discharge from a hose-pipe rapidly draining through the floor. RIGHT, galvanised flow manifold within a cast solid concrete chamber connected to 25mm polyethylene heating pipes running through the floor at 120mm deep. The installation is nearly 40 years old. (Hillier Nurseries).

after over thirty years and function effectively with no loss of drainage, apparent surface damage or wear. No-fines concrete has been tested as a road pavement, and it showed good resistance to low levels of traffic*. Hand operated fork lift trucks and mini-tractors with grass tyres should present no danger of damage to the floor. Cost is 10–20% lower than conventional concrete, but this may be offset by the desire to use thicker slabs.

Drainage is so good that the surface may need to be 'blinded' with sand to provide sufficient water retention to maintain humidity levels (Figure 15.4). No-fines treated this way performs like a deep sand bed, so it has excellent features for environmental conditions but also facilitates materials handling. After some years, there is a danger of the surface becoming blocked with organic debris from cuttings trays, compost from pots, etc. This can be addressed by use of high-pressure hoses or covering growing beds on an ad hoc basis with light grade capillary cloth, which can be removed, replaced or re-positioned as required.

A further use of no-fines concrete, known as a heated flooded floor system[†], rarely seen in the UK but more frequently in the USA, is to build the no-fines floor above a layer of sand-free gravel in a polythene-lined tank as for a capillary bed. After construction, this is filled with water to a depth which can be adjusted using capillary bed systems. This construction has the advantage of a solid working surface, rather than the non-load-bearing surface of sand. By flooding the beds to the surface on an occasional basis, any tendency for blockage with organic material can be prevented by flushing it away.

GRAFTS HOUSED ON BENCHES

Benches provide a convenient height for inspecting grafts and have the advantage of less energy use because of higher temperatures at bench level compared with those on the ground. The disadvantage of low-heat storage capacity for benches, compared with the temperature mitigating effect of

---

* Harber (2005).
† Aldrich & Bartok (1994) pp.74–77 and Hartmann, Kester, Davies Jr. et al. (2014) pp.61–63.

ground level beds, has been discussed previously. A further disadvantage is its significant additional cost compared with ground level beds.

Traditional bench construction using concrete blocks as supporting walls enclosing steel heating pipes beneath is now rarely seen; modern benches are usually aluminium or a mixture of aluminium and galvanised steel sections, with an incorporated heat source. Within the available area, mobile benches provide much higher useable bench space than fixed benches, potentially 81% compared with 59% and are the preferred option for most new construction.

Bench Heating Systems

Modern mobile benches require a power supply for conventional heating cables or electrical base heating systems, often referred to as heating mats, or alternatively narrow bore, hot water heating systems. New materials and equipment are constantly being developed and discussion with specialist suppliers is recommended.

To ensure good heat distribution and prevent over-heating when traditional heating cables are installed, they require to be laid in sand. A 75–100mm depth is necessary, resulting in a significant loading on the bench, often making choice of mobile benches impracticable and heating mats the only sensible alternative.

THERMOSTATS

These vital pieces of control equipment are concerned with the control of heating and ventilation. Two major types exist:

* Electro-mechanical thermostats.
* Digital (electronic) thermostats.

Thermostats – Electro-Mechanical

Rod-type thermostats are reasonably accurate, but response to temperature change is slow, and the differential between on and off can sometimes be as much as 5°C. To obtain best results, it is important that the temperature sensing rod is placed well above and across the run of heating elements; a badly placed sensing rod will distort readings.

Thermostats – Digital (Electronic)

These rely on the use of thermistors or semi-conductors for activating the thermostat. They normally incorporate a digital display providing a read-out of temperature and differential settings. Programmes are sometimes available, allowing control according to perceived daily fluctuations, for example different day or night time temperatures.

Electronic thermostats provide a faster response to temperature changes and smaller differentials than the electro mechanical types. Temperature sensing is normally via a probe attached to a wander lead. Maximum electrical loading capacity is lower with this type of thermostat, and a heavy-duty relay is required for large numbers of electrical heating cables or motorised valves.

HUMIDIFICATION

An automated system involving dry fog ensures a high humidity environment without free water and obviates a need for sealing and enclosing the grafts. It can also significantly lower air temperatures by up to 10°C compared with non-fogged environments. Despite this, additional environmental support in the form of shading will be required during the summer months. Sophisticated automated control using a humidistat or computer is necessary for efficient operation.

Two systems are currently available: high pressure fog jets or ultrasonic fogging heads. Capital cost of installing a high-pressure system is likely to be greater than the ultrasonic option, but it is suitable for large-scale use. Ultrasonic systems are best for smaller units but require substantially more energy to operate than equivalent high-pressure systems*.

Despite the apparent advantages, the use of dry fog in grafting operations appears restricted to very few examples. To maintain high humidity, most grafters favour a simple enclosed system relying on shading, little ventilation and plenty of moisture, as described in Chapter 13. Limited personal experience on a one-off basis has shown the potential for use of dry fog for summer grafts. Dry fog used in the winter has also proved successful for a range of species of *Araucaria*†. It is surprising that it appears to be so little used in grafting procedures.

An alternative wet fog or ventilated fog system has proved particularly effective for cuttings propagation. The free water deposited on stems and leaves using this equipment means that sealing the graft union is essential to prevent water ingress between the interface of the graft components.

---

* Hartmann, Kester, Davies Jr. et al. (2014) pp.410–411.
† A. Luke (pers. comm.).

16 Hot-Pipe Facilities: Design and Construction

The hot-pipe system is designed to operate during the late autumn, winter or early spring months, though cold storage of scion wood and rootstocks would permit some extension of use. It depends upon a small, well-insulated, closed, heated space (heating chamber) sited within a contained area holding a suitable laying-in material for the roots of bare-root grafts, or to provide a bed for holding root-balls of potted or de-potted rootstocks. The heating chamber is designed to enclose only the graft union and makes it possible to provide optimum temperatures for graft union development.

The scion, the rootstock root system and, if side grafts are used, the rootstock top project beyond this chamber into the surrounding structure and are therefore subject to external ambient temperatures, normally low when the system is in use. Heat transfer through woody plant tissues is restricted; as a result, despite relatively high temperatures at the site of the graft, areas of the plant outside the heating chamber remain much cooler and retain their dormancy. This combination provides excellent conditions for promoting callus formation, simultaneously reducing premature bud-break and root activity, ensuring food reserves are conserved and root pressure curtailed.

Management aspects of the system will be discussed in Part Five: Grafting Systems.

HOT-PIPE – HEAT SOURCE

For the heat source, two major options are available and need consideration regarding advantages, disadvantages and design features:

ELECTRICITY

Electricity is generally more flexible and versatile in use than hot water. With appropriate thermostats sited in close proximity to the heat source, electrical heating cables are readily controllable, and because of low thermal inertia, quickly responsive and accurate.

Heating cables can be either traditional-screened and plastic coated or self-regulating cables, commonly used to protect water pipes against freezing. The latter have advantages, upper heating temperatures are limited, reducing danger of burn-out, and cables may be cut to the required length and capped-off.

The cable may be laid along the length of the hot-pipe heating chamber, but bends and kinks can occur and cause uneven temperature distribution. This can be reduced by attaching it to a 15–35mm diameter rigid pipe or rod using cable ties or clips. The pipe may be filled with water and capped; the resulting column of warmed static water is thought to provide a more uniform temperature along the length of the run.

Power requirement is greatly dependent upon efficiency of insulation of the heated chamber, particularly at points where portions of the rootstock and scion project outside, where much heat can be lost. With good insulation, a single cable with an output of ±12 watts per metre run (recommended to check with supplier for confirmation, as outputs vary) should provide sufficient output to maintain temperatures of 20–22°C when ambient temperatures are at freezing point or a little below. For further security, a double run of cable is often suggested*, fixed to a pipe by cable ties or similar.

---

* Ramsbottom & Toogood (1999).

Hot Water

Heat may be supplied by a suitable pipe (usually 15–20mm diameter) containing hot water. Sometimes this is placed within a larger diameter pipe filled with static water, running the length of the heating chamber (Figure 16.1). A mains pipe supply from a heat source is required and, if poly-ethylene water pipes are used, the temperature of the water leaving the boiler must be lowered to a maximum of 40°C. Various methods have been adopted for controlling hot water supply to obtain the target air temperature in the heating chamber. Most popular is the use of thermostats to control on/off valves (solenoid valves), influencing water flow to the chambers.

THE HEATING CHAMBER

Insulated heating chambers were originally designed to enclose grafts placed vertically; horizontal placement is now more usual. For vertically orientated grafts the heating chamber needs to be held above the standing-down/laying-in substrate (usually composted bark) by fixing it to a wooden or steel platform (Figure 16.1). For many horizontal systems, the chamber has slots into which indi-vidual grafts are laid (Figure 16.1), and after placement the whole run is enclosed by a top cover, usually a flexible pipe insulation material (shown in Figure 16.4). Alternative designs depend upon the grafts being sandwiched between layers of soft, spongy, insulating material, attached to a solid cover (Figure 16.2). An increasingly popular design, because they take up far less space, is to use chambers of this type stacked in layers above each other (Figure 16.3).

Slotted or solid covers are the two more usual options, but a third 'economy' version to be described later has been adopted by some nurseries.

Slotted Cover

These patterns are popular and involve use of outer rigid plastic pipes 75mm+ in diameter. These hold individual grafts in access slots between 15 and 25mm wide and 25 to 30mm apart cut into slightly less than half of the circumference on the upper surface of the pipe (Figure 16.1 & 16.4). The heat source is provided using a hot water pipe or electric heating cables placed within and run-ning along the entire length of the interior (Figure 16.4 & 16.5).

To retain heat within this chamber, the chamber wall below the slotted portion is permanently insulated, usually with elastomeric pipe insulation, such as Armaflex®, Climaflex®, etc. To complete heat retention, individual grafts, after placement within the slot, are carefully covered with flexible insulation material, usually of the same material as the base, to close off any exposed surfaces. This is important to prevent heat loss through the slots (Figure 16.4).

FIGURE 16.1   LEFT, hot-pipe vertical placement. A system using bare-root rootstocks laid in sand. The slot-ted hot-pipe runs are attached to cross members (Botanica Nursery). RIGHT, slotted hot-pipe run at ground level for grafts placed horizontally. Grafts and flexible insulating top cover are not yet in place. A water-filled pipe containing a heat source (hot water pipe) is visible in the centre (F.P. Matthews Ltd.)

FIGURE 16.2   Hot-pipe system heated by electric cable; grafts sandwiched between foam insulation in this case, fixed to square rain water gutters. Run on left has wooden boards weighted down to ensure a tight seal; fleece is packed beneath the board to improve insulation (Penwood Nursery).

FIGURE 16.3   Hot-pipe runs arranged in stacked layers. Foam insulation material compressed tightly around projecting portions of graft to trap heat in the heating chamber. Ends of grafts sealed with red grafting wax are visible protruding through the vertical wall of the heating chamber (Verhelste Nursery).

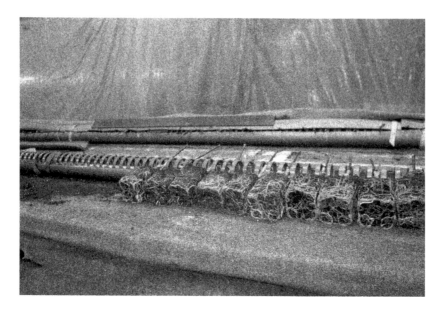

FIGURE 16.4   Hot-pipe system – placement of grafts in slots insulated with pipe insulation. Permanently insulated beneath the slots, as seen on the upper run of grafts in the foreground. To ensure a heat seal between upper and lower insulation, after placement, grafts are covered with pipe insulation weighted down with steel angle (seen in the back row). To retain moisture pot root-balls are covered with capillary cloth (back row). Those in foreground are awaiting similar treatment once all slots are filled (Yorkshire Plants).

Layered Slotted Cover

When slotted pipe designs are used, further layers of hot-pipes above the original can only be achieved by using shelves (Figure 16.5). Suitable shelving can be purpose made or purchased from companies specialising in shelving systems.

SOLID COVER ('SANDWICHED' COVER)

This design provides a heating chamber made in two component trough-shaped halves, one containing the heat source, the other providing the cover. The components meet at the edges, which are covered with a suitable foam-insulating material. The grafts are placed crossways and 'sandwiched' between the two (Figure 16.2 & 16.7). Provided sufficient pressure is applied by weighting the upper cover, heat retention within the chamber is very good.

Numerous options exist for fabrication of the two chambers. These are influenced by convenience, layout (whether single level or stacked layers) and insulation efficiency. For single level layers, the following prove practical and effective:

- Extruded polystyrene sheets, fabricated to shape.
- Half-round plastic rainwater gutters.
- Square section plastic rainwater gutters.
- Plywood and/or insulating boards, fabricated to shape.

Extruded Polystyrene Sheets Fabricated to Shape

Extruded polystyrene sheets of standard size 2400mm × 600mm × 50mm are excellent for construction of hot-pipe runs. They can be readily cut with a sharp hand-saw or circular-saw. Care is required for small sections used as edging strips. Dimensions can be adjusted as required for electric heating cable, hot water and static water heating configuration. Suitable dimensions are shown in a cross section in Appendix R. The material can only be glued using non-solvent based adhesives,

FIGURE 16.5   Shelved hot-pipes using 15mm hot water supply pipe running through 32mm static water pipe. This is within an outer 75mm slotted pipe, holding the grafts. An insulated pipe cover on the underside is visible on the top shelf (top left-hand run). Further pipe insulation is ready to close off slots when grafts are in position (alongside root-balls, left-hand side of top shelf). Each hot-pipe run has an isolating hand valve (red handle) and a solenoid valve controlled by thermostats to enable temperature control. In this design capillary matting is used to cover the roots and is seen hanging down from the top shelf before placement over the de-potted root-balls alongside the upper pipe run (Yorkshire Plants).

and suitable choices are PVA formulations, ideally choosing those which are water proof. PVA glues require two to three days under pressure before adhesion is completed, best achieved by holding the glued sections firmly in position with weights on a solid floor (Figure 16.6). Ideally, the heating chamber should be lined with aluminium foil glued into place (kitchen foil is satisfactory).

Extruded polystyrene has among the highest insulation properties of the commonly available materials*, and is so water resistant that it can be used as a float for control valves in water tanks. Good heat retention is further guaranteed by generous amounts of insulating foam at the meeting edges, with sufficient weight on the upper channel to ensure tight closure.

Extruded polystyrene has strong compressive strength, but tensile strength is relatively poor. Additionally, the material is liable to damage if handling lengths over 3.2 metres. Glueing 4mm thick plywood, or a suitable plastic substitute, to the exterior faces of the covers will significantly reduce potential damage caused by bending and knocking.

Half-Round Plastic Rainwater Gutters

Convenient for construction of hot-pipe channels; generally, all that is required is to insulate the lower portion using moisture resistant materials. The top cover may be insulated with the same material or pre-formed fibre-glass pipe lagging. Edges of each component can be either plastic foam of the type used for upholstery cut into suitably sized strips (Figure 16.7), or commercial insulating strips, or smallest available pipe lagging pushed into position (Figure 16.2 & 16.8). These materials can be glued in place using suitable adhesive.

For 112mm-wide, half-round gutter, heat is best provided by electric cables; a width of 150mm permits use of hot water. Ideally, the inside of the gutter should be lined with reflective aluminium

---

* National Mechanical Insulation Committee (2016).

FIGURE 16.6 LEFT, extruded polystyrene hot-pipe runs under construction, carried out on a flat, dry concrete floor. RIGHT, fully assembled including foam plastic edging strips. An electric heating cable is positioned at the base of the heating chamber. The openings left at each end of the run must be closed-off to prevent heat loss, using a polystyrene or plastic foam plug (visible in the chamber but not yet positioned). The chamber assembly is placed within a simple gravel board-edged bed, filled with suitable material to contain the root system of bare-root grafts or hold potted rootstocks.

FIGURE 16.7 LEFT, half-round water guttering used to produce a heating chamber. Grafts, viewed from above, placed in position across the plastic-foam-covered edges. At the top of the picture a solid cover is in place with rubber faced fibre-glass pipe lagging above for heat retention. Grafts are placed on the upper edges of the lower chamber immediately above a heating cable. When in position the upper cover is held firmly against the lower section, by placing weighted wooden boards above, 'sandwiching' the grafts between. Shown here are *Magnolias* side grafted to potted rootstocks of *M. kobus*. RIGHT, strips of plastic foam are fixed to the edges of the lower half of the heated chamber. This is contained inside a bed filled with composted bark and edged with gravel boards, onto which the root system of bare-root rootstocks or pot ball of pot grown rootstocks can be placed.

foil to improve heat retention. To help aid uniform heating, electric cables are best buried in sand placed in the base of the lower gutter.

Square Section Rainwater Gutters

Essentially the same as half round pattern but the flat side facilitates fitting of different insulation materials. Use of mineral fibre Block-Board, having slightly better insulation than elastomeric materials, can be considered and more easily cut to fit a square profile (Figure 16.8).

Plywood or Insulating Boards Fabricated to Shape

Water-resistant plywood 6 or 8mm thick or insulating boards can be cut to dimensions given in Appendix R and nailed or glued together to make the desired channel shape, with insulation fixed externally. Insulating board, e.g. Block-Board, is a convenient material for heat retention and can also be glued or screwed. As with other designs, the meeting edges have foam-insulating strips, between which the grafts are placed. Ideally internal faces should be lined with aluminium foil before fixing the source of heating.

HOT-PIPE – STACKED ASSEMBLIES

Several layers of hot-pipe assemblies can be stacked directly above each other. Two designs are currently in use; one allows assembly of each layer separately, and on completion they are brought together for stacking. This can be done consecutively as each layer is completed or delayed until convenient. The other option is to build the layers one upon the other as an integral unit.

Separate Assembly Stacking

A design for separate assembly comprises a metal pallet base with a steel box-section post in each corner (Figure 16.9). The posts support the next loaded pallet assembly to be placed above.

Material for laying-in roots of bare-root rootstocks is contained by wooden side boards attached to the posts to create a pallet tray. Within this are located pre-formed plastic heating troughs, designed to hold the grafts in position, with insulating material on the base, sides and above to create a heating chamber.

FIGURE 16.8   Insulating Block-Board cut to the correct profile and glued in position. The resultant chamber is lined internally with aluminium foil and electric heating cables, held in position by inverted cable clips. Chamber edges are fitted with pipe lagging, or similar, to provide a closed, heat retentive heating chamber for grafts when 'sandwiched' in position. Suitable weights are placed above to provide a tight heat seal (Penwood Nursery).

FIGURE 16.9 Separate assembly hot-pipe runs on a metal pallet base and with a steel box section support post in each corner. Heating chambers containing electrical heating cables within sand are surrounded by pre-formed plastic moulds to form a heating chamber which holds the grafts. Elastomeric insulation material is visible as corrugated strips to enclose the grafts and retain heat. Flat sheets of the same material are placed above the chamber, held in place by plastic 'lids', visible on the left-hand side of the picture and one section in place, top right-hand corner. Between the heating chambers can be seen the laying-in medium of composted bark (F.P. Matthews Ltd.).

Heating within the chamber is by self-regulating heating cables set in sand, often further controlled by a thermostat (Figure 16.10).

As each pallet tray is a separate unit, this is a more flexible design than the integrated and stacked alternative described below. More than one tray can be assembled at once, and once completed, individual trays can be 'plugged in' to a heat source and left until stacking is convenient. Because of the corner posts and individual pallet bases, this design occupies significantly more space than the alternative integrated stacking.

Integrated Assembly Stacking

Integrated assembly uses a pallet base on which is placed a pallet sized tray, 150–200mm deep. The assembly is built up into layers by placing (stacking) further trays above, once the one below is filled with grafts (Figure 16.11). Each tray provides two heating chambers by means of a pair of suitably spaced vertical boards on either side of the centre. A heating element (electricity or hot water) is placed in position within each pair of boards. Grafts are laid horizontally, next to each other, across the boards and above the heating element. The scions are faced inwards, the ends projecting into the empty gap between the pairs of heating chamber boards at the centre of the assembly; roots (or pots) extend into the void between heating chambers and exterior boards of each tray. The grafts are held firmly in position by trays below and above (Figure 16.11 & 16.12). Insulating strips attached to the edges of the vertical boards surrounding the heating chamber 'sandwich' the grafts between

FIGURE 16.10   LEFT, layout/work area of the separately assembled hot-pipe runs, based on a pallet system. Once filled with grafts, pallets are loaded above each other, supported by the box section corner posts. In the foreground a pallet is shown being filled with grafts above the electrical heating cable buried in sand beneath the graft union. The team of grafters is partially visible in top-middle of the picture. RIGHT, grafts have been positioned across the heat source. Profiled insulation strip is being placed in position before the flat insulation sheet and covering lid (visible between grafts and insulation sheet) is positioned above and over, to cover the grafts and heating chamber, enclosing and insulating the whole. A thermostat box, yet to be positioned, can be seen in the foreground (F.P. Matthews Ltd.).

FIGURE 16.11   LEFT, integrated hot-pipe trays stacked directly above each other on a pallet base. Each layer contains grafts, heating chambers and laying-in material. Hot water pipes are visible projecting from the end wall. Visible on the top layer are the insulating covers over the top chamber held down with concrete slabs and the rootstock root covering material kept moist with a polythene sheet cover. With this design fewer, pipes are necessary towards the top as heat rises from one chamber to the chamber above (John Richards Nursery). RIGHT, similar but not identical integrated stacking equipment with additional steel reinforcing. Showing grafts sealed with Red Wax packed in rows between insulating layers of the internal vertical walls (Verhelste Nursery).

and serve to retain heat in the heating chamber. As each layer is completed, the tray has laying-in material (peat/composted bark, etc.) placed between the heating chambers and exterior boards. This serves to fill the void and at the same time surround and 'lay-in' the rootstock root systems for bare-root gafts or those potted. The next tray is placed in position above and the procedure repeated. Once all layers are stacked into position, the whole assembly is transported to the standing-down area by a fork lift truck and 'plugged in' to the hot water or electricity supply.

FIGURE 16.12   A single-sheet end wall and steel sections strengthen the assembly allowing many layers of grafts to be safely accommodated. A reduced number of heating pipes is necessary towards the top. Flexible pipes carrying hot water and connected to each assembly may be seen on the floor. These integrated stacked hot-pipe units have been moved from the assembly area into an un-heated glasshouse corridor by fork lift trucks (Verhelste Nursery).

Work procedures are less flexible in this system because all layers are assembled in sequence, ideally having sufficient grafts available to allow the whole stack to be completed in one session. In the design pictured on the left of Figure 16.11, each layer has no additional metal reinforcement and consequently eight layers high is a safe working limit. Some designs which have steel section built-in to the construction can contain up to 12 layers of double hot-pipe runs (Figure 16.11 & 16.12).

ALTERNATIVE DESIGN FOR ALIGNED CHAMBERS

Integrated assembly stacked designs result in aligned heating chambers stretching from top to bottom. This configuration provides opportunities to heat the entire chamber from a single heat source. Options for the provision of heat include a heating element at the base which relies on convection aided by small fans to provide good mixing from top to bottom. Another option might be an externally generated hot air supply, ducted to circulate within the chamber. Pressure and output could be gauged to supply several hot-pipe 'stacks' simultaneously.

A further development based on this concept might permit a system of hot-pipe shelves held in a chest-of-drawers or side entry design. This would allow individual units to be slid in and out and permit independent packing of separate shelves, facilitate close monitoring, timely handling and reduce problems with the complexities of temperature and duration of treatment. The engineering and design demands for such a system do not appear to be insuperable.

'ECONOMY' DESIGN

Grafters on the US west coast have devised a simple design for providing focussed heat to the graft union. This involves use of a good insulating material, in this case coarse sawdust, which is in plentiful supply, but fine wood chips, even composted bark, should be acceptable. A 100–150mm × 50mm pressure treated timber plank is placed narrow edge-on, running along the length of the grafting bed at a suitable distance from the path (300–400mm). An electric heating cable or narrow bore hot water pipe is placed to run along the top edge of the plank. Small profile wooden edging (40 × 15mm) fixed to this will help hold the cable or pipe in position and prevent direct graft contact with the heat source. The bed is filled to the top of the plank with sawdust and grafts laid-in with the union directly over the plank. To hold the grafts in position, a batten (50 × 50mm) is laid on top, taking advantage of the insulating value of the batten and sawdust.

Heat cannot be contained within the vicinity of the graft as effectively as when modern insulating materials are used, but in practice it has provided good results in commercial situations. When cheap timber and wood waste materials are available, it offers a simple and economic alternative to conventional hot-pipe facilities (Figure 16.13).

For design drawings for this system, see Appendix S.

SUMMER STORAGE

Single layer systems, particularly solid cover designs, can be lifted from the grafting position and stored when not in use, releasing areas for growing on nursery plants (Figure 16.14).

TEMPERATURE CONTROL FOR HOT-PIPE SYSTEMS

Options for heat control of base heated beds using thermostats have been discussed previously; digital (electronic) thermostats with a wander lead probe are recommended. Probe placement should reflect the precise position of the graft union and some adaptation of the probe may be required to achieve this; a simple method is shown in Figure 16.15.

FIGURE 16.13 Economy hot-pipe system showing layout of two runs in a grafting bed housed within an unheated structure. Graft unions are placed over pressure-treated planks and heat source. Timber battens used to cover and hold the grafts in position have been removed to gain access to the grafts and are shown in the top left-hand corner. Note the close spacing, at approximately 25mm centres, possible with the tap rooted *Quercus* rootstock used for *Quercus* x *bimundorum* 'Crimson Spire' grafts (J. Frank Schmidt Nursery, photo Guy Meacham).

FIGURE 16.14   A solid cover design, single layer, ground level layout, consisting of insulating boards fabricated to shape, used in an unheated growing area from December to April. After this period the installation is lifted and packed under benches to release the previously occupied growing area for the duration of the non-grafting season (Penwood Nursery).

FIGURE 16.15   Placement of thermostat probe in a hot-pipe run may be achieved by attachment to a cross piece (split bamboo cane is suitable), which can be positioned to coincide with that of the graft unions, providing precise temperature measurement.

17 Growth Chambers for Grafts

Recent developments in the provision of economical lighting make it possible to consider replacing warm callusing systems by growth chambers or growth houses.

A suitably insulated building supplied with heating, cooling and humidification equipment lacks only light to efficiently provide all environmental conditions for grafts. Until recently, high electricity consumption has meant such systems are unlikely to justify the running costs of replacing naturally lit structures with artificial illumination. This view may need reappraisal in the light of recent developments in irradiation technology, notably use of LED (light emitting diode) lighting. Power consumption of the LED lamp is about half that of alternatives and has transformed energy costs for horticultural applications.

A further advantage of this lamp is the ability to customise spectral quality to meet requirements for photosynthesis and other plant responses. It has been shown that blue light stimulates stomata to open. By removing blue, transpiration is reduced. No reports have been found on the influence of light quality on graft union development, but it seems possible, given the correct spectral mix, that, at least for leafy grafts, there could be a positive effect. Growth-extending deciduous grafts or leafy grafts, both requiring light levels in the photosynthetically active range (PAR) can be provided with the correct spectral mix by LED lamps*.

Buildings based on those for refrigerated cold stores converted with necessary heating, humidifying and lighting equipment may fulfil the basic requirements for large-scale climate-controlled growth chambers. A more economical option may be found in the use of polytunnels modified for use as mushroom houses. These are well-insulated and vapour sealed but will require heating, humidifying and lighting for conversion to a suitable structure with the potential to offer optimum environmental control for grafts.

This concept may mean that it will be feasible to provide conditions for grafts in energy efficient, non-traditional growing structures with the possibility that in the future such facilities become commonplace.

---

* Davis (2015).

Part Five

Grafting Systems

18 Features and Management of Different Grafting Systems

Building on previous information, the following descriptions are of features and basic procedures for managing a range of grafting systems. These may need some modifications, as outlined in the specific requirements of the genera in Parts Seven and Eight.

The terminology used is an attempt to unify names given to the main systems, sub-cold callusing; cold callusing; warm callusing; and hot-pipe, currently in use. The others described might be regarded as categories or strategies utilising the main systems but for convenience are included here.

SUB-COLD CALLUSING (HIBERNATION CALLUSING)

For some deciduous species, particularly those within Rosaceae, 'sub-cold callusing' utilising refrigerated cold storage facilities has provided a grafting system with potential to augment, even in some instances replace, summer field budding.

It is important that the bare-root rootstocks and scions are not water stressed and totally dormant when grafting commences. Apical methods – normally splice, whip and tongue, apical wedge or inlay – are used. Hand grafting, grafting tool, grafting machine or combinations of these provide high productivity which can, for some genera, exceed that achieved in field budding. Grafts are sealed by dipping in sealant or use of impermeable enclosing ties, often the self-adhesive type. Ties and sealants must be able to withstand the low temperatures and subsequent treatment implicit in the system; Buddy-tape®, Medelfilm® or similar enclosing ties are particularly suitable. Spraying with a fungicidal solution before final packing ensures that, in addition to lowering disease pressures, the risk of dehydration is reduced. However, good judgement is required to avoid water-soaked wood caused by too much free water within the packed grafts.

Once prepared, tied and sealed, grafts are bundled and packed into large polythene bags or polythene-lined boxes, carefully closed to prevent moisture loss. Some grafters use smaller bags (7.5 × 25 × 50cm) which are carefully sealed after filling and packed into wooden crates. These permit easy inspection through the bag if desired*. Grafts are then cold stored at temperatures between −1°C and +0.5°C; although, some suggest temperatures as low as −1.5°C are suitable. At these low temperatures the development of the graft is arrested and placed in a state of 'hibernation'.

Maintaining health of the bare-rooted rootstock relies entirely on maintaining correct environmental conditions. Because of low storage temperatures, moisture requirements are very low and desiccation is avoided by enclosing the grafts in polythene. Cold storage equipment and monitoring must be totally reliable; temperatures significantly lower than those stated will cause irreparable damage and higher will allow unwanted development of the grafts.

When open-air ambient temperatures are suitable for callus and union formation, the next stage is usually direct planting into the field; although, some species may be potted up for container production. In northern latitudes, field planting is usually carried out from late April to mid-May. If necessary, grafts may be retained in store for longer periods and cold, wet, spring weather can delay operations until as late as July. Satisfactory planting conditions require soils with excellent structure and texture; equipment to provide irrigation is essential.

---

* Roller (1973).

COLD CALLUSING – FOR DECIDUOUS SPECIES

Carried out during winter/early spring, the most successful species for cold callusing are those which are able to develop callus tissue and slowly form a union at temperatures as low as ±6°C. Below 5°C, most species produce callus very slowly, or scarcely at all, over a period of some weeks or months*. Many having a higher optimum will often take longer but can still produce good results. Temperatures below 4°C should be offset by installing mobile heaters or an integral heating system. Favoured locations for cold callusing are a north facing shed, an unheated polytunnel or glasshouse (Figure 18.1). The system is equally effective for pot grown or bare-root rootstocks.

Because the grafts are housed in unheated structures, usually a cold greenhouse, cold callusing treatments enable scions to retain complete or near complete bud dormancy throughout the grafting cycle. Consequently, potential water loss caused by soft, extending growth is eliminated. Bark is an effective barrier to moisture loss; sealing only the edges of the union interface and exposed cut surfaces is sufficient to ensure that water content of the scion is maintained. The usual practice of immersion (dipping) of the scion and graft area in sealant is adopted out of convenience rather than need.

After grafting development of the union proceeds slowly and, dependent upon prevailing temperature, completion normally takes six to eight weeks. Some species which callus easily, including many in Rosaceae (Apple, Pear, etc.), may be ready for the next stage after five weeks. Following callus/union formation, most can be cold stored to delay further development until other handling procedures, planting or containerising can be conveniently carried out.

Where long, straight main stems are not required, e.g. *Syringa* and *Viburnum*, cold callusing systems using bare-root grafts are gaining in importance as an alternative to field budding.

Bare-Root Rootstocks

December to March is the normal time for grafting, with early dates usually producing the best results; cold storage of rootstocks and scions extends the period. Grafts are normally placed in cold glasshouses or polytunnels, often on ground level beds. Very rarely the completed grafts may

FIGURE 18.1 Single span polytunnel for cold callusing deciduous grafts. Note the open ends for ventilation which can be closed in cold weather (Witch Hazel Nursery).

---

* Shippy (1930).

be laid-in soil beds directly within the structure, but grafts are usually placed vertically in boxes (Boskoop trays) or pallet bins with the rootstock root system plunged in a suitable, moist, but well-drained, 'laying-in' medium of fibrous peat, with or without added 'opening' material such as perlite, bark or bark peelings* (Figure 18.2).

A wide range of species is involved, often with relatively small numbers; consequently, there is far more emphasis on grafting by hand than machine. Apical grafting methods are normally used: splice, whip and tongue or wedge grafts are most popular. For the more demanding species, apical short or long tongue veneer are sometimes employed, though these are generally reserved for pot grown rootstocks.

Sealing options include enclosing self-sealing ties, which are convenient and becoming increasingly popular, but the majority of grafters favour sealing the graft by dipping in hot wax and laying-in to leave the graft union above the medium. To avoid the need for sealing, some grafters favour plunging the union along with the rootstock root system. To prevent desiccation of the scion for grafts treated this way, the terminal cut will require waxing or painting with cold sealant. Provided the laying-in material is sufficiently 'open' and well drained, grafts can unite well in these conditions. To avoid damage to soft external callus cells after removal, a short period of heavy shading and light misting over may be necessary to acclimatise these soft tissues and allow sufficient suberisation.

POTTED ROOTSTOCKS

Potted rootstocks, with the advantages of a complete undisturbed root system, permit more difficult and demanding species to be successfully grafted, despite being in sub-optimum temperatures. Small capacity pots (9cm or 1 litre) are generally used; large plugs can be considered as an alternative.

Grafts are usually housed in an unheated polytunnel or glasshouse (cold frames can also be used) and require only occasional monitoring. There is always the possibility of rodent attack; in the UK field voles (*Microtus agrestis*) being the main culprit. Routine baiting or trapping is recommended.

FIGURE 18.2   Bare-root apical grafts laid in peat in boxes (Boskoop Trays) in an unheated structure. Note that grafts have been dipped in sealant (paraffin wax) as a convenient method of sealing. Boxes are stood-down on ground cover fabric on well-drained soil substrate (Yorkshire Plants).

* Bradley (1982) and Davis II (1982).

Apical grafting methods are favoured, often splice, but wedge or veneers are possible. Sealing will be required.

With some species, potted rootstocks increase the risk of root pressure. However, low compost temperatures together with no bottom heat causes water uptake to be lower than for plants in warm callus systems, consequently, in most cases, root pressure is sufficiently controlled by withholding water for four to six weeks before grafting takes place. In addition to drying-off, some grafters believe root extension is necessary and, during this drying period, advocate bringing the rootstocks into a warm area at 15–18°C to stimulate root growth.

Temperatures below freezing can cause losses; when temperatures exceed 8–10°C ventilation should be given*. In warm periods, ambient temperature can quickly rise to surprisingly high levels, causing unwanted premature bud development, these may be lowered by additional shade and light spraying or misting.

MODIFIED COLD CALLUSING

This system, involving the use of summer side grafts, has been described for *Acer palmatum* cultivars. After completion the whole graft, including the de-leaved scion, is entirely covered in wax sealant. This is then stood down in a non-humidified, ventilated polytunnel. It is claimed that good results are obtained. It may be successful for other summer grafted species with similar requirements to *Acer palmatum*, such as *Cornus*.

WARM CALLUSING

Warm callusing is the term used when grafts are placed within a high humidity, enclosed, supportive environment, normally provided by poly-tents, walk-in enclosures, or free standing walk-in polytunnels; these conditions are essential for the following types of grafts:

- The grafts are dormant but due to high temperatures, induced to produce extension growth before union formation.
- Deciduous species in leaf.
- Evergreen.
- De-leaved.
- Left unsealed.

During the low ambient temperatures and light levels of winter and early spring, temperature differences between warm and cold callusing may be marginal. Nevertheless, due to the influence of light, enclosed conditions always have elevated temperatures compared with those unenclosed, hence use of the term.

Most warm callus systems have an added heat source, often provided in the form of basal (bottom) heat. For many species, warm callusing systems providing higher ambient temperatures, at or close to optimum, give better results than are possible with cold callusing.

Rather than enclosing the grafts, humidity may be maintained by using equipment such as humidifiers (dry fog), misting jets or irrigation sprinklers. In this publication, to distinguish these methods from conventional warm callusing, they are called supported warm callusing, to be discussed further.

---

* Lane (1982) and Upchurch (2009).

Dormant Hardwood Species

Placing dormant deciduous grafts in warm conditions invariably stimulates extension growth. To avoid desiccation of the soft, un-ripened shoots until union formation is complete, a supportive environment using warm callusing is the best strategy. The alternative of providing no enclosure means frequent monitoring involving shade, spraying-over, etc., is more labour intensive and usually less successful.

Normal timing for the system is between December and mid-March. Species which break dormancy late in the season, such as *Cercis*, *Fagus* and *Quercus*, allow grafting dates to be extended to late March or early April. As a general rule, late collected scion wood is less successful, but cold storage allows scion wood to be collected at an optimum time when completely dormant and, if necessary, grafting can be delayed to allow busy schedules to be completed.

Apical grafts are the predominant method, but some favour side grafts on the assumption that root pressure is reduced. Experience indicates that rootstocks insufficiently dried-off are able to 'drown' graft unions regardless of whether apical or side grafts are used. However, a stem above the graft union may have some mitigating effect.

Until the advent of hot-pipe, warm callusing was the most successful system for those species which develop callus most effectively in the 15–28°C temperature range. Some species including *Cercis*, *Juglans* and *Vitis* perform best at 25–27°C. Others such as *Corylus*, *Liriodendron*, *Magnolia* and *Quercus* respond favourably to temperatures between 22 and 24°C. Many species achieve optimum callus development at temperatures of 20–22°C. Shading and ventilation will require careful management during the winter/early spring months when temperatures can increase surprisingly quickly under protected conditions. Well-controlled shading allows more flexibility in ventilation timing and duration. Good callus development, combined with adequate suberisation, will allow continuous night time ventilation and eventual hardening-off.

For a limited but important range of species including Grapevines, a supportive environment may be achieved by burying the graft to the top bud of the scion in a moist, open, freely drained medium, maintained at temperatures between 25–28°C. Only the exposed cut surface at the apex of the scion requires sealing. Grafts are often packed in purpose made 'callusing boxes' designed to provide easy access for packing and removal of the grafts* (see *Vitis*, Chapter 55 for further details).

Deciduous Hardwoods in Summer Dormancy

Despite the advent of hot-pipe systems, grafting of summer dormant deciduous species (summer grafting) is a favoured method to achieve good takes for many of the more difficult types. This is due to a combination of seasonal reduction in root pressure, good physiological activity and suitable temperatures. For some species, such as Japanese Maples, summer grafting remains an important system for commercial producers.

Summer dormant grafts require a supportive environment, provided by warm callusing or supported warm callusing, and temperatures above a minimum of 15°C. In enclosed environments, sunlight can cause excessively high temperatures, a potential problem which must be controlled by heavy shading. Grafting methods are predominantly side grafts, usually short or long tongue veneers.

Two major decisions to be made are whether or not to retain scion leaves, and whether or not to seal the graft union. Removal or retention of scion leaves has been discussed in detail in Chapter 10. Many species, such as Japanese Maples, respond well to non-sealing combined with leaf removal.

Supported warm callusing systems, utilising mist or fog, can avoid, to an extent, the problem of excessive temperatures, always a problem with conventional warm callusing systems. Unless dry fog is available, they do require grafts to be sealed to prevent water ingress into the union.

Consideration must be given to handling and transporting grafts from the work station to the final propagation standing down area. Safest procedure for all summer grafts, but particularly those left

---

* Garner (2013) p.208.

unsealed is provision of a high humidity 'holding tent' adjacent to the grafting bench, as described in Chapter 14.

A supportive environment is normally provided by contact polythene or a poly-tent, but when large commercial runs are involved, polythene enclosures, or walk-in grafting polytunnels, provide opportunities for easier handling and monitoring. Dry fog humidifiers appear to have received little attention but potentially offer an efficient means of achieving controlled humidity.

Once fully united and acclimatised, grafts are maintained under normal growing conditions until leaf-drop and dormancy. For many species, overwintering involves protection from freezing as relatively young tissue within the graft union may still be vulnerable. Experience may determine that some species are sufficiently hardy to be over-wintered in unheated structures.

Deciduous Hardwood Species in Active Growth (Early/Mid-Summer Grafting)

This procedure is carried out in early/mid-summer (late May-early July) for a few species, notably deciduous Azaleas (Rhododendron subgenus Pentanthera). There is a requirement for skilful management of the environment, notably shade, to avoid excessive temperatures. Ambient temperatures are normally sufficient but supplementary heating set at a minimum of 18°C may be required during cool weather and at night-time. Side grafts are the best choice, as transfer of metabolites from rootstock to scion will aid rapid growth and development.

Pot grown rootstocks are in an active state, and 1+1P 4–6mm are normally most suitable. Scion wood is collected in a soft, semi-mature condition and requires immediate use, or only short-term storage in polythene bags under cool conditions. The wood has comparatively low food reserves, and some foliage must be retained to enable continuing photosynthesis. Once on the grafting bench, scions are best kept in moistened polythene bags until grafted.

Completed grafts must not be left standing on the grafting bench but immediately transported in covered carriers to their housing position or more conveniently held in a holding tent before subsequent transportation in bulk numbers. Holding tent and housing position must be heavily shaded and maintained at virtually 100% humidity. When combined with high temperatures under these conditions, callusing and union formation occur rapidly.

Dormant Evergreen Hardwoods

Grafting is carried out in the period of late December to March, the earlier/mid-dates generally giving best results. Side grafts are the most popular method as it is considered disadvantageous to remove the evergreen top from the rootstock. In the majority of systems, pot grown rootstocks are used.

All species require a supportive environment, normally provided by poly-tents. In warm temperatures above 15°C, humidity levels must be maintained close to 100%. Optimum levels have not been established but in cool conditions (<15°C), after an acclimatisation period of five to ten days, lower humidity and increased ventilation may be safe for those species which respond to such conditions, e.g. *Elaeagnus pungens* cvs; *Mahonia fremontii*; *M. nevinii*.

Supplementary heat sources will be required in temperate climates and usually bottom heat is installed. Shade levels should follow those given previously, but for many species grafted in cool conditions, callus formation is slow. During warm, sunny conditions later in the season, adjustment of shade and ventilation will become more demanding.

Provision of drip-free conditions avoids the need to seal the graft union, but some favour sealing or burying the union in a suitable medium such a damp peat, or mixtures such as peat/perlite.

Summer Dormant Evergreen Hardwoods

Certain hardwood evergreens respond well to grafting using current year's scion wood which has entered the summer dormant phase, best described as well ripened, semi-mature material.

Timing is normally from late summer to early autumn. As with summer grafting of deciduous species, the system merits wider use. Some of the more difficult genera (e.g. *Kalmia*) can often produce much improved results, compared with conventional methods involving grafting dormant material in late winter (Figure 18.3). Further examples are given in Parts Seven and Eight.

Relatively high light levels and temperatures during the summer demand more attention to environmental management than for dormant material grafted earlier. Use of side grafts is normal; the long tongue side veneer incorporating a 'stimulant bud' is recommended. Grafts are usually left unsealed.

Shade levels are as described previously. It is important, with unsealed grafts, that 'non-drip' polythene covers are used. Humidity levels need to be high, close to 100%. Ambient temperatures are normally sufficient, and limiting rather than increasing the temperature is often the main problem; avoid sub-optimum temperatures (<18°C) by provision of supplementary heat. Callus development and graft union formation are faster with this system than for fully dormant material. Even species which normally develop slowly may show signs of callus in 14 to 21 days. After callus and union formation, care is required to harden-off successful grafts which may have commenced bud break and extension growth. The following winter should be spent in frost-free conditions.

SUMMER ACTIVE EVERGREEN HARDWOODS

This system is restricted to very few genera, *Daphne* being the most important. Requirements are close to those described previously for deciduous species in active growth.

Grafting commences early, normally late June to mid-July. Scions are selected using only current season's growth while still active at the soft, 'semi-ripe' stage. The extending tip is removed to leave slightly firmer wood. Pot grown rootstocks are also active and usually side grafted, the extending terminal growth trimmed into firmer wood.

Scion collection is best in the early morning prior to grafting, scions being stored in dampened polythene bags tightly closed once full and placed in a moistened sack, or similar material. Any

FIGURE 18.3   *Kalmia latifolia* 'Pinwheel' grafted to *K. latifolia* seedling in mid-August. LEFT, pictured after partial heading-back the following March. RIGHT, 2-year-old graft of the same variety.

delay requires them to be kept in cool conditions at temperatures of ±5°C where they will remain usable for two or three days. Rootstocks must not wilt and attention to watering is necessary. Root pressure does not appear to be an issue.

After grafting and before placement into a holding tent, it is advisable to mist over with hand-held 'misting bottles' to prevent leafy scions showing any sign of wilting. When transporting these grafts from grafting area to grafting structures, well-covered, heavily shaded carriers must be used.

Due to the warming effects of light, high temperatures, in the 28–32°C range, can occur. Heavy shading is essential to reduce these to a more acceptable 25–27°C. Once placed in these conditions, some callus formation may appear after only five days. Little ventilation is required for the initial five to ten days; following this, early morning or evening periods should be chosen for ventilating, initially lasting for 30–60 minutes. Eventually, after suberisation, night-long ventilation is appropriate and subsequent morning and evening ventilation periods can be significantly extended. Hardening-off requires careful management and is a demanding procedure unless automated dry fog facilities are available.

SUPPORTED WARM CALLUSING SYSTEMS

This term is used to describe situations where an enclosed structure is replaced by a supportive environment provided by hand misting, wet or dry fogging, automated misting, or irrigation sprinklers. Certain nurseries in the USA have success using wide-span glasshouses or polytunnels with summer dormant grafts standing under mist or irrigation jets.

Some grafters favour burying the union in a well-drained 'open' medium with overhead mist or fog for certain hardwood species, and selected conifer species (*Pinus* etc.), grafted in the late winter/spring or summer period. Unless graft unions are buried, methods involving wet fog, mist jets or irrigation sprinklers require careful sealing; left unsealed, water ingress between the graft components is unavoidable and invariably causes losses. Dry fog heads provide high humidity without free water and for these sealing is not necessary.

CONIFER GRAFTING

Conifers grafted in temperate areas contain seven families and over 50 genera. Systems divide between those carried out in the winter/early spring using dormant material and those grafted during the summer dormancy stage (August–October). The families and genera together with suggested timings are listed in Table 18.1.

Evergreen conifer scions are more resistant to desiccation than evergreen hardwoods but still require a supportive environment. This is usually provided by a poly-tent, and they are therefore categorised as warm callused, despite many having a low-temperature requirement. The large majority of evergreen species are side grafted; a few grafters, particularly in Scandinavia, favour apical grafts, often using bare-root rootstocks.

There is a divergence of opinion among grafters on the need to dry-off conifers. Some see no necessity[*] while others disagree preferring to control sap flow[†]. In the Netherlands it is standard practice when winter grafting *Picea pungens* cultivars to dry-off the rootstocks and, when summer grafting, to have them 'on the dry side'. It seems wise to avoid extremely wet or dry conifer rootstocks during the dormant grafting period. Summer grafted conifers should not be excessively dried, as there is always a danger of desiccation

The majority of grafters state that conifer rootstocks should be showing active roots before grafting takes place[‡]. Others recommend root activity on a selective basis so that winter grafted

---

[*] Hatch (1982) and Gregg (2013).
[†] Garner (2013) p.241 and Alkemede & van Elk (1989) p.114.
[‡] Sheat (1948) p.414, Hartmann, Kester, Davies Jr. et al (2014) p.454, Macdonald (1988) and Klapis Jr. (1964).

TABLE 18.1

Conifer Families and Genera – Grafting Times*

CUPRESSACEAE	CUPRESSACEAE	PINACEAE	PODOCARPACEAE
Arthrotaxis (S&W)	Libocedrus (S&W)	Abies (W&S)	Afrocarpus (W)
Austrocedrus (S&W)	Metasequoia (W)	Cathaya (W)	Dacrydocarpus (W)
Callitris (W)	Platycladus (S&W)	Cedrus (W&S)	Dacrydium (W)
Calocedrus (S&W)	Sequoia (W&S)	Keteleeria (S&W)	Halocarpus (W)
Chamaecyparis (S&W)	Sequoiadendron (W&S)	Larix (W)	Lagerostrobus (W)
Cryptomeria (W)	Taiwania (W)	Picea (W&S)	Lepidothamnus (W)
Cunninghamia (W)	Taxodium (W)	Pinus (W)	Podocarpus (W)
Cupressus (S&W)	Tetraclinis (S&W)	Pseudolarix (W)	Prumnopitys (W)
x Cuprocyparis (S&W)	Thuja ((S&W)	Pseudotsuga (W)	Saxegotheca (W)
Fitzroyia (W)	Thujopsis (S&W)	Tsuga (W)	AURUCARIACEAE
Fokenia (S&W)	Widdringtonia (S&W)	TAXACEAE	Aurucaria (W)
Glyptostrobus (W)		Cephalotaxus (W)	Agathis (W)
	Xanthocyparis (S&W)	Taxus (W)	Wollemia (W)
Juniperus (W)	SCIADOPITYACEAE	Torreya (W)	GINKGOACEAE
	Sciadopitys (W)		Ginkgo (W)

*The following code is used: W = winter grafting; S = summer dormancy grafting, with the first letter shown taking priority. Despite suggestions for summer grafting, many restrict their activities solely to the winter period.

Chamaecyparis, *Juniperus* and *Pinus* are pre-warmed to encourage root activity, whereas *Picea* are grafted on totally dormant rootstocks*. Personal experience has not indicated a need for root activity before grafting commences; although, due to raised temperatures, it is normal for root extension to begin at an early stage during graft union development.

DECIDUOUS CONIFERS

Regardless of their family, deciduous conifers are grafted in the dormant state, normally using apical grafting methods, and environments described for cold callusing deciduous hardwoods.

EVERGREEN CONIFERS

For evergreen species there are broad differences in temperature requirements between the families. Pinaceae and Sciadopityaceae are best in lower temperatures while those in Cupressaceae are best at higher. Aurucariaceae, Podocarpaceae and Taxaceae are an intermediate group.

Pinaceae and Sciadopityaceae

Pinaceae and Sciadopityaceae winter/early spring grafts are placed either in poly-tents with covers raised 45–90cm above the grafts, or in well-sealed walk-in tunnels. Despite the major genera having low requirements, temperatures should not be allowed to fall below 6°C for extended periods. Side grafts are invariably used for the evergreen species. The high resin content of the wood and resulting exudation means that sealing the graft union is not normally necessary.

Pinaceae (particularly *Pinus*) have a high light requirement and shading levels are best using those previously recommended for cuttings. Sciadopityaceae will tolerate lower light levels. Frequency of ventilation will depend upon conditions discussed previously but most will withstand more

* Decker (1998).

ventilation than other species. Some recommend use of ventilation whenever temperatures escalate. Experience has shown this can result in excessively low humidity and grafts with an uncompleted union slowly lose moisture, leading to losses. To avoid excessive shading, misting or fogging and damping-down surrounding areas to reduce temperature is a better option.

At low temperatures, graft union development is slow and callus development is normally not visible for 40–60 days, or longer. A further 30+ days is required for further significant development. Once this is completed, hardening-off procedures are used as described previously.

Cedrus, *Cathaya*, *Keteleeria* and *Tsuga* respond better when temperatures do not fall below 10°C, but recommendations regarding raised poly-tunnel covers still apply.

Cupressaceae

Cupressaceae species may be split into three groups:

1. Deciduous species (*Metasequoia*, *Taxodium*) requiring cold callusing as for hardwoods.
2. *Cupressus*, *Cryptomeria*, *Juniperus*, *Sequoia*, *Sequoiadendron*, and *Taiwania*: these have requirements for higher temperatures, minimum 15°C. Some grafters suggest much higher temperatures (24°C) for certain species within the group, e.g. *Juniperus*[*]; however, at these temperatures, problems can quickly develop. Sealing is optional, and sometimes favoured is the alternative of burying the union and maintaining supported warm callusing conditions by overhead misting controlled by a time-clock[†]. To avoid excessive humidity and drips *Cupressus* and Juniper grafts can be placed under low covers of thermal fleece rather than polythene[‡] (Figure 18.4).
3. Species with flattened, spray-like branches with scaly leaves (*Chamaecyparis*, etc.) can be grafted using dormant material with similar environmental requirements, though at temperatures 2 or 3 degrees lower than group (2). Grafts are normally sealed. Temperatures

FIGURE 18.4 Grafting benches containing winter/early spring *Cupressus* and *Juniperus* grafts covered with contact thermal fleece (John Richards Nursery).

---

* Savella (1977).
† Hall (1977).
‡ J. Richards (pers. comm.).

should not be allowed to fall below 10°C. These genera are susceptible to disease caused by excessive humidity and condensate drips. Ventilation must be provided on all possible occasions in tandem with light levels recommended for cuttings.

An alternative and successful strategy for this group is to graft in the summer dormancy phase during mid-August to late September (see below: Summer Dormant Species – Summer Grafting). Dutch grafters favour later timing, October/early November are suggested.

Aurucariaceae, Podocarpaceae and Taxaceae

The remaining families (Aurucariaceae, Podocarpaceae and Taxaceae) are grafted using dormant scion wood in late winter/early spring. If possible, a minimum temperature of 12°C should be maintained. They are more tolerant of higher humidity than the other families and grafts are normally sealed.

SUMMER DORMANT CONIFERS

Due to labour demands during the summer, this system is now less often used, but is very successful for achieving good results for the genera listed above with priority 'S'. It has been the favoured method for producing Blue Spruce (*Picea pungens* cultivars) in the Boskoop nurseries of the Netherlands. The grafting period is normally between mid-August and mid-September. Side grafts are usual, and sealing is not essential provided suitably close conditions can be maintained.

For species in Pinaceae having well-defined growth phases and prominent over-wintering buds (e.g. *Abies*, *Picea*, etc.), extension growth from the apical and lateral buds in the spring following summer grafting is said to be stronger and more uniform than when produced from late winter/early spring grafts.

HOT-PIPE (HOT CALLUSING) – HORTICULTURAL MANAGEMENT – PLANT MATERIAL

The term hot callusing is commonly applied to this system, but "hot-pipe" reflects more accurately its design and construction (described in Part Four); it also avoids any confusion with the term warm callusing and is adopted in this publication.

Aspects covered here are concerned with the practicalities of managing the system and plant material and intended to compliment the previous discussion of hot-pipe design and construction in Chapter 16 and Appendices R and S.

The plant material used in hot-pipe systems comprises dormant rootstocks and dormant scions of deciduous species. A real advantage of the system is that drying-off and inducing pre-grafting activity in the rootstock is of less importance and can be reduced, or ignored, for many species.

The hot-pipe system has improved the growth and development of plants grafted to bare-root rootstocks to a point where they perform almost as well as their pot grown equivalents. It has provided a reliable means of achieving good results for many difficult deciduous species, often being superior even to those obtained from warm callusing. Because of the quick response and fast 'turn-round' it also can have a place in the production of easily grafted subjects.

Satisfactory structures for housing hot-pipe installations may be one of several options: open-air conditions protected from rainfall, storage sheds, unheated growing structures or cold stores (see Temperature and Timing and Figure 18.11). To fully maintain dormancy, it is important that they are located in cold conditions. External temperatures at a minimum of +5°C are suggested*, but lower temperatures of ±1.0°C would better retain dormancy and conserve food reserves. Experience here is that temperatures slightly above freezing have proved satisfactory for many species (Figure 18.5 & 18.6).

---

* Ramsbottom & Toogood (1999) and Hartmann, Kester, Davies Jr. et al. (2014) pp.548–549.

FIGURE 18.5   A hot-pipe system placed on outside of a nursery shed (north side). With polythene curtain lifted in one section to show construction. Insulation is provided by strips of glass fibre wool overlying the heating chamber. Grafts are placed on either side of the heating chamber to increase capacity (Sandy Lane Nursery).

FIGURE 18.6   Hot-pipe system installed in open-sided structure, photographed after snow-fall; ambient is 1.0°C. Hot callus run set at 18°C for *Betula*. Temperature on the thermometer is reading 17°C but may well be 19°C on occasion. This is within normal parameters. Results were not adversely affected by low external ambient temperatures.

It is important that in low ambient temperatures, insulation of the heating chamber is efficient; loss of heat from this area results in failure to maintain optimum temperature for union formation and higher heating costs. Provided insulation is good, most would agree that energy costs are lower for hot callus systems than for warm callusing facilities.

Apart from the capital cost of building and utilising equipment for specific use over a rather short time period, the major disadvantage of hot-pipe (hot callusing) is a lack of suitability for evergreen species. Strategies for overcoming this limitation have been the use of polythene bags for separate wrapping of individual grafts; however, adequately sealing the graft union is difficult. The small volume of air within each bag makes management of environmental factors problematic.

Housing the whole hot-pipe installation within a dry fog environment may provide an alternative solution for evergreens, but as enclosed conditions are required, difficulties arise in maintaining suitable low temperatures outside the heated chamber.

OUTPUT AND CAPACITY

Commercial considerations determine the need for maximum output to achieve necessary returns on the investment. For most species, time required for development of a reliable graft union is less than with conventional grafting systems. Graft unions for some species can be adequate in 14–18 days[*]; for others 21–28 days are required. Placing the first batch of grafts in position in mid-December for removal in early January offers time for a further three batches. Inclusion of late breaking species, such as *Fagus* and *Quercus*, permits a total of five batches from mid-December to late March. Cold storage of rootstocks and scions can further extend this period and output.

Spacing in the hot-pipe channel is the important factor in determining capacity. Distances between grafts in chambers with slotted placements are pre-determined; for designs allowing the graft to be sandwiched between foam insulation, spacing is mainly influenced by rootstock specifications. Three options are available: bare-root, potted or plug grown. Hard-pruned, bare-root grafts permit spacing as little as 25mm, but with normal trimming 50mm is more usual. A system using large transplanted bare-root rootstocks of *Betula* and *Fagus* (8–10mm diam.) gave a distance of 42mm apart in a 160-metre pipe run[†].

9cm square pots theoretically give 90mm centre to centre, but in practice, unless the pot is removed, rarely less than 100mm. Pot removal and some squeezing together can produce a spacing of 87mm centre to centre. If construction and design allow, when rootstocks in large pots are used, extra capacity may be achieved by placing grafts opposite each other on either side of the heating chamber (Figure 18.5). Spacing for plug grown rootstocks reflects plug size but is usually closer than is possible with potted rootstocks.

Total area occupied is often an important issue. To keep requirements to a minimum, stacked or shelved designs greatly improve holding capacity. A six-shelved design gave a Figure of 5000 grafts occupying 40 square metres[‡] (125/sq. metre). Integral stacked designs are more space efficient and graft capacity is approximately 250/sq. metre. Graft capacities based on volume occupied provides an accurate measurement, but few examples are available. One given is 120 grafts per cu. metre based on two tiers high[§].

ROOTSTOCKS

More uniformity of growth, better establishment and vigour are obtained from scions worked on pot grown or plug raised rootstocks compared with those bare-rooted[¶], but the differences are less using hot-pipe than traditional cold callused systems.

7cm or 9cm square pots are usual, but plug grown rootstocks are becoming more popular because they combine the potential for close spacing with desirable characteristics of pot grown plants. Plug cell sizes are normally in the 25–45mm × 100–120mm deep range.

Use of bare-root rootstocks is a popular and important strategy for hot-pipe systems. Optimum sizes are influenced by species. To enable close spacing for maximum capacity, easily transplanted

---

[*] Ramsbottom & Toogood (1999) and Hewson (2012) p.70.
[†] Dunn (1995).
[‡] Berg (1999).
[§] A. Wright (pers. comm.).
[¶] C. Verhelst (pers. comm.).

types can be heavily root pruned; *Actinidia, Cotoneaster, Crataegus, Euonymus, Hibiscus, Ligustrum, Lonicera, Populus, Rosa*, Salix, *Sambucus* and *Vitis* are all suitable. For tree forming species, root pruning should be limited to allow best spacing but leaving sufficient to support vigorous growth.

Bare-root rootstocks in hot-pipe runs require protection to ensure roots do not dry out. Hot-pipe system design must provide available space for roots and covering medium (Figure 18.7). This is normally achieved by burying root systems in a moist but free-draining medium, such as composted bark. To save weight and speed-up handling, some grafters opt for covering roots in lightweight capillary cloth (Figure 18.8). This requires routine damping once or twice weekly or a water trough/ wicked system to keep it sufficiently moist. Polythene is also suitable, particularly for de-potted pot grown rootstocks (Figure 18.11 – see Temperature and Timing below).

Water requirements for potted rootstocks will be influenced by species and ambient temperatures outside the heated channel. Species susceptible to root pressure problems are always a concern, particularly when external ambient temperatures rise to 10°C or above. For these species, rootstocks should be dried back to 'just moist' and maintained at that level for the duration of hot-pipe treatment. Pot grown rootstocks lying on their side are virtually impossible to water. Control of pot root-ball moisture is most easily achieved if pots are removed and the root-balls protected from excessive drying, as for bare-root grafts, by covering with light grade capillary cloth or polythene film.

This procedure enables close monitoring of the root-ball, which can be further moistened or dried-out by applying water or removing covers as required. For less demanding species, the usual procedure is to water the pots 7–14 before grafting.

Grafting Methods

Apical grafting methods, such as splice, whip and tongue or wedge, are normally used. Apical methods enable easy graft placement in facilities using slotted holding positions and layouts allowing opposite graft placement on either side of the heating chamber. The technical advantages of

FIGURE 18.7   Hot-pipe design to accommodate root system of bare-root grafts. After positioning the grafts, roots are covered with composted bark, as shown in the foreground. Scions and graft union have been dipped in FPM green grafting wax (F.P. Matthews Ltd.).

FIGURE 18.8 Hot-pipe runs located on a shelving system within an unheated shed. Bare-root and de-potted root-balls are covered with damp capillary matting, kept moist by regular spraying. De-potted root-balls are just visible on the second shelf from bottom (bottom left-hand side) (Yorkshire Plants).

apical veneer grafts over splice have been previously discussed and are particularly relevant with demanding species such as *Acer* and *Cercis*.

Chamber designs using grafts sandwiched between foam rather than slotted insulation more easily permit the use of side grafts, even multi-grafts, and for some demanding species this can prove advantageous. When considering the use of side grafts, the importance of scion/rootstock placement, to ensure a close fit of the hot-pipe chamber components as described in Chapter 4, must not be overlooked.

All scions will need to be placed slightly higher on the rootstock to take into account the width of the surrounding insulation. This will allow placement of the union centrally within the chamber (see Figure 16.2 in Chapter 16). After tying in the graft, moisture loss through the union and other cut surfaces is prevented in apical grafts by dipping into heated wax sealant to a point about 10–20mm below the union. Some grafters may prefer to use an alternative such as polythene tape or self-adhesive enclosing ties. For side grafts and certain genera, particularly species with large buds such as *Magnolia*, grafters may apply wax sealants by brush.

Rootstock and scion buds will become active and often extend in the heated chamber (Figure 18.9). Some see this as no problem provided grafts are potted after removal and kept frost free[*]; others prevent this by removing buds located in the area before sealing[†]. Presence of soft, sucker growth is a danger if the next stage after removal from the hot-pipe involves direct field planting.

TEMPERATURE AND TIMING

It is important that heating chamber runs are level; slight rises or falls can cause temperature differentials along the run. Systems using slot holders must have the slot carefully covered or heat will be lost through any gap. The same limitations apply to systems involving sponge interface closures, which may require additional pressure to close gaps formed by large diameter rootstock/scion wood.

---

[*] Dunn (1995).
[†] Byrne & Byrne (2004).

FIGURE 18.9   Buds within heating chamber often break dormancy and grow out. LEFT, a range of species showing bud extension in the heated channel. Note the close spacing of heavily root-trimmed bare-root grafts. RIGHT, *Prunus cyclamina* with extending rootstock bud.

Detailed recommendations are given in the literature for precise temperature requirements and lengths of exposure for a range of species in hot-pipe systems. It is important to realise that speed of callus formation is greatly influenced by temperature*. Grafts in low temperatures will take longer for callus to be produced than in higher temperatures. The overriding influence on callus production appears to be the interaction between temperature and time of exposure. Suggestions that temperatures and times of exposure must be closely adhered to are not borne out by practical experience.

Conventional design of hot-pipe heating facilities makes it difficult to provide the recommended wide range of different temperature regimes. A practical approach is to select a suitable mean temperature and apply this to virtually all species; ideally, some will need to remain in place for longer than others. The optimum temperature for the widest range of species is open to debate but for most 21°C would be a good choice.

For systems utilising grafts stacked in integral layers, extending or reducing timings for different layers may present difficulties. A possible solution may be the use of 'chest of drawers' construction discussed in Chapter 16. At present a pragmatic approach is to group together subjects with the most likely similar temperature/timing requirements and treat the whole batch identically. Grafting a wide range of species required in small numbers is better achieved by single layer installations where grafts can with relative ease be inspected, removed or left longer, as appropriate (Figure 18.10).

Temperatures external to the heating chamber have been discussed previously, and the low temperatures recommended can be provided most accurately by housing the hot-pipe system in cold storage facilities (Figure 18.11).

COLD STORAGE POST-GRAFT UNION FORMATION

Grafts produced in hot-pipe systems are well suited to subsequent cold storage. If this is the intention, it is best to avoid premature bud extension within the heated channel by removing all scion and rootstock buds in close proximity to the graft union before tying and sealing. A short period of acclimatisation (at least four days) may be advisable before placement in the cold store.

TOP-WORKING

Added value of grafted plants may be achieved by placing scions high on the rootstock stem, a system known as top-working. For various selections including pendulous types, it provides a means of achieving production of shapely specimens in a fraction of the time required to train low-worked

---

* Hartmann, Kester, Davies Jr. et al. (2014) p.453.

FIGURE 18.10 Simple, single-run hot-pipe layout for small scale production. Grafts (pot grown rootstocks) sandwiched between foam insulation layers within an extruded polystyrene heating chamber. A moveable plastic junction box in the foreground houses an electronic thermostat for controlling the heating cable.

FIGURE 18.11 Hot-pipe system installed inside a cold store. Note pot grown rootstocks de-potted enclosed in polythene to maintain moisture content (John Richards Nursery).

or cutting-raised plants to a similar size and shape. In recent years, fashion has created a significant demand for this type of plant.

Popular among these is the so-called 'Patio Plant', often a top-worked specimen on a relatively low stem of 400–800mm. Suitable species include relatively easily grafted deciduous hardwood species, e.g. *Cotoneaster*, *Photinia*, *Pyracantha*, evergreen *Euonymus*, dwarf and weeping *Salix* and certain deciduous conifers, *Gingko* and weeping *Larix*. Some growers extend their range to include more demanding types such as evergreen conifers, normally dwarf, prostrate or weeping forms (Figure 18.12), as well as species such as Japanese Maple (*Acer palmatum* cvs) and *Rhododendron*. For these, a high selling price limits numbers required. See also, Genera Specific Requirements in Parts Seven and Eight.

Choice of grafting method depends on vigour of the scion variety and required specifications for the finished article. For strong growing species, a single apical graft at a specified height is sufficient. Higher quality, full-headed plants may be more quickly produced by using multi-grafting techniques. This may be extended to include so-called 'novelty patio plants' where more than one variety is grafted onto a single stem.

To add value and because the finished article tends to be top-heavy and unstable, some growers favour finishing plants in large pots*. Producing plants of this quality can involve an apical graft with one or two side grafts (usually side wedge graft) all placed to leave a clear stem height to required specification.

The easily grafted deciduous species respond to cold callusing but completely or partially evergreen hardwood species (*Cotoneaster*, *Euonymus*, *Elaeagnus*, *Viburnum*, etc.) require a supportive environment. A tightly closed polytunnel, ideally double skinned, supplemented during sunny periods by overhead spraying and appropriate shading provides suitable conditions for the relatively easily grafted types (Figure 18.13). Demanding species such as *Rhododendron* require fully supportive enclosed conditions. Top-worked evergreen conifers within the winter grafting group will need tall poly-tents or walk-in facilities; some grafters enclose grafts individually in polythene bags (Figure 18.14).

FIGURE 18.12   Top-worked weeping *Cotoneaster*, *Picea* and *Pinus*, produced as 'Patio plants' (Yorkshire Plants).

---

* Berg (1999).

FIGURE 18.13   Semi-evergreen, weeping *Cotoneaster* top-worked onto stems of *C. frigidus* or suitable alternative. Enclosed conditions in polytunnel achieved by use of 'bubble poly' to close off vents and doors at far end. Some additional damping-down and/or shade may be necessary during sunny periods (Yorkshire Plants).

FIGURE 18.14   Top-worked *Pinus* grafts individually enclosed in polythene bags within a structure (Kilworth Conifers).

FRAME-WORKING

A technique known as frame-working, normally used in the field for fruit trees and very similar in most respects to top-working, can be utilised under protection to speed-up production of large high value specimen plants, such as weaker growing *Acer palmatum* cultivars, Filigree, Ukigomo, etc. This involves use of large, well-furnished rootstocks side grafted with an appropriate number of scions, often between 5 and 15, suitably placed to provide a balanced result (Figure 18.15).

FIGURE 18.15   *Acer shirasawnum* 'Autumn Moon' grafted to a large *A. palmatum* rootstock. Frame-working using approximately 10 scions was carried out late summer 2013; photographed after two seasons' growth, May 2016. By 2017, the plant was of substantial size.

ENHANCED SCION PRODUCTION

Shortages of scion wood, especially problematic with new cultivars of commercial potential, may often be overcome by top grafting onto large vigorous rootstocks. This produces scion wood more rapidly and plentifully than alternative methods. As plant growth is cumulative, large plants bulk-up more quickly than small ones. Figure 18.15 indicates the potential for scion wood production of a Japanese Maple cultivar, using a frame-worked branched rootstock.

DOUBLE-WORKING

A portion of inter-stem, normally 50–100mm long, is grafted to the rootstock; subsequently, the scion variety is grafted to the inter-stem. Double-working has an important role for certain fruit and nut crops, where it may be used to aid the production of specific tree sizes (usually dwarf or semi dwarf types) or overcome difficulties allied to compatibility while at the same time retaining a robust, effective root system.

Use of double-working for addressing the problems of incompatibility is fully discussed in Chapter 12: Future Progress - Strategies to Overcome Incompatibility, as well as specific in examples in Genera Specific Requirements in Parts Seven and Eight. For commercial use, an important aspect will be whether success can be achieved if the graft the components are grafted at the same time or delay is necessary by grafting the scion variety to the inter-stem a season later. Experience will determine the outcome; most likely species will differ in their response.

MULTI-GRAFTS

Multi-grafts are relevant for potentially difficult species, uncertain combinations with regard to compatibility or poor-quality scions, when a number of scions may be placed on a single rootstock to keep wastage to a minimum. Those which succeed can subsequently be removed and re-grafted next season to produce single plants (Figure 18.16).

Use of multi-stem grafting can be applied to bottom-worked species grown to provide a bush form. For fast-growing species the extra time taken in producing additional grafts may not be warranted, but for weak, slow growing, high-value types, there is an under-exploited potential.

Side veneer or wedge grafts positioned more or less opposite on the rootstock stem has the potential to produce a saleable, well-balanced plant in half the time of conventional systems (Figure 18.17). Opposite placement may require simultaneous tying-in of both scions.

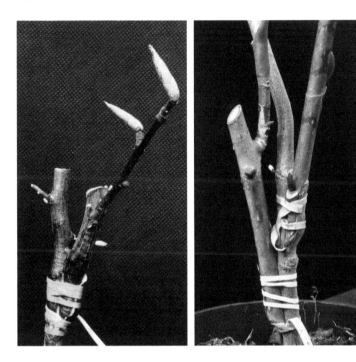

FIGURE 18.16   LEFT, *Carpinus fangiana* 'Wharton's Choice', to conserve rootstocks of an unknown combination scions were originally multi-grafted onto *C. betulus*. Successful 'takes' can then be re-grafted individually to achieve quick bulking-up of numbers. RIGHT, multi-graft of *Magnolia* 'Summer Solstice'; both grafts can be left in situ to produce a multi-stemmed plant. Alternatively, the upper one can be detached and re-grafted to produce a tree form.

FIGURE 18.17   LEFT, *Rhododendron* 'Moonstone', a slow-growing hybrid 'bulked up' by multi-grafting two scions on a branched rootstock. RIGHT, *Stewartia ovata* (left) and *S. malacodendron* (right) are both difficult species to bulk-up in containers but multi-grafts more quickly add substance to the resultant plants.

A variant of multi-grafting is to graft scions spaced along the length a stem from a low position to a height of up to 200cm. Correctly employed, the results can significantly speed-up production of specimen plants*. It is important to select suitable combinations, or eventual growth of the root-stock stem may cause an unsightly effect and prejudice opinion against grafted plants.

Multi-grafts can also be used to place more than one variety on the same rootstock. This may include combining suitable varieties for cross pollination, as in the so-called 'family tree' which incorporates several apple varieties. The same technique can also be used in enhancing seed production by increasing opportunities for cross pollination in seed mother orchards for timber trees.

ROOT GRAFTING

The general principles of root grafting have been discussed in Chapter 8.

Apart from its use with a limited number of species such as *Wisteria* and Tree Peonies, this system is somewhat neglected in the UK. Reasons for this are unclear, probably a of lack of tradition and little accumulated experience, added to the comparative difficulty in obtaining suitable roots for use as rootstocks. For some species, *Wisteria* being an exception, personal experience of root grafting is that growth after a successful union is slow compared with a conventionally grafted equivalent. This is hardly surprising in view of the relatively small root system involved and may be another factor in its lack of popularity. The system has an unexploited potential to provide a means of overcoming incompatibility[†] and for this reason alone deserves further investigation.

Root grafting may be used to propagate species compatible on the chosen root system where scion rooting may or may not be the objective. However, in selected genera it is used to 'nurse' the scion, which is encouraged to produce adventitious roots arising from the buried portion – known as nurse root grafting.

Source, Specifications and Storage

Rootstocks can either be a complete root, with stem and hypocotyl removed, known as a whole-root graft, or a portion of an existing root system, i.e. the piece-root graft. Normally, whole-root systems are derived from one- to three-year-old seedlings, the age difference determining root girth at the point of grafting (Figure 18.18). Older seedlings, having a well-developed root system, sometimes produce additional, suitably sized pieces which can be used separately as rootstocks.

In one example, when cut to include the hypocotyl with some root, and further cut to yield more root portions, 100 top-grade apple seedlings provided sufficient material for 250 grafts[‡]. Saleable plants produced in the field may yield appropriate material during lifting when excess roots are exposed and become available. A limited range of species, *Paeonia*, etc., are grown in beds specifically to provide roots for use as rootstocks. These are normally grown 'in house' and rarely, if ever, offered for sale, possibly another reason why the system is not widely practised.

Root size is best matched to scion diameter, but allowance must be made for the considerably thicker cortex tissues of the young root, adding a further 4–8mm to the required overall diameter. Large unbranched or old roots are not favoured, and most authorities suggest that some degree of root branching significantly improves results[§]. Length of root should not be less than half that of the scion and most recommend an equivalent length. Young roots may not provide sufficient girth at the required length; in this situation two undersized roots may be used and placed on the scion more or less opposite each other. Extra time is required, but the chances of success are increased as loss of one root does not totally preclude good results from the other.

---

* Mezitt (1973).
† Leiss (1988).
‡ Fillmore (1951).
§ Leiss (1995) and Bazzani (1990).

FIGURE 18.18 *Wisteria floribunda* 'Multijuga' ('Macrobotrys') grafts to roots of *W. sinensis* 2+0 seedlings. After completing the scion and graft, it is dipped in wax sealant (here coloured green) to just below the graft union (John Richards Nurseries).

Polarity of the root piece must be identified, and to ensure this, it is usual to cut the proximal end (nearest to the stem) square and the distal end sloping (Figure 18.19). Orientation decides the long-term role of the root system. Maintaining polarity by attaching the proximal end of the scion to the proximal end of the rootstock results in long-term survival; the roots function normally and remain as part of the grafted plant. Inverted root grafting by reversing polarity of the root (proximal end of the scion attached to distal end of the root) causes eventual root failure; the objective is to sustain the scion for a period until it is able to produce adventitious roots and become self-supporting.

Treatment of roots after collection follows different procedures influenced by tradition, grafters' opinions and past experience. Storage of roots before grafting is often linked to subsequent grafting methods. Roots from species which produce callus in low temperatures, e.g. most of the Rosaceae, are stored at low temperatures (0–1°C) packed in moisture retentive material such as mixtures of peat and perlite. Species which require warm temperatures for callusing are said to need pre-grafting storage at 18–25°C. Duration of this stage may vary between three to four weeks and eight to ten weeks*.

Reasons for suggesting prolonged periods at comparatively high temperatures are not obvious and again raise the issue of how quickly stem and, in this case, root tissues respond to heat. Despite roots initially containing higher levels of stored carbohydrates than stems, long periods at relatively warm temperatures result in significant carbohydrate depletion. Reported losses of grafts using pre-stored roots were due to death of the rootstock rather than scion, most likely caused by severe carbohydrate loss.

---

* Leiss (1995), Leiss (1988) and Stoner (1974).

FIGURE 18.19   *Paeonia* 'Gansu Hybrid' roots prepared for grafting, proximal ends uppermost. Note the sloping cuts on some distal ends. This hybrid has proved particularly suitable as a rootstock for *P.* 'Highdown' (P. 'Joseph Rock').

Based on personal experience, recommendations are for no pre-grafting root storage or limited storage (<14 days) at low temperatures <4°C before grafting takes place.

GRAFTING METHODS AND TYING-IN

Grafters need to take care to ensure good cambial matching, particularly as the root cambium is less obvious and more deeply seated than in the stem. Recommendations are made for apical grafting methods, e.g. splice, whip and tongue, wedge, apical veneer, cleft, inlay or saddle*. Of these, splice or wedge would meet requirements, but a short tongue veneer would give best cambial matching (Figure 18.20). When grafting roots, it is particularly important to ensure cleanliness both to reduce the possibility of disease-causing organisms entering the graft interface, as well as to prevent any possibility of the knife blade engaging grit or other debris resulting in damage to the cutting edge. Extra dirty roots may be brushed using a narrow carpet tile. Washing off roots in running water is normally adopted, but some grafters dip the apex of the root to a depth of 10–15mm in diluted sterilant such as peroxyacetic acid. For *Paeonia*, recommendations are made to give the entire root a 1–2 hours soak in bleach at 1:10 dilution[†]. Use of soil or compost-incorporated fungicides, such as Carbendazim, may also be helpful.

As the graft union is buried, the choice of tie is important. Ties which do not readily degrade in darkness, rubber ties, etc., are appropriate when it is intended to allow or encourage scion rooting. For grafts depending on the rootstock to provide a permanent lasting root system, degradable ties are essential unless the graft union is accessed and the tie cut. Light grade fillis string, heavy cotton thread, string derived from hemp or good quality raffia twisted for extra strength and durability are all suitable. Non-degradable, self-adhesive enclosing ties such as Buddy Tape®, Medelfilm®, etc. are used if there is concern over possible infection of the graft interface. These might be considered for species having the original nurse graft removed, e.g. *Paeonia*, when roots of herbaceous species are used for tree-form types[‡]; for further discussion see Chapter 43: Paeonia.

---

* Hartmann, Kester, Davies Jr. et al. (2014) pp.511–530, Bazzani (1990), Leiss (1988), Leiss (1995), Jaynes (2003), Garner (2013) pp.203–207 and Sheat (1948) pp.22–24.
† Bennison (2010) pp.28–31.
‡ Bennison (2010) pp.28–31.

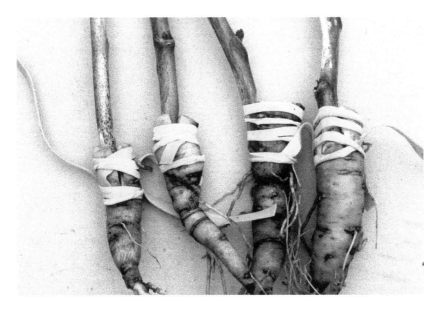

FIGURE 18.20 Root grafts of *Paeonia*, using *P.* 'Gansu Hybrid' roots. Left to right; Nos. 1–2 are Apical wedge grafts. Nos. 3–4 are Apical short-tongue veneers. On right-hand graft, note that rind is so thick that, to ensure good cambial matching, it projects well beyond the scion.

AFTERCARE

Grafts may be immediately planted, potted or alternatively, as occasionally recommended for *Paeonia**, stored in polythene bags at 20°C for 3–4 weeks. With relatively easy species, the grafts may be directly field planted, usually after a period in cold store, until soil conditions are suitable. More demanding species should be potted-off individually and stood down on heated beds, alternatively bedded-up into boxes or beds with facilities for supplying additional heat. Temperatures chosen are dependent upon species but normally 10–18°C. This may result in extension growth before union development is completed; therefore, a supportive environment such as a poly-tent is necessary.

NURSE GRAFTING

This system involves grafting the scion variety to a rootstock with known short- or long-term incompatibility. The rootstock may comprise a root with stem portion or, more often, root only. The objective is to induce scion rooting with the intention of eventually producing the scion variety on its own root system. Rooting may be encouraged by the use of non-degradable ties which have the effect of constricting growth at the union; this may be accompanied by deliberately wounding the scion just above the tie and the application of growth substances (rooting hormones) such as indolyl-butyric acid (IBA). Two further alternatives for stimulating scion rooting are: use of rootstocks which show delayed incompatibility, e.g. Californian Privet for Hybrid French Lilac; or grafting the rootstock roots in an inverted position.

Syringa vulgaris hybrids are the prime example of use of this system; for further discussion, see Chapter 52: Syringa.

---

* Bennison (2010).

NURSE-SEED GRAFTING

With this system, scion rooting is the objective. Only species which produce large seeds with hypogeal germination characteristics (cotyledons remain below soil surface) are suitable and experience is restricted to relatively few genera: *Aesculus*, *Camellia*, *Carya*, *Castanea*, *Juglans*, and *Quercus*. To date, there appears to be little commercial uptake of the system, but it may have potential for overcoming incompatibility problems. Since the method was devised and described, a modified method has been used for some species and is described in Chapter 23: Aesculus.

GRAFTING METHOD

Seed of appropriate species germinated in the spring is used as the rootstock. Once the radicle (root) and plumule (shoot) have fully emerged, they are cut away from the seed by severing the cotyledon petioles close to the cotyledons which remain enclosed within the seed coat (testa) (Figure 18.21).

Scions are selected from current year's wood of the same genera as the prepared seed. Cold storage of scion wood is recommended to ensure fully dormant material is available when the nurse-seed rootstocks (seeds) are ready for grafting.

Insufficient investigation has taken place to establish whether known compatible combinations within the genera must be chosen, or whether juvenility factors present within the newly germinated seed can overcome incompatibility. At present it is wise to group known compatible combinations together. For example, germinated *Quercus palustris* acorns would be the sensible choice for scions in the Lobatae section, and *Quercus robur* in the Quercus section.

A wedge-shaped cut is made at the base of a 75–100mm long scion having a diameter which ensures good cambial matching with the rootstock. Once prepared, the scion is pushed into a slit

FIGURE 18.21 *Aesculus hippocastanum*. Top seed has petioles severed and the knife point inserted. Bottom seed has radicle and plumule still attached to cotyledons by petioles.

created by inserting the knife point between or at right angles to the severed petioles (Figure 18.21). Which position is optimum cannot be guided by results as comparisons do not appear to have been made.

The scion is often held in position by tension within the seed, but relatively flimsy acorns of some Oaks, such as *Q. robur*, are best held in place with a rubber graft tie placed around the acorn. *Aesculus* are normally strong enough to ensure that the scion is held without assistance. After assembly, the grafts are placed individually in pots or alternatively in trays or beds. The graft is placed 25–40mm deep in a suitable medium with the scion projecting above (Figure 18.22).

Aftercare

Well-drained and aerated media are essential to provide suitable conditions for union formation and subsequent rooting. Enclosed environments are required with temperatures maintained in the 20–22°C range.

Once a union is developed, juvenility factors present in the petioles promote rooting in the scion*. However, observations indicate that rooting from the severed petioles often occurs before scion rooting. Both provide support for the graft until full development is achieved.

Growth following union and rooting is often very slow, a reason for the comparative lack of commercial interest in the technique.

CUTTINGS-GRAFTS

PRE-FORMED ROOTS

For a limited range of species, the technique of using rootstocks of unrooted stems, with internal pre-formed root initials, has been standard practice for many years, *Salix* being an important example.

FIGURE 18.22 *Aesculus* nurse seed graft. LEFT, scion of *Aesculus* 'Autumn Splendour' inserted into prepared seed. Note matching of cambium with cut face of petiole. RIGHT, assembled graft to be placed individually 30–40mm deep into well-drained medium (perlite/peat/bark mix) in a deep propagation pot or tray.

---

* Moore (1963).

Such plants will rapidly produce a new root system and are able to be treated as rootstocks, not requiring additional handling to accommodate bare-root or root-ball. After grafting, the base of the stem is pushed 150–200mm deep into a floor-level soil bed or suitable compost held within a container.

As the graft union develops, extension growth of the pre-formed roots soon produces a functional root system, capable of supporting scion growth. With this system, subsequent growth and development of the grafted plants is comparatively slow. To overcome this disadvantage, some grafters favour stems with an established root system produced the previous year rather than cuttings-grafts (Figure 18.23).

Unrooted Stems

For discussion on the principles of using unrooted stems for rootstocks, see Chapter 8. In practice, the improvements in rooting techniques made over past years have made possible the concept of using stem cuttings of a broad range of species as rootstocks. Many of those which have potential for simultaneous rooting and graft development can be rooted and grown-on satisfactorily without the need for grafting. However, sometimes use of a suitable rootstock provides better vigour and health than plants having their own roots. An important example of this is seen in Grapevines, where many millions of plants used for fruit production are grafted and rooted simultaneously using unrooted rootstocks of a type resistant to the soil borne pest Phylloxera (*Daktulosphaira vitifoliae*). A further example using leafy cuttings is seen in *Rhododendron*, where German producers and consumers consider, with some justification, that grafted plants perform better than those on their own roots (Figure 18.24). This has resulted in cuttings-grafts being a major method of propagating this genus in Germany (see Chapter 50 for further details).

An important aspect of rootstocks produced from unrooted stems is the selection of appropriate forms and the establishment of suitable stock mother plants to provide sufficient cuttings material. As the range of genera using this system is limited, choice of rootstocks will be discussed in Parts Seven and Eight, and provision of stock mother areas has been discussed in Chapter 9: Rootstocks for Cuttings-Grafts.

FIGURE 18.23 *Salix* x *smithiana* stems rooted the previous season, awaiting top-working with *S. caprea* 'Pendula' (Yorkshire Plants. Photo by Steen Berg).

FIGURE 18.24   A cuttings-graft of *Rhododendron yakushimanum* previously grafted to an unrooted cutting of *R*. 'Cunningham's White' now at the well-rooted stage.

"LONG JOHN" GRAFTS

A system – so-called 'Long John' grafting – developed in the USA, utilises extra-large scions (over 1 metre) of two- to four-year-old fruit tree branches carrying flower buds and short flowering spurs. Advantages claimed are very rapid fruit production for the following season providing an 'instant' return*.

The system has been developed for field use but could be adapted as a bench grafting technique. So far it appears to be restricted to Rosaceae species grown for fruit production: apples, cherries, pears and plums. This would undoubtedly be applicable to ornamental selections of these types and allied genera such as *Sorbus*. Careful selection of appropriate species will be required to exploit the full potential. Most important will be development of sufficient stock mother plant material to yield adequate top-quality scion wood.

Such large scions require an extra-large graft. Whip and tongue with a splice length of 100mm or more is suggested. In the field, the method was used with de-leaved scion wood grafted in the late summer and protected by a shaded, enclosed, polythene sleeve. Humidity within the sleeve under these conditions would be best achieved by the addition of a small quantity of hydrated polyacrylamide gel granules (in the UK, Broadleaf P4® granules) rather than the addition of free water. For bench grafting, enclosed conditions would afford the necessary supportive environment; cold callus systems maintained at minimum 6°C during the dormant season should prove to be effective.

The concept is worthy of further investigation and may provide the basis for an accelerated tree production system, particularly suited to container growing.

---

* Gilbert (2005).

APPROACH GRAFTING; GRAFTING WITH ATTACHED SCIONS

Approach grafting, or grafting with attached scions, is substantially different to methods for detached scions. It has features which greatly increase chances of successful takes. Normally used in the field, the procedure may also be carried out under protection and is therefore appropriate for inclusion in bench grafting systems. Grafting takes place while both rootstock and scion are attached to their respective parent plants and remain in this state until a successful union is achieved. A high percentage of successful grafts can be expected because there is no limit on the length of time taken for a union to develop.

Rootstocks are grown in pots and prior to grafting placed close to the scion mother plant. Grafts of the side veneer type are invariably used, the simplest being a slice taken out on the side of a well-placed branch on the rootstock to match a similar slice on the scion. If large enough, several grafts per mother plant are possible. Grafts need tying firmly and sealing. Mother plants and rootstocks must be kept in good health until union formation is completed.

To achieve a viable new plant once the union has formed, all that is required is to sever the mother scion branch just below the graft and the rootstock above.

Approach grafting has received very little attention from commercial growers because of high labour costs, relatively low output per given area of production and availability of scions only when attached to an established plant. There may be a case for revisiting the technique commercially as results are predictably high for difficult and high-value species. Successful utilisation in commercial production of Red Flowering Gum (*Corymbia ficifolia*) has upgraded results from the norm of 30% for detached scion grafts to 90% or more*. Use of relatively cheap structures and provision of good growing systems, such as capillary beds, are further reasons for re-assessment.

It has a place in less commercial situations such as private gardens, botanic gardens and specialist nurseries.

DECIDING UPON THE BEST GRAFTING STRATEGY, METHOD AND SYSTEM

Two very different types of grafting facilities are described, and how these influence the choice of strategy, methods and system:

SMALL GRAFTING UNIT CONCERNED WITH SPECIES DIVERSITY

Small grafting unit used to propagate small numbers of individual species across a very wide range, linked either to a small specialist nursery producer, or botanic or private garden maintaining a diverse collection of rare woody plants.

The aim is to achieve highest takes possible, more or less regardless of economic considerations. The range of species grafted is likely to vary from the easiest to the most difficult. A wide range of pot grown rootstocks is required, but where appropriate these can be supplemented by bare-root rootstocks for specific purposes.

Grafting strategies utilising side grafts allowing re-use of the rootstock in the event of failure, can be significant in keeping costs down. Scion material may be of poor-quality requiring use of the most technically superior method, side veneer grafts on potted rootstocks. A small-scale, single run hot-pipe system may provide opportunities for improving results over traditional warm callusing, especially for less experienced staff. For particularly rare, difficult or valuable material, approach grafting can be considered.

---

* Gleeson (2007).

LARGE, SPECIALISED GRAFTING UNIT CONCERNED WITH HIGH OUTPUT

Large grafting unit often specialising in a narrow range of species, with the objectives of high output and efficient materials handling.

Ease or difficulty of grafting the chosen species will be the overriding factor affecting features of the unit. Species producing callus in low ambient temperatures are normally cold callused using bare-root rootstocks. For difficult subjects, combinations of pot grown rootstocks and summer grafting will be strong candidates. Techniques using the hot-pipe system may change decisions in favour of winter grafting.

For easy subjects, the grafting method will assume less importance and emphasis will be placed on high output, possibly involving partial or complete mechanisation. Hot-pipe systems may be designed to incorporate materials handling systems such as pallets and forklifts.

Cold storage facilities may be usefully employed to improve flexibility and scheduling.

Part Six

Role of Grafting in Conservation

19 Conservation

The broad issues of conservation, the philosophical and practical reasons for its importance have been discussed in detail in other publications.

There are two major aspects of plant conservation strategy: those concerned with conservation in cultivation and those in the wild. The former will include many garden varieties but only certain species. The latter is mainly concerned with naturally occurring species, subspecies and natural hybrids.

As we enter what many are terming the "Anthropocene Age" – a period in which human activity is the dominant influence on the planet's environment and climate – some authorities argue that traditional approaches towards conservation are becoming, or have become, invalid. It is suggested instead that the influence of mankind on the Earth is so profound, the loss of true wilderness so great, that long term attitudes between in-situ and ex-situ conservation should be merged, to provide conservation in any suitable location.

In this publication, the traditional, more conventional view is taken.

CONSERVATION OF PLANTS IN CULTIVATION (EX SITU)

Arguably, conservation of cultivated plants requires less significant input than for plants in the wild. The National Trust is the largest membership organisation in the UK with a broad mandate to preserve heritage sites in all aspects, including their associated plants. To this end, a production nursery has been established with the task of maintaining, enlarging and reinstating plant collections. Other nation-wide collections have been set up under the auspices of a body known as Plant Heritage to promote and conserve plants in Britain.

Also in the UK, a number of specialist groups exist within the orbit of the Royal Horticultural Society (RHS), the Rhododendron, Camellia and Magnolia Group (RCMG) being an example. In a continuation of assessment work started by Plant Heritage and following its sphere of interest, this group aims to re-propagate and establish plants considered to be endangered.

The route taken by the RCMG highlights a basic requirement for any preservation initiative; this is the need to identify plants in cultivation which are in danger of becoming lost and take steps to ensure their propagation and re-establishment in gardens and arboreta.

CONSERVATION OF PLANTS IN THE WILD (IN SITU)

Habitat loss and, more recently, climate change are major reasons for reduction or demise of species in the wild; consequently, conservation often involves broad issues of retention or re-establishment of in situ ecology. This is likely to require political involvement, significant support from organisations involved in this sphere of activity, often with substantial financial support from governments, businesses and private sources. Organisations concerned with conservation in these situations vary from quite small local initiatives to co-operating international institutions, of which Botanic Gardens Conservation International (BGCI) is the largest, comprising joint cooperative conservation programmes based in many of the world's major botanic gardens and affiliated groups.

In situ plant conservation problems are complex and varied but as with plants ex situ, an essential element is the need to make an assessment of the status of the species being investigated. This may vary from small-scale investigations carried out by volunteers, to large-scale international surveys. Significant among these are the World Conservation Union (IUCN), Species Survival Commission/ Global Tree Specialist Group (SSC), United Nations Environment Programme (UNEP), World

Conservation Monitoring Centre (WCMC) and others. A major result of this collaboration is the production and publication in hard copy and on-line of the Red List of Threatened Species (shortened to Red List). The purpose of this is to provide a detailed assessment of the conservation status of families or genera, often within given geographical areas.

Results from these assessments influence subsequent strategies. Sometimes improvement or control of habitat, or other factors affecting loss, such as eliminating illegal collecting may be sufficient to reverse the decline. Frequently there is a need for additional plants to supplement existing numbers, and at this stage procedures for propagation must be considered (Figure 19.1).

PROPAGATION FOR CONSERVATION

Very few books deal specifically with the subject of propagation related to conservation. None appear to focus specifically on a wide range of woody species, and most information appears to be in the form of scientific papers mostly relating to procedures in vitro (micropropagation). One publication (perhaps the only one to date) giving wide coverage of the topic, including a number of woody species, devotes 23 pages to vegetative methods and 62 to in vitro propagation; unfortunately, for reasons which are not clear, grafting receives only one quarter of a page of text (126 words). Included within this is a coloured photograph of a *Rhododendron* cultivar, saddle grafted to *R. ponticum*. Neither the grafting method nor choice of rootstock can be recommended.

Further mention within the context of in vitro techniques states that, "if the plant can be conveniently propagated by grafting the plants obtained are of better quality and more convenient to harvest". An additional advantage is said to be that tissue surfaces from grafted plants are easier to disinfect. An example is cited where only three plants were remaining and grafting was used to establish suitable material for subsequent micropropagation*. It may be pertinent to ask why grafting was not continued.

FIGURE 19.1 *Magnolia omiensis* grafts in a nursery bed in Emeishan Botanic Garden, China. The grafts are from a range of genotypes from the natural population. These plants are due to be planted back on Mt. Omei to supplement those in existence on the mountain. Graft union is just visible on 1st and 2nd plants from the right in the front row (photo: Xiangyin Wen, BCGI China Programme).

---

* Bowes (1999) p.105.

Other commentary on the role of grafting takes a dismissive approach. In a book on Birch taxonomy containing a small section on conservation, reference is made to conservation of *Betula calcicola*, *B. delavayi*, *B. potanini*, *B. insignis*, and *B. bomiensis**. Grafting is written off as: "Lack of suitable close relatives to use as stocks for grafting, or shrubby habit, make grafting an unsuitable method of propagation for all but *B. insignis*, and cuttings of mature plants may not root easily". *Betula calcicola* has thrived well when grafted to *B. pendula* 15 years previously and has retained its shrubby habit, why would it not? In a commercial situation, *B. delavayi* has been reliably grafted using the same rootstock. There is no reason to suppose that with some degree of investigation the other species mentioned will not respond satisfactorily. In the same publication the authors admit to having no personal experience of grafting.

Understandably, the widely preferred method of propagation in virtually all conservation situations is by seed; if this is not feasible, cuttings may be considered but usually in vitro methods are regarded as the best alternative. Many of the senior personnel involved are from academic scientific backgrounds where, rather than horticulturally based activities, technologies involving chemistry, laboratory work and associated procedures are the obvious natural choice. Decisions by senior management on how plant material can best be produced are made well before the planting/re-establishment phase. While in vitro propagation techniques may be ideal in a number of situations and cuttings propagation can have an important role, grafting has qualities making its use at certain stages an extremely useful tool. A limitation is that, with a few exceptions, grafting techniques can be used only for woody dicotyledons.

ATTRIBUTES OF GRAFTING FOR CONSERVATION

Arguments for grafting as a significant propagation method for use in conservation initiatives are listed below. The various points are grouped under their major influence.

Genetic Integrity; Diversity; Selection

- Appropriate scion selection can ensure retention of genetic diversity.
- Trueness to type is guaranteed with grafting procedures.
- The genetic integrity of the scion variety is unchanged and mutations (relatively frequent with certain genera in some in vitro methods) are no more likely than in natural populations.
- Individual plants within the natural population selected for desirable traits, can be grafted to ensure rapid establishment of elite mother plants.
- Choice of suitable rootstocks widens the tolerance of a chosen subject to adverse growing conditions such as soil type.
- The possibility of hybrid characteristics cannot be excluded from the progeny of seed-raised sources, unless the provenance is carefully selected and monitored.

Transport – Plant Health

- Relatively easy, economical and rapid transport of scion wood across great distances[†].
- Scion wood is transported without any requirement for soil, avoiding numerous regulations related to the transportation of soil or substrates.
- It is often possible to surface sterilise scion wood, reducing possibilities for transfer of plant pathogens and pests.

Successful Outcomes – Rejuvenation

- Species can be grafted which cannot be rooted, or only with great difficulty.
- Poor quality or otherwise unsuitable material can be grafted when rooting from cuttings would be impossible.

---

* Ashburner & McAllister (2013) p.99.
† Bowes (1999).

- Senile material lacking the potential to root can be induced after grafting to reassume juvenile properties and subsequently make possible the production of rootable cuttings.

Output

- If required, easily grafted species can be quickly bulked up to produce large numbers of plants. With more input and a longer timescale, the same can be said of difficult-to-graft species.

Speed of Response

- Grafting procedures can provide a quick response to shortages of endangered plants.
- Timescale between receipt of material and successful re-establishment of a new plant is faster than for most other propagation techniques. Choice of large, vigorous rootstocks enables completed grafts to establish quickly.

Facility Requirements

- Grafting facilities can be provided at much less cost than for micropropagation.
- Many of the facilities required for grafting will also be required for establishing propagules (weaning) produced by in vitro techniques and, therefore, may already be present.
- Facilities may be shared between centres providing much more flexibility in use.

Technical Input – Artisan Involvement – Skills Transfer

- Less high-level technical input is required for grafting than for micropropagation.
- Grafting offers more opportunity for artisan involvement than in vitro methods.
- The skills acquired in grafting procedures can be transferred to benefit local initiatives which may be adjusted to produce additional income for the resident community. A good example of this is seen with a tropical species, the Wild-aromatic Pickle Mango (*Mangifera indica*). Expertise has been built up to train communities in the Western Ghats, India in the techniques of grafting the plant. As a result, superior clones have been identified and conserved, and the procedure has resulted in improved income for the participants[*]. Participation of this sort is likely to foster an interest in conservation by the total population.

PRACTICAL APPLICATIONS FOR GRAFTING IN CONSERVATION

Expanding on some of the issues raised above, the following highlights a number of situations where grafting has demonstrated, or could demonstrate, potential for successful outcomes in different conservation scenarios:

Recovery of Endangered Species from the Wild

A number of examples exist of material collected in the wild sent to the UK from areas as distant as remote parts of China and Taiwan, and subsequently established using grafting techniques. Such material, mostly leafless and in poor condition, could not have been propagated with any degree of certainty by any other method. *Tilia endochrysea*, *Fagus hayatae* and *Carpinus omiensis* are examples of species recently 'rescued' in this way. *Ostrya rehderiana* has also been received but grafted to a *Carpinus* rootstock, all that was available, and failed after producing restricted growth and foliage. This failure illustrates an important point, which is the need for suitable rootstocks at the reception area.

For an opportunistic approach to be successful, there should be close liaison between the collector and recipient, highlighting the need for pre-planning and, as far as possible, a schedule of target

---

* Vasudeva & Reddy (2015).

species for collection. Given these provisos grafting has great merit for securing species under severe pressure by land clearance, logging or a multitude of other events. The disadvantage, common to all vegetative propagation, is the normally small genetic base of plants produced when only limited amounts of parent material are available.

Recovery of Juvenility

Grafting senile scion wood onto rootstocks with juvenile characteristics (seedlings) induces juvenility in scion extension growth; juvenility is further increased by repeated grafting of further material collected from successful grafts. As is well known, the effect of juvenility is to enhance cutting rootability. Recent discovery (2016) of an ancient Wentworth Elm (*Ulmus* 'Wentworthii Pendula')* at Holyrood Palace in Scotland is a good illustration of the potential of this technique, which could be used to rejuvenate vegetative material from the tree and provide cuttings material more or less guaranteed to root. The genus *Ulmus* responds very well to grafting methods and invariably produces high takes, even in the field. Present published strategy for this very rare tree is to attempt to root cuttings, grafting to be used as a last resort, an indication of the current status of grafting in the awareness of senior botanic garden managers and conservationists.

Retention of Genetic Integrity

Lack of awareness regarding genetic integrity is sometimes apparent: a recent account concerns attempts to preserve the apple tree variety, *Malus domestica* 'Flower of Kent', said to have helped inspire Sir Isaac Newton's theory of gravity[†]. It is stated that seedlings have been raised from 'pips' collected from the original tree, still present at his birthplace in Lincolnshire. These have been distributed to various centres, including the Royal Society's headquarters in central London, to create a "Newtonian apple orchard". Unfortunately, resultant plants from the Newtonian initiative will be seedlings of 'Flower of Kent', not truly representing Newton's original plant, but rather just *Malus domestica* seedlings. To ensure the genetic integrity of the resultant plant's progeny, the existing original 'Bramley's Seedling' apple tree has to be grafted. It is equally certain that to ensure the continuation of a true Newtonian apple orchard only grafting or other less practicable vegetative methods, such as air-layering, should be considered.

Speed and Quantity of Production

It would be foolish to suggest that grafting can compete with micropropagation in output potential for very large numbers of plants. However, it is important to recognise the lengthy timescale involved in micropropagation. This includes isolating clean cultures, inducing proliferation and subsequent rooting, and finally weaning and establishing vigorous, healthy but very small plants, which require sufficient periods of growing on before becoming suitable for planting out. The difficulties of the last horticulturally based phases are often underestimated by those in the laboratory. If a broad genetic base is required, the procedures outlined will need to be duplicated to account for each genotype, further extending the timescale.

In the early stages, grafting can compete with micropropagation in output. The main limiting factor in many situations will be a shortage of scion wood. Given sufficient material, easily grafted species can be produced in very large numbers, certainly in thousands. Above this figure, micropropagation potentially has a clear lead; although, this has been demonstrated only on a fairly limited range of woody species, represented by a narrow range of genotypes.

Need for Rapid Response

The ability of grafting procedures to quickly produce plants has particular relevance in collections under cultivation containing a number of accessions in a dangerously low state, for example only

---

* The Plantsman (2016).
† The Garden (2017).

one plant in poor condition. A visit by a skilled grafter can improve the situation, producing by the following season modest numbers of the endangered plants for eventual growing on.

A specific example may be cited for a visit made to a private estate housing an extensive collection of *Rhododendron* species and hybrids. A number of specimens in the collection had reached a senile stage with poor extension growth and signs of failing vigour and health. Two-and-a-half hours spent collecting 38 scions of 12 species resulted in 28 successful takes from ten species. After two seasons, the plants were well-branched vigorous specimens, growing well in 3-litre pots and at a stage which made them suitable for planting out. A similar exercise for endangered *Rhododendron* cultivars has produced equivalent or better results.

In this example, because of poor material, short runs and the wide dispersal of donor plants, the time required during and after collection to prepare and make the grafts is substantially longer than in commercial practice. The six to eight weeks required for a successful take would in practice be 'rolled-in' with other grafts having the same environmental requirements.

The same approach could be employed for in situ situations, for example, where only few of an almost certainly graftable species exists as a small population in the wild. Even genetic diversity could be maintained by taking scion material from each of the plants present. *Mespilus canescens* (now thought to be x *Crataemespilus*) is an example of such an initiative, and several plants from the remnant population in Georgia (USA) are growing successfully in the UK on rootstocks of *Crataegus oxycantha*. Unfortunately, they may comprise only one genotype, but assuming others exist, this omission could be redressed by further collection.

Rootstock Selection to Overcome Environmental Challenges

The established horticultural practice of using rootstocks to overcome specific environmental problems, for example higher pH-tolerant Rhododendrons utilising INKARO® rootstock, could be adapted for use in conservation projects. Success would undoubtedly rely on precise reasons for failure and whether they relate solely to soil conditions or other environmental factors. Alleviation of problems caused by soil conditions for oaks has been suggested*, an example might be *Quercus hintonii*, critically endangered and reputedly difficult in cultivation. Difficulties may be associated with a requirement for very sharp drainage and low rainfall. This species, in Section Lobatae, appears to be closely related to *Quercus crassifolia*, another Mexican species. *Quercus crassifolia* has grafted well onto *Q. palustris* rootstocks, and a specimen under cultivation in the UK (coastal East Anglia) has made 2.5 metres of extension growth in one season. It may be that *Q. hintonii* would succeed equally well on this rootstock but, as with so many oaks, a high percentage of incompatibility is likely, implying specific rootstock/scion and/or inter-stem combinations will need identification.

GRAFTING POTENTIAL FOR RED LIST SPECIES

Looking through Red List entries provides opportunities for identifying critically endangered candidate species for grafting. The possibility is considerable: in the Central Asia Red List, 16 species would be graftable; in Betulaceae Red List, 7 species; in Quercus Red List, many of the Cyclobalanopsis subgenus are critically endangered and early indications are that some species will successfully graft onto relatively hardy species such as *Q. myrsinifolia*, *Q. schottkyana* or *Q. glauca*. Taxonomic and phylogenetic work may provide useful guidelines for compatible partners in this genus. Some endangered species in the Red List of Rhododendron (*R. asterochnoum*, *R. hemslyanum*) have already been successfully grafted here.

---

* Baldwin (2017).

SETTING UP A GRAFTING FACILITY

Specific grafting facilities can be placed within existing nurseries; ideally, these should be strategically placed to link with conservation site(s). The type of facility required is very dependent upon its location and the species involved. Provided ambient temperatures are at suitable levels, 15–23°C+, sites located in rain or cloud forest areas, even for evergreen species, may rely heavily on grafting procedures carried out in the field. In these instances, some provision of shelter in the form of shade houses or polythene tunnels will provide opportunities to amend environmental conditions and improve results. Hot-pipe systems based on the economy version might also be very successful in this type of location.

Judgement will be required to decide how much basic field systems need enhancement to meet the more demanding needs of difficult species. In less favourable areas, the grafting facilities described in preceding chapters will be required. Where only very scarce material is available, every graft is of vital importance and facilities must be sufficient to meet the requirements for achieving high success rates.

Use of the hot-pipe system can be recommended and will provide best results for many species (see Chapter 16: Hot-pipe, and Chapter 18: Grafting Systems, for full details). The system has further advantages in that demanding horticultural techniques (drying-off, management of enclosed environments, etc.) are less crucial to success. If finances are restricted, the 'economy' design is an option.

Hot-pipe is best suited for grafting deciduous species; sub-tropical or tropical hardwoods which have leaf fall during 'rest periods' can be included. It should also be possible to defoliate some evergreen species in order to use this system. A large number of sub-tropical and tropical species will be evergreen and where defoliation of scion wood is considered unsuitable; hot-pipe runs could conceivably be constructed inside a supportive environment such as dry fog. In some instances, this will not prove satisfactory, and there will be no alternative to establishing warm callusing procedures using poly-tents or similar equipment to provide enclosed conditions.

To achieve good results, grafters require dexterity, attention to detail and commitment. In the likely absence of trained staff, those already involved in local craft work are often the best candidates to be taught the skills. Training along the lines discussed previously will be required (Chapter 4). It may be possible to train grafters who will be prepared to work on a part-time or seasonal basis, resulting in savings on costs.

Grafting knives may prove a problem. For the majority of grafting material and for lady grafters, the lightweight and economical Tina 685 or more costly Tina 605 will provide the best option (Chapter 5). Heavy or very hard wood will prove too much for this relatively small knife and the expensive 600 or 600A is the ideal alternative. Unfortunately, there does not appear to be a cheaper, fixed blade option for the 600 pattern, but large conservation initiatives requiring many knives may persuade the manufacturer to fabricate a 'special' run, especially if attendant publicity could be arranged. Use of poor-quality, blunt knives will jeopardise good results from the outset.

THE FUTURE OF GRAFTING IN CONSERVATION

Grafting should have an important role in propagation initiatives for conservation. For this potential to be realised, better recognition and implementation is required than has been evidenced to date. Grafting skills can be taught to local workers, providing essential rapid action to combat imminent species extinction. This may be achieved far more effectively than when reliance is placed on centralised skills, utilising technological solutions, far removed from the knowledge and skills of the potential work force on the ground. The value and potential of decentralised initiatives to the people involved should not be underestimated.

Adequate recognition of the role of grafting as another tool for use in conservation strategies is long overdue.

Part Seven

Genera Specific Requirements

Introduction

Part Seven contains detailed descriptions for grafting a range of genera, and their species and cultivars as appropriate. The choices made are those considered to be of commercial significance and/or with a particularly interesting or complex set of requirements.

To conserve space and avoid repetition, detailed descriptions of standard procedures dealt with in previous chapters, particularly Chapter 18, are not discussed in depth. Specific requirements, common to more than one genus, are not repeated but cross-referenced as appropriate.

When considering observations made by the author, readers should be aware that these are made under conditions in the southern half of the United Kingdom. This is a northerly maritime climate with comparatively cool, cloudy summers, temperatures not often exceeding 26°C, mostly in the 18–21°C range, with annual sunshine hours of 1500–1700 per year (summer average 6.5 hours per day). During mid-summer, daylight hours are long (±16 hours). Winters are relatively mild (US zone 6, occasionally 5; UK range H4 to H5). Temperatures during the winter are rarely below –5°C, but lower does occur at –10 to –12°C, usually only for a short period, very occasionally even lower temperatures of –14°C or below can occur.

Under these conditions, growth is prolonged well into the autumn and the degree of ripeness in the wood, particularly of species adapted for a continental climate, is much less than for those growing in more extreme conditions. These factors undoubtedly have subtle but significant influences on the response and performance of grafts.

There is no point in grafting incompatible combinations. For some genera, compatibility of rootstock and scion is a fundamental and crucial issue. Attempts are made to identify compatible groupings based on past experience. In many instances, when results are not conclusive or speculation on unknown combinations is made, these are supported by taxonomic and phylogenetic information. For many genera, throughout Parts Seven and Eight, discussions on taxonomy, possibly also linked to phylogeny and their likely influence on graft compatibility, are a continuing theme.

20 Abies (Pinaceae) – Fir

INTRODUCTION

Most species are propagated by seed; vegetative propagation by cuttings is suitable for only a few, mostly dwarf types. Selections have been made for growth habit (dwarf or prostrate), foliage colour (blue-grey, silver-grey, yellow and gold) and combinations of these characters. For propagating these horticultural selections and species for which true-to-name seed is unobtainable, grafting remains the most important method.

COMPATIBILITY

Opinions vary on compatibility and best rootstock/scion combinations. Some suggest there is no compatibility problem[*]; this view is supported by the experience of many grafters. From work carried out, some combinations may affect growth rates, longevity and ultimate size; consequently, they are of particular concern for forestry[†].

Choice of rootstock species for nursery and amenity purposes is best based on availability and performance under nursery conditions. In Northern Europe and the USA, *Abies alba*, *A. koreana* and *A. nordmanniana* are prime candidates; in Southern Europe and Southern USA, the choice would be *A. firma* with resistance to high soil temperatures sometimes coupled with high moisture content.

TIMING

Grafting can take place between late August and early April. In the UK most are grafted in the mid-January to late February period. In the Netherlands, especially in the Boskoop area, many favour September to December.

GRAFTING METHODS

Conventional short tongue side veneer graft is preferred, because the thick rind makes good cambial matching difficult when using the long tongue veneer.

Low graft placement, facilitated by the use of plug grown rootstocks (Figure 20.1), aids production of shapely plants, with the first whorl of branches close to soil level.

Tying-in is best done using the double finger tying technique described in Chapter 6. The rather stiff horizontal needles and side branches of most *Abies* can be more easily held and controlled with this method. Sealing *Abies* grafts is not necessary and can be injurious as it prevents exudation of the resinous sap. Some grafters favour spacing tie loops widely to allow exudate to flow away[‡].

ROOTSTOCKS

Pot grown rootstocks are usual, but bare-root rootstocks are used by grafters in Scandinavia[§]. For potting up, 2+0 seedlings are satisfactory for many types; three-year seedlings (3+0) are preferred

---

[*] Dirr & Heuser (2006) and Hatch (1982).
[†] Karlsson & Carson (1985).
[‡] G. Meacham (pers. comm.).
[§] A. Koosgard-Laursen (pers. comm.) and Thomsen (1978).

FIGURE 20.1   LEFT, *Abies pinsapo* 'Aurea' graft with current year's wood scion. RIGHT, *Abies pinsapo* 'Aurea' graft using a basal branched scion. Both grafts are placed low down close to the root system, easily made possible on plug grown *Abies koreana* rootstock.

for thick wooded scion varieties such as *A. pindrow* and *A. procera*. Consequently, 2+1P 5–8mm or 3+1P 8–10mm rootstocks should be specified if they are obtained from an external source. Pot sizes used are usually between 9cm square and 1-litre deep pattern; standard compost mixes are suitable. Seedlings should be well established in their pots at grafting time.

Increasing production and availability of plug grown seedlings has led to greater use by some grafters* (Figure 20.2). Grafts may be made directly onto plug grown seedlings which are then either returned to the plug with some gaps left for spacing, laid in trays of peat or peat/bark mixes or potted-on.

SCION WOOD

The principles of scion selection and handling for conifers has been discussed in Chapter 10. Terminal shoots selected from good lateral growths may be successfully induced to assume apical dominance after grafting, providing they are staked, tied early and kept free from competition. Use of branched scion material with a two-year-old wood base can provide an early low branch structure (Figure 20.3).

SYSTEMS

All require the use of warm callusing systems; although, the temperatures employed may be quite low. At the beginning and end of season, i.e. August to October and March to April, grafts require housing in shaded structures designed to keep ambient temperatures as cool as possible (Figure 20.4).

In the mid-winter period, *Abies* are slow to form callus. Once the first signs of callus formation are detected, normally after 6–9 weeks, daily ventilation is recommended. Some grafters favour routine application of fungicides at intervals of 14–21 days.

* R. Ward (pers. comm.).

FIGURE 20.2   *Abies koreana* understocks in plugs.

FIGURE 20.3   LEFT, on the left of picture *Abies cilicica* branched with leading shoot; centre, *A. bracteata* single terminal shoot; right, *A. equitrojani* branched with leading shoot. RIGHT, picture, *Abies pinsapo* 'Aurea', left-hand terminal shoot scion and right-hand branched scion with a two-year-old basal stem.

FIGURE 20.4   LEFT, conifer grafts including *Abies* in a poly-tent on a mobile bench in a glasshouse (Kilworth Conifers). RIGHT, conifer grafts (including *Abies*) housed in poly-tent on ground level bed within a polytunnel (Golden Grove Nursery).

The grafts are hardened off to full exposure over a minimum period of three to six weeks. Rootstocks should not be allowed to dry excessively, and watering may be necessary.

Grafts placed on bare-root rootstocks are rarely used except by some Scandinavian growers*, trials in the UK have produced significantly lower results than with the use of potted rootstocks. Once the union has established, successful bare-root grafts are potted up.

PHYSIOLOGICAL AND ENVIRONMENTAL FACTORS

Opinions vary on the beneficial influence of drying off, but most consider prolonged drying is unnecessary while grafting very wet rootstocks is not advisable. Many consider rootstocks are at the ideal stage for grafting when white roots are visible[†]. Experience here is that this is not necessary.

Higher ambient temperatures in late summer and spring provide conditions for fairly rapid union development. Late autumn/winter temperatures in the 8 to 15°C range are adequate for slow, steady union formation. Temperatures at freezing point or below must be avoided by the use of supplementary heating. Managing the grafting environment for *Abies* follows that described in Chapters 13 and 18.

AFTER CARE

Except for dwarf, pendulous and prostrate forms, the dominant terminal leading shoot must be trained up vertically; when side grafted this can be conveniently achieved by tying it to the upper portion of the rootstock (Figure 20.5). After heading-back to the graft union, a suitable stake to replace the rootstock stem is required.

FIGURE 20.5   LEFT, *Abies concolor* 'Sherwood Blue' before heading-back, the scion comprising leading shoot plus laterals. RIGHT, *Abies pinsapo* 'Aurea' showing the use of rootstock as early support for the terminal shoot. A 'spacer' cane used for positioning lateral branches to provide eventual branch symmetry is visible on the right.

---

* Thomsen (1978).
† Larson (2006).

If a branched scion is used potential laterals can be arranged to provide symmetry using 'spacer canes' (Figure 20.5) (see *Picea*, Chapter 44 for a full description of the use of 'spacer canes').

GROWING ON

Light shading generally enhances growth rates and leaf colour. The central lead should be tied to a cane and any developing rival vertical shoots removed. Spacer canes may be required for more than one season. Well-judged use of space canes to improve branch placement can substantially accelerate symmetry and advance the stage to saleability.

21 Acer (Sapindaceae) – Maple

INTRODUCTION

Acers are mostly restricted to the Northern Hemisphere, where they are widely distributed throughout Europe, North Africa, Asia and North America, an exception is a small incursion by *A. laurinum* across the equator into Indonesia. All are woody species ranging in size from small shrubs to large trees. Most are deciduous, but some species from the warm temperate regions are evergreen. High value timber is obtained from certain species and *A. saccharum* has an important role in the production of maple syrup. A number are important for ornamental use in temperate climates. The Japanese maple (*Acer palmatum*), in a multitude of varieties, is among the most popular of all woody ornamentals and propagated in very large numbers by bench grafting.

COMPATIBILITY

The simple rule is milky types (containing latex) are compatible between themselves, but not with those having non-milky sap; the non-milky must likewise be kept together. This distinction provides general indications but is by no means comprehensive. To achieve better matching it is important to follow guidelines provided by previous experience and the taxonomy of *Acer*.

Acers have a complex taxonomic classification, subject to amendment, involving the recognition by recent authorities of approximately 16 distinct sections, further subdivided into 22 series*. In virtually all cases, grafts should be made from rootstocks and scions in the same section and, for many, only partners from the same series are suitable. This emphasises the link between taxonomy and compatibility, discussed in Chapter 12. An interesting feature of Maples is the parallel between compatible graft unions and existing natural or artificial hybrids. In nearly all cases both are from within the same section or series[†]. A notable exception is the remarkable hybrid *A.* 'Purple Haze', a cross between two sections, Trifoliata (*A. griseum*) and Acer (*A. pseudoplatanus* 'Atropurpureum'). Although not in the front rank of ornamental trees, when used as an intermediate stem in the double-working technique, this hybrid has the ability to form a compatible graft union between the rootstock *Acer pseudoplatanus* and the scion, *Acer griseum*. So far, these double-worked grafts have provided very high takes and appear to be healthy and sucker free.

Tables below set out guidelines for making broad decisions based on sections and series (Table 21.1), and for deciding more precise, suitable combinations (Table 21.2). Specific examples of compatible unions are given and suggested. These are based on a literature search, discussions with other grafters and personal experience.

In Table 21.2. where more than one option for a rootstock is shown, those listed on the same line indicates an equal chance of success; those listed on line(s) below are increasingly less likely to be successful.

Recent work on phylogeny[‡] re-emphasises taxonomic difficulties in *Acer* and shows many unresolved issues. Some results support suggestions made for compatible combinations, for example a link between *A. sieboldianum*, *A. elegantulum* and *A. pubipalmatum*. The linkage between graft compatible *A. campestre* and *A. miyabei* is one of the most highly supported relationships within the phylogenetic work. However, other close connections, such as between the incompatible species,

---

* van Gelderen, de Jong & Oterdoom (1994).
[†] van Gelderen (2001).
[‡] Harris, Frawley & Wen (2017).

TABLE 21.1

Acer – Graft Compatibility by Section and Series*

Section	Series	Suggested Rootstock (In Order of Preference)
Parviflora	Parviflora	*?pseudoplatanus*
	Distyla/Caudata	*? palmatum ?spicatum*
Palmata	Palmata	*palmatum*
	Sinensia	*palmatum*
	Penninerva	*palmatum*
Wardiana	–	*?davidii* or other Macrantha
Macrantha	–	*davidii* or other Macrantha. *A pseudoplatanus* has proved satisfactory for *A. pectinatum* and its forms
Glabra	Glabra	*?spicatum/?pseudoplatanus*
	Arguta	*? palmatum/?pseudoplatanus*
Negundo	Negundo	*negundo*
	Cissifolia	*? negundo*
Indivisa	–	*carpinifolia*
Acer	Acer	*pseudoplatanus* for some sp.
	Monspessulana	*monspessulanum/?pseudoplatanus*
	Saccharodendron	*saccharum/?pseudoplatanus*
Pentaphylla	Pentaphylla	*pseudoplatanus/rubrum*
	Trifida	*buergerianum/?palmatum*
Trifoliata	Grisea	*griseum/rubrum/triflorum/saccharum*
	Mandschurica	*mandschuricum/rubrum*
Lithocarpa	Lithocarpa	*sterculiaceum/?pseudoplatanus/?macrophyllum*
	Macrophylla	*macrophyllum/?pseudoplatanus*
Platanoidea (species groups in the adjoining column)	*A.campestre/miyabei*	*campestre*
	A.platanoides	*platanoides*
	Most other species in series	*cappadocicum*
Pubescentia	–	*pseudoplatanus*
Ginnala	–	*tartaricum*
Rubra	–	*rubrum/saccharinum/saccharum*
Hyptiocarpa	(Tropical species)	unknown

*Amended from van Gelderen (2001).

A. pseudoplatanus and *A. platanoides*, highlight uncertainties in the relationship between phylogeny, taxonomy and graft compatibility. At present, taxonomic, rather than phylogenic factors, may be more important for assessing *Acer* graft compatibility. Hopefully future phylogenetic work may further resolve relationships in *Acer* and provide guidance toward better compatible graft combinations.

The role of double-working *Acer griseum* to *A.* 'Purple Haze' on an *A. pseudoplatanus* rootstock has been described in Chapter 12: Compatibility, and can provide excellent plants (Figure 21.1). There may be other satisfactory combinations using this technique, which will overcome the problems of grafting the more difficult sections and series.

Once the graft has united, delayed incompatibility symptoms in *Acers* are comparatively rare. There are exceptions, notably with *Acer rubrum*, when after time grafted cultivars of this species can fail; consequently, *A. rubrum* cultivars are normally propagated by cuttings. Occasionally, apparently compatible grafts can show adverse interactions and reduced growth rates may sometimes result, an example being seen with *Acer wilsonii* in Figure 21.2.

TABLE 21.2

Graft Compatibility by Species*

Species	Rootstock (In Order of Preference)	Comments
acuminatum	*palmatum, spicatum, pseudoplatanus, buergerianum*	Although not in the same series, has survived over 7 years on *palmatum* but slow growing. Worth trying on spicatum.
argutum	*spicatum, pseudoplatanus, palmatum*	Said to be easy on spicatum. Failed on *pseudoplatanus*.
binzayedii	*saccharum/rubrum*	Not confirmed.
x bornmuelleri	*monspessulanum, pseudoplatanus, campestre*	Has failed on *pseudoplatanus* despite recommendations. Failed on *campestre*.
buergerianum & cvs	*buergerianum*	–
caesium, caesium ssp. giraldii	*pseudoplatanus*	Difficult and often subsequently fails.
calcaratum (osmastonii)	*palmatum, campbellii*	More closely related to campbelii but this has proved less satisfactory than *palmatum*.
campbellii subspecies, *flabellatum, sinense, wilsonii*	*palmatum, campbellii*	*A. palmatum* much more satisfactory than *campbellii*.
x boscii	*tartaricum*	No evidence of success.
campestre cvs	*campestre*	–
capillipes cvs	*davidii*, or other Macrantha	–
cappadocicum forms and cvs	*cappadocicum, platanoides*	*A. platanoides* less successful than *cappadocicum*.
carpinifolium cvs	*carpinifolium*	No other known compatible rootstock as it is in its own section (Indivisa). A primitive species therefore *Dipteronia* may be worth trying.
caudatifolium	*davidii, pseudoplatanus*	Has thrived many years on *pseudoplatanus*.
caudatum and *subspecies*	*spicatum, pseudoplatanus, palmatum*	*A. spicatum* produced from cuttings if no seed. No other reliable rootstocks available.
chapaense	*cappadocicum*	Successful on *cappadocicum* which was selected as it has milky sap. Originally listed in section Palmata but now placed in Platanoidea.
circinatum cvs	*palmatum*	*A. palmatum* superior to *circinatum*, which has produced lower takes and is slow to callus.
cissifolium	*negundo*	Failed in many trials.
x conspicuum cvs	*davidii* or other Macrantha section, *pseudoplatanus*	Growth rate and ultimate size less when grafted to *pseudoplatanus*.
cordatum and forms	*palmatum*	Has grown well on this rootstock.
coriaceifolium	*buergerianum, pseudoplatanus*	In Buergerianum Section, but *pseudoplatanus* has also been suggested. Failed here on this with *pentaphyllum* used as an inter-stem.
x coriaceum	*monspessulanum, pseudoplatanus*	Not successful here on *pseudoplatanus*. Surviving for some years on *monspessulanum* but significant suckering.
crataegifolium forms and cvs	*davidii* or other Macrantha section	*A. palmatum* is also said to produce a compatible union.
ceriferum (robustum),	*palmatum*	–
davidii forms and cvs	*davidii* or other Macrantha section	Will succeed rather poorly on *pseudoplatanus*.
diabolicum and forms	*rubrum, pseudoplatanus*	Not generally successful.
x dieckii	*platanoides*	–

(Continued)

TABLE 21.2 (CONTINUED)

Graft Compatibility by Species

Species	Rootstock (In Order of Preference)	Comments
discolor	*buergerianum* or *palmatum*	Short-lived on *palmatum*.
distylum	*palmatum?*	Occasional success on this rootstock is claimed.
divergens	*platanoides, cappadocicum, campestre*	Possibly short lived on *cappadocicum*. Not good on *campestre*.
duplicatoserratum and forms	*palmatum*	–
x durettii and forms	*monspessulanum*	–
elegantulum and forms	*palmatum*	–
erianthum	*palmatum, campbellii*	Not easy on *palmatum*, very difficult on *campbellii*.
fabri	*palmatum* *buergerianum*	–
x freemanii cvs	*rubrum, saccharinum*	May eventually show incompatibility on some seedling rootstocks.
ginnala cvs	*ginnala, tartaricum*	–
glabrum and forms	*pseudoplatanus*	Unconfirmed.
griseum cvs and hybrids	*griseum,* 'Purple Haze', *rubrum, saccharum*	*A. rubrum* has been successful. *A. saccharum* claimed to be successful for the type and some hybrids. Claims that *triflorum* or *maximowiczianum* are compatible have not been supported by experience here.
heldreichii forms *(trautvetteri)*	*pseudoplatanus*	–
henryi	*negundo*	All attempts have failed here.
x hillieri and forms	*cappadocicum*	
hyrcanum and forms	*monspessulanum, pseudoplatanus*	*A. pseudoplatanus* less satisfactory than *monspessulanum*. Both difficult.
japonicum forms and cvs	*palmatum*	–
laevigatum,	*palmatum*	–
subsp. *lobellii*	*cappadocicum, platanoides, pseudoplatanus*	Much better on *cappadocicum* than *platanoides* and particularly *pseudoplatanus*. The latter is not a recommended combination, with poor takes and many later failures.
longipes forms and cvs	*cappadocicum, platanoides*	Much better on *cappadocicum* than *platanoides*.
macrophyllum cvs	*macrophyllum, pseudoplatanus*	*pseudoplatanus* not successful here.
mandschuricum	*rubrum, buergerianum, griseum*	Results always poor. Most find those listed incompatible.
maximowiczianum and hybrids	*rubrum, saccharum, buergerianum, griseum,*	Results always poor. Most find those listed incompatible.
micranthum	*davidii* or other Macrantha section	–
miyabei	*campestre, platanoides* or *cappadocicum*	Results best on *campestre*. Failed here on *platanoides/cappadocicum*.
mono and forms	*cappadocicum, platanoides*	Results better on *cappadocicum*.
monspessulanum and forms	*monspessulanum, campestre, pseudoplatanus*	Neither *campestre* or *pseudoplatanus* have succeeded here.
morifolium	*davidii* or Macrantha series	–
negundo forms and cvs	*negundo*	–

(Continued)

TABLE 21.2 (CONTINUED)
Graft Compatibility by Species

Species	Rootstock (In Order of Preference)	Comments
nipponicum	?	*A. pseudoplatanus* and *palmatum* suggested; both have failed here.
oblongum and forms	*palmatum, buergerianum*	Has grown well for 2 yrs. on *palmatum*.
obtusifolium (orientale)	*monspessulanum, pseudoplatanus, campestre*	*A. pseudoplatanus* and *A. campestre* have been suggested but unlikely to succeed.
oligocarpum	*palmatum*	–
oliverianum and forms	*palmatum*	–
opalus	*monspessulanum or pseudoplatanus*	*A. pseudoplatanus* has failed here, *monspessulanum* a better choice?
palmatum forms and cvs	*palmatum*	–
pauciflorum and forms	*palmatum*	–
paxii	*buergerianum, palmatum*	–
pectinatum forms and cvs	*davidii* or Macrantha series, *pseudoplatanus*	This species performs well on *pseudoplatanus*.
pennsylvanicum forms and cvs	*davidii* or Macrantha series, *pseudoplatanus*	Grows slowly on *pseudoplatanus*.
pentaphyllum	*pseudoplatanus, rubrum, saccharinum or saccharum*	Has made successful lasting union on *pseudoplatanus*.
pentapomicum	*pseudoplatanus*	Poor results.
platanoides forms and cvs	*platanoides*	–
x pseudo-heldreichii	*pseudoplatanus*	–
pseudoplatanus forms and cvs	*pseudoplatanus*	–
pseudosieboldianum and forms	*palmatum*	–
pubipalmatum	*palmatum*	–
pycnanthum cvs	*rubrum, pseudoplatanus*	Satisfactory on *pseudoplatanus*, longevity not known.
robustum and forms	*palmatum*	–
x rotundilobum	*monspessulanum, pseudoplatanus*	Poor results on *pseudoplatanus*.
rubescens and forms	*davidii* or Macrantha series	–
rubrum forms and cvs	*rubrum*	Long term compatibility doubtful.
rufinerve forms and cvs	*davidii* or Macrantha series	–
saccharinum forms and cvs	*saccharinum, rubrum, pseudoplatanus*	Long term compatibility not guaranteed. Results on *pseudoplatanus* poor.
saccharum forms and cvs	*saccharum, rubrum, pseudoplatanus*	Long term compatibility not guaranteed.
sempervirens and forms	*monspessulanum, pseudoplatanus*	Results poor especially on *pseudoplatanus*.
shirasawanum forms and cvs	*palmatum*	–
sikkimense and forms	*davidii* or Macrantha series, *pseudoplatanus*	Has formed a lasting union on *pseudoplatanus*
sinense	*palmatum*	–
sinopurpurascens	*pseudoplatanus*	Results poor.
skutchii	*saccharum/rubrum*	–
spicatum forms	*spicatum*	Type species propagated from cuttings.

(Continued)

TABLE 21.2 (CONTINUED)

Graft Compatibility by Species

Species	Rootstock (In Order of Preference)	Comments
sterculiaceum and subsp. *franchetii*	*rubrum* *pseudoplatanus, macrophyllum*	Some success on *rubrum* Results poor on *pseudoplatanus. A. macrophyllum* not tested here.
tartaricum forms	*tartaricum (ginnala)*	–
tegmentosum and forms	*davidii* or Macrantha series, *pseudoplatanus*	Best on Macrantha series.
tenellum and forms	*platanoides*	*A. cappadocicum* worth trying.
tibetense	*platanoides, cappadocicum*	Successful on *platanoides*, should also succeed on *cappadocicum*.
tonkinense	*palmatum, davidii* or Macrantha series	Should succeed on *palmatum* but Macrantha series worth a try.
triflorum and hybrids	*triflorum, saccharum* or *griseum*	Results unreliable.
truncatum forms and cvs	*platanoides*	–
tschonoskii and forms	*davidii* or Macrantha series, *pseudoplatanus*	Successful on *pseudoplatanus* but growth rate slow.
tutcheri	*palmatum*	–
velutinum and forms	*pseudoplatanus*	Results unreliable.
wardii	*davidii* or Macrantha series,	–
wilsonii	*palmatum*	–
x zoeschense and forms	*cappadocicum, campestre*	Poor takes and growth on *campestre*.

*Sources for the above table derived from: Alkemade & van Elk (1989) pp.81–82, Buchholz (pers. comm.), Dirr & Heuser Jr (2006), P.Dummer (pers. comm.), S. Pavloski (pers. comm.), K. Rushforth (pers. comm.), van Gelderen (2001), van Gelderen, de Jong & Oterdoom (1994), Vertrees (1975) and Vertrees (1991).

Phylogenic work may eventually reveal the link between differences in section Platanoidea (milky sap) where three compatibility groups appear to exist:

1. The *A. campestre/miyabei* group, which are most successful when grafted to *A. campestre*.
2. Those which are characterised by persistent green shoots, e.g. *A. cappadocicum, A. longipes, A. mono*, which are best on rootstocks of *A. cappadocicum* (Figure 21.3).
3. Species and hybrids that produce young shoots which quickly become brown, e.g. *A. platanoides, A.* x *dieckii* and *A. truncatum*; these are most successful when grafted onto *A. platanoides*.

If there is any doubt regarding the group to which the plant belongs, experience has shown that as a general rule *A. cappadocicum* produces better takes and provides a better longer lasting union for groups 2) and 3) than *A. platanoides*.

Some species, notably *A. pseudoplatanus**, have the ability to unite successfully with some unrelated species (*A. pectinatum, A. pentaphyllum,* etc.) (Figure 21.3). Other unlikely groupings exist, for example two American species, *A. rubrum* and *A. saccharum*, have produced successful although unlikely combinations, i.e. *A. griseum*, successfully grafted to *A. saccharum* and *A. rubrum; A. triflorum* to *A. rubrum*.

* Teese (1979).

FIGURE 21.1 LEFT, *Acer griseum* 'Fertility' double-worked; the rootstock is *A. pseudoplatanus* with an inter-stem of *Acer* 'Purple Haze'. RIGHT, *Acer pectinatum* ssp. *pectinatum*, on the left showing two grafts on *Acer rufinerve* (Macrantha section) rootstocks. On the right, two grafts of the same species on *A. pseudoplatanus* rootstock. Excellent initial growth is evident when *Acer pectinatum* is grafted to this species.

FIGURE 21.2 *Acer wilsonii* grafted onto *A. palmatum* showing rootstock/scion interaction. Graft union at the same height. Grafts are the same age. The left-hand individual is significantly smaller than that on the right, with some sucker production (bottom left). However, though small, it appears healthy and compatible.

FIGURE 21.3   LEFT, successful use of *Acer pseudoplatanus* rootstock for *A. pentaphyllum* on the left, and *A. pycnanthum* on the right. RIGHT, *Acer zoeschense* 'Annae'; on the left a graft using *A. campestre* as the rootstock, with atypical small, green leaves. Compare this with graft on the right using *A. cappadocicum* rootstock, showing good extension growth and appropriate foliage colour. The graft on *A. campestre* subsequently failed.

Sole species within a particular series, *A. nipponicum* in Parviflora and *A. carpinifolium* in Indivisa, mean that choice of a reliable, compatible partner is not yet possible.

Dipteronia is a genus closely allied to *Acers* and is suggested as a potential rootstock for some species (for instance, *A. carpinifolium*) which, due to problems of incompatibility, have so far failed as grafts. DNA data has shown that some species of *Dipteronia* are closer to some *Acer* species than they are to each other[*]. Attempts have been made to graft *Dipteronia* to *Acer platanoides*, but after leaf emergence the grafts subsequently died, almost certainly due to incompatibility[†]. Further work is required to establish the links between *Acer* and *Dipteronia*.

TIMING

Acers respond well to grafting from mid-January to early spring and again in summer from mid-July to early September. These dates may be extended during the winter period by the use of cold storage and in the summer by individual preferences for an early start, possibly by late June for early breaking species. In Japan, working with *A. palmatum*, some grafters use wood which has not yet hardened during the period late May to early/mid-June, a system described as softwood grafting[‡]. With several species, takes are lower for winter grafting than those in the summer, but the winter period causes less disruption of other essential cultural work and is therefore favoured by many grafters.

Several species commence growth early in the season and for these grafting often starts in January. If the rootstock is pre-conditioned by raising ambient temperature to induce root activity, it is important not to delay grafting until extension growth is seen because in warm callusing conditions the resultant soft growth is liable to become diseased and infect the graft.

* Harris, Frawley & Wen (2017).
† W. Barnes (pers. comm.).
‡ Kawarada (2008).

GRAFTING METHODS

To produce best results, *Acers* demand neat, accurate knifesmanship. Winter grafting usually involves apical grafts; a few use splice*, but generally, veneers are most popular. To mitigate the effects of bleeding, side veneer grafts are sometimes recommended[†] and offer the possibility of re-grafting any failures the following summer. Summer grafting usually involves the use of side veneers, if possible, incorporating a stimulant bud. Some favour placing the graft between two buds on the rootstock. This means that after heading-back a bud is present on either side of the graft union which, it is claimed, provides better and quicker healing[‡] (Figure 21.4).

Grafting height may be varied, from close to ground level, especially for those with attractive bark, to 150–200mm for dissected or weeping forms. Plants produced to decorate patio areas such as *A. palmatum* cvs or selected cultivars of other species, such as *A. pseudoplatanus* 'Esk Sunset', may be grafted onto suitable rootstock stems at 70–100cm high. Some Japanese maples are grafted to produce so-called 'novelty plants' involving several rootstock stems plaited together, each one being top-grafted with a dissected or weeping form at 500–800mm high. Eventually the stems form natural grafts and meld into each other, showing an attractive characteristic woven pattern for many years. Using a splice, two side veneers, or combinations of these, top-worked plants of a range of species and cultivars, e.g. *A.platanoides* 'Globosum' and *A. pseudoplatanus* 'Brilliantissimum', may be grafted onto suitable stems at 2 metres high.

A winter side grafting method, known as 'Stick Budding', has been described for use when there is considerable disparity in diameter between rootstock and scion[§]. The rootstock is induced into early re-growth in late winter by placing it in warm, moist conditions under protection. Normally, after 3–4 weeks, cambial activity makes it possible to detach the rind from the wood and grafting can commence. A T-cut is made in the rootstock, and a scion, comprising a dormant shoot of

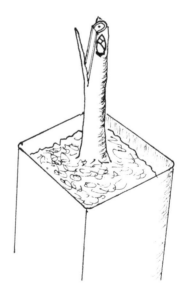

FIGURE 21.4   LEFT, position of scion between two axillary buds after heading-back on a summer side graft of *A. palmatum* 'Shidava Gold'. RIGHT, showing the position of a cut on the rootstock for an apical long tongue veneer graft made in late winter; this allows scion placement between a pair of axillary buds.

---

* Intven & Intven (1989).
[†] Lane (1976).
[‡] D.van der Maat (pers. comm.).
[§] Hottovy (1986).

1-year wood carrying several sets of buds, is prepared by producing identical cuts to those for short tongue side veneers. It is then pushed under the lifted rind as for T-budding. Tying-in is achieved using polythene tape which encloses the union and the scion. Experience with summer grafts, using the same technique, have proven less successful than conventional short tongue side veneer grafts. However, to obtain good cambial matching on such small scions, care and skill is required to perform this graft.

Sealing is essential for hot-pipe and cold callusing systems. When mist or ventilated fog is used during the summer (supported warm callusing), many grafters adopt the use of impermeable enclosing ties and seal any uncovered cut surfaces to prevent free water ingress into the graft. Also, under these conditions, if side grafts are used, it is particularly important to seal the divergence point of rootstock and scion at the top portion of the graft.

Provided condensation drips are prevented, grafts in warm callusing systems can be left unsealed and tied using non-enclosing ties. Personal experience is that *Acer* grafts unite more successfully when left unsealed.

Summer grafts are headed back when dormant the following winter (Figure 21.5 & 21.6), normally carried out using secateurs and leaving a short, 5–7mm, of stem to wall-off the rootstock stem below. Side grafts made in the winter are sometimes headed back in two stages: by approximately half when the union appears well established and 3–4 weeks later, when completed.

ROOTSTOCKS

Most *Acer* rootstocks are raised from seed, and choices on what to grow are made on the basis of known compatible combinations. For *Acer palmatum*, there is an opportunity to choose between subspecies. The majority favour the so-called Little Leaf form, representing plants raised from *A. palmatum* sub species *palmatum*, characterised by relatively small leaves and small fruits. Seed of *Acer palmatum* collected from nursery, arboretum or garden sources often comprises a mixture

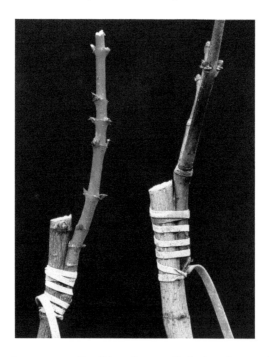

FIGURE 21.5   Headed-back summer grafts of *A. palmatum*. Left, *A. palmatum* 'Winter Flame'. Note axillary buds on either side of the graft as in Figure 21.4. Right, *A. palmatum* 'Ruby Ridge', headed-back. 5–7mm of rootstock stem is left above both graft unions to allow walling-off.

FIGURE 21.6   *Acer palmatum* cultivars overwintered in a frost-free structure, following grafting the previous summer. Having been headed back a few weeks previously, they are just commencing active growth. In the back left-hand corner, some top-worked Patio plant stems are visible.

of all three subspecies (Amoenum, Matsumurae and Palmatum). Collection from strong growing clones of *A. palmatum* subsp. Matsumurae, such as 'Osakasuki', is claimed to produce larger seedling rootstocks with resultant additional vigour in young grafts. Vigorous seedlings are best selected to be grown on for the production of stems for top-working.

Some growers have selected extra vigorous individuals from the mix. These are propagated by cuttings to produce what are considered to be superior rootstocks[*]. Others have selected clones which can be rooted from hardwood cuttings and have produced equally good results when used to supplement seedlings[†].

Stems for working as Patio trees, half-standard or full-standard trees, are usually produced from seedlings graded-out from field plantings. Suggestions for novelty selections have been made, among them use of the rough-barked maple *A. palmatum* 'Arakawa', as a stem for suitable *A. palmatum* cultivars. To encourage additional branching of top-worked *A. pseudoplatanus* 'Brilliantissimum', work in the Netherlands has demonstrated that it is best to use the clones 'Rotterdam' or 'Negenia' as an inter-stem[‡]. The same choice would almost certainly apply to *A. pseudoplatanus* 'Esk Sunset' and 'Prinz Handjery'.

POTTED, PLUG GROWN AND BARE-ROOT ROOTSTOCKS

For most *Acer* species, including *A. palmatum*, grafters favour use of 1+1P rootstocks in the 5–7mm or 7–10mm diameter range. The larger grade being suitable for *A. palmatum* cvs grafted higher at 150–200mm. For summer grafting, rootstocks are lifted and potted in the dormant season but, because of the reduced growing period, larger seedling grades (7mm+) are selected for potting.

---

[*] Wolff (1973).
[†] Curtis (1978).
[‡] van Doesburg, Schalk & Stam (1977) p.119.

Those for high working are often grown in the field and potted or root-balled for subsequent graft-ing, when they become 2+1P or 2+2P with a girth of 7–12mm at grafting height.

Pot sizes for I yr. seedlings are normally in 9 × 9 × 9cm, 9 × 9 × 12.5cm, or 1-litre deep. Rootstocks for top-working are usually potted into 2- or 4-litre deep patterns.

For dormant season grafting of *A. palmatum*, some growers favour lifting suitable rootstocks from open beds before hard winter weather and potting immediately, or after a period of cold stor-age. Grafting is carried out either directly at, or shortly after, potting*. Advantages of this system are no requirement to cultivate the rootstocks for a season before grafting, as well as a likely reduc-tion in problems with root pressure. Disadvantages include the possibility of lower takes compared with established rootstocks and reduced subsequent growth[†].

Use of plug grown rootstocks is viewed with increasing interest. Well-grown 1 yr. seedlings of *A. palmatum* in plugs can provide 80% graftable rootstocks by the following winter and virtually all are useable by the summer. It is claimed that plug grown understocks provide better growth than pot grown or bare-root rootstocks after potting[‡].

Bare-root rootstocks 1+0 or 1+1 6–8mm or 8–10mm stem girths may be used for some systems involving forms of *A. pseudoplatanus*, *A. platanoides*, *A. campestre* and *A. cappadocicum*.

SCION WOOD

One-year-old shoots and, for summer grafts, current year's growth is usually considered best; firm, ripened basal portion should always be included. Scion length is usually between two pairs and three pairs of buds. In normal environmental conditions, cutting the scion 7–10mm above the top buds provides the opportunity for the shoot to wall-off the tissues below[§].

Given sufficient material, larger scions appear to perform well and personal experience, sup-ported by the view of others[¶], favours scions incorporating a portion of 2-year wood for *A. palmatum* and most other species. Catapult shaped scions, the two 'arms' comprising 1-year shoots trimmed to length, and a 'handle' of 2-year-old wood, provide a well-balanced, substantial scion (Figure 21.7). The graft cuts are made on the 'handle'. Easily grafted cultivars of *A. palmatum* can utilise even larger scions which quickly out-perform those with a single shoot of 1-year-old wood (Figure 21.8). Dwarf or densely branched cultivars of various species inevitably need to have scions which include some proportion of 2-year-old or older wood. Some grafters opt for encouraging scion wood of some species, e.g. *A. buergerianum*, to produce longer internodes by growing the mother plants under protection**.

SYSTEMS

Dormant Season Grafting

Cold callusing can only be attempted in the winter/early spring. However, top growth often occurs before union formation is complete, and without the support of an enclosed environment, emerging shoots collapse, and results can be poor. Some of the later breaking, more easily grafted species, such as *A.campestre*, *A.pseudoplatanus* and *A. platanoides*, are successful, but the enclosed condi-tions of warm callusing produces better results.

Warm callusing at temperatures between 15°C and 20°C are generally recommended; higher temperatures, especially when bottom heat is used, can cause root pressure and bleeding. To avoid

---

* Foster (1992), Omar (1962) and van Gelderen, de Jong & Oterdoom (1994) p.30.
[†] Intven & Intven (1989).
[‡] Thompson (2005).
[§] D. van der Maat (pers. comm.).
[¶] Harris (1976).
** Buchholz (2016).

FIGURE 21.7   LEFT, showing two grafts using large scions with 2-year-old basal wood and current season's shoots. RIGHT, *A. palmatum* 'Shidava Gold' grafted using a catapult shaped scion with 2-year-old base. After two growing seasons this compact cultivar has reached 1-metre high and is suitable for retail sale.

FIGURE 21.8   *A. palmatum* 'Seiryu', substantial leafy scions incorporating 2–3-year-old basal growth. Following grafting, the scions dropped their leaves in a poly-tent after 2–3 weeks, which is not unusual, subsequently producing Lammas shoots, usual for this cultivar. With over-wintered grafts in frost-free conditions, the leaves have dehisced by December, by which time the grafts become dormant. The best grade will be saleable at the end of the next growing season.

this, some grafters favour the lower temperature range (15°C) and rely only on space heating. Under these conditions, with unsealed grafts, the small amount of sap exudation produced is not considered a problem. Some species, particularly in the Macrantha section, bleed copiously even when ambient temperatures are as low or lower than 15°C, and severe drying-off is required.

Various responses to temperature and environment are further discussed in the section below on physiological and environmental factors. Good management of enclosed conditions is required, especially for unsealed grafts and is described in detail in Chapter 13.

Hot-Pipe System

Hot-pipe systems do not appear to be widely used for *Acer* grafts, possibly because good alternatives have been devised. Hot-pipe grafting is particularly suitable for some of the more difficult types, *A. pennsylvanicum* 'Erythrocladum'*, *A. conspicuum* 'Phoenix'[†] and those which have very thin wood and require uniting quickly, such as *A.* tenellum or those involving difficult manipulations such as double-worked *A. griseum* (Figure 21.1 & 21.9). The system encourages quick union formation; in *Acer palmatum* up to 22 days in December and, it is claimed, only 10 days in March[‡].

Sections such as Macrantha, particularly liable to root pressure problems and, as a result, posing particular difficulties, respond very well. For this section, pot grown rootstocks should be dried-off before placement and it is best to de-pot the rootstock to facilitate good monitoring of moisture levels.

In general, for hot-pipe systems, potted rootstocks are favoured, but cultivars of *A. campestre*, *A. platanoides* and *A. pseudoplatanus* respond well when grafted to bare-root rootstocks.

Summer Grafting

Popular in the UK and western USA, this system has a significant role in the production of *A. palmatum* but may also be used with good success on the full range of species. For *A. palmatum*, timing is best between mid-July and the end of August but may be delayed to late September and grafters in western North America suggest as late as October[§]. Some recommend that for species other than *A. palmatum*, summer grafting should be delayed until September[¶].

Experience is that the more demanding *Acer* species respond better to side grafting carried out before mid-September; in the UK mid-August is best. De-leaved scions show much less tendency to produce Lammas shoots if grafted late, but callus production at this time is slower than for those made earlier. The consequences of Lammas shoot production and use of leafy and de-leaved scions have been discussed in detail in Chapter 10. The importance and methods of controlling excessive temperatures and maintaining humidity, always a challenge at this time of year, are dealt with in Chapters 13 and 18.

Trials over 15 years of the more easily grafted species carried out at the Lienden Boomteeltproeftuinen Station were summarised in the Boskoop Trial Station Yearbook for 1995**. Grafts of *A. campestre* and *A. platanoides* using apical short tongue veneers made in September were compared with those made in January to March. September grafts were left unsealed and placed in poly-tents, with humidity maintained at 95% RH or above. Because apical grafts were used, leaf retention on the scions produced better results than when scions were de-leaved. The first half of September gave better results than the second. Optimum temperature was 18°C bottom heat. Comparison of results for grafting in early September rather than January/March were better

*   Twombly (1996).
†   Buchholz (2014).
‡   Hummel (1999).
§   Meacham (2008).
¶   van Gelderen, de Jong & Oterdoom (1994) p.30.
** Ravesloot (1995).

for *Acer campestre* 'Elsrijk' and *A. platanoides* 'Crimson King'; equally good with: *A. campestre* 'Columnare', *A. capillipes*, *A. negundo* cvs and *A. lobelii*; less good with: *A. negundo* 'Variegatum' and *A. pennsylvanicum*. The same trial also investigated differences between potted and bare-root rootstocks and produced results slightly in favour of bare-root. Apart from the possibility of better takes the technique has obvious potential for use as a means of balancing grafting schedules and seasonal labour requirements.

A modified cold callusing system used during summer months for *A. palmatum* cultivars is described in Chapter 18. The system has not been described for use in this way by other grafters of *A. palmatum*.

PHYSIOLOGICAL AND ENVIRONMENTAL FACTORS

All Acers show varying degrees of root pressure when grafted. In some sections, especially Macrantha, this is so significant that control of bleeding is the paramount factor in success or failure. For those in other sections, it is an important issue, and some level of drying-off before grafting is standard practice*. Re-wetting successful grafts after drying-off needs careful judgement as the effects of root pressure can remain for a surprisingly long time. *A. cappadocicum* rootstocks, apparently grafted successfully with *A.* x *hillieri*, heavily watered 5 weeks later, responded with copious bleeding from the union, resulting in callus decay, leading to the loss of many grafts. The same effect has been noted with *A.* x *conspicuum* 'Silver Cardinal' and *A.* x *conspicuum* 'Phoenix'. For *A.* x *conspicuum* forms, winter grafting using the hot-pipe system is a successful grafting system (Figure 21.9), presumably because the root system is kept cool and root pressure effectively controlled. See Chapter 11, including Figure 11.21 for further discussion, and for a comparison between the effect of warm callusing and hot-pipe on the root pressure of *A. tegmentosum* 'White Tigress', another species in the Macrantha section.

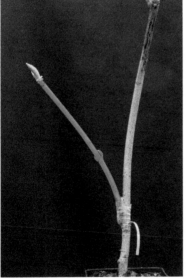

FIGURE 21.9   *Acer* grafts are successful in the hot-pipe system. LEFT, *A.* x *conspicuum* 'Phoenix'. After the formation of an established, strong union, this is a good example of where the stem of the side grafted rootstock can be used as a temporary 'supporting stake' to straighten up the scion which can be tied to it. RIGHT, *A. palmatum* 'Purple Ghost' grafted using a long tongue side veneer graft and placed in a hot-pipe until the union is complete.

---

* Vertrees (1978), Wolff (1973), Foster (1992) and van Gelderen, de Jong & Oterdoom (1994) p.30.

Experience of Macrantha section species placed in rather severe warm callusing conditions during summer grafting has produced exceptionally good results. Applied shade was less than that recommended, and humidity levels were lower than optimum because of imperfect sealing. Under these conditions, grafting takes were above normal; the problem is to gauge the level of severity to achieve good results without exceeding upper limits and causing loss.

Reported side graft treatment to influence water relationships has produced somewhat confusing suggestions. In the Netherlands (Boskoop), the practice of removing all rootstock leaves "to force the sap into the scion" follows a recommendation "to dry for a week or so to decrease sap flow"[*]. In the western USA, some growers reduce top growth of the side grafted rootstock by 50–60%. The reason is not clear, but work at East Malling UK has shown removal of rootstock growth prior to field budding has had an adverse effect on takes[†].

Temperature effects on *Acer* grafts have not been widely investigated but winter and summer grafts vary in their response. For *A. palmatum* grafted during the late winter some recommend 15°C, possibly in an attempt to reduce root pressure. Higher temperatures are recommended for the same species by others, from 25°C[‡] for winter grafts to 29°C for summer grafts[§]. Personal experience of summer grafts of *A. palmatum* indicates that temperatures in the 25–26°C range promote rapid graft development. Petiole dehiscence can occur after 4–5 days, complete after 8 days, and first signs of callus on rootstock and scion in 7 days, plentiful by 10 days. *A. campestre* cvs grafted in late summer responded to supplementary heat of 18°C with increased takes of approximately 70% higher than an unheated comparison[¶].

From work at the Trial Station in Boskoop (Netherlands) in 1968, fungicides used as scion dips for *A. palmatum* cultivars 'Dissectum Nigrum' and 'Elegans Purpureum'[**] can significantly improve takes by between 17% and 32%. For further information on this topic, see Appendix P.

TOP-WORKED PLANTS

Apart from dwarf, compact or globular forms of *A. campestre*, *A. platanoides* and *A. pseudoplatanus*, the main interest in top-working is for the Japanese maples. Cultivars from within *A. buergerianum*, *A.circinatum*, *A. japonicum* and *A. palmatum* are grafted at 70–80cm high, to produce Patio trees. Side veneer grafts using one or two scions are usual. 'Dissectum' forms of *A. palmatum*, particularly 'Baldsmith', 'Crimson Queen', 'Dissectum' ('Viridis'), 'Emerald Lace', 'Filigree', 'Orangeola', 'Red Filigree Lace' and 'Tamukeyama' are among the most popular. *A. palmatum* 'Omure-yama', a strong growing cascading form, can look very effective grafted on suitable stems at 100–120cm high.

AFTER CARE

Summer grafts should be over wintered in frost-free conditions, particularly if Lammas shoots have appeared (Figure 21.6). Very short periods at, or slightly below, freezing can be tolerated by *A. palmatum* grafts after they have become dormant, normally not until after mid-December in the UK.

Warm callused winter grafts need to be kept in protected conditions until growth has fully ripened and danger of frost is over. A period of light shading in the early part of the year is an advantage for most species but eventually many thrive best in full sun. Exceptions are the understory woodland species e.g. *A. crataegifolium*, *circinatum*, and *palmatum*, which are best given some level of shading throughout the growing season.

---

[*]   van Gelderen, de Jong & Oterdoom (1994) p.30.
[†]   Howard & Oakley (1997).
[‡]   Hottovy (1986).
[§]   Thompson (1988).
[¶]   Lienden (1985) p.258.
[**] van Elk (1968) pp.46–47.

PERFORMANCE COMPARISONS

Particular difficulties occur in sections with a limited range of species which are not obtainable as seedlings for use as rootstocks; these include Trifoliata, Glabra, Lithocarpa and Hyptiocarpa. As a result, species from other sections must be used, substantially adding to the difficulties of obtaining good results. The absence of tyloses in the xylem, responsible for controlling root pressure, leads to bleeding, evident in all species, particularly the Macrantha section and especially the *A.* x *conspicuum* hybrids. This feature makes demands on the grafter, requiring a review of grafting methods and skilful control of water content. The huge range of cultivars, especially within *A. palmatum*, demonstrates different levels of ease or difficulty. Some of the small leaved varieties, particularly *A. palmatum* 'Hanami nishiki', are always a challenge. The variegated forms, such as *A. palmatum* 'Oridono nishiki' and some of the fine leaved forms of the Linearilobum group, can sometimes produce disappointing results. Among the most dependable are the strong forms such as 'Osakazuki', and possibly the easiest of all is *A. palmatum* 'Seiryu'.

GROWING ON

Tree forms need tying-in as soon as possible to a suitable support once sufficient extension growth has developed. This is particularly important with Acers as arrangement of the buds often encourages development of a sideways pointing shoot before it can be trained to grow vertically. Side grafts permit early tying-in; before the graft is headed back the stem above can be used as a support to which the scion is tied.

Dissectum forms of *A. palmatum* are liable to produce unbalanced, one-sided growth for the first season or more, resulting in a reduction in value for retail sales. The usual method of dealing with

FIGURE 21.10 *Acer palmatum* 'Tamukeyama' showing use of 'spacer cane' to deflect lateral branches into a symmetrical balanced shape without the need for pruning. Skilfully applied, this technique can save time to reach saleability.

this is to prune to shape. Consequently, reaching saleable size of these rather slow growing forms is delayed by a period, often until the next season. Use of 'spacer canes' (see *Abies* and *Picea*), suitably placed in position to space the branches in the first year, obviates the need for pruning and provides shapely specimens the following season often at an earlier date than would otherwise be possible (Figure 21.10).

22 Actinidia (Actinidiaceae) – Kiwi Fruit

INTRODUCTION

Natives of eastern Asia, *Actinidia* are vigorous, twining, woody climbers. Most are grown for their ornamental value, but *A. deliciosa* is important for producing Kiwi Fruit, otherwise known as Chinese Gooseberry. A number require inter-planting with pollinators to produce fruit. All species root comparatively easily from cuttings, but because of disease problems development of resistant rootstocks for *A. deliciosa* has required plants to be grafted.

COMPATIBILITY

All species appear to be compatible, but there are no reports to confirm this. Further investigation into levels of compatibility and influences on cropping can be expected.

TIMING

Winter grafting using dormant scion material in December to February is normal.

GRAFTING METHODS

Actinidia are easily grafted, and mechanised methods, such as use of grafting tools which produce 'V' notch or inlay type grafts, are often employed. Comparisons between grafting methods have been made*; the less accurate alignment and poorer cuts of mechanical systems resulted in a significant percentage reduction in takes, from 100% for those produced manually, to 79–80% for tool-made grafts. Despite this, economics dictate that mechanised grafting is favoured by most growers.

Pot grown, disease-resistant, rootstocks are normally grafted at a stem height of 300mm or more. This raises the susceptible scion variety above the higher humidity and splash zone of soil level, and chances of infection are reduced. After completion, the graft union is usually tied-in with enclosing, impermeable ties and any exposed cut surfaces are sealed. Callusing is rapid and extensive. Successful grafts grow away vigorously and are available for planting in the same season.

ROOTSTOCKS

Hardiness, tolerance to adverse soil conditions and partial if not complete resistance to *Pseudomonas syringae* pv *Actinidiae* (PsaV), a major bacterial disease of *A. deliciosa*, are now factors of great importance to growers of Kiwi fruit. This has resulted in far more interest in identifying superior rootstocks.

Originally the selection 'Bruno' (1+0 6–10mm) was commonly used because of its vigour and resistance to PsaV. Best producers propagate the rootstock by cuttings (0+1+0 or 0+1+1P), but some use seedlings when genetic variation may produce less resistant or vigorous individuals. Further selections are being made and have resulted in the rootstock 'Bounty 71'. This is less vigorous than

---

* Zenginbal, Ozcan & Demir (2006).

'Bruno' and consequently requires closer planting but has the advantage of tolerance to wet, heavy soil conditions, not the case with 'Bruno', which requires light, well-drained soils.

SCION WOOD

Strong, well ripened 1-year wood is chosen for grafting. Scions are prepared in short sections carrying one or two buds.

High-health-status-licenced scion wood is available from sources in New Zealand. The cultivar Hayward is the standard variety because of its cropping reliability and hardiness, down to −10°C.

SYSTEMS

Cold callusing in unheated, protected structures is the usual system. Supplementary heating is advisable to prevent temperatures falling below 6°C.

PHYSIOLOGICAL AND ENVIRONMENTAL FACTORS

Root pressure does not appear a problem, particularly with relatively high worked grafts. PsaV is inhibited at temperatures above 25°C, consequently, young plants are best grown in warm, dry conditions. Adequate control of Spider Mites (*Tetranychoidea*) will be required in this environment.

GROWING ON

Young grafts are produced in 2–4-litre pots staked and tied to a suitable support for subsequent field planting.

23 Aesculus (Sapindaceae) – Horse Chestnut & Buckeye

INTRODUCTION

Natives of the temperate Northern Hemisphere, Eurasia and North America, the genus includes large, medium and small trees, as well as a few which can be considered as large shrubs. Some are among the most spectacular late spring/early summer flowering trees with large panicles of flowers varying from white, pink, deep pink, yellow or mixtures of these. The Eurasian and Chinese species are of considerable interest for their potential medicinal properties.

COMPATIBILITY

The genus is divided into four sections; the species included within each are listed below:

- Aesculus – within Eurasia – *A. hippocastanum, A. turbinata*.
- Calothyrsus – within southeast Asia & western USA – *A. assamica (A. wangii), A. californica, A. chinensis, A. chinensis* var. *wilsonii, A. indica*.
- Pavia – within southeast USA – *A. arguta, A. glabra, A.flava, A. pavia*.
- Macrothurothyrsus – within southeast USA – *A. parviflora*. Some authorities recognise a further section Parryanae in California which includes *A. parryi*.

With the possible exception of Macrothurothyrsus, covering *A. parviflora*, all others appear compatible between sections and species. *A. parviflora* is reported as compatible on *A. hippocastanum* and *A. glabra**, which leaves only Parryanae as an uninvestigated group.

It is suggested that incompatibility is a factor in the failure of cultivar *A. turbinata* 'Marble Chip', grafted on *A. flava*, but it was successful on *A. turbinata* and *A. hippocastanum*[†]. This raises the possibility that a few genotypes may have particular characteristics which cause problems, (discussed further in Chapter 12).

TIMING

Aesculus seem to be bench grafted only in the winter/early spring period. One grower's trial compared grafting dates for bare-root grafts combined with a hot-pipe system set at 21°C. The earliest date, mid-January, gave a final establishment of 81%, compared with the latest date, mid-March; with only 6% success[‡]; this, however, can be considered exceptionally low, as even for late timing, takes in excess of 75% can be regarded as normal.

Winter grafting is used out of convenience as field budding in the summer months is successful for the stronger growing types, indicating that summer grafting should also be successful. This would need to involve side grafts utilising de-leaved scions. Summer time apical grafts requiring leafy scions would occupy too much space within the poly-tent to be practicable.

---

[*] Dirr & Heuser Jr. (2006) pp.112–113.
[†] Brotzman (2012).
[‡] Cross (1995).

GRAFTING METHODS

Apical grafts are almost always from one of the three main methods: splice, wedge or veneer. Splice grafts can be difficult to match if a thick, large-diameter rootstock such as *A. hippocastanum* is chosen for a comparatively thin wooded species such as *A.* x *mutabilis* 'Induta'. Matching cambia on one side only should be avoided, a wedge graft is sometimes suggested, but the best option is a short tongue apical veneer which, when well executed, guarantees good matching with even the most imbalanced combinations[*].

Aesculus have a strong tendency to produce adventitious buds and shoots on exposed callus tissue of graft unions. Splice grafts reduce this reaction, as less area is present on the narrow graft interface compared with larger amounts on the surfaces of wedge grafts[†].

A modified form of nurse seed grafting has been used for a number of species. The method involves allowing the seed to germinate, and, after development, unlike the conventional system, the plumule and radicle are not removed[‡]. The graft is performed on the young plumule once it has reached a diameter of approx. 6mm. It is cut back to within 10–20mm of the cotyledons and the scion variety is grafted to this using a wedge or an inlay graft of the type often called cleft (Chapter 2). It seems doubtful if the good takes reported using this method represent any increase on those which can be obtained by conventional methods A further disadvantage is that subsequent growth will take place more slowly.

ROOTSTOCKS

A. hippocastanum is the main species chosen and considered suitable for all species[§]. The specification is normally 1+0 8–10 to 10–12mm or 1+1P 8–10mm. Others, such as *A. flava*, *A. pavia*, *A.* x *carnea*, (6–8 or 8–10mm), have been recommended as being more suitable for those with lighter wood[¶]. Difficulties of producing pot grown rootstocks, due to the problems of strong taproot development on young seedlings, may be addressed by using under-cutting procedures in open air seed beds (1/2 u 1/2 +1P); extra deep pots may be required, or producing plug grown seedlings. Bare-root rootstocks are often used.

The large spreading foliage of this genus means that wide spacing between pots is necessary. Shearing back, if required, must be carefully managed to avoid excessive loss of foliage in the growing season. *Aesculus* are greedy feeders, and the potting compost should contain good levels of fertiliser, possibly supplemented by additional top dressing or liquid feeding during the growing season.

SCION WOOD

Most *Aesculus* commence growth early, and this should be reflected in collection dates. Late collection can jeopardise results, and it is of great assistance if cold storage facilities are available to permit early collection and allow provision of totally dormant scion material at a later grafting date. By using such material, cold callusing works best if grafting is delayed until ambient temperatures have risen above the norm in late winter.

Conventional choice of 1-year-old wood is normally recommended. In maritime climates, some *Aesculus* species can produce wood which is 'pithy' and lacking in substance. This problem may be avoided by use of 2-year-old wood at the base of scion to provide material which is well ripened.

---

[*] Macdonald (1986) p.516.
[†] Brotzman (2000).
[‡] Villis (1975), Alexander III (2001) and Brotzman (2000).
[§] Krüssmann (1964) p.366.
[¶] Bean (1976) vol.1 p.250 and McMillan Browse (1979).

Leading shoots often prove too thick for seedling rootstocks and choice of thinner but well-ripened lateral shoots may be best*.

SYSTEMS

Cold callusing, warm callusing and hot-pipe all appear to work well, and choice of system is governed by available resources and expertise.

Use of bare-root rootstocks avoids the need to pot-off into pots too small for the unpruned root system. For seedlings characteristically producing a deep tap root, severe root pruning will be required to enable them to fit into a standard pot, or use of a special deep-pattern pot. Once bare-root grafts have united, potting-off can be into a larger pot capable of accommodating the root system; 2 litre deep is ideal. Hot-pipe systems work well and normally produce results close to 100% take.

PHYSIOLOGICAL AND ENVIRONMENTAL FACTORS

Most grafters consider *Aesculus* among the more easily grafted species. No problems with regard to root pressure have been experienced or reported, but for warm callusing it is wise to use reasonably dry rather than very wet rootstocks. Callus production is rapid and plentiful even at quite low temperatures. In the UK, *Aesculus* can be grafted in the open field with reasonable success. Optimum temperature is probably in the 20–22°C range, but cold callusing at temperatures fluctuating for periods between 8–18°C still provide quite rapid union formation. Higher temperatures, sometimes suggested, are not required.

TOP-WORKED TREES

Slow-growing forms, particularly *A.* x *neglecta* 'Erythroblastos', can be considered for top-working, at between 150 and 200cm high, on stems of *A. hippocastanum*. Initially looking somewhat imbalanced, such top-worked trees, once developed, create a decorative impact in the garden far more quickly than bottom-worked specimens. Interestingly, the disparity in stem girth, which might be expected between *A. hippocastanum* and *A.* x *neglecta*, appears to be diminished by the scion variety; after 20 years, the union between the two components is often barely noticeable.

GROWING ON

Root disruption caused by rootstock bare-root procedures or pot growing, combined with the usual single flush of growth, means that *Aesculus* are rather slow to grow away after bench grafting. They are best in sunny positions in the heavier, calcareous, moist soils.

---

* McMillan Browse (1971).

24 Amelanchier (Rosaceae) – Snowy Mespilus

INTRODUCTION

Restricted to the Northern Hemisphere, the Amelanchiers, otherwise known as Snowy Mespilus or Service Berry, reach their greatest species diversity in northern USA and Canada. They comprise mostly large, multi-stemmed shrubs or small trees producing abundant masses of mostly white or pink-tinted flowers in the spring. Autumn colour adds to their ornamental value.

COMPATIBILITY

Compatibility appears possible for all species within the genus and extends further to include other members of Rosaceae. Trials in Holland have tested *Amelanchier* compared with other genera as rootstocks*. *Pyrus communis* has also been suggested, but it was not part of this trial. Takes on *Amelanchier* averaged 60%; *Cotoneaster* gave 80%; *Sorbus aucuparia*, 90%; *Sorbus intermedia*, 85% and *Crataegus*, over 80%. After 2 years, *Amelanchier* rootstocks produced scion growth of 130cm; *Cotoneaster*, 107cm; *Sorbus aucuparia*, 105cm; *Sorbus intermedia*, 106cm and *Crataegus*, 90cm. Subsequently, *Amelanchier* rootstocks continued to show superior growth performance. If alternatives to *Amelanchier* are considered, Dutch recommendations were to use *Cotoneaster acutifolius*, *Cotoneaster bullatus* or *Sorbus intermedia*.

General opinion in the UK is that *A. lamarckii* or *A. canadensis* provides the best rootstocks, but due attention must be given to grafting procedures to ensure good takes.

TIMING

Grafting is normally restricted to the winter period. Field summer budding is possible which implies summer bench grafting can be considered but is rarely, if ever, used.

GRAFTING METHODS

Apical short or long tongue veneer, even side grafts, can be considered and facilitate good cambial matching. Splice grafts can be satisfactory when scion and rootstock diameters are well matched.

ROOTSTOCKS

1+1P 6–8mm grade pot grown rootstocks are to be preferred over bare-root 1+0 or 1+1 8–10mm; many grafters opt for the latter and achieve good results.

SCION WOOD

Well-ripened, previous years' wood from the basal portion of the shoot is preferred. Where only very thin, light shoots are obtainable, the use of 2-year-old wood at the base will provide satisfactory results.

---

* Joustra (1985).

SYSTEMS

Cold callusing provides a reliable outcome given good cambial matching and good management of grafting environment. Recommendations are for use of potted rootstocks and early scion collection followed by cold storage. This permits grafting dates to extend into the late spring when ambient temperatures are higher.

Hot-pipe systems invariably produce good takes. Ideally, placement should not extend longer than 21 days and high temperatures should be avoided, 18–20°C being optimum. Bare-root rootstocks are most often used with this system.

PHYSIOLOGICAL AND ENVIRONMENTAL FACTORS

As with Rosaceae in general, root pressure does not appear to be a problem. If using pot grown rootstocks, compost is best on the dry side at grafting time.

GROWING ON

Most species naturally form a large, multi-stem shrub, but particular selections may be trained to provide a tree form. Some grafters consider that grafting onto other Rosaceae genera aids this procedure; however, evidence points to the fact that most vigour is achieved when grafted onto a species within the genus.

The following are most suitable for training into trees: *A.* x *grandiflora* 'Ballerina', *A.* x *grandiflora* 'Robin Hill', *A.* x *grandiflora* 'Rubescens' and *A. laevis* 'R.J.Hilton'. Successful grafts should be trained at an early stage, a single shoot being tied-in to a short stake.

25 Aralia (Araliaceae)

INTRODUCTION

A family which includes herbaceous species, but only the woody, *A. elata* is of interest to grafters. Three forms of this species: *A. elata* 'Aureo-variegata', *A. elata* 'Silver Umbrella' and *A. elata* 'Variegata', constitute some of the most spectacular and beautiful variegated, hardy, woody plants. It appears impossible to root stem cuttings, and suckers will produce only green shoots; thus, grafting is the only option.

COMPATIBILITY

Normally grafting takes place on rootstocks of the same species. The alternatives, *A. chinensis* or *A. spinosa*, are also very likely to be compatible but no confirmation of this has been discovered.

TIMING

Aralias are grafted in mid-winter/spring, the huge, doubly pinnate leaves and strong, soft, pithy growth would seem to preclude any attempts at working with the material in the summer.

GRAFTING METHODS

Aralias are mostly grafted using bare-root* or root-balled rootstocks[†].

The structure and arrangement of *Aralia* shoots determines the type of scion material available and the grafting methods used.

Thick, pithy scion wood with widely spaced buds causes difficulty in using methods such as splice; consequently, most grafters use variations of veneer or budding techniques. Occasionally it is possible to obtain terminal shoots of a suitable diameter (Figure 25.1), and these may be splice or veneer grafted provided rootstocks of sufficiently large diameter are available. Apical grafts are popular, and for this material most favour a short tongue apical veneer graft, incorporating two buds if possible.

Aralias also produce spur (shoot) growths surmounted by a single bud (Figure 25.2). These are mostly short with very compressed internodes. With pruning, some may be induced to produce longer lengths with an apical bud and are consequently easier to deal with. In both types, the spur is cut away from the parent stem to produce a single bud trimmed to produce a short tongue veneer scion (Figure 25.3).

Budding methods normally involve a choice between T budding, chip budding or, if the bud is present on a thick stem, patch budding. For apical grafting of *Aralia*, it is always beneficial to make placement of the graft opposite the site of an axillary bud on the rootstock stem. When this bud breaks dormancy during the grafting process, it acts "as a sap drawer", to quote some authorities. A more likely explanation of its role is stimulation of meristematic activity and maintenance of stem viability.

Apical T budding is possible if a scion bud of suitable size can be obtained. Normally, this is only possible from lighter, thinner portions of donor material, often from young stock mother plants rather than older specimens. Suitable buds are inserted under two rind flaps, made by a single

---

* Jaynes (2003) p.495.
[†] van Doesburg & Pellekooren (1962) pp.66–69.

FIGURE 25.1   *Aralia elata* 'Variegata' shoots, the upper shoot is of 1-year old terminal growth carrying two buds. The lower shoot comprises 2-year-old wood with one bud, terminating in a length of 1-year old wood with buds (out of picture). Note: buds in photograph are extending and too late for use as scions.

FIGURE 25.2   *Aralia elata* 'Variegata', short spur growths from the main stems. LEFT, a reasonably vigorous spur shoot. RIGHT, a more typical short spur shoot. Note that the buds on both shoots are too advanced for use as scions.

vertical cut 25–30mm long, extending to the apex of the rootstock which has been headed-back to a suitable height of ±150mm. Before carrying out this procedure, it is important to check that it is possible to separate the rind from the wood. Despite being carried out on dormant rootstocks, the soft nature of *Aralia* wood normally permits this, but to facilitate insertion some trimming back of the bud shield including removal of wood and pith may be required.

Tying-in is adjusted to leave the bud free but firmly fixed. Sealing is said always to be required. Hot waxes should be chosen as small gaps between bud shield and rootstock flaps are inevitable; cold sealants are likely to permeate the graft interface and cause loss. Careful sealing is important as the thick rind of rootstock and scion easily desiccate. If the graft can be warm callused in enclosed, strictly non-drip conditions, it is likely that unsealed grafts will perform better than those which have been sealed.

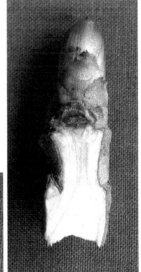

FIGURE 25.3   LEFT, single bud short tongue veneer graft scion prepared from *A. elata* 'Variegata' spur shoot. RIGHT, inner cut face of scion showing wide pith in the basal portion which is typical of *Aralia* wood. Cambial layer is just visible as a thin green line on the inner edge of the rind. The cambium is easily confused with the demarcation line between xylem and pith, also clearly visible. Skilled knifesmanship is required to achieve a good cambial match between rootstock and scion. Note: bud development on this scion is too advanced for practical use and only for demonstration.

Side grafting, involving conventional T budding as used in the field, is an alternative method. Here a T cut is made in the rootstock at an appropriate height of ±150mm. With this method, lifting the rind flaps to allow access for the bud is more difficult than for the apical T budding technique just described. Comparisons between the two methods have not been discovered, but basic principles indicate that T budding, retaining rootstock stem above the graft, should produce better results.

The alternative method, chip budding, is also most easily achieved on lighter, thinner scion wood. Producing a configuration virtually identical to the short tongue veneer, this process is done by placing the knife blade initially on the rootstock across the plane of the wood at a height of ±150mm and pushed in ±6mm to produce a downward sloping cut of 25–40 degrees. This produces a basal cut identical to that of the short tongue veneer. The blade is withdrawn and commencing 35–40mm above the base, an initially scooped cut is made downwards parallel to the stem until it meets the basal cut below; the resultant loose piece of wood (chip) is then removed. A chip bud from the scion is obtained by cuts matching the cambium with that on the rootstock, and also made to ensure the bud is positioned approximately mid-way along the chip. Once completed, the chip bud is placed in position.

Many grafters favour placing the chip bud as an apical graft, but better results might be expected from chips placed as side grafts with rootstock buds above reduced in number. For apical chip buds, tying and sealing can follow the same procedure as for T buds: for those placed as side grafts it is possible to use carefully applied enclosing ties. After grafting, bare-root and root-balled rootstocks are laid-in a peat bark mix within a poly-tent. Using the apical chip budding method, one grafter claimed callus is visible after 3 weeks at temperature settings of 20–25°C and takes of 50–80% are possible*.

Some grafters have developed patch budding to cope with the large diameter stems of mature *Aralia*. Horizontal cuts through the rind are made 10mm above and 120mm below the selected bud.

---

* Jaynes (2003).

These are linked with two vertical cuts and the resultant patch of bark with the bud is removed. The rootstock stem is headed back to 150mm, and an identical sized portion of rind is removed from the apex of the rootstock and replaced by the patch containing the cultivar bud. Tying-in using rubber ties is followed by sealing with grafting wax. Completed grafts are then laid-in sawdust and heated to 15–18°C, 2–3 weeks later suckering on the rootstock begins and grafts are potted individually into suitably sized pots. Variable results from 10–60% are reported*.

ROOTSTOCKS

Most reports describe using bare-root rootstocks, probably because *Aralia* is normally propagated from root cuttings and the straight, rather thick root system produced is not readily potted. Stem girth of plants from this source is rarely less than 15mm and mostly 20–25mm. Seed raised plants normally have thinner stems, 12–20mm in girth.

Dutch grafters favour root-balled rootstocks, generally produced from root cuttings, but seedling production is also possible and probably to be preferred. Recommendations are to prepare the root system before lifting by cutting round established plants lined out in nursery rows. This procedure is carried out in August in two stages. Plants are first cut round halfway only, the other half completed 14 days later. The plants are lifted with a rootball during the following early winter and stored in an unheated but reasonably frost-protected structure. Grafting in the period February–March completes the operation.

Potted rootstocks are not described but experience here is that pot grown seed raised rootstocks provide ideal material. Deep, reasonable capacity pots are required for what will become a substantial plant; 2- or 3-litre-deep pots are suitable, though in view of the value of the finished graft, a 4-litre-deep pot can be justified. Alternatively, use of air-pots may be considered, being particularly appropriate for this group. Compost should contain suitably high levels of feed for what is a fast-growing, large species. Seedlings normally take 2 years in the pot (1+2P) to reach graftable size. Grafts from well grown potted rootstocks outperform others in takes and growth rates.

SCION WOOD

Efficient production of the *Aralia elata* cultivars requires provision of top-quality scion material. Moist, well-drained sites with a naturally rich medium to heavy textured soil are often recommended, these tend to produce thick, rather soft shoots, not ideally suited to matching with somewhat lighter rootstocks. Personal experience is that mother plants are best grown in relatively light, not highly fertilised soils. Adequate moisture is necessary for plants which have large, wide-spreading leaves, and for this reason, wind protection is also important. In the UK, full sun helps the production of well-ripened wood and encourages branching. In sunnier climates, light shading may be helpful to prevent leaf scorch. Ripe wood and balanced growth are encouraged if mother plants are well spaced at a minimum of 1.5 metres each way.

The aim should be to produce as many shoots as possible. Some tying-down of major branches may aid this, especially when supplemented by well-judged pruning involving pinching out terminal buds and some shortening back of extra strong shoots (see Chapter 10). Of the three, *A. elata* cultivars, 'Silver Umbrella' is likely to produce the largest number of suitable shoots for use as scions for splice or short tongue veneer grafts. The thicker wood of 'Aureo-variegata' and 'Variegata' will often only provide material for chip or patch buds.

The soft wood of *Aralia* is vulnerable to attack by Coral Spot (*Nectria galigena*), and after scion wood collection a preventative spray against this disease is a wise precaution. Routine sprays against fungal infection are recommended, particularly after wet weather and humid autumnal days. Some

---

* Leiss (1977).

grafters seal cut surfaces of the mother plants with a fungicidal pruning dressing after scion collection, a wise precaution if time allows.

SYSTEMS

Warm callusing is the system usually chosen for this species. Because of soft sappy extension growth, the grafts are very susceptible to damping-off (*Botrytis*), and some grafters prefer to leave the grafts standing on the open bench (supported warm callusing) in a structure with available supplementary heat. This approach requires careful management to avoid early extension growth becoming dehydrated before the union is able to provide sufficient support. Heavy shading and some spraying over will be required. Personal experience is that the use of drip-free poly-tents and skilled manipulation of shading and ventilation have the potential to produce best results.

For most grafts, hot-pipe would not be practicable; scion buds either constitute part of the graft itself, as in conventional chip or patch budding, or are located close to the graft union and would be enclosed within the heated chamber, resulting in premature bud break. If sufficient length of scion stem can be provided to ensure buds are outside the chamber, hot-pipe systems should produce good results but have not been tested here.

PHYSIOLOGICAL AND ENVIRONMENTAL FACTORS

Thick, sticky sap flow from over-wet rootstocks can potentially cause losses. Drying-off to a moderate level seems to eliminate this problem; extreme drying is not required. The fact that most *Aralia* rootstocks are bare-root or root-balled probably helps reduce excess sap emission. Aralias are basically able to produce callus freely and unite well. Ensuring reasonable cambial matching, managing rootstocks and provision of suitable scion wood are the main problems associated with grafting the genus.

Views on temperature requirements seem to vary: some recommend temperatures in the 15°C region for patch budding,* while others suggest bottom heat temperature set at 20–25°C[†]. Personal experience indicates that a temperature of 20–22°C is optimum.

Use of short scions of spur shoots can result in a terminal bud which is totally dormant or very slow to produce extension growth. Such buds can be induced to extend if treated with a mixture of Gibberellic acid (GA4+7) and Benzyl adenine, applied as a solution at 100ppm. Occasionally, stronger dilutions of up to 500ppm are required and re-treatment may be necessary. On some plants, these spurs are produced in quantity and, if induced to extend by this method, can substantially add to the number of plants produced from a given number of stock mother plants.

GROWING ON

Successful grafts are often bedded up in the field or ideally in a polytunnel, in rich soil that is well fertilised and watered; light shading is beneficial; some winter protection is advisable. Growth from bud grafts is normally rather limited in the first season. Strong grafts can make excellent growth the first year, particularly when produced on potted rootstocks and from scions originating from shoots rather than buds. If kept under protection a significant number can be saleable by the end of the first growing season.

---

* Leiss (1977).
[†] Jaynes (2003).

26 Berberis & Mahonia (Berberidaceae)

INTRODUCTION

Berberis are a large genus of deciduous and evergreen shrubs. Also included in this account are the closely related *Mahonia*, comprising only evergreen species.

Most are propagated by seed or cuttings; some are difficult to root and in the case of some species seed is rarely available. For these and some cultivars grafting is the only practicable means of propagation. Species and cultivars included in this group are the deciduous: *B. chillanensis, B. concolor, B. hispanica, B. montana* and *B. temolaica*; and the evergreen: *B. actinacantha, B. asiatica, B. comberi, B. hakeoides, B.* x *lologensis* cvs, *B. trigona* (*linearifolia*) cvs and *B. valdiviana*. A few Mahonias, mostly species from semi-arid areas in North and Central America, respond similarly to the evergreen *Berberis* and are discussed with them. They include the following: *M. fremontii, M. haematocarpa, M. nevinii, M. swaseyi* and *M. trifoliolata* 'Glauca'.

COMPATIBILITY

No compatibility issues have been discovered in the literature. The experience of those grafting the two genera is to use as the chosen rootstock one of the more easily seed-raised deciduous species, usually *B. thunbergii* or *B. vulgaris*. The *Berberis* species listed appear to thrive and live a normal lifespan when grafted on either of these. *Mahonia* is less certain with regard to initial take or longevity. Indications are that *M. fremontii* and M *trifoliolata* 'Glauca' perform well for many years on *Berberis* rootstocks. To what extent scion rooting has taken place is uncertain; it seems likely that often scion rooting is well developed, and in many examples may well have replaced the original rootstock.

TIMING

Grafting is normally carried out in the mid-winter/early spring period; although, to spread the work load and avoid winter damage in cold areas, some favour early winter (November–December). The main commercial species, *B. trigona* (*linearifolia*) and *B.* x *lologensis* may also be grafted in late summer (August-October). The Mahonias are more demanding, normally grafted early in the season but may respond well to late summer grafting when wood is well-ripened, and any previous winter damage has repaired.

GRAFTING METHODS

Apical grafts are normal; short tongue veneers are favoured, but splice may also be used. For the more demanding Mahonias, the advantages of long tongue apical or side veneer incorporating a stimulant bud should not be overlooked.

ROOTSTOCKS

Both genera are normally grafted onto 1+1P 6–8mm rootstocks of *B. thunbergii* or *B. vulgaris*. Commercial runs of *B. trigona* and *B.* x *lologensis* cvs are occasionally grafted to bare-root rootstocks (2+0 grade).

SCION WOOD

Pruned stock plants produce plentiful 1-year shoots for use as scions. Plants in gardens and arboreta may provide suitable material. If the girth of 1-year shoots is not sufficient, the addition of 2-year wood at the base produces satisfactory results. *Mahonia* extension growth is not always vigorous enough to provide scion wood of young material and use of 2-year, or even 3-year wood, is acceptable.

In many areas the evergreen species are subject to winter damage, and collection dates and timings for grafting should reflect this. Deciduous *Berberis* scions may be cold-stored, securely packed in polythene bags for some weeks at temperatures close to freezing. Evergreen scions should be grafted while still fresh but will be useable for a period (2–3 weeks) if cold stored. To avoid loss of quality, the glaucous foliaged *Mahonia* species should not be stored for more than a few days before use.

SYSTEMS

All the evergreen species require warm callusing; for convenience, the deciduous species may be treated identically. Deciduous species can be cold callused but should not be subjected to temperatures below 6°C for long periods.

PHYSIOLOGICAL AND ENVIRONMENTAL FACTORS

Bleeding has not been recognised as a problem with *Berberis* or *Mahonia*; apart from the usual recommendations for rootstock pots to be 'on the dry side', no further drying-off is necessary. Callus production is plentiful, and with the exception of the Mahonias, the genus normally produces good results. Temperature requirements are low for most species, especially *Berberis*. Mahonias from the warm, arid areas are likely to respond to higher levels (±22°C), but comparisons between this and the more usual 15–18°C have not been made. It is likely, however, that the Mahonias may be maintained at rather lower humidity levels than many hardwood evergreens.

GROWING ON

Most species are easily grown in field or containers. Those from dry desert areas require sharp drainage and winter protection, ideally provided under glass for glaucous foliage *Mahonia* species.

27 Betula (Betulaceae) – Birch

INTRODUCTION

Distributed within the northern hemisphere, birches are important ecologically for forestry and as amenity trees. Horticultural interest for amenity purposes is for bark effect, elegant growth habit and attractive foliage; many have handsome catkins and good autumn colour. The genus hybridises freely and, unless collected from a wild stand, seed propagation for species cannot guarantee trueness to type. Some are micropropagated and a limited range can be produced from cuttings. When wild seed is unavailable, grafting, including field budding, remains the major means of propagation.

COMPATIBILITY

Recent taxonomic revisions of this genus have identified 4 subgenera, 8 sections and 2 subsections[*]. The major group containing most of the known species and hybrids of interest to grafters are found in subgenus Betula; with section Betula containing the so-called white and silver bark birches (*pendula, papyrifera, pubescens,* etc.); section Costatae, the Himalayan and E. Asian white barked species (*ermanii, utilis,* etc.); section Dahuricae, the shaggy/paper barked birches (*nigra, dahurica,* etc.); and section Apterocaryon, the dwarf birches of tundra and peaty swamps (*nana, pumila,* etc.).

Subgenus Acuminata contains some species from the warmer temperate, sub-tropical Sino-Himalayas including Vietnam and Japan (*luminifera, cylindrostachya, maximowicziana,* etc.).

Within both the above subgenera and sections, all species appear compatible, and choices of suitable rootstock species are largely dependent upon availability and suitability for cultivation in a given area. *Betula pendula, B. pubescens* or *B. nigra* appear the most often used and reliable choices.

The remaining subgenera Aspera and Nipponobetula present a more complex picture. Aspera, section Asperae, subsection Asperae contains species *B. calcicola* and *B. delavayi*. These appear compatible with *B. pendula*. However, *Betula schmidtii*, in the same subsection, and *B. chinensis*, in subsection Chinensis, have both produced poor results when grafted to this species, providing only 20% success for the former as well as poor growth and likely delayed incompatibility for *B. chinensis*[†].

In the other section of Aspera, namely Lentae, *B. alleghaniensis, lenta, insignis* and *insignis* subsp. *fansipanensis,* all appear incompatible with *B. pendula*. Surprisingly, within the same group, *B. medwediewii* remains compatible when grafted to this rootstock (Figure 27.1), still healthy and successful after 25 years.

If these findings are confirmed, the simplest option for intractable species in this subgenus should be to use rootstocks of species such as *B. lenta* or *B. alleghaniensis*, or consider the use of *B. medwediewii* as an inter-stem. Personal experience is that even these combinations cannot guarantee success. *B. insignis* subsp. *fansipanensis* is particularly difficult to accommodate with a high level of incompatibility and poor results when grafted to *B. alleghaniensis*; the closely related rootstock species, *B. lenta,* has not been tested.

---

[*] Ashburner & McAllister (2013) pp.20–21.
[†] Tubesing (1988) p.184.

FIGURE 27.1   LEFT, *Betula medwediewii* side grafted onto *B. pendula*. RIGHT, *B. medwediewii* (left) and *B. insignis* (right) grafted onto *B. pendula*. Some live buds just visible on *B. insignis* but the graft eventually failed.

In the subgenus Nipponobetula, attempts at grafting *B. corylifera* failed in the 1960s, possibly due to incompatibility*. This has not been further investigated; unfortunately, a search for other suitable rootstocks may prove fruitless as the species is the sole member within this Subgenus.

TIMING

Dormant season grafting normally takes place from January to early March; although, some favour an earlier start in mid- to late December. Early dates generally produce best results; the reason may well centre on the deep dormancy of the rootstock linked to the possibility that root pressure may not develop until late winter/early spring. Unless prevented by drying-off, pre-conditioning root-stocks to encourage root activity will promote the development of root pressure.

Cold stored dormant scion material permits grafting to be delayed up to late May. By this time, potted rootstocks have developed leafy shoots and problems associated with root pressure are reduced. Grafting may also take place in late summer using potted rootstocks. The advantage of this is the significant reduction of root pressure problems. Surprisingly the method is not often used though it can produce reliably good results.

GRAFTING METHODS

For many grafters, an apical graft such as the splice is favoured, but side is popular because it is thought it may offset the adverse effects of root pressure. Where scion and rootstock diameters are similar, best choice is the long tongue side veneer method, ideally with a stimulant bud. If a heavy grade, ±10mm diameter, bare-root rootstock is grafted with small diameter scion wood, i.e. *B. pendula* 'Dalecarlica', apical short tongue side veneer is a better choice than splice.

Trials in Holland of different grafting methods for *Betula pendula* 'Laciniata' compared wedge (called 'cleft' in the report), apical and side, long and short tongue veneers. Best results were

* P. Dummer (pers. comm.).

achieved with those side grafted, which were marginally better than apical veneers. Wedge (cleft) grafting produced significantly lower takes: 44% compared with an average of 86% for the others*. The sloping side veneer can be used where root pressure is a potential problem due to insufficiently dried-off rootstocks. It is essential that it is not sealed after tying-in (Figure 27.2).

Tying and sealing procedures have been covered in Chapter 6. For species such as Birch, which bleed copiously, the importance of leaving grafts unsealed cannot be over-emphasised.

ROOTSTOCKS

In Europe the favoured rootstock is *B. pendula*. *Betula pubescens* is an obvious alternative but more susceptible to the disease Grey Mould (*Botrytis cinerea*)[†] and more likely to produce suckers[‡].

In areas characterised by heavy, poorly drained soil, high rainfall and high summer temperatures, *Betula nigra* is the preferred rootstock[§] but less tolerant of high pH and dry soils than *B. pendula*.

The growth rate and size of *B. pendula* and *B. nigra* are very similar. *B. lenta*, required for species in subgenus Asperae, may require a further season's growth to achieve similar size.

Normal pot size for pot grown rootstocks is 9 × 9 × 9cm. Larger 9 × 9 × 12.5cm is preferable; larger sizes of 1.0-litre-deep pattern are favoured by some.

FIGURE 27.2   Side sloping veneer. LEFT, sap oozing from the graft. The sloping cut acts as a drip guide. RIGHT, sap is seen running down the rootstock stem. A conventional veneer graft would have resulted in sap 'damming up' at base of graft and resultant loss. Note: twist in the tie wraps, not deliberate but of no concern other than aesthetic effect.

---

* Kloosterhuis (1977).
† Lane (1993).
‡ A. Wright (pers. comm.).
§ Bir & Ranney (2000), Upchurch (2008) and Barnes (2006).

A seedling collar size of 3–5mm in diameter when potted will provide suitable rootstocks in the 6–8mm grade after one growing season (1+1P 6–8mm) (Figure 27.3). Well-grown, 1-year seedlings can also produce a graftable grade. Because of the reduced growing period before grafting takes place rootstocks for summer grafting will require larger sizes at ±6mm Ø for potting-up.

Comparisons have been made between successful takes of Birch grafts on well-established potted rootstocks grown for a full season with others severely root pruned and potted in the autumn prior to grafting in February. Results gave 'takes' of 100% for the former compared with 68% for autumn potting*.

Plug Grown Rootstocks

With the aid of protection, good compost and nutrition, seedlings of 6–8+mm in diameter can be produced in a single growing season from >150cc. cells (Figure 27.3). The plug also simplifies drying-off because in use the root system and surrounding compost are visible. In hot-pipe systems, moist but not over-wet plugs laid-in damp peat or covered with damp capillary matting are recommended.

FIGURE 27.3   LEFT, *Betula pendula* 1+1P rootstock in 9 × 9 × 12.5cm rootstock pot, ready for grafting. RIGHT, *Betula pendula* 1-year, plug grown seedling, for light wooded types suitable for grafting immediately, or for stronger growing species requiring to be pot grown a further season.

* van Doesburg (1961).

BARE-ROOT ROOTSTOCKS

Rootstocks will need to be between 6 and 10mm in diameter at the collar. The heavier grade may not be ideal for thin scion wood and grafting methods will need to be adjusted accordingly. Larger grade rootstocks are usually able to promote stronger scion growth in the first season compared with light grade. Pot grown rootstocks normally outperform bare-root in takes and subsequent growth, even when the hot-pipe system is used.

SCION WOOD

MOTHER TREES/STORAGE

The provision of ripe, strong, relatively thick shoots by managed scion mother trees can have real advantages for grafting this genus (Figure 27.4). For details of establishment and management of scion mother trees, see Chapter 10.

During late winter, Birch trees are particularly liable to 'bleed' when shoots are pruned to supply scion wood for winter grafting. This response is to be avoided; if possible, by collection in early December/January followed by cold or cool storage*. Temperatures during cold storage should not fall below −2°C, or there is the possibility of development of ice crystals in the wood, rendering them useless as scions.

SIZE/QUALITY

For winter grafting, scions of dormant, mature, basal, 1-year wood, furnished with well-developed vegetative buds are recommended[†]. Inclusion of a basal portion of 2-year wood is acceptable and can sometimes produce better results than poorly ripened 1-year old wood (Figure 27.5).

FIGURE 27.4   *Betula* scion mother trees before scion collection (Witch Hazel Nursery).

---

* Lane (1993).
[†] van Doesburg (1962), Lane (1982) and Cave (1982).

FIGURE 27.5 *Betula*, selection of scions for various uses. First from left, collected from arboretum source with 2-year wood basal portion carrying 1-year shoots. Second has a 2-year base with well-spaced 1-year shoots suitable for forming a multi-stemmed tree. Third is a scion of 1-year wood with a 2-year base. Four and five are conventional scions of 1-year wood. Fifth, on extreme right, is from the large growing *B. maximowicziana*.

Suggestions for optimum length range from two buds, 50mm–200mm*; a sensible length is in the region of 120mm. When warm callusing systems are used, certain selections, e.g. *B. utilis* subsp. *jacquemontii* 'Grayswood Ghost', are prone to producing buds which become diseased. The use of long scions increases the number of buds present and the chances of some healthy shoot development.

Grafts made in the late August to late October period use late summer dormant, ripened scion wood, normally taken from the basal portion of current season's shoots, but basal 2-year wood could be included. Scions for summer grafting are best de-leaved before use.

Alternative Scion Wood/Scions for Multi-Stemmed Trees

To obtain sufficient girth and substance, scions collected from un-pruned Birch species, and varieties found in arboreta and gardens, often require wood with a 2- or 3-year-old base. Some 1-year-old wood should always be included to ensure healthy buds are available for extension growth (Figure 27.5 extreme left). This type of scion may also be used to help in the production of multi-stemmed trees, popular for Birch. Best quality 'multi-stem scions', comprising a 2-year-old base furnished with vigorous, healthy, ripe, 1-year shoots suitably arranged on the stem, if possible avoiding a central straight stem, provide a good start for multi-stemmed trees (see scion 2nd from left Figure 27.5).

SYSTEMS

Grafts made with dormant scion wood can be dealt with in various ways:

---

* Lane (1982) and Cave (1982).

COLD CALLUSING

Cold callusing procedures follow those described in Chapter 18 for dormant, deciduous species, carried out in unheated structures, cold glass, ventilated polytunnel* or a cold frame. Potted rootstocks are often used, but bare-root or plugs are alternatives.

Prolonged temperatures below 6°C should be avoided. Normally, apical grafting methods are used and sealing all cut surfaces is essential. An early start, combined with low ambient temperatures, will somewhat supress root pressure and only modest drying-off is required.

WARM CALLUSING

Warm callusing requires a minimum air temperature of 18°C, maintained by supplementary heat, further boosted by sunlight, giving an average of 20–24°C. These conditions result in callus development accompanied by bud burst and possibly extension growth after 3–4 weeks. This may be before the union has completely formed, so a supportive environment is required. Grafting early in the year, December-early February, produces better results than those done later, emphasising the need to commence drying-off sufficiently early.

Potted rootstocks are frequently used, but bare-root rootstocks, laid-in a suitable medium, are an alternative and lessen the risk of 'bleeding'. Bare-root rootstocks rarely provide extension growth comparable with pot grown rootstocks.

Pre-drying of potted rootstocks is vital for success; without this, in warm callusing systems, sealed grafts decline into a slimy, mould-covered disaster. Unsealed grafts with some bleeding can sometimes drain themselves and succeed.

HOT-PIPE

The hot-pipe system has proved highly successful when used for Birches and has to some extent redressed the imbalance in subsequent growth between potted and bare-root rootstocks. Use of plug-grown rootstocks in conjunction with hot-pipe is becoming increasingly popular, as they share many of the advantages of potted rootstocks but can be closely packed within the heated channel.

LATE SPRING/EARLY SUMMER – LATE SUMMER GRAFTING

Late spring/early summer grafting using scions which have been cold stored takes place in unheated polytunnels, cold glasshouses or cold frames. Investigations in the Netherlands over a four-year period (1958–1962) demonstrated that late May grafting, using cold stored scion wood, produced better results than conventional early grafting†.

Late summer grafting is best in warm callusing conditions using de-leaved scions left unsealed. Cold callusing with de-leaved scions at this time is possible and simplifies management but sealing the graft union is then essential.

PHYSIOLOGICAL AND ENVIRONMENTAL FACTORS

ROOT PRESSURE/DRYING OFF

Sap bleeding from wounds during scion collection will alert the grafter to a major problem in this genus which has the potential, once grafted, for excessive rootstock bleeding. Until this was recognised, Birches were difficult to graft and often unsuccessful‡. Methods of overcoming this problem

---

* Lane (2012).
† van Elk (1967).
‡ Krüssmann (1964) pp.379–383.

have been discussed previously. Significant drying-off is required for warm callusing but less critical for cold callusing and hot-pipe systems (Figure 27.6).

Temperature

Many suggest that Birch grafts respond best to rather low temperatures. For winter/spring grafting, recommendations vary from 3–4°C above ambient to a maximum of 14–18°C[*]. Higher temperature, 22–24°C, is used by some, especially in hot-pipe systems[†].

A temperature differential of ±2°C was produced in a hot-pipe heating channel by removing insulation from one side. After 3 weeks, callus development on the 'warm' side (22°C) was significantly more advanced than on the cool side (Figure 27.7), indicating that Birch respond to temperatures higher than those often recommended. Personal experience indicates a hot-pipe temperature of 20–22°C is optimum for most, if not all, species.

Graft Infection Treatments

Birch grafts appear susceptible to infection. In the Netherlands in the 1960s and 70s, soil-borne diseases affected grafts, and investigations demonstrated the benefits of dipping the cut surfaces of scions in fungicide before attaching them to the rootstock[‡]. Grafts treated with carbendazim and thiophanate methyl often produced significantly better results than those untreated. Routine use of fungicidal dips is sensible if there is any possibility of graft infection.

FIGURE 27.6 Rapid drying off 'de-potted' *Betula* rootstocks destined for warm callusing. Those considered suitably dry are left in the pots.

---

* van Doesburg (1961), Lane (2012) and C. Verhelst (pers. comm.).
† A.Luke (pers. comm.) and Hewson (2012).
‡ Perquin (1975).

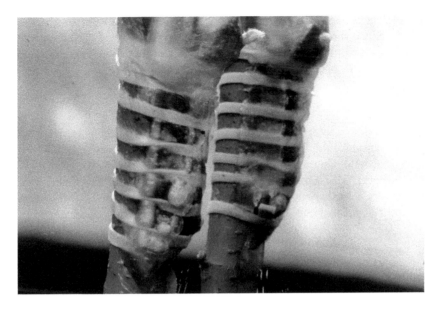

FIGURE 27.7   Effect of temperature on callus formation on *Betula* in a hot-pipe system; on the left the warm cable side, extensive callus formation; on the right, cooler side away from the cable with less callus development.

CARE OF GRAFTS BEFORE, DURING AND AFTER UNION FORMATION

All systems require careful and controlled watering before the union has completely formed. Irrigation applied too early increases root pressure and can result in losses. Allowing some development of sucker growth reduces this risk, but if these become too vigorous, pinch back before final removal once the graft is well established.

Side grafts should not be headed back until good union development and some scion extension growth is evident, even at this stage watering must be controlled. Experience and judgement are needed to decide when removal of suckers and full irrigation is appropriate, but it is always best to err on the side of caution. Droplets of sap or, worse still, sap running down the stem from the cut end of the headed-back rootstock indicates root pressure is a likely impending problem. In worst cases, the only remedy is to de-pot the grafts in the hope of drying them back sufficiently. Even laid-in, bare-root rootstocks can produce excessive root pressure after re-establishment of an active root system if watered too soon. This can result in dramatically lower success rates*, and, for any hope of success, the grafts must be pulled out of the laying-in medium to dry them back.

Successful grafts are sometimes cold stored for a period before subsequent handling and before any extension growth appears. A few grafters have reported problems with cold storing birch following hot-pipe treatments. It is likely that some days in cooler conditions before full cold storage will avoid this problem.

TOP-WORKED AND MULTI-STEMMED TREES

Some weeping forms, notably Young's Weeping Birch (*B. pendula* 'Youngii') and dwarf types such as 'Trost's Dwarf', or *B. nana*, *B. delavayi*, etc., may be top-worked onto stems in the late winter period. *Betula pendula* is usually chosen for this but *B. utilis* can provide a more decorative effect. Stems are normally pot grown and require drying-off, but, as discussed previously, this is less critical when grafting high on the stem.

---

* Deering (1979).

Birch are intolerant of hard pruning, and production of multi-stems is not easy. Use of 'multi-stem scions' as described above can help, but maintaining equal vigour of each stem is always difficult, and requires careful monitoring and pruning during the formative stages. Some selections, notably *Betula* Edinburgh, appear suited to multi-stem production.

PERFORMANCE COMPARISONS

As a general guide, best results can be expected from most *B. pendula* cultivars when using *B. pendula* as the rootstock; although, 'Dalecarlica' and 'Gracilis' are more demanding; *B. nigra* and *B. ermanii* cvs also fall within the easy group. *B. utilis* cultivars can be considered intermediate in difficulty except *B. utilis* 'Grayswood Ghost', 'Silver Shadow' and 'Inverleith', which are less successful. From personal experience *B. pendula* 'Trost's Dwarf' and the *B. albosinensis* forms, particularly 'China Ruby', are the most likely to produce poor results.

GROWING ON

Successful grafts, usually under protection in containers using modern potting compost, and copious watering, respond well to subsequent growing on procedures. After potting-on from potted rootstocks, many of the main commercial varieties of *B. utilis*, *B. ermanii* and *B. pendula* reach 2 metres or more in the first growing season.

28 Camellia (Theaceae)

INTRODUCTION

A large and important genus of evergreen shrubs and trees from east and southeast Asia, found particularly in southern China. Recently, a number of species of potential value for ornament and commerce have been introduced from Vietnam; undoubtedly more will be discovered in future years.

In the west, they are mostly grown for their ornamental value but one species, *C. sinensis*, has major economic importance for the production of tea. Camellia seed oil is produced on a large scale in China from a number of species including *C. oleifera* and *C. chekiangoleosa*. Having numerous uses from culinary and cosmetics to lubrication, 3 million hectares of land produced 164,000 tons of oil in 2011–2012*.

Most Camellias are propagated by cuttings, but a number prove difficult to root, and grafting is necessary.

COMPATIBILITY

Taxonomy of *Camellia* is in a state of flux with authorities disagreeing on the basic number of species, from 90 to 400, and number and descriptions of sections within the genus, from 5 to 14. A recent publication lists 14 sections[†]. The following five sections contain species at present likely to be encountered by grafters:

- Section – Camellia: *C. chekiangoleosa, C. japonica, C. pitardii, C. reticulata, C. saluensis, C. tunganica.*
- Section – Paracamellia: *C. oleifera, C. sasanqua.*
- Section – Protocamellia: *C. yunnanensis.*
- Section – Thea: *C. sinensis.*
- Section – Theopsis: *C. cuspidata, C. lutchuensis, C. transnokoensis, C. tsaii.*

Experience here of grafts of those listed above is that *C. japonica* cvs have proved suitable rootstocks for all, and after five to ten years there has been no indication of incompatibility symptoms. A recent introduction from Vietnam, *C. dongnaiensis* (section unspecified) proved compatible on *C. japonica* and *C. sasanqua* rootstocks[‡]. An IPPS discussion group in 1974 has provided the only report suggesting that incompatibility may occur, one member of the group noting that symptoms of delayed incompatibility were visible around the union of *C. japonica* on rootstocks of *C. reticulata*.[§]

TIMING

The majority of Camellias are grafted between January and early March, a period from August to early September is also possible. Propagation by cuttings-grafts is carried out in the summer period (July–August). Modified nurse seed grafting takes place on recently germinated seedlings in the spring.

---

* Lu, Jiang, Ghiassi et al (2012).
† Parks (2017).
‡ Orel & Curry (2017).
§ Macdonald (1974).

GRAFTING METHODS

Splice, veneer and wedge grafts are used variously by grafters within and between the different grafting systems. An IPPS GB&I (Great Britain & Ireland) discussion group in 1974 reported that splice grafting was favoured by most British grafters. This choice may not be a result of any comparisons made between the methods but rather continuation of a traditional approach.

It is suggested that Camellias, especially *C. reticulata*, are poor callus producers, a reason for some to recommend approach grafting for this species rather than attempting normal detached scion methods[*]. Reluctant callus production is supported by personal experience and that of others[†], emphasising the need for accurate cambial alignment and contact. The conclusion is to use veneers as the preferred graft, and to aid good matching there is a case for cutting the scion before the rootstock.

The characteristics of side grafts for supporting rootstock and scion health is an important factor in deciding the best strategy to follow. The short tongue side veneer is often advocated[‡], but the advantage of the additional cambial contact and callus production of the long tongue side veneer should not be overlooked[§]. If a stimulant bud and associated leaf, preferably reduced by a half, can be included on the tongue results may be further improved (Figure 28.1). Well-matched cambial alignment may be achieved more easily by using the short tongue side veneer; grafters must decide their best option. Side veneer grafts are also normally recommended for cuttings-grafts, a procedure described fully below in Grafting Systems.

FIGURE 28.1   LEFT, *Camellia* 'Royalty', a *C. reticulata* x *japonica* hybrid. Long tongue side veneer, extension growth seen on rootstock and scion and the graft is ready for heading-back. RIGHT, *Camellia reticulata* 'Valentine's Day', showing a long tongue side veneer with stimulant bud and associated leaf which has been reduced by half, extension growth just commencing.

---

[*] Hartmann, Kester, Davies Jr. et al. (2014) p.453.
[†] Hellriegel (1982).
[‡] Sheat (1965) pp.72–73.
[§] Wells (1955) p.218.

Apical wedge grafting (in some papers described as cleft) is a popular method for Camellias*. A grafter in the USA has described a variation of the wedge graft involving two cuts, one conventionally down the centre of the headed-off rootstock stem, into which the scion is placed; the other bisects it at the apex but cut at an angle to the original to provide a drainage channel for any excess sap formation around the graft interface[†].

Side wedge grafts are recommended by others but because of the inability to be certain of a perfect match, it may be necessary to align the cambia on one side of the graft only, and 'church windows' are required[‡]. Wedge grafts are also used for modified nurse seed grafting, to be later discussed in Systems.

As a successful method of grafting small, thin scion wood to stouter thicker rootstocks, the so-called 'bark graft' is described by Chinese authors[§]. Close-up photographs show an apical wedge graft placed off-centre very closely to the rind. Cambium of the wedge-shaped scion cuts are well-matched to the cambium of the rootstock on the outer face only. Successful grafts can be expected to grow away strongly due to the influence of the large rootstock.

Rind grafts are used by Chinese grafters when old mature plants are headed-down and used as rootstocks. The method was used to convert *C. oleifera* plantations to more commercial varieties for amenity use after a sharp reduction in demand for Camellia seed oil (Figure 28.2).

ROOTSTOCKS

When possible, seedling rootstocks should be chosen over those produced from cuttings because they have the advantages of juvenility and vigour. They are also unlikely to be infected with virus disease, which is always a potential problem with vegetative propagation. Views on the choice of best species vary; *C. sasanqua* has been suggested as having a healthier root system than the

FIGURE 28.2   Using rind grafts, *Camellia oleifera* was grafted the previous season with an ornamental variety (photo: Maurice Foster).

---

* Macdonald (1986) p.557, Zhang, Li, Guo et al (2016) and Dirr & Heuser Jr. (2006) p.133.
† Wells (1986).
‡ Clark (1976).
§ Zhang, Li, Guo et al. (2016).

alternative, *C. japonica*. It is suggested that *C. sasanqua* is resistant to Phytophthora root rot*, a view supported by work at the North Carolina Cooperative Extension Service[†]. *Camellia hiemalis*, which may be a hybrid of *C. sasanqua*, is a possible substitute[‡], *C. hiemalis* 'Kanjiro' being the favoured clone.

Personal experience has not confirmed these suggestions as the best choice. In the UK, problems are concerned with the difficulty of obtaining sufficient stem girth with *C. sasanqua* or *C. hiemalis* to match thick wooded species, such as *C. reticulata* or strong forms of *C. japonica*. To achieve equivalent size the plant must be held in the rootstock pot for one or two seasons longer than for the strongest selections of *C. japonica*. After such extended periods, *C. sasanqua* has proved to be as susceptible to root health problems as any of the alternatives.

Setting aside the desirability of seedlings, the usual alternative is to root selected types from cuttings and grow these on to produce rootstocks. The requirements are for a vigorous, upright, strong, thick wooded variety able to produce the necessary girth as quickly as possible in the 6–8mm diameter range. Other attributes are no evidence of virus infection, relatively easily rooted from cuttings and the ability to produce a healthy disease-free root system. This requires the correct choice of species and cultivar, good propagation procedures, suitable growing medium and the best growing conditions.

Comparatively few alternatives are available. *C. japonica* cultivars are the obvious choice and any vigorous form can be considered. Ideally, the selection should be from one of the following: 'Berenice Boddy', 'Grand Slam', or 'R. L. Wheeler'. Another recommendation is *C. japonica* 'Debutante'[§] but this does not provide sufficient stem girth in the UK to warrant inclusion among the first choice. The primary candidate, fulfilling most of the requirements, has proved to be 'R. L. Wheeler'. The *C. japonica* cultivar 'Kumagai' is gaining in popularity in the USA. Its main attribute is said to be its healthy root system[¶]. An alternative which should be particularly suited as a rootstock for *C. reticulata* cultivars and hybrids is *C.* 'Leonard Messel' (*C. reticulata* x *C.* x *williamsii* 'Mary Christian'). Unfortunately, it is less easily rooted than the *C. japonica* forms, and because of the influence of *C.* x *williamsii*, it has rather less stem girth at an equivalent age.

Conventional grafts are always made on potted rootstocks, usually $9 \times 9 \times 9$cm though $9 \times 9 \times 12.5$cm is a better choice; some favour a 1-litre-deep pot which may be even more preferable. The plants are normally produced as 0+1+2P 6–8mm, i.e. held in the rootstock pot for two seasons, and periodic additional nutrition in the form of top dressing or liquid feed is required.

Compost and growing conditions are crucial to success. Camellias are prone to root disease, especially when held for a period in the comparatively small pots suggested. In the UK, a high percentage of pine bark incorporated into the mix reduces the incidence of this problem. Skilful use of slow release fertilisers and liquid feeding allows use of up to 50% bark with resultant improvements in root health. Capillary bed culture, light shade during the growing season and winter protection to avoid any possibility of freezing temperatures at the root zone all contribute to the production of top-quality rootstocks.

The favoured side grafting method normally requires some light trimming during the growing season to maintain correct stem height to facilitate tying-in. If possible, a single stem only should be produced, but light branching may be unavoidable. Routine monitoring for pest and disease is essential. Rootstocks of appropriate size, vigour and good health, especially of the root system, are a crucial factor in successful grafting of Camellias.

---

* Trehane (1998).
[†] Benson & Jones (2015).
[‡] Clark (1976).
[§] Wells (1986).
[¶] B. Green (pers. comm.).

SCION WOOD

Most *Camellia* scions comprise the basal portion of 1-year-old shoots. Length of scion is normally in the range 75–120mm; for most varieties this will involve removal of an upper stem portion. This is particularly important with summer dormant scion wood when scions may not have become fully ripened along their entire length. Leaf area on the scion should be reduced by removal of all but 2–3 leaves, and these may have their size reduced by up to a half.

Use of a 2-year basal portion of scion material is permissible for dwarf varieties such as *C. sasanqua* 'Tanya', where extension growth is limited and grafting on a short stem is required to produce a novelty plant (Patio Plant). Scions from senile plants may also require use of older wood because of limited extension growth.

Healthy, well-ripened, 1-year wood with no evidence of winter damage is the ideal choice to achieve best results.

SYSTEMS

WARM CALLUSING

The most important system for *Camellia* production is warm callusing of side grafts on potted root-stocks. Procedures for all evergreen species have been fully described in Chapter 18. Differences in techniques between winter and summer grafting centre on the need to control excessive summer temperatures while at the same time maintaining high humidity levels, see Chapter 13.

Enclosed conditions are usually provided by conventional poly-tents. Unsealed grafts are best, but Camellias do not respond well to any drips falling on the union; therefore, it is important that poly-tents are drip free. Some growers in the USA* and South Africa cover individual grafts with glass jars (wide mouth ±1 litre capacity). It is claimed that these provide ideal conditions for high humidity and cleanliness. Covering grafts individually with a polythene bag or plastic lemonade bottle with screw top replaced and base removed is another option.

CUTTINGS-GRAFTS

Less often used is the technique of cuttings-grafts. This system has been reported by a number of grafters[†] and is claimed by some to have the advantage of producing a budded plant in 3 years. This compares with the traditional grafting system involving use of a rootstock which is 4 years old after completion of the whole grafting procedure[‡].

The method involves side grafting the scion to a suitable, easily rooted rootstock previously prepared as a cutting. Choice of appropriate variety can follow those suggested for rootstocks. In the USA, the clone *C. sasanqua* 'Kanjiro' (regarded in the UK as *C. sasanqua* 'Hiryu') appears to be the main candidate for this purpose. Basal portions of 1-year shoots are severed from the stock plant close to the junction with the previous season's growth and cut to approximately 125mm long, leaving 2–3 leaves at the apex. IBA hormone treatment, ±8000ppm powder or preferably 2000ppm aqueous solution or quick dip, aids rooting. The scion variety is grafted using a short or long tongue side veneer placed ±30mm above the base of the cutting. Placement should avoid any possibility of IBA contamination of the graft area, and it is advisable to treat after, rather than before, making the graft. It is important to leave the union unsealed.

Upon completion, the graft union is buried approximately 25mm deep in the rooting medium. To minimise subsequent disturbance, it is recommended to use plug cells for each individual cuttings-graft rather than setting out in boxes. Rubber ties are normally suggested, but as the union is buried,

---

* Wells (1986).
† Clark (1976), Dirr & Heuser Jr. (2006) p.133 and Trehane (1998) pp.85–86.
‡ Clark (1976).

there is a danger of non-degradation and girdling. Provided the scion variety produces roots within a suitable timespan, this will present no problem, but if not, cutting the ties may be required, usually a difficult and slow procedure. A degradable tie is therefore preferable and un-waxed cotton or hemp ties may be substituted for rubber. Once well rooted, cuttings-grafts may be lifted, the rootstock headed back to the union and completed grafts potted-on into larger containers. These procedures normally take place during the spring following cuttings-grafts preparation the previous summer.

An interesting variation of the above method has been described by workers from the Tea Research Foundation in Malawi*. The rootstock comprises a single leaf-bud cutting to which a single bud scion is grafted more or less opposite the bud of the rootstock, using a chip bud technique. Retention or removal of the leaf of the chip bud is not discussed, but it would seem wise to retain a small portion (say one-third) of this leaf to maintain chip bud viability. The stem below the union is inserted into the rooting medium within a poly-tent equipped with bottom heat. The graft union itself is retained in a position above the rooting medium; the rootstock therefore equates to an inter-nodal cutting. Rooting takes place after 4 months, and after a further 2 months, the rootstock bud is removed. Whether this includes removal of the rootstock leaf is not clear, but removal of the bud ensures rootstock sucker production is prevented. Total removal of rootstock cutting leaf and bud at the time of grafting resulted in total loss.

A further trial investigated the effect of chip budding the scion variety to single bud cuttings some months after they had rooted. Results were reduced from 84% for the method described to 49% for the trial, probably due to disturbance of the young root system.

Nurse Seed Grafting

Large seeds produced by Camellias make them suitable candidates for nurse seed grafting. Suggested procedures follow the same modified method described for *Aesculus*[†]. Workers in China use *C. fraterna* as the chosen nurse seed, but this species is likely to be too tender for use in cooler temperate areas. Availability of viable seed of suitable species such as *C. japonica* is likely to be a problem. Many commercial seed sources cannot be guaranteed to provide seed which germinates reliably, so hydration and dormancy breaking pre-treatments may be required, and even then, results can be disappointing.

A wedge graft is used once the extending plumule from the germinating seed has reached sufficient size, ±6mm diameter. It is also suggested that the radicle of the germinated seedlings is trimmed back. The scion, taken from dormant material, comprises a single bud and leaf from a portion of the terminal growth of a 1-year shoot; the leaf area is reduced by half. Completed grafts are tied-in with a self-adhesive, non-porous enclosing tie. No details are given but presumably the completed grafts are laid-in a suitable free-draining medium and placed in a supportive environment until a graft union is established. Because such juvenile material is used, callus formation is almost certainly accelerated and enhanced. This, coupled with necessary skill to achieve good cambial matching on such light, easily manipulated material, indicates that graft takes may well be very high. Unfortunately, subsequent growth is likely to be slow.

PHYSIOLOGICAL AND ENVIRONMENTAL FACTORS

Sluggish and comparatively little callus production in many of the *Camellia* species and hybrids makes Camellias among the more difficult temperate woody plants to graft. Cuttings may be rooted with reasonable success for many important commercial varieties; consequently, comparatively few are involved with grafting, adding to a lack of widespread knowledge and experience.

---

* Kayange & Scarborough (1976).
† Zhang, Li, Guo et al. (2016).

There can be little doubt that top-quality, vigorous and healthy rootstocks and scions are a major contributory factor for success. Use of seedling rather than the more usual cutting raised rootstocks should be a factor in raising the number of successful takes. There remains the problem of obtaining reliable supplies of well-grown and graded *Camellia* seedlings.

The influence of root pressure has been rarely discussed but features significantly in some accounts from California*. Personal experience has not identified root pressure as a major problem, but it possibly has an important effect under optimum conditions of growth and vigour, such as might be encountered by Californian growers. Normal recommendations are for using potted rootstocks 'on the dry side' rather than wet. Grafts are best left unsealed.

Views on optimum temperature vary. Some prefer cool conditions and recommend applying external heat only when the temperature falls below 4°C. Most opt for temperatures in the 18–20°C range. Personal experience supports the view that high temperatures can have adverse effects due to promotion of premature bud break of dormant scions. Too high temperature in the summer makes maintenance of appropriate humidity levels more difficult. Cool temperatures in the 4–8°C range seem too low and must significantly delay callus formation; despite this, evidence of callus was claimed after only 3 weeks at these temperatures, surely requiring further substantiation. In the UK, temperatures of 18–20°C have produced the early stage of union development after 4–6 weeks.

TOP-WORKED PATIO PLANTS

There is a limited demand for this high-value product, and most Patio Plants on a 45–75mm-high stem are produced by training a cutting-raised liner into the required shape. To achieve this in an economic time span implies that reasonably vigorous and therefore, in the long term, rather unsuitable varieties are chosen. The few true dwarf types (*C. sasanqua* 'Tanya', *C. x williamsii* Contribution, etc.) are best grafted by top-working onto a trained stem.

PERFORMANCE COMPARISONS

C. reticulata is considered among the most difficult of the genus to graft. The *C. japonica* x *reticulata* hybrids 'Royalty' and 'Valentine's Day' have proved easier, as has *C. reticulata* 'Mandalay Queen', considered a *C. reticulata* cultivar but probably a hybrid with *C. japonica*. Some of the species, while demanding, produce good results, particularly *C. tsaii*, which appears among the easiest.

GROWING ON

Invariably grown in containers, young grafts are potted on into an ericaceous mix and placed under protection throughout the production cycle. Regular pruning is required to ensure shapely plants.

---

* Wells (1986).

29 Carpinus (Betulaceae) – Hornbeam

INTRODUCTION

The genus comprises about 50 species, found mostly in the temperate regions of the northern hemisphere, particularly in China. One warm, temperate/tropical species extends southwards into Central America (*Carpinus caroliniana* var *tropicalis*). They appear to comprise mostly small- to medium-sized trees, often multi-stemmed, but some species and individuals can reach large sizes. Their ornamental value is not always considered to be in the first rank, but *Carpinus betulus* does produce a fine dense hedge. Best forms of *Carpinus fangiana*, the Monkey Tail hornbeam, produce dramatic and highly decorative pendant fruit clusters. Many selections of the European hornbeam are used in amenity situations and these, along with some desirable species for which seed is often unavailable, are propagated by grafting.

COMPATIBILITY

Carpinus are divided into two sections: Carpinus and Distegocarpus, the largest numbers of species being within the Carpinus section. Distegocarpus species, distinguished by large pendant fruit clusters, are *C. cordata*, *C. fangiana*, *C. japonica* and *C. rankanensis*.

An investigation into the performance of *Carpinus* on various rootstocks at the Morton Arboretum has provided guidelines for future combinations*. Rootstocks used were *C. betulus* and *C. caroliniana* from Carpinus section, and *C. japonica* from the Distegocarpus. The plants were grafted in 2001 and conclusions drawn over a short time span of 18 months, therefore give no indication of long term compatibility.

On the basis of results from this trial using relatively small numbers of grafts, *Carpinus betulus* was not considered suitable for species in Distegocarpus but acceptable for all in the Carpinus section. *Carpinus japonica* was considered satisfactory only for those in the Distegocarpus section. *Carpinus caroliniana* was not tested for Distegocarpus species, but proved best of the three tested for the intersectional *C. caroliniana* x *cordata* hybrid. *C. caroliniana* did not confer as much vigour on the other species as *C. betulus*.

Personal experience of Distegocarpus species, *C. cordata* and *C. fangiana* grafted to *C. betulus* has proved successful. On a limited number of grafts tested, *C. rankanensis* either failed immediately or after 12–18 months. In the absence of *C. japonica* or similar rootstocks this species would warrant further testing on *C. betulus*. It has succeeded well on *C. fangiana* x *C. betulus* hybrid rootstocks from a botanic garden source.

Compared with rootstocks of *C. betulus*, enhanced performance of Distegocarpus species can be expected from those grafted to rootstocks of *Carpinus japonica*, or seedlings raised from seed of any in the Distegocarpus section (Figure 29.1). However, after some years on *C. betulus* rootstocks, grafts of *C. fangiana* 'Wharton's Choice' and *C. cordata* are showing reasonable extension growth and virtually no sucker production.

In the absence of *C. japonica* for the Distegocarpus section, recommendations at this stage are to use *C. betulus* as a rootstock for all *Carpinus*[†] except *rankinensis*. If they exist, the shortcomings

---

* Wiegrefe (2003).
[†] Alexander III (1998) and Nelson (1968).

FIGURE 29.1 *Carpinus fangiana* 'Wharton's Choice', a successful side veneer graft on *C. japonica*. The rootstock stem on this side graft could be used to tie-in and straighten-up the lead of this young, extending graft until caning is possible.

of *C. betulus* as a rootstock for species in Distegocarpus section may become evident after a long timescale. In dry soils, *Carpinus caroliniana* is suggested as a possible alternative for *C. betulus**.

TIMING

Hornbeams are usually grafted in mid- to late winter using cold callusing or hot-pipe techniques. Before the introduction of hot-pipe, some in the UK grafted *C. betulus* cultivars to pot grown rootstocks of *C. betulus* in August–September. Work in the Netherlands has demonstrated early September as an optimum time for grafting *C. betulus* cultivars.

GRAFTING METHODS

Splice, wedge and apical short tongue are favoured and normally are all that are required for *C. betulus* cultivars; more demanding species always respond to the additional support and cambial contact afforded by long tongue side grafts.

---

* Wiegrefe (2003).

ROOTSTOCKS

Usual grades of *C. betulus* for *C. betulus* cultivars are 1+1 8–10mm, 1+1P 6–8mm or 1+2P 8–10mm. For species having lighter wood, such as *C. kawakamii*, *C. omiensis*, *C. viminea* and *C. fangiana*, pot grown rootstocks 1+1P 6–8mm in girth is preferable.

Bare-root rootstocks may be used for late winter cold callusing. A system developed in the Netherlands has shown that bare-root rootstocks of *C. betulus*, lifted from the field just before grafting in early September, are equally or more successful than late winter/spring timings. For this purpose, transplanted or undercut seedlings of *C. betulus* (8–10mm) are ideal. During the lifting procedure care should be taken to prevent drying-out, and total soil removal from the root system is best avoided.

Plug grown seedlings of suitable girth can be grafted directly; small plugs require moving on into larger ones or potting-on into 9 × 9 × 12.5cm rootstock pots before grafting 9–12 months later (Figure 29.2).

SCION WOOD

Ripened current year's wood is normal. For apical grafts made in September, recommendations are to have leaves retained on the scion, which has been trimmed to length, with the less ripened terminal portions removed. Side grafts made at this time can have scion leaves removed. The very dense, short extension growth of *Carpinus betulus* 'Columnaris' may require inclusion of some 2-year-old wood at the base to obtain sufficient length.

SYSTEMS

Warm callusing of early September grafted *C. betulus* 'Fastigiata' is popular with some grafters in the Netherlands. It involves use of recently lifted bare-root rootstocks (potted can be used) and unsealed apical short tongue, splice or wedge grafts, which when completed are placed in poly-tents or under contact polythene. Temperature is set at 18°C, callusing and union development is usually complete in six weeks. Results for the tested species (*C. betulus* 'Fastigiata') were better

FIGURE 29.2 LEFT, *Carpinus betulus* after 1 year in plugs. This grade (4–6mm) is suitable for immediate use as a rootstock for some species with light scion wood. Moving on into a larger plug or potting-on will be required to produce a 6–8mm or 8–10mm grade for grafting stronger growing types the following September or Jan–March. RIGHT, *Carpinus viminea* grafted directly to a plug grown seedling.

for September grafting than late winter grafting. Although the system was developed for *C. betulus* 'Fastigiata', there is no reason to suppose it would not also be effective for other reasonably strong-growing *C. betulus* cultivars such as 'Frans Fontaine', 'Purpurea', etc. It may not be successful with some of the lighter wooded species and those in Distegocarpus section. Prior to hot-pipe, a similar system for *C. betulus* cultivars using potted rootstocks was in use in the UK.

The cold callus system follows usual procedures but is less successful than the hot-pipe system which now widely replaces it for grafting forms of *Carpinus betulus*. Hot-pipe systems have particular application for some of the more demanding species such as *C. fangiana* (Figure 29.3).

PHYSIOLOGICAL AND ENVIRONMENTAL FACTORS

Despite being a member of the Betulaceae, *Carpinus* appear not to show root pressure problems. However, it is best to apply the usual recommendations for potted rootstocks to be on the dry side at grafting time. Some species, especially in the Distegocarpus section, show bud infections similar to the Birches (Figure 29.4). This problem is undoubtedly exacerbated by conditions of warm callusing and much less evident in hot-pipe systems.

Temperature requirements for warm callusing in the late winter are 15–18°C and in the September period 18°C; although, it is likely that ambient temperature on sunny days will be significantly above this. Hot-pipe recommendations for *C. betulus* cvs are for 21–22°C for 16 days*. However, as discussed previously, length of time in the hot-pipe may be adjusted by altering the temperature to provide a requirement for the standard 21 days in the heating chamber.

FIGURE 29.3   *Carpinus fangiana* long tongue side veneer grafts. LEFT, graft is shown in position in the hot-pipe with top cover removed. Note: no obvious signs of extension growth as the bud is just outside the heated channel, first signs of callus development on the long tongue is visible. A take of 90% was achieved with this system. RIGHT, the graft on the left has been taken from the hot-pipe and is showing no extension growth but good callus development, seen breaking through the wax seal between tie loops. On the right, the graft is from a warm callus poly-tent system with buds bursting into extension growth; no sign of callus development at the unsealed graft union. Great care is required to ensure the survival of this graft and a number of others in the same batch failed.

---

* Dunn (1995) and MacDonald (2014) p.182 and p.214.

FIGURE 29.4   *Carpinus fangiana* grafts in warm callusing conditions. LEFT, showing a healthy apical bud but with a diseased lateral bud below. RIGHT, badly diseased buds are present on two scions, resulting in the death of both grafts. Note: on the right-hand scion infection has spread into blackened wood visible at the centre.

After removal from the hot-pipe, grafts can be cold stored at ±2°C for a period followed, for bare-root grafts, by potting or bedding-up. For grafts produced in the September period it is safest to over-winter under frost-free conditions.

PERFORMANCE COMPARISONS

The *C. betulus* forms perform reliably, but some of the Distegocarpus section presents a challenge, notably *C. rankanensis* and *C. fangiana*, particularly using warm callusing when bud infection can result in the loss of many grafts.

GROWING ON

Normal training procedures apply, involving timely staking, tying-in and pruning to produce a tree form.

Carpinus betulus and its cultivars are susceptible to Bryophid Mite (Acari family), which by mid-summer can cause significant foliage bronzing and reduction in vigour. Infection is less prevalent or absent in most other species. The mites are very tiny and are only visible with the aid of a hand lens. Control is difficult as many of the most effective acaricides are no longer available.

30 Cedrus (Pinaceae) – Cedar

INTRODUCTION

Among the most impressive and stately conifers in size and appearance, Cedars have been culti-vated for centuries in large gardens and arboreta in the UK and mainland Europe. Only four species and one subspecies have been identified, three of which are closely related and considered by some to be geographic variants of *Cedrus atlantica*.

Horticulturists have selected a number of cultivars of each species; these exhibit various differ-ences in form, habit or foliage colour and, because cuttings have proved difficult to root, grafting is the main method of propagation.

COMPATIBILITY

All species within the genus appear compatible. Opportunities for hardier alternatives to *C. deodara*, the rootstock species of choice, would be useful for cold areas. A number of attempts to broaden the range of rootstocks by utilising other hardier genera within Pinaceae, e.g. *Picea, Abies*[*] and *Pinus strobus*[†], initially resulted in reasonable graft takes but subsequently failed, almost certainly due to incompatibility[‡]. This is perhaps not surprising in view of the phylogenic work carried out on the genus which has indicated that cedars occupy their own distinct position within Pinaceae. Attempts to utilise *Larix decidua* has also been suggested but no details given on its long-term reli-ability[§]. As no further accounts of this combination have been discovered, it seems unlikely that it was successful.

TIMING

For many, the preferred timing is summer/early winter (late July–November)[¶], but the alternative, late winter/early spring, is also frequently chosen[**]. No comparative figures for results between the two timings have been discovered, but in cold areas the possibility of winter damage to scion wood and/or rootstocks may influence decisions in favour of summer/early winter.

GRAFTING METHODS

Without exception, side grafts are chosen, the long tongue side veneer, called in some papers side wedge, is favoured. When there is significant disparity between the diameter of rootstock and scion, short tongue side veneer is an obvious alternative.

ROOTSTOCKS

Most agree that *C. deodara* is best choice, some suggesting that *C. atlantica* cvs grafted onto *C. atlantica* rootstocks make weak unions[††]. *C. deodara* is also said to have a better root system than

---

[*]   Barnes (2005).
[†]   Barnes (2008).
[‡]   W. Barnes (pers. comm.).
[§]   Krüssmann (1964) p.572.
[¶]   Krüssmann (1964) p.572, MacDonald (2014) p.183, Macdonald (1988) and Sheat (1965) p.419.
[**] Thomsen (1978) and Wells (1955) p.138.
[††] Nelson (1968).

C. atlantica. Potted rootstocks 1+1P 6–8mm are favoured over bare-root, an alternative choice is plug grown seedlings, claimed to provide a more even grade than bare-root seedlings[*]. A very few favour bare-roots[†], one even suggesting that after lifting the roots should be washed and trimmed to 75–100mm long[‡].

SCION WOOD

Scions usually comprise current year's wood for summer and winter grafts. Terminal growth should be chosen to provide material easily trained into an upright leading shoot; although, for spreading forms such as *C. deodara* 'Feelin Blue' this is less important. Experience here is that scion material can include older wood without any adverse influence, a view supported by many Dutch grafters[§]. Top-worked Patio Plants may reach saleability more quickly by using large-branched scions incorporating a basal portion of 2-year-wood.

SYSTEMS

Warm callusing is essential to provide supportive conditions. The rather soft foliage of *C. deodara* is susceptible to fungal infection (*Botrytis*), and if poly-tents are used, it is important that the polythene cover is raised well above the grafts. Foliage should be kept as dry as possible by a fleece lining to prevent drips, and the use of frequent ventilation.

Bare-root grafts are laid in a free draining media: 75% pumice, 25% peat has been suggested if the union is buried, or pure fibrous peat when exposed. A system in New Zealand utilising bare-root rootstocks, root pruned, grafted and placed in open beds, occasionally syringed over, claimed 90% success[¶].

PHYSIOLOGICAL AND ENVIRONMENTAL FACTORS

Cedars are more tolerant of warmer temperatures during grafting than a number of other genera in Pinaceae. However, despite heavy shading, grafts in enclosed conditions during July and August may experience temperatures in excess of 25°C, which is above optimum requirements. Later in the autumn and early winter, recommendations for bottom heat settings vary between 10–12°C. Personal experience is that a maximum of 18°C provides optimum conditions for callus formation and easiest management of the graft environment. Some suggest lower temperatures are preferable.

Active roots at the time of grafting are recommend by many, but this is considered less relevant in the late September to November period. Using bare-root grafts, especially when the roots are trimmed, obviates the possibility of root activity immediately post grafting but good results are claimed.

TOP-WORKED PATIO PLANTS

Selected dwarf and weeping clones can make attractive Patio Plants grafted onto rootstocks of *C. deodara*, which have been trained to form straight stems 600–900mm high. These are usually grown in containers but can be substituted by root-balled plants. Normally, two scions are side grafted to the rootstock more or less opposite each other (Figure 30.1).

---

[*] Thompson (2005).
[†] Thomsen (1978).
[‡] Richards (1972).
[§] Alkemade & van Elk (1989) p.112.
[¶] Nelson (1968).

FIGURE 30.1   *Cedrus libani* 'Sargentii' makes an attractive Patio plant top-worked at 600mm high on a suitable *C. deodara* stem.

GROWING ON

Tree-forming species such as *C. atlantica* 'Glauca' require training and staking at an early stage to develop suitable, straight leading growths.

31 Cercis (Caesalpiniaceae) – Redbud, Judas Tree

INTRODUCTION

Distributed mostly in the warmer temperate areas of the Northern Hemisphere, *Cercis* are usually seen as small trees or large shrubs. They are characterised by handsome rounded heart-shaped leaves; a number of selections, particularly of *C. canadensis*, produce spectacular violet/purple or golden leaves. From April onwards many have attractive deep-pink or white, pea-shaped flowers wreathing the stems, often extending to the older branches. *Cercis racemosa* has flowers in short racemes hanging from the branches.

A number of species and selected forms must be vegetatively propagated. Cuttings have proved difficult to root, and grafting is the main method of propagation.

COMPATIBILITY

Botanic investigations are ongoing but there is general agreement that species from the eastern side of North America (*C. canadensis*) and those from western Eurasia (*C. siliquastrum*), central Asia and eastern Asia have close phylogenic linkage. *Cercis chinensis*, *glabra* and *racemosa* are thought to be fairly closely related to *C. siliquastrum**.

Experience has demonstrated mostly reliable compatibility between *C. canadensis*, *C. chinensis* and *C. racemosa* when grafted to *C. siliquastrum* rootstocks. Individual exceptions occur due to seedling variation rather than widespread incompatibility. Examples may be seen in the occasional *C. chinensis* 'Avondale' clone which fails due to delayed incompatibility (see Figure 12.3, Chapter 12). Dutch grafters have long maintained that *C. siliquastrum* is an unsuitable rootstock for *C. canadensis* cultivars, presumably based on the assumption of incompatibility. Personal experience is that this is not the case, supported by several 30-year-old specimen mother plants grafted to *C. siliquastrum* rootstocks, and the production of several large commercial batches with no sign of incompatibility.

Only *C. occidentalis* may demonstrate consistent incompatibility symptoms with *C. canadensis* or *C. siliquastrum* rootstocks. It is possible that, if obtainable, *C. chingii* will provide a universally compatible rootstock or inter-stem.

TIMING

Grafters in the UK and USA graft *Cercis* mostly in the mid-/late winter period. In the USA, field budding is popular in areas where summer ambient temperatures can be relied upon to reach 32–37°C. Good takes at this time obviate the need for bench grafting. A wet summer can significantly reduce field budding takes, suggestions for this have centred on the possibility of excessive sap flow 'drowning the union'[†], but lower ambient temperatures may be the underlying cause, particularly as it was noted lower temperatures in the 21–26°C range during cool, dry summers reduced the success rate.

---

* Davies, Fritsch, Li et al (2002).
[†] Warren (1973).

One grafter in the USA recommends delaying grafting until May when the rootstock has made significant extension growth. Grafts at this time using dormant cold-stored scion wood have outperformed those made earlier. It is suggested this is because the rootstock is active*; as the grafting is carried out in cold glasshouse facilities, the difference in take between early and late grafting may be due more to higher ambient temperatures than active rootstocks.

Trials in the UK investigated grafting methods and timing. Using bare-root rootstocks in a hot-pipe system set at 23°C for 14 days, best results were produced from grafts made at the earliest date of 14 December, compared with those on 4 January and 15 February[†].

In the Netherlands, late winter grafting is noted as being possible using a root piece (root graft) or bare-root rootstocks. Takes appear to be generally poor and summer grafting in July–August on potted rootstocks is favoured.

GRAFTING METHODS

Splice grafts are often suggested, but as cambial alignment and good matching are important for this difficult to graft species, veneer grafts are a better choice (Figure 31.1). Side veneer in a hot-pipe system is best; long tongue side veneers are only possible when top quality young vigorous rootstocks are used. With this type of rootstock, inclusion of a stimulant bud will improve takes further. In hot-pipe systems, the advantage of the long tongue side veneer over the short tongue alternative or splice is illustrated using *Cercis* as a specific example in Chapter 3, Figure 3.16.

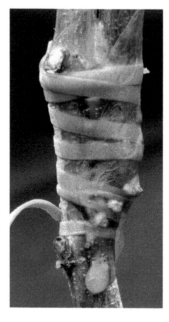

FIGURE 31.1 *Cercis canadensis* 'Forest Pansy'. LEFT, a well-callused short tongue side veneer graft of 1-year scion wood placed on young vigorous *C. canadensis* rootstock. RIGHT, showing emerging shoots of *C. canadensis* 'Forest Pansy', the right-hand graft with normal coloured extension growth, the left-hand graft with green/bronze emerging shoots.

---

* Meacham (2003).
[†] Murphy (2005).

ROOTSTOCKS

The maritime climate of the UK is not well suited to producing vigorous well-ripened, young seedlings of any *Cercis* species, so obtaining suitable rootstocks is always a problem. *Cercis siliquastrum* has been the rootstock of choice, mainly because good quality transplants have been available from European suppliers. More recently, the advantages of *C. canadensis* have become evident and supplies from home and continental producers are now more readily available. This species is the best rootstock for many of the important commercial varieties, which are mostly forms of *C. canadensis*. It also has the advantage of producing stronger, straighter stems than *C. siliquastrum* and is more vigorous, at least in the early stages. In the USA, producers from the southern states can produce excellent seedlings suitable for potting to produce the required 1+1P 6–8mm rootstocks.

Good quality vigorous young rootstocks are an important requirement for successful grafting of *Cercis*.

SCION WOOD

1-year, well-ripened wood is a crucial factor in grafting *C. canadensis*, *C. siliquastrum* and probably most, if not all, others (see Chapter 10, Figure 10.3). Experience here is that the sometimes suggested use of scions with a basal portion of 2-year wood is generally less satisfactory. In maritime areas such as the UK, it is essential that mother plants are well pruned annually, normally by cutting back to induce vigorous shoot production. Stock plants should be sited in a position which is as sunny and warm as possible. Pruning should be carried out sufficiently early to allow rapid development of buds during the following growing season. *C. canadensis* is characterised by very late bud break, limiting the growing season and reducing time for wood ripening.

Stout, well-balanced shoots should be chosen; any excessively large shoots (water sprouts) and thin material are discarded. Under UK growing conditions, normally only the basal portion of the shoot is suitable for use. Material collected when completely dormant may be cold stored for some weeks.

SYSTEMS

Warm callusing is possible during two periods, mid-/late winter or July–August. In the Netherlands summer grafting is favoured. Dutch recommendations for grafts made during winter time are for use with bare-root rootstocks or root pieces*, but well-established, pot grown rootstocks are a much better option.

Summer grafting involves decisions on whether or not to remove scion leaves. Cercis foliage is very susceptible to Botrytis, and if leaves are retained, skilful manipulation of the enclosed environment is essential to avoid significant leaf decay and loss. Leaf removal avoids this problem, but no comparative figures are available for effects on results for leafy or de-leaved scions. Experience indicates that leaf removal should improve results but should only be used for side grafts. Summer grafts involving apical methods will require scion leaves to be retained.

Until the advent of hot-pipe systems, warm callusing during mid-December to the end of January was the best option. Hot-pipe systems now provide, by far, the best opportunity for obtaining good takes, particularly if the heating chamber design allows the option of side veneer grafts (Figure 31.2).

* Alkemade & van Elk (1989) p.123.

FIGURE 31.2   *Cercis canadensis* 'Forest Pansy' side veneer grafts placed in a Boskoop tray 30 days after removal from a hot-pipe with a thermostat setting of 25°C, over 90% were successful.

PHYSIOLOGICAL AND ENVIRONMENTAL FACTORS

High temperatures of 25–26°C for 21–28 days are required to provide optimum conditions for callus formation.

Using the hot-pipe system allows control of premature bud-break by keeping scion and rootstock in low ambient temperatures, while permitting optimum temperature to the graft union. The combined effects of this on the physiology of the grafts have been to significantly improve results over other systems. An interesting feature of *C. canadensis* 'Forest Pansy' produced in warm callus and hot-pipe systems is the tendency for individuals to produce extension shoots with a greenish or bronzy hue, rather than the normal rich, ruby colour. This corrects itself as shoots continue to extend and mature (Figure 31.1).

With a combination of well-ripened scion wood, healthy vigorous rootstocks, well-matched grafts and adequate temperature, good results are guaranteed (Figure 31.2).

GROWING ON

In the UK, a combination of nutritious compost, good light, copious water and high temperature in protected structures can produce single-stemmed (feathered) plants of *C. canadensis* 'Forest Pansy' ±2.4 metres high by the second season (Figure 31.3).

FIGURE 31.3   *Cercis canadensis* 'Forest Pansy' grafted plants representing a sample of 1-year and 2-year grafts. These are being staked and tied up to canes for subsequent sale. Best 1-year grades can make 2.0+ metres high in the first year. Two-year plants make 2.4 metres plus. All these plants were grafted to *C. siliquastrum* rootstocks.

32 Chamaecyparis Including Xanthocyparis (Callitropsis) (Cupressaceae)

INTRODUCTION

All are found in North America, Japan and Taiwan. Five species are known, excluding *nootkatensis*, which was until recently included but has now been split off into a separate genus, *Xanthocyparis*. This classification is still under review as some authorities claim it should be known as *Callitropsis*.

Despite few species, an enormous number of cultivars have been selected, especially from *C. lawsoniana*. Most of these can be propagated by cuttings, but some are grafted, either because of difficulties in rooting, slow growth when cultivated on their own roots or an attempt to overcome inherent problems with susceptibility to the soil-borne disease *Phytophthora sp*, which is particularly evident in *C. lawsoniana*.

COMPATIBILITY

Reported information on the compatibility of *Chamaecyparis* is sometimes confused and conflicting.

C. lawsoniana cultivars are compatible on *C. lawsoniana*. *C. pisifera* may also be successful and it is stated that *C. pisifera* 'Boulevard' cuttings taken in October can be rooted and ready for grafting the following spring*. Some have suggested that growth on this rootstock is limited. *C. thyoides* has been the subject of trials in the Netherlands over a period of some years, but it has produced rather poor growth rates for *C. lawsoniana* cultivars[†]. *Juniperus* x *pfitzeriana* 'Hetzii', another member of *Cupressaceae*, is considered a suitable rootstock by some growers in the USA[‡].

C. obtusa cultivars are mostly propagated by cuttings but rootstocks of *C. pisifera*[§] or *C. lawsoniana* have been proposed. *C. lawsoniana* rootstocks demonstrate long-term compatibility but, because of significant differences in growth rates between rootstock and scion, an unsightly union will result, (see Chapter 12: Compatibility Figure 12.5). *Thuja occidentalis* is advocated by several as a suitable rootstock, and *T. occidentalis* 'Smaragd' stems have been recommended as the best rootstocks for top-worked Patio tree forms of *C. obtusa*[¶].

C. thyoides is normally propagated by cuttings but can be grafted to rootstocks of *C. lawsoniana*. In view of the trials in the Netherlands, long-term prospects for this combination may not be good.

C. formosensis has been suggested as a suitable understock for *C. lawsoniana*** and should provide a compatible combination when the orientation is reversed. *C. formosensis* is endangered in the wild because of over-cutting and loss of habitat. Consequently, protocols for its propagation should be of interest to those concerned with conservation.

---

* Dirr & Heuser Jr. (2006) p.147.
† Ruesink & van Kuik (1991).
‡ Decker (1998).
§ Nelson (1968).
¶ Buchholz (2017).
** Holland, Warren, Ranney et al. (2001).

Xanthocyparis nootkatensis cultivars may be grafted to seedlings of the parent species if they are available; unfortunately, this is rarely the case. The obvious alternative, *C. lawsoniana*, has consistently produced low takes and poor subsequent growth. Trials in the Netherlands to test x *Cuprocyparis leylandii*, *Chamaecyparis pisifera* 'Aurea' and *Platycladus (Thuja) orientalis* produced results in the region of 100% for all three species; regrettably, their long-term performance was not established*.

Several grafters recommend *Platycladus orientalis*[†], compatibility with *Xanthocyparis* seems to remain lengthy and probably permanent.

Juniperus x *pfitzeriana* 'Hetzii' has potential as a possible understock for *Xanthocyparis nootkatensis* 'Pendula'[‡]. Subsequent experience has confirmed the compatibility of this combination. Trees of 4.5 metres, 150mm stem girth and excellent fibrous root systems have been produced on this rootstock[§].

Recently, x *Cuprocyparis leylandii* has been further investigated as a rootstock for *X. nootkatensis* 'Pendula' in the USA and produced promising results. Good initial takes have been supplemented by healthy normal growth, at least for the succeeding 12 years[¶]. No reports have been discovered of tests for the other x *Cuprocyparis* hybrids, x *C. notabilis* and x *C. ovensii*. There is every reason to think that they may prove compatible. x *Cuprocyparis notabilis* may prove a better choice than x *C. leylandii* because the other parent used for the hybrid was *Cupressus arizonica* var. *glabra*, which can be considered hardier than *C. macrocarpa*.

TIMING

For *Chamaecyparis* and *Xanthocyparis*, grafting during February into March is popular; Dutch grafters extend the winter period into late spring (April/May). Some favour the summer/late summer to early winter period (mid-August to mid-November). Any of these should provide good results and choice will largely depend upon the need to organise schedules for the annual grafting programme.

GRAFTING METHODS

For pot grown rootstocks during the summer to early winter period, side veneer grafts are favoured (Figure 32.1), for late winter, splice or wedge grafts are also occasionally recommended**. Grafts made on bare-root rootstocks in the late winter may also be splice grafted. The long tongue side veneer graft works well for *Chamaecyparis* and *Xanthocyparis*, when rootstock diameter does not exceed 8mm.

ROOTSTOCKS

Potted rootstocks, 1+1P, in 9 × 9cm square pots with a stem 6–8mm Ø are the usual choice. One-year plug grown, seedling rootstocks potted on into a 9 × 9cm pot for the final year or grown in a larger plug for two seasons will make excellent rootstocks. Bare-root rootstocks of *Chamaecyparis lawsoniana* 1+1 or 2+0 are suggested for late summer (late September/October) grafting[††] and are suitable for the late winter period; higher takes can be achieved using potted rootstocks.

---

*   Verhoeven (1985).
†   Buchholz (2016), Dirr & Heuser Jr. (2006) p.147, Krüssmann (1964) p.575 and Holland, Warren, Ranney et al. (2001).
‡   Hendricks (2013).
§   B. Hendricks (pers. comm.).
¶   W. Barnes (pers. comm.).
** Sheat (1965) p.420.
†† Krüssmann (1964) p.575.

FIGURE 32.1 *Chamaecyparis lawsoniana* 'Winston Churchill', grafted to 1+1P *C. lawsoniana* rootstock using long tongue side veneer graft.

x *Cuprocyparis leylandii* is raised from cuttings and should reach graftable size for *X. nootkatensis* cvs 1 year after potting-on from an established, rooted cutting, 0+1P, 6–8mm.

In the Netherlands, trials of *C. thyoides* as a Phytophthora-resistant species have been carried out over a three-year period. They concluded that *C. thyoides* has no strong resistance to *Phytophthora cinnamomi**. A reputedly resistant clone of *C. lawsoniana* is available in the USA[†]. Oregon State University has released a clone, 'DR', which claims to be resistant to the disease, whether or not prolonged resistance can be achieved will eventually be revealed.

Some growers adopt a programme of routine drenching of rootstock pots with a suitable fungicide as a precaution against Phytophthora root rot. Furalaxyl is probably the most effective control but has been withdrawn from use in Europe. Metalaxyl is currently available as an alternative product; elsewhere, others may be available. It should be kept in mind that fungicides in this group have a fungistatic action and do not eradicate but only halt progress of the disease for a period. An alternative is to add copper ions to the irrigation water (±2 ppm) via a copper ionising unit. Trials of this system in Europe have reduced infection of a susceptible *C. lawsoniana* cultivar by a significant amount.

---

* Ruesink & van Kuik (1991).
[†] Buchholz (2015).

SCION WOOD

A terminal portion of current year wood is normally chosen. Any unripe wood at the tip should be removed. To obtain a suitable size for dwarf varieties, inclusion of a 2-year-old basal portion may be necessary. All species seem to callus equally well on 1- or 2-year wood.

It is always important to select scion material which reflects the true habit of the source plant; for dwarf cultivars, strong growing atypical shoots should not be chosen, and similarly upright scion wood should not be collected from those grown for their pendulous habit.

In the winter/spring period, scions may be stored for some days in cool conditions and for some weeks in cold storage, packed in polythene bags at ±2°C. Storage of summer-collected material is less satisfactory, but prolonged storage is possible towards the end of the period in late September/October.

SYSTEMS

Warm callusing is required. A few growers adopt supported warm callusing by burying the union in suitable, freely-draining media and placing the grafts under misting systems. Some recommend burying the union when using bare-root rootstocks, which is done conveniently as the rootstock is laid-in. Top-worked grafts are normally placed in tall poly-tents while callusing takes place; alternatively, polythene bags placed on and tied-in over each graft may be used.

PHYSIOLOGICAL AND ENVIRONMENTAL FACTORS

Temperature requirements for the whole group are low and most grafting in summer and autumn is achieved without any supplementary heat other than sunlight. Bottom heat temperature for the late winter/early spring is sufficient if set at 10–12°C; from mid-spring, sunlight is usually sufficient.

Scion wood carrying flowers (strobili) is liable to develop fungal infection; non-drip conditions should be created in the covering enclosure and ventilation applied as often as possible. Routine spraying of fungicides, particularly against Botrytis, is recommended; under poly-tents, dust formulations are better than aqueous sprays.

TOP-WORKED PATIO PLANTS

Dwarf, pendulous, filiform and golden forms, or combinations of these, can make spectacular novelty Patio plants when top-worked at 600–750mm high. Parent species can be run-up to make stems for *C. lawsoniana* and *C. pisifera*. *Thuja occidentalis* 'Smaragd' is suitable for *C. obtusa* varieties.

Top-working normally takes place in the late winter period and tall poly-tunnel structures are recommended; individual polythene bagging may be used but can result in foliar disease problems.

C. lawsoniana, *pisifera* and *obtusa* all have suitable cultivars to make attractive Patio plants. These are comprehensively illustrated and described in the RHS Encyclopaedia of Conifers*.

GROWING ON

The weaker-growing, variegated, golden and filiform cultivars respond to wind protection and light shade. Pendulous types require staking and conscientious tying-in to produce a shapely plant with a reasonably straight central stem. Light trimming and training of side branches will enhance the symmetry of many types

---

* Auders & Spicer (2012).

33 Cornus (Cornaceae) – Dogwoods, Cornels

INTRODUCTION

Widely distributed throughout the northern hemisphere, from the sub-arctic to the equator, with a small incursion by one or two species into South America and Africa, the Cornels are an important genus of ornamental plants for temperate climates. Their ornamental value is very diverse, from those grown primarily for their winter bark effect to some having beautiful showy flowers. A number contribute to the autumn display of fruit and 'fall colours'. Grafting is an important procedure for some species and cultivars.

COMPATIBILITY

Subgenera within three groups contain species which are grafted and can be recognised by the following characteristics*:

1. The blue- or white-fruited group contains three subgenera: the biggest, Kraniopsis, comprising species which flower in the summer and are grown for their winter stem colour. They include the shrubby species *C. alba*, *C. sanguinea* and *C. sericea* (*stolonifera*). A small subgenus, Mesomara, within the same group has two tree-forming species, *C. alternifolia* and *C. controversa* of importance to grafters.
2. The Cornelian Cherries, characterised by large red fruits, contains subgenus *Cornus*, with *C. mas* and *C. officinalis*, and subgenus Sinocornus containing the rather tender *C. chinensis*.
3. The big-bracted Dogwoods producing large, showy flowers have two subgenera of importance: Cynoxylon comprising *C. florida* and *C. nuttallii*, and Synocarpa with *C. capitata*, *C. hongkongensis* and *C. kousa*.

Choice of rootstocks for *C. alternifolia* and *C. controversa* cultivars might be expected to centre on the parent species, both of which are available as seedlings. Surprisingly, the experience of many grafters is that better rootstocks are provided by *C. alba* and *C. sanguinea*, species from within the same blue- or white-berried group. Work in 1966 in the Boskoop Trial Ground demonstrated that for C*ornus controversa* 'Variegata' and *C. alternifolia* 'Argentea', C*ornus alba* 'Elegantissima' produced marginally better takes (90%+) than the alternatives, *C. alba* 'Gouchaultii' and *C. alba* 'Sibirica'[†]. Rather than rely on cutting raised clones, many would opt for seed raised rootstocks of the species C*ornus alba* or *C. sanguinea*. In the UK, when both species are raised from seed, *C. sanguinea* may be more prone to sucker production than *C. alba*. An advantage of cutting raised, variegated cultivars of *C. alba* is that they possibly produce less sucker growth than seedling equivalents.

The improved performance of *C. alba* and *C. sanguinea* rootstocks is linked to the negative effects of bottom heat and high ambient temperature on the root systems of *C. alternifolia* and *C. controversa*. This is often seen as significant root rotting in warm callusing systems. Hot-pipe

---

* Fan & Xiang (2001) and Xiang, Thomas, Zhang et al. (2006).
[†] van Elk (1965).

systems are likely to avoid this problem and may well allow the use of these ideal rootstock species, but this has not been tested here.

All the Cornelian Cherry Group, within the two subgenera noted, appear compatible on *C. mas*.

C. florida and *C. nuttallii* are in separate subgenera from the other big-bracted Dogwoods, and this may account for the unreliable performance of *C. nuttallii* grafted on rootstocks of *C. kousa*.

When grafted together, *C. florida* and *C. kousa* provide reliable unions and there are reports of *C. kousa* surviving for at least 25–30 years when grafted to rootstocks of *C. florida**. In the UK, *C. kousa* would be the favoured rootstock for the group because of its wide tolerance to a range of soil types. Clones of *C. florida* ('Cherokee Chief', 'Cherokee Sunset', 'Daybreak', etc.) have grown well here for 15 years when grafted to *C. kousa*.

Despite the benefits of *C. kousa* it seems wise to use *C. florida* rootstocks for *C. florida* cultivars and crosses between *florida* and other species. Also best on *C. florida* are *C. nuttallii, nuttallii* cultivars and hybrids between *C. florida* and *C. nuttallii*. The group of hybrids between *C. florida* and *C. kousa*, raised at Rutgers University and released in the early 1980's, represent a genetic mix between the two species and pose less clear options. *Cornus* x *rutgersensis* 'Stellar Pink'® and 'Celestial'® have performed well when grafted to C. *kousa*. After a slow start, 'Ruth Ellen'®, 'Constellation'® and 'Aurora'® have made large plants; 'Stardust'® failed after two years. Trials comparing results between *C. florida* and *C. kousa* rootstocks for C. x *rutgersensis* hybrids would be useful.

C. *kousa* is the best choice of rootstock for *C. kousa* cultivars, *C. kousa* hybrids, *C. capitata* and *C. hongkongensis*. Hybrids between *C. kousa* and *C. nuttallii, Cornus* 'Venus'® and 'Starlight'® have produced good results on *C. kousa* rootstocks and are growing well after ten years. Subsequent grafts of C. 'Venus'® on *C. florida* have produced plants with a more upright growth habit, but after time the differences may be undetectable.

TIMING

Most grafters favour the late January–March period for all species. Summer bench grafting during July-early September is also suitable for all the big-bracted dogwoods, *C. controversa* and *alternifolia* cultivars. The Cornelian Cherry group are almost always grafted in late winter.

GRAFTING METHODS

Warm callused grafts during the late winter period are mostly side grafted, often retaining only a short (75–100mm) portion of rootstock above the graft (Figure 33.1); a number of grafters recommend apical grafts using splice or whip and tongue for *C. florida*[†].

Any suckers which develop are allowed to remain until the scion reaches the leafing out stage when they are rubbed off. For all species, increasing the use of hot-pipe systems will place more emphasis on the use of apical grafts.

Apical long or short tongue veneer grafts will provide best matching for a genus which has a rather thin and indistinct cambium (see Chapter 3, Figure 3.1).

Side grafts are used for summer grafting, normally veneers, but side cleft is possible. Warm callused summer side grafts are best left unsealed. The extra time and scion wood required to multigraft the large flowered Dogwoods may be warranted as the saleable stage is reached more quickly than when only a single scion is used (Figure 33.1).

---

* Dirr & Heuser Jr. (2006) p.154.
[†] Dirr & Heuser Jr. (2006) p.154 and Curtis (1962).

FIGURE 33.1   LEFT, side grafted *Cornus controversa* 'Variegata' from a hot-pipe system. The rootstock used is *Cornus alba* 'Sibirica' raised from cuttings. RIGHT, multi-grafted summer grafts of *Cornus kousa* 'Miss Satomi' on *C. kousa* rootstock, photographed the following winter before heading-back.

ROOTSTOCKS

Many species make successful rootstocks if required; *Cornus nuttallii* cannot be recommended as the root system invariably succumbs to root disease in the container and/or after planting out. As discussed, *Cornus controversa* and *C. alternifolia* are unreliable once subjected to bottom heat and the stresses of the grafting procedure.

Potted rootstocks are normally well-established 1+1P. Stem diameter at grafting time is best in the 6–8mm region, but larger grades can be used with success, especially with the hot-pipe system. Summer grafts are normally grafted during the summer following potting and therefore have less time to establish in the pot. This needs to be taken into account when deciding upon the best grade. A minority of grafters pot the rootstocks after lifting in the autumn and graft the following late winter/ early spring. No comparisons between this and conventional established rootstocks have been made, but this technique may have a beneficial influence in reducing bleeding. Potting just before or after grafting would have even more effect on bleeding, but takes may be reduced due to other factors.

Cornus are noticeable for the extreme branching and fibrous nature of their root system. Provided feeding and irrigation are supplied as required to maintain plant health, they may be safely left in rootstock pots for longer periods than many other species. The resulting rootstocks, held in 9 × 9 × 12.5 or 1-litre pots, can be maintained in graftable condition for at least three growing seasons.

The hot-pipe system produces excellent results and subsequent strong growth for all the *Cornus* species. Even heavy-grade rootstocks, which might be considered too old for other methods, are successful, growing away strongly (Figure 33.2).

SCION WOOD

Heavy, one-year wood is satisfactory; reflecting features of the rootstocks, several grafters stress the value of heavy-grade scions and recommend wood up to three to four years old for *C. florida* and others in the big-bracted group*. In the UK, *C. florida* and *C. kousa* do not make such vigorous,

---

* Wells (1955) pp.224–227 and Upchurch (2008).

FIGURE 33.2   *Cornus* x rutgersensis 'Celestial'® grafted in a hot-pipe to heavy 1+3P rootstock of *C. florida*.

heavy growth as those in climates with warmer summers; to obtain scions of sufficient girth and ripeness, a basal portion of two-year wood is often necessary. For these, scions may be chosen with some degree of branching (Figure 33.3).

Because of the growth habit of *C. alternifolia* and *C. controversa*, branched scions including a basal portion of two-year wood are much superior to one-year alternatives. Examples of the selection and type of scion wood required are shown in Figures 33.3, 33.4 & 33.5).

The Cornelian cherry group, *C. mas*, *C. officinalis* and *C. chinensis* produce strong one-year wood suitable for scions. Some of the weaker clones of *C. mas* such as 'Aureoelegantissima' and 'Variegata' may require the addition of basal portions of two-year wood to provide material of sufficient size.

For summer grafting the question arises of whether or not to remove leaves from the scion. Some have suggested that total removal of *C. kousa* scion leaves has reduced takes by 45%; instead, reducing leaf area by one-half to two-thirds is suggested. Grafts treated this way often resulted in complete leaf drop after three weeks, but it was noted that this should be due to buds swelling, possibly a precursor to extension growth*. If this occurred, it would appear that retention of leaves in this instance did not prevent production of Lammas shoots.

---

* Thompson (1998).

FIGURE 33.3   LEFT, strong 1-year shoot of *C.* x rutgersensis; the black line surrounded by white circle indicates where the scion may be detached. RIGHT, black line surrounded by white circle indicates where the 2-year wood may be cut to provide a branched scion of *C. kousa* 'Miss Satomi'.

FIGURE 33.4   A typical branch of *Cornus controversa* 'Variegata'.

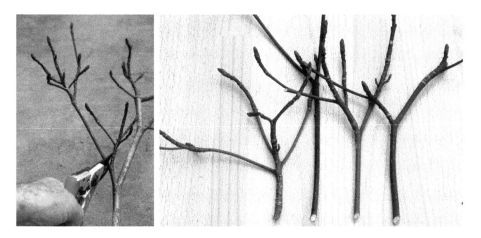

FIGURE 33.5   LEFT, selected scion being removed from a branch of *Cornus controversa* 'Variegata'. RIGHT, a selection of suitable scions of *Cornus controversa* 'Variegata'.

Experience is that the removal of scion leaves enhances rather than reduces takes. Leaves retained often become infected with Botrytis, to which *Cornus* appear susceptible, and are rarely held for more than two weeks, often less. Even the evergreen *C. hongkongensis* can have substantial leaf removal in the summer. Summer grafted *Cornus* are rapid callus producers, normally visible on unsealed grafts after 10–14 days. Apart from *C. hongkongensis*, Lammas shoot production is comparatively rare after summer grafting on cultivars of *C. florida*, *C. kousa* and their hybrids.

SYSTEMS

Cold callusing may be suitable for *C. mas* and *C. officinalis*, but warm callusing will provide more certain results in cooler areas. For most grafters, summer or winter warm callusing or hot-pipe are the preferred systems. Because of its evergreen nature, *C. hongkongensis* is unsuitable for hot-pipe and is warm callused in summer or late winter.

PHYSIOLOGICAL AND ENVIRONMENTAL FACTORS

Apart from one reference*, no mention of root pressure has been discovered in the literature. Surprising, as heavy bleeding is a feature of the big-bracted Dogwood group and also *C. alternifolia* and particularly *C. controversa*. If *C. alba* or *C. sanguinea* are used as rootstocks root pressure is less of a problem and may in part account for their success as rootstocks. Warm callusing during mid-winter using rootstocks of *C. florida*, *C. kousa*, *C. alternifolia* or *C. controversa* requires them to be dried-off quite severely. Even for a hot-pipe system, some level of drying back is advisable, particularly if ambient temperatures outside the heating chamber are often above ±10°C. Summer grafting generally alleviates the problems of root pressure and can consistently produce good results.

To avoid excessive 'soft' callus and poor strength of the union when late winter grafting, several grafters stress the importance of keeping temperatures on the cool side (15–18°C)[†]. A hot-pipe setting of 20–22°C in the heated chamber appears to be optimum. Summer grafts respond well to higher temperatures of ±25°C.

PERFORMANCE COMPARISONS

The big bracted group are more demanding than the rest. In the UK, *C. nuttallii* and its cultivars have proved to be the most difficult, *C. kousa* cvs and *C. hongkongensis*, the easiest.

GROWING ON

Although the root system of *Cornus* is well-suited to field production and root-balling, modern production is usually in containers. Most species are medium-to-large shrubs and grown as multi-stemmed plants. Types naturally forming a tree, such as some *C.* x *rutgersensis* clones, require staking.

---

* Brotzman (2012).
† Wells (1955) pp.224–227 and Standbrook (2005).

34 Corylus (Betulaceae) – Hazel

INTRODUCTION

Found in the temperate regions of the northern hemisphere, this genus comprises large, often suckering shrubs, and small- to medium-sized, frequently single-stemmed trees.

A number of forms, mostly of *C. avellana* and *C. maxima* have been selected for their ornamental value, purple foliage, weeping or contorted habit. Requiring propgation by grafting procedures, these may be grown as bottom- or top-worked specimens. The genus also has importance for hazelnut nut production which is estimated at 600,000 tons of in-shell nuts world-wide.

The genus also has importance for nut production which is estimated at 600,000 tons of in-shell nuts world-wide. Hazelnuts are currently produced as bush grown plants in plantations, involving significant labour inputs for pruning and harvesting. Interest in growing hazelnuts as non-suckering trees, preferably on a stem, has presented possibilities for more efficient plantation management. In parallel with this is an interest in the development of superior cropping varieties and non-suckering tree forming rootstocks, which may be used to provide controlled growth and enable mechanical harvesting. At present, in the early stages, this requirement appears to be met by micropropagated clones. However, future developments may involve large scale grafting procedures.

COMPATIBILITY

The genus is divided into four sub-sections: comprising those which produce fruit having leafy husks, *C. avellana*, *C. americana*, *C. maxima*; those with bristle covered husks, *C. cornuta*, *C. sieboldiana*; those with a spiny (*Castanea*-like) husk, *C. ferox* and *C. tibetica*; and those which grow into trees, *C. chinensis* and *C. colurna*. There is also an attractive tree-forming species with papery bark, *C. fargesii*, not yet assigned to a sub-section, though it appears to be close to *C. chinensis*.

It has been nursery practice to graft all but the tree types to *C. avellana* rootstocks. The extreme suckering habit of *C. avellana* has led to numerous examples of grafted plants disappearing beneath a mass of *avellana* shoots. Whether there is a compatibility issue involved in such losses is uncertain. Experience using *C. colurna* as an understock stem for top-worked cultivars of *C. avellana* and *C. maxima* appear to show no problems with respect to compatibility, at least while the plants are grown to saleable size on the nursery.

The tree types, *C. chinensis*, *C. fargesii*, forms of *C. colurna* as well as *C.* 'Te Terra Red' are most likely to be compatible with *C. colurna*. *Corylus fargesii* at USDA Corvallis is showing some graft incompatibility. In the arboretum here, one of two *C. chinensis* is showing some sucker growth of the *C. colurna* rootstock. The other plant has produced no suckers, and on both grafted plants, scions are growing away well. In Belgium, grafts of *C. fargesii* performed better on *C. avellana* (70–80%) than on *C. colurna* (55%) while others have reported successes of 90% for the *colurna/fargesii* combination*.

Experience here and of others, of fruiting hazelnut varieties grafted to short stems (300mm) of *C. colurna*, showed that after a period of some years the scion variety lost vigour and gradually died off[†]. A popular commercial variety, 'Barcelona', grafted to *C. colurna*, declined in nut production over a 20-year period compared with plants on *C. avellana* rootstocks[‡].

---

* Camelbeke & Aeillo (2016).
[†] R.Vernon (pers. comm.).
[‡] Dirr & Heuser Jr. (2006) pp.57–159.

TIMING

Hazels are graftable in winter or mid- to late-summer. Winter grafting temperatures must be raised well above ambient to obtain acceptable results. Grafting times may be delayed by the use of cold stored scion wood; however, scion collection must not be delayed because late collection can significantly reduce takes.

Summer grafting is successful and particularly suitable for top-worked grafts for which the provision of hot-pipe conditions is difficult, and use of large poly-tents to house them during winter grafting increases heating costs.

GRAFTING METHODS

Given appropriate temperature, *Corylus* produce copious callus and the knifesmanship involved is relatively straightforward. Bottom-worked grafts are normally splice grafted. Because of bark thickness, some difficulties may be experienced in achieving a good match with *C. colurna*; when grafting to this rootstock, a short tongue apical veneer may be a better choice. If multi-grafts are used, the side graft variant of this graft is essential for top-worked stems.

ROOTSTOCKS

Seed raised *C. avellana* is frequently grafted bare-root as 1+0 or 1+1 6–8, or 8–10mm grade. Usually 1-year seedlings reach sufficient girth for grafting, but transplanted 2-year rootstocks produce larger, heavier plants by the end of the first growing season. The less common species *C. cornuta*, *C. ferox*, *C. sieboldiana*, *C. tibetica*, etc. may be grafted more reliably using potted rootstocks 1+1P 6–8, 8–10mm grade.

Tree species *C. chinensis*, *C. fargesii*, forms of *C. colurna* as well as *C.* 'Te Terra Red' are best grafted to potted understocks of *C. colurna* 1+1P 8–10mm. When potted rootstocks are required for bottom-working, the 1-litre deep pot is ideal. For top-worked stems, 2-litre deep pots normally accommodate stems grafted at 60–80cm high; taller stems of 120cm will require 4- or 5-litre deep pots. Root-balling stems for top-working is always an option for winter grafting.

SCION WOOD

Usual recommendations for current year wood apply. To provide material of sufficient size and weight types such as *C. avellana* 'Contorta' and 'Red Majestic' may need to include some 2-year wood at the base.

Scion wood collection should take place early in the season, ideally December/January. Grafting results are closely related to the time of collection, when this is after mid-January takes are depressed in direct proportion to the postponement. Early collected material should be cold stored at 0.5 to 1°C if grafting dates are delayed.

SYSTEMS

Winter-grafted, dormant material making use of the hot-pipe system has significantly and consistently improved grafting success*. Trials comparing no hot-pipe with periods in the hot-pipe at ±24°C for 2–4 weeks, gave 0% takes for no treatment, 70% for 2 weeks, and 93% for the 4-week period.

Use of enclosed structures for warm callusing dormant material always results in premature extension growth and a need to carefully manage the grafting environment until the union has

---

* Lagerstedt (1981).

sufficiently developed. Carefully managed, cold callusing systems can produce acceptable results, but graft union development is slow and less certain than with hot-pipe systems.

Summer grafting from late July to early September is successful for *C. avellana* cultivars (Figure 34.1), but *C. maxima* appears less successful if grafting is delayed until September. Summer grafting is carried out in a heavily shaded, warm callusing system, when apical or side grafts may be used. Personal experience favours scion leaf removal for side grafts; they should be retained for apical grafts.

PHYSIOLOGICAL AND ENVIRONMENTAL FACTORS

Callus and union formation in *Corylus* are positively influenced by temperature. General consensus is that optimum temperature lies between 22–24°C. They are fairly amenable and will respond well to any temperatures within this range.

GROWING ON

Light shade, warm growing conditions, high humus content and nutrient-rich growing medium at pH levels above 6.2 favour best growth. Under protection, *Corylus* are susceptible to Spider Mite (*Tetranychus urticae*), which must be controlled to avoid substantial loss of growth.

FIGURE 34.1 *Corylus avellana* 'Red Majestic'. A batch of summer-grafted plants in final pots and growing on before sale.

35 Cupressus (Cupressaceae) – Cypress

INTRODUCTION

Occurring in the Old and New World, many Cypresses are notable for their statuesque columnar shape, exemplified by the iconic Italian cypress of the Mediterranean region. Several species are tolerant of a wide range of soil conditions, and the Monterey cypress is of value as a windbreak in coastal areas because of its resistance to salt-laden winds. Their main disadvantage is a lack of hardiness, making them only suitable for maritime climates in northern temperate areas. The hardiest is possibly *C. arizonica* var. *glabra*, but once past the early years, *C. macrocarpa* is sufficiently hardy in the southern half of the UK.

A number grown for their attractive, 'feathery' foliage in the juvenile phase may be rooted from cuttings, but many, including the glaucous forms of *C. arizonica* and selected clones of *C. sempervirens*, are propagated by grafting.

COMPATIBILITY

Some authorities split the species into two subgenera: Cupressus, the Old World species, and Hesperocyparis in the New World. All appear to be compatible, and so choice of rootstock is reliant on availability and hardiness.

TIMING

Summer/early autumn (early August to mid-September) are favoured by many but late winter/early spring often fits in with working programmes more conveniently and is a popular choice.

GRAFTING METHODS

Side veneer grafts using short or long tongue are invariably chosen.

ROOTSTOCKS

For most, 1+1P 5–8mm is normal. *Cupressus macrocarpa* is generally available but *C. arizonica* var. *glabra* should prove marginally hardier if suitable material can be sourced; it has been suggested that scion growth on *C. arizonica* is less vigorous than on *C. macrocarpa*[*]. *Platycladus* (*Thuja*) *orientalis* has been found unsatisfactory by some; others found no problems for 3 years, after which the scion variety was killed by frost, so no long-term compatibility could be established. *Thuja occidentalis* 'Smaragd' has provided a suitable rootstock for the tender *C. cashmeriana*; however, despite suggestions otherwise, it is inconceivable that any of its hardiness will be transferred to the scion variety[†]. A few horizontal forms, notably *C. macrocarpa* 'Greenstead Magnificent', lend themselves to top-working as Patio plants at 60–90cm high. Rootstocks for these can normally be

---

[*] Dirr & Heuser Jr. (2006) p.167.
[†] Buchholtz (2016).

produced as 1+2P plants of *C. macrocarpa*, but in the USA, because of concerns over hardiness, *Platycladus orientalis* or possibly x*Cuprocyparis leylandii* may be suitable alternatives*.

SCION WOOD

Where possible, healthy current year shoots 150–200mm long should be chosen. An advantage of summer grafting is that winter-damaged scion wood is avoided. Scions of *C. macrocarpa* 'Greenstead Magnificent', required to form a spreading, many-branched head, can include branched scions with 2-year basal growth.

SYSTEMS

Warm callusing is essential. Summer grafting up to mid-September often avoids the need for bottom heat; as a result, condensation and potential root pressure problems are reduced. *Cupressus* are particularly vulnerable to Botrytis during the winter grafting period; consequently, some grafters opt for the use of thermal fleece rather than polythene as a cover for grafts. Careful monitoring is required to ensure humidity levels do not become too low and slowly desiccate the scions before they have united.

PHYSIOLOGICAL AND ENVIRONMENTAL FACTORS

Temperature requirements are higher than for many conifers and, if possible, 15°C should be regarded as a minimum. During the summer period, grafts respond well to temperatures in the 20–25°C range. *Cupressus* are adapted to growth in high light levels and, therefore, avoid heavy shading, if possible. Heading-back the side grafted rootstock should take place once a good union is established and can normally be accomplished in one operation.

GROWING ON

Best-quality, container-grown plants are produced in sunny open beds. In cooler, temperate areas, *Cupressus* grown in containers will require winter protection. The genus transplants so badly that container production is normally the chosen growing on system.

---

* Buchholz (2016).

36 Daphne (Thymelaeaceae)

INTRODUCTION

Mostly small evergreen or deciduous shrubs noted for their beauty, these are found in the Northern Hemisphere and are confined to the Old World, Europe, Asia and North Africa. Species which do not regularly produce seed, some hybrids, slow-growing types and those which are not easily propagated by cuttings are normally grafted. There is significant interest in the micropropagation of some types; however, it has been suggested that plants produced by this method do not flower as prolifically as those grafted. From a commercial standpoint, grafting speeds up growth rates compared with cuttings raised or seedling plants.

COMPATIBILITY

Daphnes have a complex taxonomy open to interpretation and amendment involving 15 subgenera and ten sections*. Opinions differ widely on choice of the most suitable rootstock to ensure compatibility and good performance. No investigation into long term compatibility in *Daphne* has been discovered, but a number of combinations are known to have survived for decades.

One of the most commonly available rootstocks is *Daphne mezereum*, but this deciduous species is considered by many to be unsuitable for evergreens, which, when grafted to it, become semi-evergreen and eventually fail to thrive. Of the evergreen species, *D. acutiloba*, *D. longilobata* and *D. tangutica* are mostly used; *D. retusa* (Tangutica group) is suitable for the dwarf growing types.

Table 36.1 below shows botanic links with suggestions for what are considered likely to be the best rootstocks. If these are not shown, grafters must follow current practice and select from those listed, if possible from the same subgenus. The broadest guide is to always graft evergreen scion varieties to evergreen rootstocks.

TIMING

At a plant propagators' discussion group, it was reported that the period December to February was favoured for grafting[†]. Experience here favours summer grafting because of the possibility of some level of winter damage on some of the most commercially important species, *D. bholua* cvs and *D. odora* cvs Early growth and use of relatively soft, semi-mature scion wood allows a start commencing in early July.

GRAFTING METHODS

Wedge – often termed cleft or inlay – grafting is frequently recommended. With these methods, the relatively soft, stringy wood of *Daphne* promotes what can only be described as a crude grafting technique. Apical grafting is usually proposed; considering the difficulties of maintaining health and vitality in the rootstock, this is a surprising suggestion.

Recommendations based on personal experience are for the use of side grafts, short or preferably long tongue veneer. These offer the possibility of accurate cambial matching and retention of rootstock foliage above the graft helps maintain health of the root system. Grafts in a supportive environment make sealing unnecessary and is best omitted.

---

* Halda (1998).
[†] Carter (1979).

TABLE 36.1

Botanic Links and Suggested Best Graft Combinations for Daphne[*]

Species/Cultivar/Hybrid	Subgenus (Section Is Not Shown)	Comments – Suggested Best Rootstock
Species		
D. mezereum cvs	Daphne	D. mezereum
D. giraldii, D. jezoensis	Pseudomezereum	D. mezereum?
D. alpina, D. altaica*, D. caucasica, D. longilobata*	Sophia	D. longilobata or D. tangutica
D. aurantiaca	Dielsia	–
D. oleiodes*, D. jasminea*	Keisslera	–
D. arbuscula*, D. cneorum*, D. juliae*, D. petraea cvs*	Celkovskya	D. mezereum has been used with D. cneorum with claimed good success. D. longilobata would seem preferable.
D. laureola*, D. pontica*, D. glomerata*, D. blagayana*	Pseudolaurus	–
D. tangutica (incl. D. retusa)*, D. bholua*, D. odora cvs*, D. acutiloba*, D. papyracea*	Rehdera	D. acutiloba, D. tangutica, D. longilobata is a suitable alternative.
D. collina (D. c. napolitana)*, D. sericea*	Vahlia	–
D. gnidium*	Spachia	–
Hybrids		
D. x burkwoodii	Sophia x Celkovskya	D. longilobata
D. x hendersonii*	Celkovskya x Celkovskya	–
D. x houtteana	Pseudolaurus x Daphne	D. mezereum
D. x hybrida*	Rehdera x Vahlia	D. tangutica
D. x mantensiana*	Sophia x Celkovskya x Rehdera	D. longilobata or D. tangutica
D. x rollsdorfii*	Celkovskya x Vahlia	–
D. x rossetii	Celkovskya x Pseudolaurus	–
D. x susannae*	Celkovskya x Vahlia	–
D. x transatlantica*	Sophia x Vahlia	D. longilobata or D. tangutica
D. 'Valerie Hillier'*	Celkovskya x Sophia	D. longilobata
D. x whiteorum*	Keisslera x Celkovskya	–

*= evergreen or semi-evergreen in cold winter.

ROOTSTOCKS

Pot grown rootstocks are the usual preferred choice. Root-balled *D. mezereum* could possibly be considered for those listed as suitable for grafting to this species. Some grafters use young seedlings of *Daphne mezereum* up to 1-year old (other species could be considered). These are headed-down, the remaining stem split through the middle between the cotyledons at the point of the hypocotyl and a wedge graft used to place the scion*.

Practical experience of raising the main species for rootstocks, *Daphne longilobata* or *D. acutiloba*, emphasises the difficulty of successfully producing healthy rootstocks. If sufficient seed of *D. tangutica* or cuttings material of *D. odora* can be obtained, their cultivation as rootstocks in pots will prove less demanding than most others. The advantage of *D. acutiloba* and *D. longilobata*

* Carter (1979).

is that they are evergreen and can be relied upon to produce good quantities of viable seed each year (Figure 36.1). Seed mother plants are best grown in large containers under protection such as unheated poly-tunnels.

Precautions must be taken against birds, particularly Blackbird (*Turdus merula*) and, Wood mouse (*Apodemus sylvaticus*) as a determined individual can clear many grams of fruits daily. Wood mice are excellent climbers and among the most voracious of mice, being particularly attracted by *Daphne* bushes in fruit. Fruit ripens over a period and must be protected throughout.

Seed of these species is best sown soon after collection. It should be remembered that *Daphne* fruit, certainly of some species, is toxic and gloves are best worn during collecting and handling procedures. Cold storage permits delay until harvesting can be completed, but germination percentage may be slightly reduced. Seed should be sown individually in deep plug trays as it is important that the root system is not damaged or kinked in subsequent handling procedures. Air pruning of roots is an advantage if the seedlings are held until the following spring for potting. Elaborate precautions must be taken against mouse attack after sowing; mouse proof wire across the bottom, sides and lids of frames holding the sown trays provides the only certainty of preventing serious or total loss.

Potting-on is usually into deep 9cm pots (9 × 9 × 12.5) using a compost with a high bark and grit content. The plants must be kept as cool as possible, especially the root system, as root pathogens are encouraged by higher-than-normal soil temperatures. A better solution is to use clay pots, which are plunged in sand beds, when the root health of pot grown rootstocks is much improved. *Daphne* roots adhere very strongly to pot sides. Use of a root-inhibiting treatment, such as Copper hydroxide dissolved in latex paint, applied to the inside of the pot, will reduce root damage when the plants are knocked out for moving on (approximately 250gm of Copper hydroxide per 4.55 litres of aqueous-based latex paint is required).

Seedlings of *D. acutiloba* or *D. longilobata* should be grown in well-drained shaded beds and provided with a reliable water supply (Figure 36.2). 1+1P or 1+2P 5–7mm rootstocks are required or, if *D. odora* is chosen, 0+1+1P grade. Root grafting using root pieces 5–7cm long is said to be possible. It is claimed that this technique has been used in the past with good success for grafting *D. cneorum*.

FIGURE 36.1 *Daphne longilobata* in fruit at the first collection stage, photographed in a garden in early August. Note the presence of ripe (red) fruit and those still developing in the green stage.

FIGURE 36.2 *Daphne longilobata* rootstocks 1+1P 5–7mm in 9 × 9 × 12.5cm plastic pots. Growing on a capillary sand bed with some overhead shade, they make good rootstocks for summer grafting or for grafting the following winter/early spring. To achieve an even grade of this sort, home production is recommended. The difficulties of subsequent culture of *Daphne* after grafting means that it is crucial only best quality rootstocks are used.

SCION WOOD

Winter/spring grafting involves the use of scions usually comprising 1-year wood. To obtain sufficient size of small, compact species such as *D. petraea* 'Grandiflora', it is normally necessary to incorporate some 2-year wood at the base. Scions 2.5–7cm long are used, trimmed as required to provide a balanced start to the young graft.

Summer grafting of species such *D. bholua* cvs can commence once the shoots have started to firm. They constitute what might be regarded as semi-ripe material still producing some extension growth but with a maturing basal portion. The soft tip is removed at collection and the scions must be processed quickly and protected from damage. If unavoidable, short-term storage at 4–5°C for a few days is possible.

SYSTEMS

For the evergreens grafted in late winter, warm callusing is essential. For deciduous species, cold callus is satisfactory, provided that prolonged periods of temperatures below ±6°C are avoided. Cold callused grafts will require sealing, but it is preferable to leave *Daphne* unsealed when possible.

Because of the use of relatively soft scion material, heavy shading and a very supportive environment is required for warm callused summer grafts.

Daphne cneorum has been grafted using root pieces (probably of *D. mezereum* but not confirmed). After late summer grafting, the unions were buried in ground level beds comprising well-drained, organic soil within unheated Dutch Light glasshouses. Good results and excellent growth were claimed for this system.

PHYSIOLOGICAL AND ENVIRONMENTAL FACTORS

Daphnes are among the easiest of all woody plants to induce to form a graft union. Callus production is rapid and profuse; difficulties are associated with ensuring root health of the plants before and after grafting takes place. Root health is most probably aided by avoiding, as far as possible, application of bottom heat. If heating is unavoidable, it is preferable, economics permitting, to use space heating.

During the late winter/early spring high temperature is not required, 15–18°C being sufficient. At lower temperatures callus development takes place more slowly but satisfactorily.

Summer grafting inevitably involves high ambient temperatures. Fortunately, to the advantage of the rootstocks, bottom heat can be avoided. Heavy shading is necessary when grafting takes place during the early/mid-July period; a careful watch is essential to ensure the maintenance of very high humidity levels. Temperatures of 25°C or more are normal during daytime. Combined with the type of material being grafted, these temperatures promote rapid callus formation. The first signs of development are seen after 5–7 days, and the union is complete after 14 days. Takes are very high, normally close to 100%. Successful grafts should be returned to cool conditions as soon as possible (Figure 36.3). See also Chapter 18: Features and Management of Different Grafting Systems – Summer Active Evergreen Hardwoods.

GROWING ON

Growing conditions should be cool; during the summer months growers in northern districts have the advantage over those based in the south. Use of shade beds sited in woodland conditions will provide best conditions for many of the species, particularly the *D. bholua* cultivars. In all but the most favoured temperate areas, winter protection is required for the evergreens which are best over-wintered standing on capillary beds in cold but frost-free structures. Regular close inspection for vole (*Microtus agrestis*) damage is advisable, particularly after cold weather.

FIGURE 36.3   Weaned summer *Daphne* grafts in situ, grafted to seedlings of *D. longilobata*. *Daphne bholua* 'Jacqueline Postill' (left) and *D. odora* 'Geisha Girl' (right), in a walk-in grafting polytunnel. Successful grafts are being boxed-up in Boskoop trays for transport to a frost-free house for overwintering. The nearly 100% take is an indication of how easily this genus is grafted. Heading-back will be carried out the following spring. Careful monitoring is required to ensure the young grafts are not attacked by field voles which are particularly partial to young plants of this type. If nursery cats are present it is a wise precaution to move their feeding station to close by this crop. Unlike wood mice, field voles appear relatively disinterested in mouse bait and much prefer *Daphne* stems and shoots.

37 Fagus (Fagaceae) – Beech

INTRODUCTION

All are found in the northern hemisphere – Eurasia and North America; the Beeches contain some of the largest and most imposing deciduous trees in the native forests of their homelands. These attributes have been used to create spectacular plantings in landscaped schemes. Numerous selections, particularly of the European species *Fagus sylvatica*, have been made to further enhance possibilities for ornamental use.

Fagus are notoriously difficult to propagate by means of cuttings and the use of micropropagation does not seem so far to have reached significant levels. Layering, grafting in the field and bench grafting are the alternative propagation methods, of which, in northern areas, bench grafting is the most important.

COMPATIBILITY

Members of this small genus appear to show close taxonomic affinity with no clear evidence of consistent incompatibility linked to particular combinations.

Fagus sylvatica has proved a universal rootstock for all *sylvatica* cultivars and forms and the following species: *F. crenata, engleriana, hayatae*, (Figure 37.1), *japonica, longipetiolata, lucida*, and *orientalis*. *F. sylvatica* is recommended as a suitable rootstock for *F. grandifolia*, which is said to grow better with this combination than on its own roots*. This view is supported by others who also suggest *F. sylvatica* as a rootstock for *F. orientalis*, a pairing which has worked well for over twenty years[†]. In the UK *Fagus grandifolia* grows rather poorly and occasionally suckers, making it a questionable rootstock compared with alternatives, despite suggestions for this use in its native USA.

Beeches are not entirely free of compatibility problems, hardly surprising as they are in the same family (Fagaceae) as the Oaks. Rootstocks are seed-raised and the resultant variability produces individual grafts exhibiting incompatibility. This may take the form of union failure a few months or years after grafting, or sometimes dwarfed growth, often combined with early precocious flowering and fruiting (Figure 37.1). Generally, the numbers involved are quite small, but occasionally a high percentage may fail. This is particularly the case with top-worked *F. sylvatica* 'Purpurea Pendula' grafted in the field. In these examples it is possible that poor knifesmanship, notably matching the cambia on one side only, leads to an inadequate cambial linkage resulting in a weak union, prone to failure under windy conditions.

TIMING

As with so many temperate woody species, winter/early spring or mid- to late summer/early autumn are the favoured times for grafting. Some are adamant that only late winter/spring timing is effective while others take the opposite view. Cold storing scion wood is feasible and allows grafting to be delayed until late April. At this time in maritime climates such as the UK, field grafting is often successful. Work in Holland comparing grafting dates found that late August/early September produced better results for *F. sylvatica* 'Zlatia' when compared with grafts in late winter; for

---

* Upchurch (2008).
[†] Brotzman (2012).

FIGURE 37.1   LEFT, *Fagus hayatae* successfully side grafted to *F. sylvatica* 1+1P 6–8mm potted root-stock. RIGHT, *Fagus orientalis* showing delayed incompatibility of this large growing species grafted to a *F. sylvatica* rootstock. Overall height of this 10-year-old specimen is merely 50cm, which, coupled with profuse fruit production, visible in the picture, is demonstrating classic symptoms.

F. sylvatica 'Riversii', results were equal*. Personal experience is that either season can produce equally successful results.

GRAFTING METHODS

In the late winter period apical grafts, usually splice, tend to dominate; at other times, side veneer grafts are often favoured. The Dutch system of grafting in early September involves apical veneer grafts, with the scion leaves retained. Side veneer late summer grafts are best with scion leaves removed.

ROOTSTOCKS

Normally one-year-old *Fagus sylvatica* are lifted from seed beds to produce 1+1P 6–8mm potted rootstocks. Beech tends to produce a significant tap root and mid-season undercutting substantially improves seedling quality. Failing this, if undesirable heavy root pruning is to be avoided, deep rootstock pots are recommended. An alternative strategy is to obtain plug grown plants for potting-on.

Bare-root rootstocks, 1u1 or 1+1 6–8mm or 8–10mm, are successful for systems involving early September grafting and hot-pipe systems. They are less suitable for dormant season grafting using cold or warm callusing. As a general guide, potted rootstocks usually perform better than bare-root.

Close spacing of beech seedlings in growing beds makes them vulnerable to severe attacks by Beech Woolly Aphid (Phyllaphis fagi). This can have a debilitating effect on rootstock growth, causing an adverse influence on grafting takes. Should it appear, it is important that this pest is rapidly controlled. Most aphicides are effective but diligent spraying is required to ensure the insecticide penetrates the wholly protective covering.

SCION WOOD

During the winter grafting period, many grafters recommend collecting scions of 1-year wood, 200–300mm long. This is significantly longer than for many other species but is said to promote subsequent growth. Inclusion of a basal portion of older wood allows this length to be achieved without leaving the softer unripe, terminal portion attached. A number of grafters specifically recommend that a basal portion of 2-year-old wood is included†. Scions collected during the winter

---

* Ravesloot (1995).
† Creech (1954).

period can be cold stored until late March or beyond when cold callusing procedures take advantage of higher ambient temperatures. For summer grafting scions, 150–200mm long are sufficient.

Choice of scion material is important for the fastigiate cultivars, *Fagus sylvatica* 'Dawyck', *F. orientalis* 'Iskander', etc. Only good, upright material should be taken, repeated use of flopping or semi-horizontal branches, which may be easier to collect, results in loss of tightly fastigiate habit and lowering of quality. Similarly, if the habit of weeping forms such as *F. sylvatica* 'Purpurea Pendula' is to be maintained, only scion material from strongly weeping branches must be selected.

SYSTEMS

In Europe *Fagus sylvatica* and its cultivars are relatively easily grafted during the dormant period using cold callusing. The Oriental and American species and some *F. sylvatica* cultivars such as *F. sylvatica* 'Aurea Pendula', 'Mercedes' and 'Rohanii' are rather more demanding, and for these cold callusing may be replaced by warm callusing or hot-pipe systems.

Summer grafting is very successful and commences in late July/early August, continuing until early September. Normally, pot grown rootstocks and side grafts are used. Work in Holland at Lienden Boomteelproeftuinen (Trial Station) has investigated the use of bare-root rootstocks with an apical graft in early September. This system has worked well when grafts were left unsealed, placed in poly-tents at 95%+ relative humidity and provided with bottom heat set to 18°C. Comparisons were made between grafts with or without bottom heat. Differences were recorded of 23% higher takes in favour of providing bottom heat*.

PHYSIOLOGICAL AND ENVIRONMENTAL FACTORS

Personal experience has not revealed significant problems of root pressure and bleeding. At an IPPS discussion group event, it was noted that *Fagus* are susceptible to root pressure especially when grafted late[†]. It was suggested that stocks should only be grafted when very dry and that a cut should be made below the graft "to drain the sap". This procedure, if effective at all, takes some days to influence sap flow and slows down grafting output. Others take the opposite view with regard to root pressure and recommend leaving the graft unsealed, plunged in sphagnum moss, and the root-stock "soaked" before grafting. It was noted that if the sphagnum moss was too wet, blackening of the stems and loss occurred[‡].This would seem to indicate classic graft loss due to root pressure and bleeding. For most situations, the usual requirement is to graft potted rootstocks on the dry side, particularly if subjected to bottom heat. Potted rootstocks used in hot-pipe systems do not require drying-off but should not be excessively wet.

High temperatures are not necessary and cold callusing provides only somewhat elevated ambient temperatures. Warm callusing requires higher temperatures with a bottom heat setting of 15–18°C, and hot-pipe settings are best in the 18–20°C range.

TOP-WORKED TREES

Top-working is not required generally for *Fagus* with the exception of *F. sylvatica* 'Purpurea Pendula', which has strongly weeping branches, incapable of making an upright leading growth without tying to a stake. For convenience, this cultivar is sometimes top grafted in the field, but results are unreliable. Strong straight stemmed 2+2 trees may be lifted and rootballed for grafting in cold callusing conditions. Alternatively, suitable stems may be lifted and potted into 7-litre-deep pots and, after being held in the pot for one growing season, grafted the following late winter/

---

* Ravesloot (1995).
[†] Weguelin (1972).
[‡] Carville (1970).

spring. Grafts are usually side grafted with two scions per stem at 150–200cm high. *F. sylvatica* 'Aurea Pendula' is also suggested for top-working as it is slow growing and when bottom-worked takes time to reach a saleable size. Unlike *F. sylvatica* 'Purpurea Pendula', it does produce a leading terminal growth and makes a far more elegant and attractive tree when bottom-worked, staked and trained.

Patio plant production can be considered, but the range of suitable cultivars is limited. A good candidate is *F. sylvatica* 'Mercedes', which when grafted at 60–80cm high makes an attractive specimen.

GROWING ON

Normal tree training procedures of staking and tying-in apply for most species and cultivars. For fastigiate forms, some training and thinning double leads will ensure more shapely specimens for the long term. The pendulous types, *F. sylvatica* 'Pendula', 'Aurea Pendula', 'Black Swan', etc., which naturally produce a lead, require tying-in to a stake in the early years to encourage this habit. To prevent foliage scorch, *Fagus sylvatica* 'Aurea Pendula' will require some light shade, 20% in the UK, more in sunnier climates.

38 Hamamelis (Hamamelidaceae) – Witch Hazel

INTRODUCTION

Hamamelis are large shrubs or small trees providing a beautiful late autumn or winter flowering display. They are restricted to the northern hemisphere and occur in eastern Asia and eastern North America with a small disjunct population in Mexico. In addition to their ornamental value, *H. virginiana* produces an astringent known as witch-hazel, used as a treatment for skin and eye disorders.

The numerous cultivars can be propagated with difficulty by cuttings, and grafting is still the main method of propagation.

COMPATIBILITY

The species, plus their varieties and hybrids, all appear to be graft-compatible. Choice of combinations therefore depends upon the required characteristics of each component, namely the most ornamental scion varieties and as rootstocks, those most vigorous and easily produced.

The suckering habit of *Hamamelis* means that none of them are the perfect rootstock. To avoid this disadvantage, grafters have looked at other genera in the Hamamelidaceae family. Those most closely related to their are *Fothergilla*, *Parrotiopsis* and *Parrotia* – in that order. Of these, *Fothergilla*, another strongly suckering genus, would not be a suitable choice and *Parrotiopsis* is comparatively rare and unlikely to be available.

Parrotia persica appears a possible candidate for use as a rootstock, and some grafters are using or trialling it as a replacement for *H. virginiana*[*]. Certain scion varieties induce significant sucker production, which it is thought may be stimulated by a degree of incompatibility[†] (Figure 38.1). Also, some suggest that growth rate is slowed to about half that of plants on *H. virginiana*[‡]. Reduced size is not necessarily a disadvantage, especially in many small UK gardens and might be promoted as a desirable feature, equating to dwarfing rootstocks for fruit trees. An indication of longevity of the combination is given in an example of *H.* x *intermedia* 'Arnold Promise', which has been grafted to stems of *Parrotia persica* as top-worked standards for some years[§]. Further investigative work is warranted on the use of *Parrotia* as a rootstock for *Hamamelis*.

Other members of Hamamelidaceae have been considered. *Distylium racemosum* has the advantage of being evergreen and making it possible to extend summer grafting dates beyond August/early September. The evergreen habit makes the comparatively few suckers it produces immediately identifiable, and cuttings may be rooted relatively easily; seedlings appear not to be available. Resultant grafts have reduced vigour, which may be an advantage for small gardens; however, the longevity of grafts on this rootstock is open to doubt[¶]. At an IPPS discussion panel in 1979, it was noted that *Hamamelis* grafts with *Distylium* rootstocks "produced an inferior plant with poor

---

[*] Upchurch (2008) and Buchholtz (2017).
[†] S. Lynde (pers. comm.).
[‡] B.Upchurch (pers. comm.).
[§] Buchholtz (2017).
[¶] Lane (2005) pp.174–177.

FIGURE 38.1   *Hamamelis* x *intermedia* grafted to *Parrotia persica* rootstock, a substantial plant over 2m × 2m with some non-aggressive suckering at the base (Carlton Nursery, photo: Tim Brotzman).

growth"*. Its other disadvantage is lack of hardiness making it unsuitable for use in much of the USA and parts of the UK.

Other attempts to graft *Hamamelis* on rootstocks of *Corylus colurna* and *Liquidambar styraci-flua* resulted in complete failure[†].

TIMING

Three periods have been identified: i) summer grafting from late July to mid-September with the optimum period being August[‡]; ii) winter/early spring; iii) late spring using cold stored scion wood often used as a single bud scion. Summer grafting is generally the favoured period.

GRAFTING METHODS

Summer grafting invariably involves side grafting, either short or long tongue veneers. Some graft-ers favour the use of a single bud scion, often cut as a chip bud.

Late winter/spring grafting of potted rootstocks for cold or warm callusing are grafted either using a splice, or preferably, an apical short tongue veneer. Alternatively, they may be side grafted using a short or long tongue veneer. Because root pressure is a possibility, experience of many favours side grafts over apical grafts.

ROOTSTOCKS

Two of the three American species, *H. vernalis* and *H. ovalis*, sucker freely, a significant disad-vantage for a rootstock. *H. ovalis* is a rare plant, unlikely to be available in the near future. It may prove substantially more drought resistant than other species and have potential for hybridisation to produce a rootstock more tolerant of dry soils. The third species *H. virginiana* also produces suck-ers from stem and roots, but not as prolifically, and the foliage is sufficiently distinct to make any

---

* Humphrey (1979).
† Verhoeven (1985).
‡ Humphrey (1979) and Lane (2005) pp.174–177.

sucker growth identifiable. Add to this good vigour and relative ease of propagation from seed and *H. virginiana* becomes the rootstock species of choice.

When *H. virginiana* suckers badly it may be the result of hybridisation with *H. vernalis**; it is therefore incumbent upon rootstock producers to ensure their seed source is from stands of pure *H. virginiana*. Seed sources from gardens and arboreta where both species are present should be regarded with suspicion. Winter grafting of *Hamamelis* can sometimes be disrupted by rootstocks of some strains of *H. virginiana* which retain their dead leaves until grafting time, as these need to be removed the whole procedure is slowed down. There is a case for raising rootstocks from own sourced mother plants which display timely leaf-fall and no excessive suckering[†].

H. mollis, the Chinese species, and *H. japonica* from Japan and the hybrid between the two, *H.* x *intermedia*, can also be used as rootstocks. Because of almost identical foliage shape and colour, distinguishing sucker growth from the scion variety is difficult and failure to do so eventually leads to the desired variety becoming buried in a mass of rootstock growth.

Due to the suckering habit of *H. vernalis* it should always be avoided.

Well grown 1+0 or 1u1 seedlings of *H. virginiana* at 4–6mm diameter and 15–25cm high are ideal for potting to produce 1+1P 6–8mm for grafting after one growing season. For summer grafting a heavier grade, 5–7mm in diameter should increase in girth sufficiently during the reduced growing period to be graftable at 6–8mm from late July onwards. Upon receipt, usual trimming procedures discussed in Chapter 9 will apply. Normal choice of pot size is 9 × 9 × 12.5cm or 1-litre deep.

Bare-root *Hamamelis* rootstocks are best as 1+1 or 2+1 8–10mm seedlings.

The other main rootstock candidate, *Parrotia persica*, has the disadvantage of increased cost of production because at present it appears that no reliable source of *Parrotia* seedlings is available. The alternative method of propagation by cuttings requires a degree of skill, attention to detail, additional facilities and added cost but is routinely carried out by some nurseries in the USA[‡]. An alternative, potentially more economical approach would be to attempt grafting to roots of *Parrotia persica*[§]; no reports of results for this strategy are available.

Parrotia persica should be available as a 0+1+1P or 2P with a diameter of 6–8mm at the grafting stage. A reliable seed source for *Parrotia* may increase its use as a rootstock.

SCION WOOD

Well-ripened, 1-year wood from the lower portion of the shoot containing three or four buds is most suitable for winter or summer grafting. Scion wood collected in the summer must be gathered to ensure only firm ripe material is collected; soft unripe upper portions should be discarded. To achieve necessary girth and length, material collected from gardens and arboreta may need to include a basal portion of 2-year wood; this does not normally jeopardise results. A suggestion is made that very old thin and weak material has a better chance of success if top-worked to stems at over 60cm high[¶].

Scions collected for grafting in the summer or autumn are carrying leaves and retention or removal of these is open to debate. Removal is said to encourage scion buds to shoot prematurely. Experience is that this rarely happens before callus development is extensive enough to allow sufficient ventilation, consequently avoiding foliage disease problems. One report stated that takes were reduced by 45% due to leaf removal[**]. This has not been the experience here or during large-scale

---

*   Brotzman (2012).
†   C. Lane (pers. comm.).
‡   S. Lynde (pers. comm.).
§   Brotzman (2012).
¶   Lane (2005) p.176.
**   Thompson (1988).

Hamamelis production over several decades at Hillier Nurseries. Management of de-leaved scions is significantly easier than those with leaf retention.

SYSTEMS

Summer grafting, mid-July to mid-August, is favoured by most as results are considered to be better than for dormant season grafting*. Problems of root pressure, aided by the lack of a need for bottom heat, are significantly reduced during the summer. Ambient temperatures, further boosted by sunlight, are sufficient and may require some control by use of heavy shading. Warm callusing is essential, but some grafters replace this with a supported environment provided by overhead mist.

Sealing is not required in warm callusing systems. Supported systems using overhead mist or irrigation sprinklers require sealing or an enclosing tie, and for side grafts, careful sealing at the junction of rootstock and scion to prevent water ingress at this vulnerable point. When using this system, some check grafts hourly to avoid over-watering†. Dry fog would solve these problems.

Summer grafts are overwintered in a frost-free greenhouse (Figure 38.2).

Winter grafting can succeed using cold callusing systems, but ideally, *Hamamelis* require some additional warmth of 15–18°C to develop callus, and warm callusing in poly-tents is preferable. Results are generally less good than those achieved by summer grafting.

No mention of the use of hot-pipe systems has been discovered; however, it seems reasonable to assume its use may substantially improve takes during the winter grafting period. Premature bud

FIGURE 38.2   Summer grafts of *Hamamelis* grafted to pot grown rootstocks, packed in Boskoop trays and overwintered in a polytunnel (Witch Hazel Nursery).

---

* Lane (2005) pp.174–175, Leiss (1969), Sheat (1965) pp.204–206 and Thompson (1998).
† Thompson (1998).

break in the heating channel is likely; the resultant etiolated and often strongly pubescent leaves are vulnerable to leaf infection. Consequently, before sealing it may be wise to remove buds which will be within the heated channel.

Bare-root grafting is used by some grafters mainly in the USA. A splice or whip and tongue is normal, the graft union being sealed, well buried in boxes of peat and cold callused in temperatures maintained well above freezing (>6°C). Once fully callused they are cold stored until planting or potting*. A similar method, providing a 50–60% take, is advocated by a grower in Tennessee where, after grafting followed by cold storage, the grafts are field planted[†]. Personal experience of bare-root grafting using rootstocks of rather light-grade material intended for potting has produced very poor results. It is possible that heavier material, 2+0 or 2+1, ideally produced 'in house', would perform more satisfactorily.

PHYSIOLOGICAL AND ENVIRONMENTAL FACTORS

During the dormant season, root pressure is a danger to grafting success, and drying-off to just moist is recommended. While union formation takes place, careful control of rootstock pot moisture is important until extension growth is well advanced.

Temperature requirements for *Hamamelis* are modest, especially in the dormant season. Bottom heat settings should be set to 15–18°C with the expectation that temperatures will rise to a maximum of 21–22°C on sunny early spring days. During the summer temperatures, reaching 25–26°C is tolerated, but it is sensible to reduce or eliminate bottom heat to assist in controlling excessive temperatures.

GROWING ON

Where suitable, highly organic, moist, well-drained soils are available, best plants are grown in the field. Light shade will promote growth in the early stages, though eventually in the maritime climate of the UK, most flowers are produced when the plants are exposed to full sunlight. An important factor is protection from late frost, to which *Hamamelis* are particularly susceptible.

Some level of formative pruning by cutting back extra strong shoots may be required after planting out.

Hamamelis are increasingly grown in containers; composts without additions of lime or limestone but fortified with kieserite (magnesium) and gypsum (calcium) provide best results, especially if they contain at least 33% granulated pine bark in the mix.

---

* Flemer III (1986).
[†] Neubauer (1998).

39 Juglans (Juglandaceae) – Walnuts

INTRODUCTION

Juglans are mainly restricted to the northern hemisphere; however, there is an incursion southward via South America into the southern hemisphere, where species such as *J. australis*, *J. boliviana*, *J. neotropica* and *J. venezuelensis* are found.

Many form stately, fast growing trees including the Persian or Common Walnut, *J. regia*, and the Black Walnut *J. nigra*. *J. regia* is grown on a vast scale for nut production, notably in California, but increasingly in other locations, including the more coastal areas of Australia. Thousands of hectares of Walnut forest exist in Iran. Worldwide nut production is estimated to be 3 million tonnes.

Walnuts are notoriously difficult to root from cuttings; consequently, most of the nut producing varieties, and some species for which seed is rarely obtainable, are propagated by grafting. In areas with high ambient temperatures this is often carried out in the field, but for more northerly temperate climates bench grafting is the only viable option.

COMPATIBILITY

Taxonomic studies of the genus have identified four sections:*

1. Cardiocarya – this includes *J. ailanthifolia* (*sieboldiana*), *J. ailanthifolia* var. *cordiformis*, *J.mandschurica* (*cathayensis*).
2. Juglans – comprising *J. regia* and the recently brought into cultivation *J. sigillata*.
3. Rhysocaryon (Black Walnuts) – *J. nigra*, *J. hindsii*, *J. californica*, *J. elaepyren* (*major*), *J. macrocarpa*.
4. Trachycaryon – this includes *J. cinerea*.

Using *J. regia* as the rootstock, grafts with the following have been made In the UK: *J. regia* cultivars, *J. californica*, *J. mandschurica*, *J.* x *intermedia* 'Vilmoriniana' (*nigra* x *regia*), *J.* x *notha* (*ailanthifolia* x *regia*), *J.* x *sinensis* (*mandschurica* x *regia*). These combinations, covering most species in the sections listed above, have produced apparently successful takes and resultant plants reached saleable size. Their long-term compatibility has not been established with certainty.

J. nigra has been tested in the USA as a rootstock for short-term compatibility and was successful for the following: *J. cinerea*, *J. major*, *J. macrocarpa*, *J. regia* and the following hybrids: *J. microcarpa* x *nigra* and *J.* x *intermedia* (*J. nigra* x *J. regia*). It was less successful for *J. ailanthifolia* and the hybrid *J. cinerea* x *J. ailanthifolia* var. *cordiformis*[†]. In New Zealand *J. nigra* was used as the rootstock for *J. regia* 'Franquette' in various investigative trials[‡]. *Juglans hindsii* and Paradox Hybrid (*J. hindsii* x *J. regia*) have been used with long-term success as rootstocks for selected nut-producing clones of *J. regia* in California. In recent years other species, including *J. macrocarpa* and *J. mandschurica* (*cathayensis*), have been incorporated into hybridisation programmes to provide compatible rootstocks with desired properties of vigour and resistance to disease.

---

[*] Aradhya, Potter, Gao et al. (2007).
[†] Kaeiser, Jones & Funk (1975).
[‡] Deering (1991).

From the above results, it may be inferred that compatibility between the sections is good if not complete. However, grafters would be wise, when possible, to use only rootstock and scion combinations from within the sections, hopefully to avoid symptoms of delayed incompatibility. An alternative approach could be the use of inter-stems from hybrids between sections.

In an attempt to broaden disease resistance, trials have been undertaken using *Pterocarya stenoptera* (Chinese Wing Nut) with known resistance to *Phytophthora sp.* root disease*. Results after some years of trial have shown some, but not all, Walnut fruiting varieties are compatible when grafted to this genus. No reports of trials with other *Pterocarya* species and hybrids have been discovered. *Pterocarya macroptera* is a non-suckering species worthy of investigation.

TIMING

Apart from budding in the field during the summer, which is successfully carried out in areas with high summer temperatures, grafting is restricted to the dormant season.

In an attempt to avoid root pressure problems, early grafting before the wood had been sufficiently chilled resulted in limited and slow bud extension the following spring. This appears to emphasise the importance of a cold period to ensure that bud dormancy is broken and guarantee good extension growth after union formation[†].

Trials in Australia[‡], with *J. regia* 'Franquette' on rootstocks of *J. nigra*, used a hot-pipe system to compare the effect on takes of different grafting timings and dates. Some scions were grafted within 72 hours of collection, others cold stored for varying periods. Best results with an 82% take were obtained by storage until just before natural bud break. A further trial compared fresh scions with those collected and cold stored in sealed bags at 0°C. Cold storage increased takes compared with those grafted immediately, giving figures of 67% for the shortest storage period and 92% for the longest. Work in the USA has supported the view that a period of cold storage from collection until late spring produces the best takes. The results indicate that prolonged storage of scion wood at low temperatures enhances grafting results.

Under UK and northern European conditions, a strategy of scion collection in January/early February followed by grafting immediately or cold storage at 0.5–1°C for a period appears to provide best timing. For potted rootstocks, timing must also be related to when drying-off is considered to be at optimum level.

GRAFTING METHODS

Walnuts are characterised by wood which can contain a large volume of pith, within which are chambers or voids (see Scion Wood to follow). A procedure, said to mitigate the effects, is use of a complicated and slow grafting method known as the double tongue graft[§]. In practice many grafters opt for splice or whip and tongue grafts, the latter used especially in the field where the double tongue method would be impractical.

Personal experience is that veneer grafts offer a sensible technique for providing good matching and, with care, avoid too much exposure of potentially vulnerable central pith. Appropriate slices removed from rootstock and scion can be placed to expose pith only at the severed base of the scion. The long tongue veneer with all the advantages of additional cambial contact can be adopted by the skilled knifesman (Figure 39.1). Veneer grafts also have the advantage of being easily produced as side, rather than apical grafts, providing the added bonus of possible reduction in root pressure.

* Burchell (2002).
† Harrison (1978).
‡ Deering (1991).
§ Garner (2013) pp. 173–175.

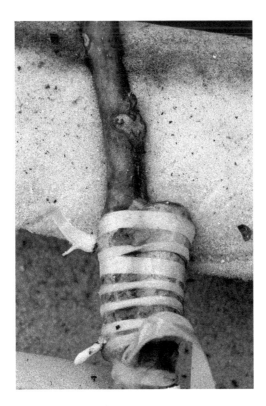

FIGURE 39.1   *Juglans* x *notha*, apical long tongue veneer graft placed in hot-pipe conditions. The long tongue of the rootstock can just be seen on the left; the stimulant bud situated at the top of the tongue has extended due to the warmth of the heating chamber. A further bud at the base of the tongue is also extending. Beneath the stimulant bud, the tongue can be seen to have slightly lifted away from the flat face of the root-stock cut. This is due to additional callus production within that area.

Young, well-grown rootstocks, bare-root or potted, may permit inclusion of a stimulant bud on the long tongue, as can be seen in Figure 39.1.

ROOTSTOCKS

In the UK and much of northern Europe, bare-root *Juglans regia* seedlings 1+0, 1u1, or 1+1 8–12mm are traditionally used as rootstocks for Walnut grafting. Some recommend larger un-transplanted 2+0 12–22mm rootstocks with the main root cut back to 20cm and side roots to 2cm*. Bare-root rootstocks have advantages in that root pressure is substantially controlled but extension growth the year after grafting is poor. This is particularly the case with un-transplanted seedlings when, combined with hard root pruning, often two seasons must elapse before grafted plants reach a size suitable for planting out from nursery beds.

Rootstocks grown in pots (1+1P 10–12mm) can boost growth rates substantially, but to gain full advantage deep pots able to contain most of the root system "as lifted" are essential to avoid the need for severe cutting back. Extra-deep-pattern, rigid pots or poly-bags, ideally with an overall length of 30–35cm are required. Suitable pot types are discussed in Chapter 9 and Figure 9.1.

It is likely that plug grown *J. regia* seedlings with air pruned roots will fulfil the need for suitable intact root systems. There will be a requirement to use the largest, deepest plugs available. After

* Lantos (1995).

the first season, potting-on into a larger pot will be necessary; in this instance a deep pattern air-pot may be a good choice.

In major nut producing areas in the USA, a number of clonal rootstocks are produced by micro-propagation and may be offered for sale in pots or plugs. The usual objective being field planting for subsequent field grafting. Bench grafting procedures could take advantage of pot grown plants as rootstocks for immediate use. For this grade of material, light-weight scions (<8mm) will be required. Alternatively they may be potted-on to achieve an increase in stem girth able to accommodate conventional scion wood (see Scion Wood to follow).

A requirement for improved cropping performance, control of tree size, shape, resistance to adverse soil conditions and diseases has promoted extensive development of rootstocks to replace seedling *J. regia*. Over many decades in the nut producing areas of California, the Californian Black Walnut, *Juglans hindsii*, has been popular for this purpose. In northern areas, its hardiness is suspect and winter temperatures much below −12°C can be expected to cause damage. Consequently, hybrid rootstock seedlings between *J. hindsii* and *J. regia* known as 'Paradox' have been in use since the 1950s*, but in recent times more effort has gone into rootstock breeding, selection and vegetative propagation.

An alternative strategy to control growth rates and influence cropping has been to double-work inter-stems on Paradox hybrid seedlings. Results showed that only *J. nigra* and *J. ailanthifolia* var. *cordiformis* had a significant effect on cropping, doubling nut production when compared with inter-stems of *J. hindsii, J. regia, J. microcarpa* (*rupestris*), and various hybrids. *Juglans ailanthifolia* var. *cordiformis* produced a much thinner stem than the rootstock below and scion variety above, and where the length of this exceeded 30cm, some breakage occurred.[†]

A number of rootstocks continue to be developed in California at the University of California, Davis, and by commercial growers. These are selected to meet requirements of varying soil conditions, vigour and resistance to soil borne diseases and pests[‡]. In most cases, only rootstocks which impart increased cropping capacity (normally a doubling of crop weight) have been selected.

Table 39.1 below lists varieties currently of significance, with propagation methods and, where known, their specific characteristics:

ROOTSTOCK PROPAGATION

The main method of vegetative propagation for walnut rootstocks appears to be by micropropagation, and a number of tissue culture laboratories in the USA are actively producing a range. Many of these are under licence from selection and trials work carried out by University of California, Davis. Specific in vitro protocols for walnuts have been developed, and most, if not all cultures, seem to be initiated from excised embryos taken from the nut. This procedure should ensure that 'off-type' mutations are eliminated or kept to a minimum.

To date, less investigative work on traditional propagation methods such as layering or cuttings appears to have taken place by researchers or nurserymen. In practice, horticultural rather than laboratory technology has obvious advantages in the propagation of rootstocks for use in the field and for bench grafting. The difficulties associated with inducing the species and hybrids to form adventitious roots on cuttings, continues to provide a significant challenge. However, the horticultural science and technology developed at East Malling Research Station, UK for use with fruit tree rootstocks might, with advantage, be applied to Walnut rootstocks.

---

* Serr (1964).
[†] Serr (1968).
[‡] Sacramento Valley Orchard Source (2016).

TABLE 39.1

Juglans Rootstocks for Walnut Production*

Name	Prop. Method		Soil Type	Tree Form/ Vigour	Pest/ Disease Resistance			Nematodes	
	Seed (sd)	Microprop (mp)			Phytophthora	Crown Gall (Agrobacterium)	Verticillium	Root Lesion	Root Knot
hindsii	sd	–	–	–	susc	susc	resist	susc	resist
Paradox	sd	–	heavy wet low salt	v. vigorous	some resistance	v susc	susc	resist	–
Vlach Paradox	–	mp	–	vigorous	–	susc	–	resist	susc
VX 211** Paradox hybrid	–	mp	poor	v. vigorous	some resistance	mod susc	–	tolerant	tolerant
RX 1** Paradox hybrid (*J. regia* x *J. nigra* selection)	–	mp	heavy wet	moderate vigorous / vigorous	tolerant	mod susc	–	–	susc
UCD 84-121** (*hindsii* x *regia*)	–	mp	–	–	? resistant	–	–	–	–
J. regia (unspecified selection) Tolerates blackline.	–	mp	–	less vigorous than Paradox	susc	fairly resistant	–	–	–
Pterocarya stenoptera	sd	–	wet	vigorous	tolerant	–	–	–	–

*Sources: McGranahan, Leslie, Hackett et al (2007–2012), Burchell Nursery (2002), Agromillora California (2017), Hasey (2016) and Sierra Gold Nurseries (2017).

**= on licence from UC Davis; susc = susceptible; resist= resistant; soil type = indication of what can be tolerated and not necessarily optimum.

BLACKLINE DISEASE

For many years, Walnut grafts using rootstocks other than *Juglans regia* failed due to the development of a black line at the point of the union and subsequent death of the scion. The cause of this response was considered to be a manifestation of incompatibility. The true cause has since been discovered to be a virus infection known as Blackline. This strain of Cherry Leaf Roll virus is transmitted from the symptomless *J. regia* cultivar scion and has an adverse effect on a number of other Walnut rootstocks including *J. nigra*, *J. hindsii* and its hybrid derivatives. After grafting, the virus travels down the scion into the susceptible rootstock causing a hypersensitive reaction. A chemical barrier to wall-off the virus is formed, resulting in graft failure and production of a characteristic black line at the union. It is essential that scion varieties of *J. regia* are virus-free until resistant rootstocks are developed.

SCION WOOD

In cool, temperate areas such as northern Europe, most grafters agree that well-ripened scion wood with minimal pith content, ideally less than 10% of the stem area is required for successful results* (Figure 39.2). In warmer climates, higher pith content may be less crucial. Trials in Australia investigated the influence on takes of varying levels of pith content†. Wood chosen comprised portions from the basal half of the one-year-old lateral shoots, where the pith occupied ±40% of the overall scion diameter, described as 'woody scions'. These were compared with those from the distal half of the shoot where the pith occupied ±60%, described as 'pithy'. Results produced a take of 59% for 'woody' versus 60% for 'pithy', showing no significant difference in takes between the two types of wood. In northern latitudes such scion wood would fail, and only basal portions of current year's wood can provide low pith to wood ratio, consequently, only one scion per shoot is possible.

Occasionally after a cool summer even basal material is insufficiently firm, and then the best option is to include a portion of two-year wood. Some authorities recommended this as standard procedure‡.

In gardens and arboreta where the trees are not regularly pruned to provide suitable material, use of scions incorporating a two-year basal portion is usually essential.

FIGURE 39.2 *Juglans regia* 'Buccaneer' scion wood showing chambered pith. Left, scion is not ideal, with too much pith/wood, but it is just useable if material is in short supply. Centre has too much pith and is unlikely to succeed. Scion on the right shows a suitable choice with a low pith:wood ratio of less than 5%.

---

* Garner (2013) p.173 and Krüssmann (1964) p.658.
† Deering (1991).
‡ Sheat (1965) pp.224–226.

In Australia, investigations were carried out comparing single bud scions with those comprising 3–4 buds. Results showed that the larger scions were significantly better, producing takes of 80% compared with those for single buds scions of 64%*.

The availability of lightweight micropropagated rootstocks grown in pots has focussed interest on producing scion wood with a stem diameter less than 8mm. Trials in California are ongoing, at present appearing to centre on pruning existing shoots in late March on previously 'hedged' mother trees. Frames erected above the mother plants are then covered with opaque tarpaulin so that emerging shoots are etiolated. These are subsequently cut-back, a procedure repeated again in June. Subsequent growth is allowed to grow in full sun or shade, to develop sufficient ripeness for use by the end of the growing season[†]. With modifications, this technique may have application in more northerly areas to produce scion wood for lighter pot grown rootstocks.

SYSTEMS

In cool, temperate areas grafting procedures take place using warm callusing or hot-pipe systems.

WARM CALLUSING

Several variants of warm callusing may be adopted:

Warm Callusing – Enclosed Conditions

Conventional warm callusing in poly-tents using completely dried-off potted rootstocks has been a traditional method of grafting Walnuts in northern Europe. Grafting takes place in January/February, and completed grafts are placed in a suitable enclosure such as a poly-tent maintained at an optimum ±26°C. Sealing is not necessary if drips are avoided and grafts are best left unsealed. Callus formation is rapid, and, as growth extension also occurs at this temperature, skilful use of ventilation is required to avoid infection of the emerging young growth. Provided this and sufficient drying-off is achieved, takes can be very good (Figure 39.3).

Warm Callusing – Callusing Boxes

For bare-root rootstocks, poly-tents may be replaced by callusing boxes similar to those used in Grapevine propagation (Chapter 55). These open-top boxes are purpose made to accommodate grafts packed in layers.

Filling is achieved by laying the box on its side and removing a detachable side portion to provide access for loading. Grafts are placed with the root ends touching the bottom of the box; as filling takes place, a free-draining moistened medium is spread between each layer to enclose the roots and graft union. Suitable materials may be chosen from moistened fine bark, coarse sawdust/fine wood shavings or peat/perlite mixtures. After filling, the removable side is replaced and when re-positioned with the open top uppermost, the box has grafts which are vertically orientated. The exposed scions may be left uncovered, but some grafters complete the packing process by covering them completely with a freely draining, moist, preferably non-organic material such as dampened perlite. This has the advantage of providing support to extending buds by surrounding them in a humid environment.

Once packing is completed, the filled boxes are placed in a controlled heated space, usually a heated storage area. The boxes may be placed in growing structures on base-heated beds or benches, but heat transfer from the base to the point of the graft union is less precise, and root pressure (bleeding) may be stimulated. Temperature is best at 25–26°C, and grafts remain in these conditions for 21 days, by which time callus and union formation should have developed.

---

* Deering (1991).
[†] Leslie, Hackett, Robinson et al. (2012).

FIGURE 39.3 *Juglans regia* 'Broadview' removed from warm callusing conditions after grafting to a pot grown rootstock. Note: extensive growth of rootstock and scion shoots due to stimulation from bottom heated base of poly-tent and high air temperature. This graft was sealed (waxed) to avoid danger of drips.

After removal from warm storage, boxes are placed in cooler conditions to allow the extending growth to harden and become acclimatised. Perlite topped boxes are upended to remove the perlite. An alternative strategy is to place the boxes in cold conditions at $\pm2°C$ for up to 4–6 weeks; careful judgement is required to avoid damage to extending shoots and soft callus.

The callus box system just described can be regarded as a variant of warm callusing as the graft is enclosed within a supportive environment.

ROOT GRAFTING

Root grafting procedures have been investigated and reported in Walnut Research Report Database papers (UC Davis) from 2007–2012*. The technique involves taking a root portion with a diameter of 16–22mm and a length of 100–120mm (unspecified – author's suggestion).The scion variety is whip and tongue grafted to this, and, after tying, the graft is placed in a plastic tub filled with wet wood wool and heated with a suspended heating cable to a temperature of 27°C. First trials heated the entire graft including the roots. Early takes were satisfactory but after field planting heavy losses occurred particularly if roots were a smaller diameter (9–10mm) than optimum.

Further trials focused heat at the graft union; although, some heat was allowed to reach the root as it was considered necessary to promote adventitious root formation. The overall effect of focused heat was to improve graft survival after planting out to an average of 74% of all types tested.

---

* McGranahan, Leslie, Hackett et al. (2007–2012).

These findings would appear to support the view expressed earlier in Chapter 18: Root Grafting, i.e. prolonged exposure of roots to high temperatures can result in a dangerous loss of stored carbohydrate reserves and subsequent death.

Reports of further investigation into this technique have not been discovered. It would appear to have potential to provide good planting material using clonal rootstocks. Comparisons of growth between root grafts and field budded stock showed that surprisingly, in the first season root grafts produced 1.75 times more growth than the budded alternative. The main disadvantage of the system would seem to be the possibility of scion rooting but suitable procedures to obviate this could probably be devised.

HOT-PIPE

Hot-pipe systems are very successful for grafting a wide range of Walnut species and cultivars. Some grafters use the 'economy design' hot-pipe beds with success and better convenience because the long root system of Walnuts is more easily accommodated*.

Bare-root or potted rootstocks can be expected to consistently provide takes in excess of 80%. After union formation, bare-root rootstocks may be cold stored with greater safety because extension growth is controlled. Consequently, grafts produced in southern areas by this system are usually field planted with minimal losses, and subsequent performance may rival that of pot grown rootstocks (Figure 39.4).

Well-grown, potted rootstocks in hot-pipe conditions can normally provide the best takes and significantly better subsequent growth than bare-root alternatives. For use in the hot-pipe system, drying-off is not required to the same level as for conventional warm callusing, but a good level of drying back is essential, particularly if surrounding ambient temperatures are allowed to exceed 5°C. Temperature for hot-pipe systems should be 25–26°C for 3–5 weeks.

Utilising the hot-pipe system may well provide a very successful means of root grafting walnuts to the roots of clonal rootstocks. Extra care will be required to prevent desiccation of the root portion housed within the heating chamber. This should be possible by the use of appropriate heavy sealing or a waterproof sleeve, such as a wide polythene tie. Successful grafts will produce a semi-dormant scion, and a rootstock still retaining plentiful stored carbohydrate, enabling good subsequent development of extension growth. It seems likely that roots will branch naturally if supported by good food reserves and stimulation from extension growth above.

PHYSIOLOGICAL AND ENVIRONMENTAL FACTORS

The two major requirements for successful Walnut grafting are control of root pressure and provision of temperatures in the region of 26°C at the site of the graft.

For warm callusing systems, particularly if bottom heat is applied, potted rootstocks must be severely dried to prevent excessive bleeding, which invariably results in graft failure. Potted rootstocks destined for grafting in January need to be brought into a dry, rain-free area when leaf-fall is more or less completed, if possible by mid-October. To achieve this, timing defoliation may be necessary. Storage conditions during drying-off are best with no additional heat other than ambient; although, in cold areas additional heat will be required to avoid very low temperatures. Potting compost can be formulated with additional bark content to aid faster drying-out.

Occasional checks should be made to ensure drying is proceeding, no rodent damage is occurring and there is no sign of any shrivelling of the younger stem tissues. Closer checks should be made 2–3 weeks before the proposed grafting date to ensure the roots and stems are in good condition

---

* Clark (1981).

FIGURE 39.4 *Juglans regia* 'Franquette' bare-root graft, removed from hot-pipe. Note the growth of buds enclosed within heating chamber; beyond this area buds are still dormant.

and there is no evidence of residual moisture in the compost. If examination reveals moisture present, grafting must be delayed, and pots may need to be removed to provide faster evaporation. Conversely, excessive drying, leading to the possibility of shrivelling, is avoided by advancing the grafting date or enclosing the rootstocks in polythene to prevent further moisture loss. Rootstocks adequately dried before required should be placed ideally in polythene bags in cold storage.

For hot-pipe systems, potted rootstocks need to be well dried-off, however this is less critical than when using warm callusing for reasons explained previously. Bare-root rootstocks are far less prone to root pressure and unless due to heat stimulation, young root regeneration occurs, bleeding is rarely a problem.

Temperature response for Walnuts is very specific: takes of 80% or more can be expected in only a narrow range between 25 and 27.5°C. The normal optimum range of 21–22°C for many species is insufficient, and poor takes, as low as 10%, are a possibility at these temperatures. Higher than 28°C results in a steady fall and at 29°C takes may be reduced to 60%*.

---

* Hartmann, Kester, Davies Jr. et al. (2014) p.453.

GROWING ON

After successful grafting, subsequent handling in southerly areas often involves field planting; in northern Europe, to avoid frost damage, this cannot take place before mid-May. For northern areas, a safer procedure is to pot-off the successful grafts and place them in growing structures often provided with supplementary heat. Deep-pattern pots are required because of the lengthy root system. The large thick root system makes it difficult to avoid severe root curling in container culture, and for this genus, Air-pots may provide the ideal container.

In field culture, transplanting must be carried out regularly to ensure the root system is maintained in good condition for sale. To provide reliable transplanting material, use of root control bags may be a practical alternative strategy.

40 Juniperus (Cupressaceae) – Juniper

INTRODUCTION

Junipers are widely distributed throughout the northern hemisphere from the Arctic Circle to Mexico and the mountains of eastern Africa. Only one species, *J. procera*, extends into the southern hemisphere. Growth habit varies enormously from low-growing prostrate evergreens to small, medium or large shrubs, and conical or columnar trees. A number are important for use in ornamental plantings. The berries of some species have culinary uses, and Juniper berry spice is an important product, used to flavour gin. Others yield oil from the wood, leaves and berries used in homeopathic medicine to aid digestion, reduce anxiety and insomnia.

Most species and important ornamental cultivars are propagated by cuttings or seed. Certain types resist attempts to induce rooting and must be grafted.

A number of changes in nomenclature have been made in recent years with some old established names no longer considered valid. New names are given here followed by the old name in parentheses.

COMPATIBILITY

Taxonomy of the genus is in a state of flux with some confusion and disagreement on the number of existing species. Several sections have been identified; however, allocation of the species within these is incomplete. The largest, by far, but likely to be further divided, is the section Sabina, distinguished by their scale like leaves. These are represented in the Old World by *J. chinensis*, *J. recurva*, *J. squamata*, and in the New World by *J. horizontalis*, *J. scopulorum*, *J. virginiana* and others. The section *Juniperus* with needle-like leaves has sub-sections of interest; sub section *Juniperus* contains *J. communis*, *J. conferta* and *J. rigida*; sub-section Oxycedrus includes *J. cedrus* and *J. oxycedrus*.

Incompatibility issues have been identified as a result of individual grafters' experience and do not follow a fixed pattern or taxonomic principles. The most popular rootstock *J.* x *pfitzeriana* 'Hetzii' (*J.* x *media* 'Hetzii') is considered by many to be fully compatible with the species and cultivars of: *J. virginiana*, *J. communis*, *J. sabina*, *J. scopulorum*, *J. chinensis* and *J. squamata*[*]. Exceptions may occur; *J. scopulorum* 'Springbank' is stated to show delayed incompatibility and Dutch grafters use *J. scopulorum* 'Skyrocket' for this cultivar[†]. Another review of Juniper rootstocks indicated poor compatibility of *J. pfitzeriana* 'Hetzii' with *J. sabina* 'Broadmoor' and poor takes with *J. virginiana* 'Hillspire' ('Cupressifolia'). Surprisingly, the same report gave poor growth and takes for *J. virginiana* 'Hillspire' as a rootstock for *J. virginiana* cultivars[‡]. Other species tested as possible rootstocks include *J. chinensis* said to inhibit growth of *J. virginiana* cvs. *J. communis* 'Hibernica', having a dwarfing effect on *J. chinensis* 'Columnaris' and producing progressive losses on other *J. chinensis* cvs[§]. Limited experience here supports the view that *J. communis* 'Hibernica' is not a reliable rootstock.

---

[*] Bakker (1992) and Hall (1977).
[†] Alkemade & van Elk (1989) p.114.
[‡] Nelson (1968).
[§] Nelson (1968).

The system of cuttings-grafts has shown other possible incompatibilities: *J. virginiana* 'Hillspire' (*J. virginiana* 'Cupressifolia') is favoured by some as a rootstock for this system because it is said to root fairly easily, produces a good root system, is a fast grower and transplants well. Surprisingly, it is said to show some incompatibility with *J. virginiana* 'Burkii', *J. virginiana* 'Caenartii' and 'Glauca'*.

Grafting to a stem for Patio trees or taller, top-worked items invariably involves the choice of *J. scopulorum* 'Skyrocket' and no problems of incompatibility have been reported with this rootstock.

Platycladus (*Thuja*) *orientalis* has been suggested as an alternative to *Juniperus* and for some species (unspecified) has proved satisfactory[†]. Other authorities have identified *Platycladus* as causing progressive losses on various *J. chinensis* forms, plus delayed incompatibility and losses on certain *J. virginiana* cultivars[‡].

The above summary contains some conflicting opinions on compatibility and must raise doubts on the validity of all conclusions. Personal experience with the use of *J.* x *pfitzeriana* 'Hetzii' across a wide range of species and cultivars has not produced any obviously incompatible combinations during their time in the nursery and arboretum. These include: *J. ashei*, *J. californica*, *J. chinensis*, *J. chinensis* 'Aurea' (growing well after 25 years), *J. chinensis* 'Keteleeri', *J. chinensis* 'Plumosa', and forms (*J. media* 'Plumosa' and forms), *J. deppeana* and *deppeana pachyphlaea* (growing well after 25 years), *J. drupacea*, *J. excelsa* (has survived well for many years), *J. monosperma* (growing well after 30 years), *J. osteosperma*, *J. oxycedrus*, *J. pinchotii*, *J. procera*, *J. recurva* (growing well after 30 years). Also, the following *J. scopulorum* forms: 'Blue Heaven', 'Hillburn's Silver Globe' 'Hill's Silver', 'Moonglow', 'Pathfinder', 'Springbank', 'Tabletop', 'Tolleson's Blue Weeping' (growing well after 30 years), 'Wichita Blue' and *J. virginiana* forms: 'Glauca', 'Globosa' and 'Schottii'.

It would be surprising if, in such a diverse and large genus, there are no examples of incompatibility, but in general most combinations involving *J.* x *pfitzeriana* 'Hetzii' appear to provide a successful long-term union.

TIMING

Grafters in the UK and USA normally graft Junipers in the December/February period. In the Netherlands grafting is generally delayed until February/ March/April.

In Dutch nurseries, rarely so in the UK and USA, autumn grafting (September–October) is sometimes used for the green and glaucous leaved forms. It might be considered that some of the foliar diseases, notably *Pestalotiopsis*, are less prevalent and aggressive at this time than in the late winter/ early spring. Grafters who consider scion wood should not be collected in frosty weather may also prefer this timing. Some, in northern Europe grafting in late winter, are unable to avoid scion collection during freezing weather but do not consider this to cause problems or reduce successful results[§].

GRAFTING METHODS

Side veneer grafting is almost universally used (Figure 40.1). In most cases the long tongue variant is favoured. Short tongue veneers are a safer option if the difference in stem girth is substantial, most likely when grafting to stems in the production of Patio Plants and standards.

One method described involves the separate production of veneer graft components: scion preparation by one team, rootstock preparation and graft assembly by another. An output of 6–700 grafts/ person /working day is claimed after two weeks' training[¶].

---

* Leiss (1966).
[†] Hall (1977).
[‡] Nelson (1968).
[§] Thomsen (1978).
[¶] Hall (1977).

FIGURE 40.1 *Juniperus deppeana* var *pachyphlaea* grafted to *J.* x *pfitzeriana* 'Hetzii' rootstock using a long tongue side veneer graft. This form of *J. deppeana* has among the best blue/green foliage of all Junipers.

Top-working is usually restricted to species and cultivars with a horizontal or pendulous growth habit. A well-balanced head is most rapidly produced by multi-grafting using two or three scions.

ROOTSTOCKS

For most grafters in Europe and America, *Juniperus* x *pfitzeriana* 'Hetzii' (*J.* x *media* 'Hetzii') is widely considered to be the best choice of rootstock for bottom-worked grafts. For top-working, *J. scopulorum* 'Skyrocket' (*J. virginiana* 'Skyrocket') is the preferred choice. Some also favour 'Skyrocket' for bottom-working, but experience here is that it is not as easily rooted as 'Hetzii', and subsequent growth of the graft is not as rapid.

The seed-raised *Juniperus virginiana* is an alternative and was for some years the major rootstock species; however, its susceptibility to Juniper blight (*Phomopsis juniperovora*) has resulted in a general move away in favour of the more resistant *J.* x *pfitzeriana* 'Hetzii', which is said by Dutch grafters also to be more resistant to Aphid attack.

Traditional methods of producing Juniper rootstocks from cuttings, particularly in the USA, involved the use of large cuttings (<25cm × 4–7mm Ø) rooted under polytunnels in the field, or bedded-up into the field after rooting in a structure. These were left in position for the remainder of the growing season, lifted late September-late October, potted-up and grafted the following late winter/early spring*. Most now root cuttings conventionally, often using mist propagation. This is followed by potting-up and growing on procedures to provide established pot grown rootstocks for grafting.

---

* Hall (1977), Bakker (1992) and Hoogendoorn (1981).

The specification for bottom-worked *J.* 'Hetzii' is 0+1P or 0+1+1P 5–7mm. The final potting stage (1P) is sometimes shortened from a growing season to a brief period to allow root regrowth before grafting takes place. A pot size of 9 × 9 × 9–12.5cm is best for bottom-worked grafts.

J. scopulorum 'Skyrocket' for top-worked grafts normally requires 0+2P or 0+1+2P, with an optimum diameter of 7–10mm at grafting height. A 2-litre-deep, final pot size is best for Patio Plants grafted at 45–75cm high. Limited requirements for tall, top-worked specimens with a 180–200cm stem may need a further year 0+1+3P, finishing in a 4-litre deep pot. A popular alternative is to use a root-balled plant lifted from the field.

SCION WOOD

Junipers grow by extension rather than in flushes, and distinguishing between 1-year wood and older is not always easy. Best scions comprise material 125–180mm long, the lower 60–100mm being mature (brown) wood. Some favour larger material, up to 300mm long; heavy gauge rootstocks are then required involving significant additional handling procedures. Upright and columnar species and varieties normally have single-stem scions, spreading and weeping types, especially if selected for top-working, are best with multi-branched stems.

Scion wood may be stored for a period of up to 3 days at 3–5°C* while winter collected scions stored at 0.5–1°C can be safely stored for several weeks. Mention has already been made of the belief by some that scion wood should not be collected in temperatures below freezing. If this is unavoidable it is suggested that the problem may be overcome by submerging the material in cold water for 15 minutes to "draw out the frost"[†]. It is hard to imagine that scion wood from hardy Juniper species suffers when collected in temperatures just below freezing.

SYSTEMS

Warm callusing using conventional poly-tents is usual but occasionally supported warm callusing using overhead mist is favoured[‡]. Open mist set to give water bursts at intervals of lengthy duration may reduce prolonged wet foliage conditions, and assist in controlling foliar diseases, to which Junipers are particularly susceptible.

Grafters traditionally plunged the graft union in the standing-down medium[§]; this avoided the need for sealing and permitted more ventilation without increasing the possibility of damaging the graft union by desiccation. In all cases, the need for a well-drained open medium to cover the union was stressed. Careful management of an enclosed environment, coupled with measures to avoid drips, has meant that graft unions are rarely plunged today, and grafts may be left unsealed. Replacing polythene film with horticultural fleece for covering Cupressus, vulnerable to foliar diseases, may prove equally successful for Junipers.

Summer grafting during August/September is carried out on a limited basis and has particular value for species subject to winter damage such as *J. californica*, *J. formosana* and *J. procera*. Warm callus conditions succeed well, and most Junipers perform favourably in high summer temperatures and long day length.

Cuttings-grafts have proved a successful technique for some, who claim by using this system saleable plants can be produced 18–24 months more quickly than by conventional systems[¶]. Others disagree and say the time from grafting to production of an established plant is nearly doubled[**].

---

* Wells (1980).
† Hoogendoorn (1981).
‡ Hall (1977).
§ Hall (1977), Bakker (1992), Stoner (1974), Decker (1998), Wagner (1967) and Hoogendoorn (1981).
¶ Chong (1981).
** Leiss (1966).

Experience suggests that a graft made on an unrooted stem portion cannot subsequently compete in growth rate with those grafted to a conventional rootstock with an established root system.

The cuttings-graft system involves grafting the chosen scion variety to unrooted cuttings. Long or short tongue side veneer grafts are placed ±40mm above the base of the cutting which has been, or can subsequently be, treated with a suitable formulation of IBA growth substance (1–2500ppm quick dip in 50% alcohol or 3–8000ppm in talc dust). Completed grafts are then inserted into a suitable rooting medium to cover the graft union. The medium normally comprises Peat/Perlite mixtures, though Peat/Bark may be a better choice (Figure 40.2).

One method claimed to avoid the need to tie the graft components together involves the use of a side graft pushed through a pre-punched Styrofoam rooting block, which serves to hold the graft components tightly together. The base of the cuttings portion of the graft ('Hetzii' rootstock) is pushed through the block, so that 5mm is protruding to allow growth substance (IBA) treatment to be applied. This promotes rooting which takes place at the base of the cutting and throughout the rooting block. The resultant grafted plant is then grown-on directly from the block by potting-off or planting-out as required*.

PHYSIOLOGICAL AND ENVIRONMENTAL FACTORS

Junipers can be placed in the 'warmth loving' group of conifer grafts with a minimum temperature requirement of +12°C. Callusing may be achieved relatively quickly in bottom heat levels of between 18–22°C. However, shading may be necessary even in the January-March period to prevent ambient temperatures rising to excessive levels.

FIGURE 40.2   Cuttings-graft of *Juniper chinensis* 'Plumosa Aurea' to *J.* x *pfitzeriana* 'Hetzii'. Note: base of cutting rootstock has been heavily wounded and IBA will be applied to aid rooting.

---

* Chong (1981).

Some grafters recommend watering the rootstocks before grafting*, and none appear to recommend significant drying-off, indicating there is no root pressure problem. Personal experience would advocate grafting with the rootstock 'on the dry side'.

Almost all recommend that the root system should be in an active state before grafting commences, some delaying grafting until there is visible root extension[†]. It is hard to reconcile this assertion in the light of the apparent success of cuttings-grafts, when no roots are present at the commencement of grafting. Personal experience has not shown a close link between graft success and rootstock root activity. Relatively soon after non-root-active grafted rootstocks are placed in warm callusing conditions, root activity is visible, but at this stage callus formation has also commenced.

PERFORMANCE COMPARISONS

Strongly blue-foliaged forms of *J. scopulorum*, particularly when linked to a weeping habit such as *J. scopulorum* 'Tolleson's Blue Weeping', and those from hot arid areas such as *J. deppeana* var. *pachyphlaea* and *J. oxycedrus*, are more challenging than most other species.

GROWING ON

Given good drainage and open-air conditions, Junipers grow well in containers or in the field. The main problems are associated with disease pressures, probably exacerbated in nurseries by close proximity of blocks of susceptible species. Regular fungicidal spray programmes, sufficient space and, if possible, separation of growing blocks by interspersing them between non-susceptible species are sensible strategies. Hard pruning should be avoided, and follow-up preventative sprays after shaping and pruning operations are wise precautions against disease outbreaks.

---

* Hall (1977), Hoogendoorn (1981) and Wells (1980).
[†] Brown (1963).

41 | Magnolia (Magnoliaceae)

INTRODUCTION

A large, very ancient genus of woody plants, they encompass some of the most spectacular and beautiful flowering trees and large shrubs in cultivation. Over 200 species are distributed in both hemispheres; their main centres in the north being the Himalayas, east Asia and N. America, and in the south, Central America to Brazil. Those from northerly regions are mostly deciduous; a high proportion of those from the Southern hemisphere are evergreen. A large number of species are found in South America, and due to habitat loss, these, along with some in eastern Asia, are becoming endangered or critically endangered. Grafting could and should play an important role in their conservation.

In temperate areas, many of the popular hardy types for gardens are propagated by leafy cuttings. Good environmental conditions provided by polythene, mist or fog, when combined with rooting hormone treatments can produce very good results in the hands of a skilled propagator. A number of important ornamental varieties are less successful, and for these grafting is essential.

COMPATIBILITY

It is not surprising that such a large genus has a complex taxonomy still being investigated and debated, often with conflicting views, by botanists across the world. For a full account of taxonomic issues readers are referred to publications by the USA-based Magnolia Society.

In recent years, a number of closely allied genera have been shown by DNA investigations to be part of the Magnolias and consequently are now amalgamated within the genus. Discovery of close linkage within Magnolias has led to the realisation that most, if not all previous separate genera, species, and hybrid varieties between them, appear to be graft-compatible (Figure 41.1). As long ago as 1952 it was noted that there was a wide range of compatibility within the genus*. Subsequent experience has confirmed these findings, and other issues, particularly concerned with cultivation, have influenced the choice of suitable rootstocks.

Despite suggestion that universal compatibility is a feature, nothing can be assumed and unexpected problems could occur. A number of new combinations, yet to be investigated, may fail and time may prove that long-term compatibility of others was not as certain as once thought. For these reasons, the following phylogenic table (Table 41.1) showing linkage of species within the two subgenera is provided. Those grouped together show the closest phylogenic affinity and might be expected to have the highest level of compatibility. Decisions on suitable rootstock/scion combinations for hybrids between the groupings may be aided by extrapolation from the list.

Table 41.1 does not include a range of tender species or those which are rarely cultivated in temperate areas. Detailed taxonomic classification is not shown beyond a broad distinction between the relevant main subgenera. Some authorities consider certain species (e.g. *M. sinica*, *M. nitida*) shown here within subgenus Magnolia should be placed within a third subgenus Gymnopodium, containing mostly tropical species and some others not listed.

Specific examples of compatible unions provide further useful guidelines. Most suggest that *Magnolia grandiflora* cultivars should be grafted to the species[†], but in Japan they are often grafted to *M. kobus*[‡]. A general suggestion is that the 'Oriental' species are best grafted to *M. kobus*,

---

* Mundey (1952).

[†] Wells (1955) p.250, Sexton (1965) and Dirr & Heuser Jr. (2006) p.238.

[‡] Baker (1976) and Knuckey (1969).

TABLE 41.1
Magnolia Phylogenic Groupings*

Subgenus Yulania	Subgenus Magnolia
foveolata	*sinica*
maudiae	*nitida var. lotungensis*
chapensis	*nitida*
champaca	*insignis*
lacei	*fordiana*
figo	*grandiflora*
doltsopa	*tamaulipana*
floribunda	*virginiana*
amoena	*delavayi*
zenii	
biondii	*obovata*
kobus	*officinalis*
stellata	*tripetala*
	rostrata
dawsoniana	
sargentiana	*sieboldii subsp. sinensis*
campbellii	*sieboldii subsp. sieboldii*
sprengeri	*wilsonii*
denudata	*globosa*
cylindrica	
liliiflora	*fraseri var. fraseri*
salicifolia	*fraseri var. pyramidata*
acuminata	*macrophylla subsp. macrophylla*
acuminata var. subcordata	*macrophylla subsp. ashei*
	dealbata

*modified from Phylogenic relationships in family Magnoliaceae. S. Kim et al. (2001).

whereas *M. acuminata* is considered best for the American species*. This is presumably based on an assumption that most 'Orientals' are from subgenus Yulania, though in fact many are in subgenus Magnolia. This view is not shared by all, and some state that takes are higher on *M. kobus* than on *M. acuminata* though the scion varieties involved are not specified[†].

Personal experience of a wide range of *Magnolia* graft combinations has identified only one possibly incompatible combination: this concerned *M. rostrata* (subgenus Magnolia) grafted to *M.* x *loebneri* 'Leonard Messel' (subgenus Yulania). Three scions gathered from the original Kingdon-Ward collection at Borde Hill garden, UK were grafted to rootstocks of *M.* x *loebneri* 'Leonard Messel' and seedling *M. obovata* (subgenus Magnolia) (Figure 41.2). One graft on Messel failed but the remaining on *obovata* and Messel succeeded. Subsequent growth the first season on both rootstocks was good, during the second season growth on *M. obovata* was very vigorous, but on 'Leonard Messel' restricted. At the start of the third season the graft on 'Leonard Messel' failed;

* Hess (1953) and Alkemade & van Elk (1989) p.97.
† Hesselein (2005) and Wells (1955) p.250.

FIGURE 41.1   LEFT, *Magnolia* (*Michelia*) *figo* var *crassipes* grafted to *M.* de Vos/Kosar hybrid Betty (both from subgenus Yulania). CENTRE, *M.* 'Nimbus' a hybrid of mixed parentage (*M. obovata* x *M. virginiana*) from subgenus Magnolia, succeeds well on rootstocks from the subgenus Yulania (*M. stellata* 'Royal Star') and has made a small tree 3m × 2m in 6 years. RIGHT, *M.* (*Michelia*) *doltsopa* grafted to *M. kobus* rootstock in February photographed late August. Showing over 30cm of extension growth.

testing the scion produced a clean break indicative of an incompatible union. Whether death was actually due to incompatibility or the original poor constitution of the scion, taken from a rather sickly mother plant, cannot be confirmed.

The group (Oyama series) containing *M. sieboldii*, hybrids within the group and *M. globosa* and *M. wilsonii*, particularly the latter, does not seem entirely happy on Yulania rootstocks. Slow growth, plentiful flowering and fruiting and a tendency to produce suckers, not profusely but more or less continuously, are indications that a further search for suitable rootstocks may be warranted, possibly from within subgenus Magnolia.

Grafters must make their own decisions on likely suitable combinations; indications are that it may be wise to ensure some *M. obovata* rootstocks are available for use for some species and hybrids within subgenus Magnolia*. In the majority of cases, *M. kobus* or others in subgenus Yulania such as *M. stellata*, *M. kobus* x *stellata* or the de Vos and Kosar hybrids make excellent rootstocks for Yulania species. However, they are also compatible partners for many species and hybrids within subgenus Magnolia. *M. grandiflora* may be added to those listed above as the most appropriate rootstock for its cultivars.

Trials in 1985† investigated the possibility of using *Liriodendron tulipifera* as a rootstock for *M. acuminata* and subsequently for *M. fraseri* and *tripetala*. If successful, the advantages of *Liriodendron* would be its hardiness and tolerance of a range of soil types. After an apparent union and significant growth for a short period, all grafts failed, showing typical incompatibility symptoms.

TIMING

Two major times during which grafting can take place are winter/early spring and mid-/late summer to early autumn. A few extend the autumn period into early/mid-winter. Experience here using warm callusing systems is that grafts after early-September show progressively poor performance up to leaf fall, and after this very poor, until mid-winter/early spring. Use of the hot-pipe system

---

* Lane (1993).
† Figlar (1985) and Hooper (1990).

FIGURE 41.2   *Magnolia rostrata*. LEFT, grafted to *M.* x *loebneri* 'Leonard Messel' which subsequently failed, possibly due to incompatibility. RIGHT, grafted to *M. obovata* (*M. hypoleuca*) not yet headed back but scion growth (right hand side) shows excellent vigour achieved during the first season from this rootstock/ scion combination.

in autumn/early- winter, rather than in the conventional period winter/early spring, has not been tested; possibly it would prove successful.

Previously, warm callusing in summer/early autumn invariably produced better results than the late winter/early spring period. However, introduction of the hot-pipe system has redressed the balance, and, using this, high percentage takes can also be expected during the winter/spring period.

GRAFTING METHODS

In warm callusing systems, side veneer grafts are preferred over splice for reasons discussed previously. More recently use of the hot-pipe system for winter grafts has produced excellent results, and splice grafts are normally used. If heating chamber design permits, long tongue side veneer grafts allowing almost perfect cambial matching are to be preferred (see Chapter 18: Hot-Pipe/Grafting Methods for further discussion). The long tongue side veneer is recommended for those evergreen species previously in the genera *Michelia*, *Manglietia*, etc.

More recently, papers on *Magnolia* grafting describe chip budding as the favoured method in protected structures, leading to a misleading impression that chip budding has replaced traditional procedures involving side or apical grafts. The chip bud was devised as a practicable system for use in the field while at the same time allowing a level of cambial matching not possible with the alternative and technically inferior rind graft ('T' bud). A chip bud is very similar to a side veneer scion but with substantially reduced wood content; this has the effect of reducing the amount of carbohydrate and protein available for supporting callus production and subsequent graft union development. Additionally, the relatively thin section of wood left above the bud is vulnerable to die-back and can be the focus of eventual failure. In enclosed environments such as a poly-tent, the side veneer graft, especially the long tongue variant (not possible with chip buds), should outperform the alternative chip bud method.

While chip buds can be considered for summer grafting, their use in a hot-pipe system during the winter is not possible. The configuration of the graft causes the bud to lie within the heating chamber; consequently it is subjected to heat designed to stimulate callus formation, causing the bud to break dormancy and elongate.

One reason given for the use of chip budding is that it results in a single leading shoot. This is obviously an advantage for the production of a plants suitable for growing-on as trees. A similar outcome can be achieved using a single bud side veneer graft with the added bonus of a long tongue and stimulant bud (Figure 41.3).

Magnolias grown in tree form are not universally popular, and a number of markets require multi-stemmed plants. Flowering at a reduced height, often with more flowers, these can create a better effect in retail outlets. A chip bud is not suitable for this purpose, and multi-budded or branched scions are best for producing this shape.

Grafts placed in enclosed environments are better not sealed. In hot-pipe systems where sealing is essential, this may be achieved by dipping splice grafts in hot wax to ±10mm below the union; side grafts will require brush application or use of non-porous enclosing ties plus sealing any exposed cut surfaces. Because of their large buds, some feel that Magnolias are better sealed using a brush rather than by dipping.

ROOTSTOCKS

CHOICE OF ALTERNATIVES

There appears little possibility of incompatibility causing problems when making choices for *Magnolia* rootstocks. Consequently, a number of other factors can be taken into account. Initially a decision must be made on whether the plants are derived from seedling or clonal sources. The obvious advantages of economy and comparative ease of production commends seedling

FIGURE 41.3   LEFT, single bud scion of *Magnolia wilsonii* 'Sandling Park' using a long tongue side veneer graft. A stimulant bud is seen growing out from the 'tongue', more or less ensuring a successful 'take'. RIGHT, *Magnolia campbellii* subsp. *mollicomata* 'Borde Hill' using the same grafting method and showing a stimulant bud growing out. Both were summer grafts warm callused. Provided the scion is long enough to ensure the bud of the single bud scion is outside the heating chamber, a hot-pipe system could be used during the late winter. Sealing the graft would then be necessary.

rootstocks to most grafters. However, there is a case for clonal rootstocks and advocates of these emphasise:

- Uniformity of growth.
- The possibility of influencing final size of the mature plant, as in fruit tree production.
- The long-term parity of growth between rootstock and scion to avoid unsightly graft unions.
- The tolerance of the rootstock root system to adverse soil conditions.
- The ability to maintain good root system health within the confines of the comparatively small pot used by most grafters for rootstock production.

Individuals place different emphasis on these factors. In ideal growing conditions, such as parts of New Zealand, where many Magnolias may be budded in the field and vigour may exceed that in the wild, considerations of differential growth rates leading to unsightly unions is of importance. Careful observations by individuals in that country have highlighted differential growth between numbers of combinations, as follows:

M. x *loebneri* 'Leonard Messel' is considered moderately vigorous with a caliper growth increment comparable to hybrids of *M. acuminata*. *Magnolia* x *soulangiana* cultivars are thought to be suitable for *M. cylindrica* and *M. sprengeri* var. *diva* but not for any hybrids with *M. campbellii* parentage. *Magnolia* x *soulangiana* 'San Jose' is considered to have similar vigour to *M.* x *veitchii* and therefore suitable for the most vigorous hybrids*. This last example highlights problems of location and nomenclature; in the UK this cultivar is much less vigorous than *M. veitchii*, and the situation is further confused by the fact that there are two clones†.

In much of northern Europe and the UK, with few exceptions, differential growth rates leading to unsightly unions is less evident, even when vigorous scion varieties are grafted to rootstocks having a small stature (see Chapter 11 and Figure 11.23: Rootstock and Scion Interactions for further discussion).

There are indications that use of *M. kobus* and smaller growing types, such as *M. stellata*, the de Vos and Kosar Hybrids, and *M.* x *loebneri* 'Leonard Messel', has a growth reducing effect on larger varieties grafted to them. Work on the influence of clonal rootstocks on *Magnolia* growth and development would be a very worthwhile subject for further long-term investigation.

Seed-Raised Rootstocks

Results from bare-root rootstocks are considered to be significantly inferior to potted rootstocks. Consequently, from the grafter's perspective, an essential factor in rootstock choice is availability and performance in small rootstock pots. *M. kobus* has the advantage of being reasonably amenable to small pot culture and has long been readily available as 1+0 seedlings suitable for potting into 9 × 9 × 9–12.5cm or 1-litre-deep pots. It is also frequently offered as 1+1P 6–8mm plants in 9 × 9cm pots. In general, 6–8mm girth is sufficient, but for some of the larger hybrids 8–10mm grade is more appropriate.

There are advantages for in-house production of *Magnolia* rootstocks from seed (Figure 41.4). In some seasons in northern Europe *M. kobus* often sets seed freely and mother plants may be planted to ensure a supply of seed for home grown seedlings. If possible, mother plants should be kept isolated from other Magnolias as hybridisation is possible and the resultant seedlings may fail to respond to pot culture as well as pure *M. kobus*.

Seed mother plants may be used to meet entire requirements or, because a good seed crop is not guaranteed every year, as a back-up to existing suppliers. This is a long-term strategy as young

---

* Hooper (1990) and Hooper (2010).
† Gardiner (2000).

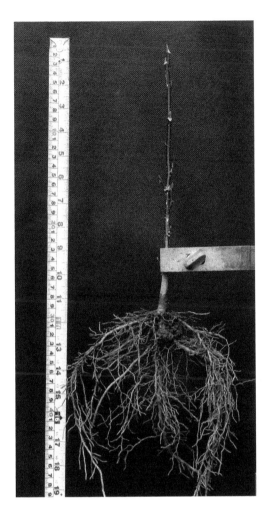

FIGURE 41.4 *Magnolia kobus,* a top grade 1+0 seedling from the open seed bed. Seedlings of this grade are best potted into 1-litre-deep pots rather than $9 \times 9 \times 12.5$cm.

plants may take at least ten years to produce worthwhile quantities of seed, but obviously this period may be shortened by planting up larger starter plants.

The plants must be accessible by long step ladders or 'cherry picker' attachments to tractors. Seed is produced in large, fleshy, cone-like fruiting aggregates, and seed itself is held tightly within carpels which split when ripe to expose the brilliantly coloured red/orange seed. Collection should not take place until the carpels are at a reasonably advanced splitting stage and the seeds readily visible, ideally some exposed on dangling threads (Figure 41.5). Early collection risks collecting immature seed and makes separation of individual seed very slow and tedious, adding significantly to costs.

Yield varies somewhat but is generally 5000–6000 seed per kilo of cleaned seed, each fruit holding 5–15 seed. After extraction, the oily seed coat (aril) is immediately cleaned by vigorous washing in water and detergent (washing-up liquid is excellent) until the black seeds are revealed. These must not be allowed to dry-off to any significant extend or viability is lost, hence problems with bought-in samples. Given the appropriate ±90 days of low temperature <5°C, germination can be up to 90%.

Seed is best sown individually in deep plug cell trays, and after a suitable period under cold conditions brought into a frost-free structure for germination and growing on (see Chapter 9: Plug

FIGURE 41.5   *Magnolia* fruit, a cone-like aggregate, the carpels split open to expose the large seeds covered in a bright red aril. One seed is already dangling on a thread-like extension from the carpels. This is the ideal stage for seed collection but requires prompt action to avoid seed being lost.

Grown for further description). Large plugs and sufficient heat can produce graftable seedlings at the end of one growing season but normally plants in plugs are removed and potted up into suitable pots using compost with up to 45% pine bark and grown on for a further season (Figure 41.6)

CUTTINGS RAISED ROOTSTOCKS

Well-rooted cuttings make excellent rootstocks grown on in rootstock pots (9 × 9 × 12.5 or 1-litre deep) as 0+1P 6–8mm or 0+2P 8–10mm (Figure 41.7). The following are best: *M. stellata* cvs ('Royal Star' is favoured), *M.* x *loebneri* 'Leonard Messel' or 'Merrill'; the de Vos or Kosar hybrids (*M. liliiflora* x *M. stellata*), particularly 'Betty', 'Pinkie' and 'Susan'; finally, *M. liliiflora*, usually only the cultivar 'Nigra' is available although the type species is rather more vigorous.

These varieties are normally reliably rooted from cuttings, given suitable environmental conditions of enclosed mist or ventilated fog plus rooting hormones, a number being grown on for sale as sizeable plants. All that is necessary, therefore, is to produce additional rooted cuttings for use as rootstocks. These selections have the important characteristic of thriving under pot culture, retaining a good root system and comparative freedom from pest and disease. Of the species listed, *M. stellata* is the most accommodating, invariably providing a vigorous healthy root system which outperforms even *M. kobus*. This may well be because

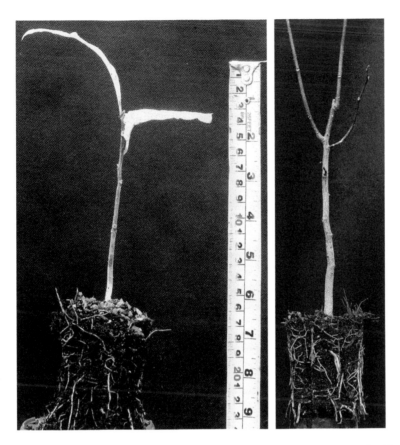

FIGURE 41.6   LEFT, *Magnolia kobus* 1-year old seedling removed from a plug. RIGHT, *Magnolia kobus* 1-year plug grown seedling. One season's growth after potting-on into 9 × 9 × 12.5cm rootstock pot. Note: extensive healthy root system evident in both examples.

of its natural occurrence throughout much of Japan in less well-drained areas on mixed and sometimes poor soils.

Species other than those listed are less certain when kept for extended periods in small pots and subjected to various manipulations in the grafting process, such as drying-off and bottom heat. Experience here would indicate the often recommended *M.* x *soulangiana* is usually a disaster with significant root death, subsequent yellowing of the foliage and little extension growth. Unsuitable types are not restricted to cuttings raised plants; several raised from seed fail to perform well. Notable among these are *M. sieboldii* and *sieboldii* subsp. *sinensis* and the closely related *M. wilsonii*. These are often recommended as rootstocks for the beautiful hybrid *M.* x *wieseneri*; however, they rarely prove successful because of extensive root death during and after the grafting procedure (Figure 41.8).

Conclusions on Rootstocks

Choice of rootstocks for *Magnolia* bench grafting is clearly a contentious issue and the choice of suitable types may vary considerably by location, climate, soil conditions and root disease pressure. In northerly areas, uneven growth rates between rootstock and scion of established plants seems to be less significant, and the main focus must be on types which thrive well under the cultural conditions existing when grown as potted rootstocks. Experience under UK conditions is that *M. kobus*, *M. stellata*, *M. liliiflora* and hybrids between them are far more successful than many suggested alternatives (Figure 41.9).

FIGURE 41.7   Trays of rooted and weaned *Magnolia stellata* 'Royal Star' cuttings to be used as rootstocks, being transported to a frost-free structure for overwintering prior to potting up into 9 × 9 × 12.5 pots the next spring.

FIGURE 41.8   LEFT, comparison of root system of *M. stellata* 'Royal' Star (left) and *M. sinensis* (right), both grafted with *M.* x *wieseneri*. RIGHT, subsequent growth of *M.* x *wieseneri* after one growing season – on left grafted to *M. stellata* 'Royal Star', and on right grafted to *M. sinensis*. After planting-out, the graft on *M. sinensis* subsequently failed.

FIGURE 41.9   Pot grown root systems comparing left, *Magnolia stellata* 'Royal Star', with on the right, *Magnolia x soulangiana* 'Rustica Rubra'. Note the healthy and extensive root system of *M. stellata* 'Royal Star' compared with less root and some signs of die-back on roots of *M. x soulangiana*. Both types potted at approximately the same time in the same high-bark content compost mix.

SCION WOOD

Magnolias appear to be amenable regarding choice of scion wood. Successful takes are normally achieved using ripened one-year-old wood; if this is unobtainable, up to four-, even five-years-old material has produced acceptable results for both winter and summer grafts. It is likely that extension growth will be more vigorous when 1- or 2-year wood is used but a number suggest 2 year is equally, or more successful, than 1-year wood*. When it is intended to produce grafted trees with a single dominant leading shoot, 1-year scion wood, possibly a single bud scion, has clear advantages over the alternatives. A selection of suitable material is shown in Figure 41.10; the larger branched scions produce a well-balanced branched plant more quickly than single bud or 1-year scion wood and often produce flower buds at an earlier stage. They therefore have advantages over alternatives for retail sales (Figure 41.11).

For summer grafted deciduous species, opinions are divided on whether to remove or retain scion leaves. Experience grafting annually several thousand Magnolias during the summer dormancy period supports removal of leaves for the reasons discussed in Chapter 10. However, many grafters favour retention of all or some of the leaves, usually reducing those remaining by third to half. Virtually all summer grafts are made using side veneers; if apical grafting is employed, some scion leaf retention of a third to half the original leaf area will be necessary.

Evergreen species also require some leaf retention; this involves removal of some leaves and reduction of three or four remaining by a third or more (Figure 41.12).

Dormant deciduous scions may be placed in sealed polythene bags and cold stored for some weeks at 1–2°C. *Magnolia grandiflora* and other evergreens can be stored for less time, probably not more than 2 weeks. Scions of deciduous species for summer grafting are suitable for use for up to 10 days after de-leafing and storing at 4°C.

---

* Alkemade & van Elk (1989) p.90, Krüssmann (1964) and Hess (1953).

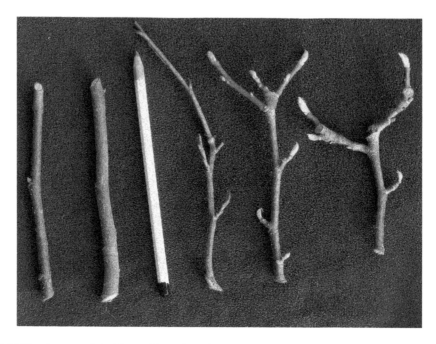

FIGURE 41.10   A range of deciduous *Magnolia* scions. A pencil has been included to provide some idea of size. On the extreme left is a 1-year, well-ripened shoot about 150mm long including the basal portion. Next is a heavier grade including a small portion of 2-year wood at the base. The following two represent branched scions comprising 2-year-old stems with short, 1-year shoots. Finally, on the extreme right, a 3-year-old stem carrying short, 2-year and 1-year shoots.

FIGURE 41.11   *Magnolia* 'Felix Jury', a hybrid with strongly arborescent habit and difficult to induce to branch. Some consumers specify the branched habit of the plants pictured because they may be more appealing than tree specimens to retail customers. Choice of branched scions combined with selective pruning enables this to be achieved in reasonable time.

FIGURE 41.12 *Magnolia (Michelia) doltsopa*, scion of an evergreen species prepared ready for grafting.

SYSTEMS

Until the advent of hot-pipe systems, warm callusing during the summer was arguably the most successful procedure for grafting Magnolias. This is continued by a number of grafters to lessen pressures on grafting programmes during the very busy winter/early spring period. Techniques and methods follow those given in Chapters 13 and 18; grafts respond best to non-sealing, but it is important to avoid drips in the poly-tent. Severed petioles left on de-leaved scions are a potential source of infection, and after ±14 days when they start to dehisce, if time permits, it is advantageous to brush them off the stems. Leafy scions are liable to develop Botrytis (grey mould) and any infected leaves should be removed. Routine spraying or preferably dusting with a suitable fungicide is a wise precaution.

Depending on location, some of the evergreens are liable to a level of winter damage and are best grafted during the late summer, using scions of well-ripened, 1-year wood rather than older.

The high ambient temperature within structures during the late summer causes rapid callusing and union formation; ventilation can be increased after 3–4 weeks from commencement of grafting. A generally low percentage of early grafts will produce extension growth (Lammas shoots); provided they are overwintered in a frost-free structure, this should present no problem and is a guarantee of a successful union. If they are allowed to even slightly freeze at this stage, significant losses can be expected.

Warm callusing during the mid-January to early March period is also possible, but normally takes are higher in the summer period*. Rootstocks will require drying-off because *Magnolia* exhibits some root pressure problems. Grafters using dormant material usually adopt the splice

---

* Wells (1955) p.250.

rather than side graft. However, problems associated with root pressure may be better dealt with using side grafts, despite some reservations regarding their efficacy in this respect.

Hot-pipe system procedures follow those outlined previously; the rootstocks should be grafted 'on the dry side', and for this reason it is sensible to de-pot the rootstock to allow good monitoring of moisture levels during the ±3 weeks taken to produce a graft union. Pot root-balls may be plunged in a slightly moist, well-drained open medium or covered with polythene sheeting or light-grade capillary matting. The fleshy roots of *Magnolia* must not be allowed to shrivel during the hot-pipe procedure (Figure 41.13).

With the exception of the evergreen species, hot-pipe systems have increased grafting takes for winter grafting to a level equal or superior to those for summer grafting.

PHYSIOLOGICAL AND ENVIRONMENTAL FACTORS

With *Magnolia*, root pressure is generally considered not to be a serious problem. In the late autumn/spring period when grafts are given bottom heat in a warm callus environment, bleeding can occur and seriously reduce takes. Drying-off rootstocks for grafts during these periods is important. It does not appear to be a problem with grafts placed in a hot-pipe when ambient temperatures are low.

Magnolias fall into the intermediate temperature requirement group, the optimum lying between 22 and 24°C, which should be the setting for hot-pipe systems. Lower temperatures delay callus formation but can be successful, particularly for the evergreens when 18–22°C may be preferable as humidity levels are more easily maintained.

GROWING ON

Magnolias demand good drainage and for many species pH levels neutral or below. For commercial production most are now grown in containers; to ensure healthy roots, high levels of pine bark incorporated into the mix are recommended. In northern areas the evergreens in containers require winter protection.

FIGURE 41.13 *Magnolia* grafts placed on hot-pipe shelving system. Light-grade capillary cloth drawn back to expose de-potted root-balls, the condition of the root system may be easily monitored and if necessary water applied to the roots or capillary cloth, which is then replaced over the root-balls. Note the extensive healthy root system of *M. kobus* (Yorkshire Plants).

PERFORMANCE COMPARISONS

The hybrids are generally more easily grafted than the species. Those in subgenus Magnolia section Oyama (*globosa*, *wilsonii*, etc.) and hybrids between them are among the more demanding, as are some forms of *M. sprengeri*, notably var. *diva* 'Lanhydrock'. The evergreens always present more of a challenge than the deciduous types. Time is required to ascertain whether all the species previously within *Michelia*, *Manglietia* and associated genera, now grouped within Magnolia, are successful on the commonly used rootstocks, but at present, results appear encouraging.

42 Malus (Rosaceae) – Flowering Crabs and Orchard Apples

INTRODUCTION

Forty or so species are widely distributed throughout the temperate regions of the northern hemisphere. The genus contains those grown for their ornamental value and for their fruit. Flowering crabs constitute some of the most decorative small- to medium-sized trees for gardens combining attractive floral display with pleasing yellow, orange or red fruits in the autumn; some also have good autumn colour.

Worldwide apple fruit production is very significant; in 2012 it was 76 million tonnes, but now has probably risen to over 80 million tonnes. By far the largest producer is China with over 37 million tonnes (48% of worldwide total), with USA sitting in second place at 4 million tonnes (6%). Poland is the largest European producer, 2.9 million tonnes. UK, at 200,000 tonnes, currently produces less tonnage than Italy, France, Germany, Spain, Austria, Belgium, Netherlands and Greece.

As with many Rosaceae, establishing a graft union for *Malus* is easily accomplished. Budding in the field during the summer or field grafting in the dormant season can be expected to provide very high takes. Bench grafting is gaining in significance because of better working conditions and higher output compared with field grafting. In climates with harsh winters, bench grafting has always been a popular option, not least because it can provide productive work for full-time nursery staff when work is impossible in the open.

COMPATIBILITY

FLOWERING CRABS

Flowering crabs may be grafted to other *Malus* species, or to virus-free clonal rootstocks originally produced for dessert and culinary apples. Incompatibility is not a problem for the vast majority. Of the 41 species and natural hybrids listed in the *Hillier Manual of Trees and Shrubs* (8th Edition 2014), all but one appears compatible on the commonly used rootstocks. An exception is *Malus trilobata* where incompatibility is sometimes evident (see Chapter 12, Figure 12.4). This is reflected in the taxonomy of the genus, which is subject to debate and the inevitable disagreement. Some authorities have proposed *Malus* should be split into 3 subgenera: Malus, Sorbomalus and Chloromeles. Others have suggested two new genera *Docyniopsis* and *Eriolobus*. Under these proposals, *Malus trilobata* from section Eriolobus would become *Eriolobus trilobata*[*].

Views on suitable rootstocks for *M. trilobata* vary; some have specifically singled out *M. Bittenfelder* as producing an incompatible union, others take the exactly opposite view and cite it as the rootstock of choice. MM 106 and MM 111 have been suggested and provided the source is from virus-free material it is very suitable[†]. Budded to MM 111, growth and takes were less satisfactory than if M25 was used and for some the latter is now the preferred rootstock[‡].

Other incompatibilities are implied: for example *Malus ioensis* 'Plena' (Bechtel's crab), an exceptionally attractive form, is often quoted as being difficult to grow on a given rootstock (*M. pumila*,

---

[*] Luby (2003) pp.9–12.
[†] D. Harris (pers. comm.).
[‡] C. Sanders (pers. comm.).

for example), and alternatives such as *M. baccata* and *M. prunifolia* are recommended*. It is likely that the poor performance of Bechtel's is due to its weak constitution rather than any problems of incompatibility.

Despite the few incompatible combinations which have been recorded, in view of the complex taxonomy and wide geographical spread of this quite large genus, it will be surprising if no compatibility issues other than *M. trilobata* occur.

For orchard apples, choice of rootstock is not significantly limited by concerns regarding compatibility, but controlled by the need to influence growth and fruiting characteristics. In some important fruit producing areas, other factors, such as hardiness, and the ability to tolerate various soil conditions, assume real significance. It is necessary to recognise that there are some examples of particular apple varieties showing delayed incompatibility on particular rootstocks. Recent hybridising programmes, with the objective of selecting new improved rootstocks, are addressing this problem.

TIMING

This genus is invariably grafted in the dormant season during the period December to early April. Cold stored scions permit a further extension to late May. The ease with which *Malus* may be budded during summer months suggests that, in early September, similar procedures as those previously outlined for *Acer, Betula, Carpinus, Fagus*, etc. could be successfully employed.

GRAFTING METHODS

Apical grafting methods are normally used, involving splice, whip and tongue, veneers or wedge. Grafting machines or tools are also sometimes favoured for large numbers of apple grafts; generally, results are less satisfactory than for manual grafting methods (see Appendix K). Well organised manual grafting procedures, making use of the whip and tongue grafting method, allows assistants to do the tying-in, sealing and ancillary work. This permits highly skilled knifesmen to concentrate their efforts on the graft, resulting in outputs which rival those achievable with mechanical systems whilst retaining the quality of manual methods.

Apple trees on clonal rootstocks produced for fruit growers must be grafted sufficiently high to ensure scion rooting does not occur, normally a minimum of 150mm from soil level. To facilitate harvesting, some fruit producers require scion placement higher; 450mm is a popular choice but chosen height is often subject to specifications from the purchaser.

It is usually convenient to dip the completed apical graft into melted wax sealants, immersing the graft to about 15mm below the union. Some favour use of a vapour-proof self-adhesive enclosing tie (Buddy tape®, Medelfilm®, etc.); the terminal portion of the scion will require sealing. A cold sealant or melted wax may be used, gathering many grafts together and dipping the cut ends as a batch.

Root grafting is possible using splice or short tongue apical veneer and subsequently placed in a heated area at 18–20°C for 3–4 weeks to encourage callus formation, or kept cold until planted or potted†. Roots may be obtained from seedlings at the point of the hypocotyl plus sufficiently large portions below; a large seedling may produce 2 or 3 useable sections 100–200mm long and 8–15mm in diameter. They may also be obtained from nursery trees during lifting operations and stored at 0°C for some weeks until required. Because of the possibility of scion rooting root grafting is more appropriate for ornamental crabs than orchard apple varieties.

---

* Krüssmann (1964) p.463.
† Leiss (1995).

ROOTSTOCKS

ORNAMENTAL CRABS

In Europe, N. America, and much of western Asia including Russia, the species of *Malus* and numerous varieties and cultivars grown for ornament are usually grafted to types chosen from the following: *M.* Antonovka, *M.* Bittenfelder, *M. sylvestris*, *M. domestica* or *M. pumila*. Precise choice will be dependent upon factors such as hardiness, tolerance of soil conditions and resistance to pest and disease pressures. In China, eastern Siberia, Mongolia, northern India, Bhutan, Nepal, Japan and Korea, a wider range seems to be utilised and, in addition to those listed above, a selection from the following *Malus* species might be used: *baccata, baoshanensis, bhutanica (toringoides), halliana, hupehensis, micromalus, prunifolia, pumila, sieversii, sikkimensis, toringo* and *transitoria**.

Of those listed *M. baccata, M. bhutanica, M. hupehensis* and *M. sikkimensis* are considered apomictic and can be expected to produce very evenly graded batches of seedlings usually with a virtually identical genetic make-up. In most areas the main interest in rootstocks from the species listed is their use as understocks for orchard fruit production. In China the main rootstock species for this purpose are *M. baccata, M. prunifolia, M. micromalus, M. hupehensis* and *M. sieversii*.

The clonal rootstocks listed for orchard apples varieties may be considered for ornamental crabs. Many producers take advantage of their influence on growth to produce trees suitable for small and large gardens. MM106 is a popular choice where reasonable but not excessively large trees are required; MM111 and M25 are favoured by nurserymen as they quickly produce saleable standard trees. The extreme dwarfing effect of M27 may be used for extra vigorous types such as 'Cowichan', or for culture in containers, grown as Patio trees.

Early attempts at using the Malling range of rootstocks for ornamental crabs were not successful. The cause was eventually identified as due to the presence of latent virus infection. These viruses are not evident by visual symptoms on many of the commonly grown orchard varieties, and for some years their presence in the ornamentals was not realised. In 1968, a variety of flowering crab apples failed on M1, M2, M4, M7, M9, and M25[†]. Following the recognition and eradication of latent viruses at Long Ashton Research Station UK, and introduction of new virus-free EMLA rootstocks, no failures were recorded[‡]. The effect of latent virus infection can cause problems for certain rootstocks when grafted with infected scions. The promising rootstock Geneva 16 is an example, proving to be highly susceptible, and can only be used in combination with virus-free scion material.

A disadvantage of some commonly used rootstocks of 'Golden Delicious' and 'McIntosh' raised from seed, is that although hardy for cold areas in northern Europe and the USA, they sucker heavily. Hardiness is not a problem with the Budagovsky and Geneva selections. Of these, B.118 (B54-118) is promising for larger ornamental trees and Geneva 11 for smaller[§]. Malling rootstocks are also non-suckering but too tender for use in these areas.

As noted previously, *Malus* roots can make a suitable rootstock. Scion rooting is the almost inevitable result of this strategy, and it is therefore only suitable for ornamentals or large apple trees in non-intensive orchards.

ORCHARD APPLES

Modern orchard systems invariably demand the use of clonal apple rootstocks. Since the early twentieth century, when the first selections were released from East Malling Research Station, there has been a constant flow of new rootstock introductions. These have been bred and selected to give

* Luby (2003) pp.2–5.
† Nelson (1968).
‡ Alexander III (1998).
§ Scott (2003).

FIGURE 42.1   LEFT, apple rootstock Budagovskij B9 showing typically reddish/purple leaves derived from the anthocyanin content of one of its parents, the extremely hardy Siberian/central Asian species: *M. neidzwetzkyana* (Boomkwekerijen Gebr. Janssen B V.; photo Paul Janssen). RIGHT, apple rootstocks (MM106) bundled ready for subsequent handling (F.P. Matthews.; photo: N. Dunn).

better growth control and cropping, endure lower temperatures, withstand adverse soil and growing conditions, improve pest and disease resistance, facilitate easier propagation and ensure they are fully compatible with the chosen scion variety.

Many rootstocks for orchard and plantation crops are derived from sources designed to ensure high health status. This often originates from government- or state-sponsored organisations that provide high health status (virus-/disease-free) material to licenced growers for subsequent re-propagation and distribution (see Chapters 8 and 9 for further discussion).

Rootstocks currently considered of importance for orchard apple production are listed in Table 42.1 (see following page). They are arranged according to their influence on ultimate tree size from small to large. A key listing abbreviations is shown beneath the table. Where known, the origin and major characteristics of each rootstock is also shown beneath the key (Table 42.1).

Double-Working

A further strategy to control growth and crop production is double-working. This involves the use of a strong growing rootstock, such as MM111 or M25, to which is grafted a dwarfing intermediate stem (inter-stem), such as M27 or M9, with the chosen apple variety above. This strategy provides a strongly anchored root system, often adapted to a wider range of soils than would otherwise be possible, coupled with the ability to significantly control growth of the fruiting variety. Dependent upon the precise combination, growth rates of 35–55% of full sized trees is possible. The length of inter-stem has an effect on growth rates and ultimate size; it is generally considered the longer the inter-stem, the more significant its influence.

SCION WOOD

Well-ripened, current year wood containing 3–5 buds selected from the lower portions of the shoot are recommended. A few slow-growing compact cultivars such as 'Adirondack', 'Coralburst', *ioensis* 'Plena' and x *scheideckeri* may make such little extension growth that a basal portion of 2-year-old wood must be included.

Collection is best during December to January, the scion wood being cold stored or laid-in well drained substrate situated in a northerly aspect, and protected from desiccation with a polythene sleeve around the bundles. Material will remain viable for up to three months if cold stored in temperatures of 0.5–1°C.

Methods involving intensive manual grafting, or machine/tool-made grafts, work far more efficiently if scion material is graded and matched to similar rootstock dimensions before grafting takes place.

TABLE 42.1
Orchard Apple Rootstocks*

% of Full-Sized *M. sylvestris*	Name	Origin/Source	Hardiness V H. Hardy Mod.	Soil W. Dry. H. L.	Staking Yes Limited No	Cropping/Precocious	Collar Rot Susc. Res.	Fire Blight Susc. Res.	Scab/Mildew Susc. Res.	Woolly Aphid Susc. Res.
15–20	M 27	E.M	Mod	Heavy	Yes	VG (small fruit)	M Res	M Susc	Mod	–
20–25	V3	Vineland	Hardy	–	Yes			M Res	–	–
25–30	P2	Polish	VH		Yes	G	M Susc	Susc	–	Susc
30–35	M 9	E.M	Mod	Heavy	Yes	VG		Susc	–	Susc
30–35	M 9 – F1 56 (widely used for Jonagold)	Fleuren	–	Heavy	Yes	VG	Res	Susc		Susc
30–35	M 4 (for non-commercial fruit use)	E.M	–	–	Limited	–	Res	Res		–
30–35	G 214	Geneva	VH	Good calcium uptake	Yes	VG	Res	Res		–
30–35	Bud 9	Budag-ovskij	VH	–	Yes	VG	Res	–	Mod	Susc
35–40	M9 – T 337	NAKB	Mod	Heavy	Yes	VG		Susc	–	Susc
35–40	M 9 Nic 29	Belgium	Mod	Heavy	Yes	G	M Res	V Susc	–	–
35–40	G 11	Geneva	VH	Dry	Yes	VG	Res	Res	–	M. Susc
35–40	G 41 (compatibility not certain all varieties)	Geneva	VH	Good nutrient uptake	Yes	VG	Res	V Res	–	Res
40–45	G202	Geneva	VH	–	Limited	VG	V Res	V Res	–	Res
40–45	M.26	E.M	Mod	Light	Limited	G	Susc	V Susc	–	Susc
50–60	M 7a	E.M Virus-free	Hardy	W-D	Limited	–	M Res	Susc	–	–
55–60	M 116	E.M	Hardy	Heavy	No	–	V Res	–	–	Res
60–70	MM 106	E.M	Hardy	Light	No	VG	Susc	Susc	–	Res
75–85	P 18	Polish	Hardy	W-H	No?	–	Res	Res	–	

(Continued)

TABLE 42.1 (CONTINUED)

Orchard Apple Rootstocks*

% of Full-Sized *M. sylvestris*	Name	Origin/ Source	Hardiness V H. Hardy Mod.	Soil W. Dry. H. L.	Staking Yes Limited No	Cropping/ Precocious	Collar Rot Susc. Res.	Fire Blight Susc. Res.	Scab/ Mildew Susc. Res.	Woolly Aphid Susc. Res.
75-85	MM 111	E.M	Hardy	W-H	No	–	Res	–	–	Res
80-85	Bud 118 (B54-118)	Budag-ovskij	VH	H-L	No	G	Res	M Res	S. Susc	–
90-95	M 25	E.M	Hardy	H-L	No	–	–	–	–	–
100	*M. sylvestris*	Nursery	Hardy	H-L	No	–	–	–	–	–
100	Antonovka	Nursery	VH	H-L	No	–	–	–	–	–
100	Bittenfelder	Nursery	VH	H-L	No	–	–	–	–	–

*Sources: N. Dunn (pers. comm.), Auvil (2016), Schupp & Crassweller (2018), Wilson (2000), Orange Pippin Ltd (2018), Frank P. Matthews Ltd (2018) and Washington State University (2017).

Table Key:

Source

- E.M: East Malling Research Station Kent, UK: work on selection commenced after the Station was inaugurated in the early 20th century through the initiative of fruit growers in the area (Figure 42.1).
- Polish: originating from the Research Institute at Skierniewice, Poland.
- Vineland: rootstocks developed at the Vineland Station, Ontario, Canada.
- Fleuren: a commercial nursery in Baarlo, N. Limburg, Holland.
- Budagovskij: extra hardy rootstocks raised at the Mincurinsk University of Agriculture, Russia. Most have red leaves derived from *Malus neidzwetzkyana*, an extremely hardy species from southwestern Siberia and Central Asia (Figure 42.1).
- Geneva: from a comprehensive rootstock raising programme based at New York Agricultural Experimental Station, Geneva, N.Y., USA.
- Belgium: a selection of M9 made by Johann Nicolai in Belgium.
- NAKB: the research station based at Roelofarendsveen in the Netherlands. They have been responsible for providing virus-free material to European growers in recent years.
- Nursery: seed-raised rootstocks from the following – seed mother plants of Antonovka, Bittenfelder, *Malus sylvestris*; the latter may possibly include a content of *M. domestica* (*M. pumila*).

Hardiness: VH = very hardy; H = hardy; Mod = moderately hardy

Soil: W = wet; D = dry; H = heavy; L = light

Staking: Yes = required; Limited = limited requirement; No = not required

Cropping: VG = very good, often also precocious; G = good

Collar Rot, Fire Blight, Scab/Mildew, Woolly Aphid: Susc = susceptible, S. Susc = slightly susceptible, Res = resistant, V Res = very resistant, M Res = moderately resistant

SYSTEMS

Grafts respond well to cold callusing and may be placed in unheated structures to form a graft union. Most cold callusing systems involve the use of bare-root rootstocks, treated as discussed in previous chapters. For modest numbers, the laying-in medium is usually contained within Boskoop trays or similar. Large-scale production for orchard planting normally involves palletised, large capacity boxes or bins. Once filled, these are held in cool sheds, climate controlled sheds or cold stores until the grafts are required for planting into nursery beds or field rows. For those growing plants to the point of sale in containers, bare-root rootstocks may be potted-up immediately grafting is completed, confident in the expectation that a virtually 100% take will result.

Temperatures should be prevented from remaining long below 6°C. Dependent upon temperature, grafts should have united after 5–8 weeks. At this stage, decisions must be taken on whether to pot-up or plant out. Those destined for planting out may require some delay, easily achieved by cold storing the grafts at 1–2°C until soil and growing conditions become conducive to establishment and growth. If there is minimal bud activity at the time of storage this may be prolonged for up to 6 weeks.

Sub-cold callusing is a suitable method for orchard apples and ornamental types and allows planting to be delayed until suitable conditions prevail, often not until mid- to late May or beyond in northerly temperate areas.

A few favour hot-pipe conditions, and, provided scion and rootstock material is in good health, 100% takes can be guaranteed. Temperature of 21°C in the hot-pipe will produce united grafts after 20–25 days. Successful grafts may be potted-up and grown on in frost-free conditions in glasshouse or polytunnel when overall production times can be shortened.

Root grafting, now a rarely used system, has been previously discussed in Grafting Methods.

PHYSIOLOGICAL AND ENVIRONMENTAL FACTORS

Root pressure has not been reported as a problem with *Malus* and this, coupled with a readiness to produce callus and unite in a wide range of temperatures, has placed the genus among the more easily grafted species.

TOP-WORKED TREES

A limited number of weeping cultivars such as 'Echtermeyer', 'Elise Rathke', 'Excellenz Thiel', 'Louisa', *prunifolia* 'Pendula', 'Red Jade' and 'Sun Rival' may be top-worked. This is often carried out in the field but may take place as a bench grafting procedure, using bare-rooted stems of suitable length, normally 1.5–2 metres. When completed the grafted stems have their roots laid-in well-drained open mix placed in tall-sided bins. When filled, these are transferred to a cool shed or unheated structure. It is usual to containerise the plants after the graft union is well established.

Dwarf and compact cultivars such as 'Coralburst' can also benefit from top-working. Normally, for this type, stem length is shorter (60–90cm), the objective being to produce Patio trees; after grafting, treatment is as for the weeping forms.

GROWING ON

Commercial orchards are invariably planted up using bare-root, field grown plants. Medium textured soils, with a pH of neutral or slightly below, maintained in a high state of fertility, irrigated when necessary and free of replant disease produce best quality planting material.

Many ornamental crabs are now offered for sale in containers. These are mostly produced by containerising field grown trees and holding them for one growing season before marketing. Some growers opt for growing in containers throughout. Standard potting compost is satisfactory for good performance.

43 Paeonia (Paeoniaceae) – Tree Paeony

INTRODUCTION

Peonies are mostly herbaceous species; those from western China and adjoining Tibet develop woody stems becoming medium-height shrubs. Among these are some of the most spectacular flowering woody plants for gardens in temperate climates, various selections and hybrids within and between the species have increased the range.

A number may be propagated by seed (*P. delavayi, P. ludlowii, P rockii*, etc.), but specific cultivars of these, *P. suffruiticosa* cvs and selected colours in the Gansu Group (including 'Highdown', the original 'Joseph Rock'), are normally propagated by grafting.

COMPATIBILITY

No specific problems related to compatibility have been discovered. Even herbaceous species appear compatible with the tree peonies at least in the short- and medium-term. Cultural issues such as poor vitality, specific plant health problems, i.e. virus infection, and a propensity to produce suckers, influence the choice of rootstock.

TIMING

Most favour summer grafting, July-early September; however, in the UK some graft in winter/early spring. Reasons for choice of summer may be linked to field production, as grafts are able develop over winter and conveniently reach a stage for planting out the following spring.

GRAFTING METHODS

Root grafting predominates (Figure 43.1), and an overview of the method is discussed in detail in Chapter 18: Root Grafting.

Most grafters recommend what they call a cleft graft*. What is described is actually an inlay graft, as defined in Chapter 2, Figure 2.6. Choice of this method increases the difficulty of achieving good cambial alignment, due to the varying disposition between the cambium of the scion stem tissues and the rootstock. Furthermore, the rather delicate small tip section left after scion cuts are completed provides a vulnerable point, potentially an area for graft failure, especially for the rather pithy wood of peony. Despite these disadvantages, the juvenile characteristics and composition of the root cortex, comprising mostly parenchyma cells, usually permits differentiation to eventually establish cambial continuity. However, results are less likely to be good when compared with well-matched grafts of a carefully cut wedge, or short tongue side veneer (see Chapter 3, Figure 3.2 and Chapter 18, Root Grafting, Figure 18.20).

Conventional apical or side veneer methods on rootstock stems, as opposed to roots, give excellent takes if seed-raised rootstocks of appropriate types are available. In the early stages, such grafts out-perform root grafts with regard to growth and development. Using these methods, long-term performance has not been evaluated, but numbers of *P.* Gansu 'Highdown' grafted to *P.* Gansu

---

* Bennison (2010), Dirr & Heuser Jr. (2006) pp.255–256, Haworth-Booth (1963) pp.85–91, Leiss (1971) and Reath (1982).

FIGURE 43.1   *Paeonia* 'Highdown' root graft to *P.* Gansu one season after grafting, removed from pot containing pure bark compost. No sign yet of scion rooting but grafting ties have been cut.

seedlings have been satisfactory over a ten-year period with no sign of suckering. There is no reason to think this will not continue.

Tying-in the graft has been discussed in Chapter 18; for peonies some favour use of self-adhesive enclosing tape, which is claimed to reduce chances of infection and be convenient to use; however, it does not disintegrate and must eventually be removed.

ROOTSTOCKS

Non-suckering or restricted suckering species, such as the Gansu Hybrids, can provide seed-raised rootstocks potted into standard pots for conventional grafting to the stem. Seed mother plants are required, and for commercial runs it may prove impractical to provide sufficient numbers for required volumes of seed.

Choices for root grafting depend upon species which produce long, straight roots with non-suckering characteristics, coupled with good health, disease-resistance and vigour. The *P.* Gansu Group has been identified as fulfilling this requirement and the hybrids between tree and herbaceous peonies, the Itoh or intersectional group, are also suitable*. Alternatively, selections from herbaceous species may be considered, of which the most often used are: *P. lactiflora* and its

* J. Bennison (pers. comm.).

cultivars and hybrids, e.g. 'Mons Charles Elie', 'Charles White', 'Early Windflower', 'Red Charm' and 'Krinkled White'.

P. officinalis and *P. peregrina* and their hybrids should be avoided as they have the ability to produce true suckers from adventitious buds on their roots*.

Herbaceous peony rootstock mother plants are lifted from the field, the roots are washed off to remove the soil, and correctly sized root portions cut away as required; good clumps should yield at least 20 suitable pieces.

Opinions on optimum size vary between unbranched portions 150–200mm long and 15–20mm in diameter, to those favouring well-branched roots[†]. Lengths of 100–150mm and a diameter of 20mm are the general consensus[‡]. Personal experience using *Paeonia* Gansu roots would favour something not above 100mm in length and 12–15mm in diameter. Larger roots eventually require more support from foliage above, which is delayed for months following summer grafting and a long period of dormancy.

A recommendation has been made to lift the roots early and allow them to feel 'rubbery' when grafted, presumably achieved by allowing them to partially dry. This seems counterintuitive but perhaps no more so than drying-off potted rootstocks. The reasons for this recommendation are not clear but centre on the statement "they will not split away from the scion"[§].

Root grafting methods result in comparatively large amounts of exposed cut surfaces, indicating a need for protection from infection. Careful cleaning with tap water starts the process, and before or after completing the graft, some recommend sterilisation by soaking with a sterilant such as bleach. Peroxyacetic acid at a suitable dilution, say 1:500 should also prove satisfactory. An alternative, less invasive approach is suggested in the section to follow on Grafting Systems.

SCION WOOD

For commercial quantities, development of stock scion mother plants is essential.

Scions taken during the summer are selected from vigorous current year wood which has matured but not completely hardened, i.e. firm, semi-mature. The choice of the basal portion is preferable as the wood can have a pithy structure. Some recommend selection of single bud scions taken from along the length of the shoot. No comparative figures for takes are available to enable comparison between scions from the basal or terminal portions.

When using leafy scions during the summer, opinions differ between partial or total leaf removal. Leaves left on the scion usually fall off after 10–20 days. Most favour scions containing at least two buds, and although only one normally grows out the first year, the other is retained and develops in the second season to provide a better-balanced plant in the early stages of growth (Figure 43.2).

Late winter grafting involves similar but fully matured wood. Although never suggested, there is no apparent reason why scions with a two-year basal portion would not be successful, thus avoiding some of the pithy structure common to tree peonies.

SYSTEMS

Most summer root grafting, July–early September, involves planting the grafts after completion. Some omit storage and plant out within a day or two; others delay for a period of between 3–8 weeks to allow the graft to "heal" (commence or complete union formation). Recommended conditions at this stage vary. One method suggests packing the completed grafts upright with dampened tissue paper in a sealed polythene bag, which is placed to ensure vertical graft orientation,

---

* Bennison (2010).
† Leiss (1971).
‡ Haworth-Booth (1963) p.86 and Reath (1982).
§ Leiss (1971).

FIGURE 43.2   Root grafts of *Paeonia* 'Highdown' showing the advantages in use of two bud scions. The scion on the left has so far failed to produce terminal extension growth, but fortunately the lower bud is growing well. On the right, the opposite is the case, the lower bud showing only slight activity.

and maintained at ±20°C for approximately 3 weeks*. Others advocate heeling-in sand, peat and perlite in a "very warm" enclosed environment at temperatures of 20–24°C. Alternatively, grafts may be placed in layers horizontally above each other, up to three layers deep, with sawdust between and retained in this position for 3–4 weeks when it is claimed "the union is generally well underway"[†].

After storage, some plant out the grafts during mid-September to October. Planting takes place in well-prepared, open land or under polythene cloche runs or cold frames. In all cases, soil conditions must be first-class with good texture and drainage. Many soils will need amelioration with additions of grit, sand, peat, granulated bark and/or wood chips.

In view of the hardiness of tree peonies, use of sub-cold callusing would warrant investigation as a more efficient method of production (see Chapter 18: Sub-Cold Callusing). Short tongue apical grafts on peony roots during December-late February could be followed by cold storage at 0.5°C, and field planting in the early May/early June period (±4 months after grafting). Adverse soil conditions during the winter are avoided and rapid graft union in suitable soil temperatures is followed by uninterrupted growth.

Techniques for autumn planting vary, some favour planting with the tip of the top bud on the scion just showing. Deeper planting is sometimes preferred, the graft union being 100–150mm below the surface. Others plant the vertically placed grafts with their top buds covered by 50–75mm of soil followed by a further 50–75mm of bark chips; finally, the whole is covered with black plastic sheeting. When this operation takes place in northern USA, the plastic is covered by a further layer of 300mm deep straw or hay to prevent the grafts from freezing[‡]. The following spring, the protection and plastic sheeting is removed, and the grafts left in position for one or two growing seasons to develop into saleable plants.

Other grafters lift the young plants after a season's growth. The lifted plant is inspected for root development as it is intended that the scion should have produced roots. If these are well developed,

* Bennison (2010).
† Haworth-Booth (1963) p.87.
‡ Reath (1982) and Haworth-Booth (1963) pp.85–91.

the herbaceous rootstock can be cut away and removed; often such roots can be reused for further grafts. At the same time, any grafting ties remaining and still intact should be cut through to avoid girdling. The root system of the herbaceous rootstock can be left when the plants are planted deeply. If for any reason this is impossible, shallow, herbaceous roots develop into a club-like structure and disrupt development of the scion (tree peony) root system. Rootstock roots of suitable tree peony species and Itoh types can be left attached to the scion variety.

Personal experience is limited to mid-winter grafting in warm callusing conditions using non-suckering roots from *P.* Gansu hybrid seedlings. Roots are cleaned as far as possible by washing off and prepared for grafting (see Chapter 18, Figure 18.19). No further treatment with sterilants is applied and grafts are tied with rubber ties. To reduce possibilities for infection, the completed grafts are potted individually in deep pots with the union placed as low as possible, leaving the upper bud projecting from the compost surface. The grafts are potted into 100% fine grade pine bark. This contains a content of bark dust and although very free-draining has some water retention. Utilising the natural fungal suppressant attributes of pine bark has proved to be an effective way of preventing infection.

To avoid drying-out the pots are stood on a sand base which is kept wet by any convenient means. Using this system, takes are high and successful grafts are potted-on during the following late spring. The bark compost may be easily removed to expose the grafting ties, which can be cut. At this stage, to encourage rooting, the scion stem can also be wounded and treated with IBA, painted on with a brush using a quick dip formulation (2500ppm in 50% alcohol). Grafts are then potted as deeply as possible in standard compost. Use of pot culture means that if required the herbaceous root may be relatively easily removed. The high sales value of tree peonies would seem to support their production in containers particularly as the system provides successful grafting results.

PHYSIOLOGICAL AND ENVIRONMENTAL FACTORS

Relatively high temperatures are likely to deplete carbohydrate reserves in roots and are best avoided. Experience has demonstrated slow but steady callus development and union formation succeeds well at temperatures in poly-tents set at a bottom heat temperature of 12°C. In the late winter period with some shading ambient temperature is not likely to exceed 15–18°C. Small trials here of conventional tree peony side grafts placed in a hot-pipe at 20–22°C using potted rootstocks have resulted in total failure. Reasons for this are not clear and reinvestigation is warranted.

GROWING ON

Some suggest Paeonies are gross feeders and use of high-nutrient levels (rotted farmyard manure, etc.) incorporated into the soil is recommended by a number of growers. Tree peonies, especially *P. suffruiticosa*, are particularly vulnerable to *Botrytis paeoniae* (Peony wilt). To help reduce incidence of the disease, every attempt should be made to keep growth as firm as possible by avoiding high levels of nutrition.

Most producers favour field culture for peonies. Whether decisions on tree peony culture is influenced by the herbaceous species is uncertain, but their relatively high value and the fact that they are often sold at a younger and smaller stage than their herbaceous equivalents, suggests that container culture can be an alternative strategy.

44 Picea (Pinaceae) – Spruce

INTRODUCTION

Found through the northern hemisphere but concentrated in eastern Asia, the genus comprises about 38 evergreen species mostly from temperate areas. Spruce are among the most beautiful and elegant conifers; many are large trees, but others are small- to medium-sized and even from among the largest species, such as *P. sitchensis*, a number of dwarf forms have been selected. Attractive golden foliage forms also occur.

As they are difficult to root from cuttings, species for which seed is not available, hybrids and cultivars are propagated by grafting.

COMPATIBILITY

It is generally agreed that all *Picea* species are compatible* and no evidence of incompatibility between any combinations have been recorded. Performance differences related to various rootstocks do occur but not directly attributable to incompatibility; these will be discussed in the section rootstocks.

TIMING

Two periods can be considered, January–March and late July-early September. Grafters in the Netherlands favour the summer period, particularly August, whereas those in the USA and UK mostly graft during January–March; a delay until April is possible if scions collected for winter grafting are cold stored.

At the Boskoop trial station, comparisons were made between grafting in late March and August/ September. *Picea abies* and *P. omorika* cultivars produced an average take of 96% for late summer grafts compared with 62% for those in March[†]. The Blue spruce varieties (*P. pungens* cvs) were traditionally grafted during August in the Boskoop nurseries, often using double glass systems within cold frames. Advantages of summer grafting have been identified by Macdonald[‡]; these were listed as less root pressure at this time, improved scion growth and subsequent plant development during the season following grafting, and a useful distribution of labour demands.

Virtually all accounts of spruce grafting in the USA involve grafting during the January–March period[§], similarly in Scandinavia[¶]. Winter grafting is also favoured in the UK[**], and improved results are claimed for grafting at this time rather than during the summer[††]. This view is supported by personal experience. In southern England it is important that grafting should take place early, preferably not after late-January, because after this period ambient temperatures in enclosed environments can become excessively warm.

---

* Larson (2006) and Mathers (1999).

† van Elk (1967).

‡ Macdonald (1979).

§ Carville (1970), Mahlstede (1962) and Wells (1955) pp.261–264.

¶ Thomsen (1978).

** Hatch (1982).

†† Willard (1968).

GRAFTING METHODS

Almost without exception, side veneer grafts are chosen, the main decision being whether these should be long or short tongue. If there is significant discrepancy in the diameter of rootstock and scion (inevitable with too large rootstocks) the short tongue is favoured. When young rootstocks are available, ideally from plug-grown seedlings, use of a long tongue veneer has every advantage. More cambial contact is possible, and this has a very beneficial effect on takes when combined with the presence of stimulant buds and active needles*.

Many grafters stress the importance of grafting as close to the base of the rootstock as possible; this goes some way to ensuring that the first whorl of branches is produced close to grown level, an important factor in the quality of many forms of spruce (Figure 44.1).

Opinions differ on whether or not to seal the graft. Grafts in enclosed conditions (poly-tent, etc.) are best left unsealed. Although sap emissions are not as extreme in Spruce as in Pine, it is wise to

FIGURE 44.1   Short tongue side veneer graft of *Picea pungens* 'Hoopsii' to *P. abies*. Rubber grafting tie has been removed. Note low graft placement, easily possible on the plug grown rootstock. The top facing lateral bud is discoloured and probably dead, the common cause of a non-symmetrical basal whorl of branches. 'Spacer canes' inserted to deflect shoots produced next season may partially redress the problem (see Growing on).

---

* Barnett & Miller (1994).

leave tying-in loops fairly well spaced to allow excess to flow away. The resinous content of the sap means that within a few days of completion the union is sufficiently sealed. Where grafts are stood on the open bench, the union must either be sealed with wax or with non-porous enclosing ties. An alternative, avoiding the need for sealing is to plunge the graft union within a well-drained substrate such as bark, peat/perlite mixtures or sawdust/shavings derivatives. Some grafters both seal and plunge the union.

Bare-root grafts are rarely recommended but some using this method bury the union at the same time as laying-in the rootstock*.

ROOTSTOCKS

For all spruce species and varieties, *Picea abies* is the favoured rootstock. Trials at the Boskoop trial station compared *P. pungens* 'Koster' grafted to potted rootstocks of *P. abies*, *P. pungens* and *P. sitchensis*; highest takes were obtained on *P. abies* (96%) and on *P. pungens* and *sitchensis* results were 90% and 89%, respectively[†]. Subsequent growth of successful grafts is more questionable, as the work in Boskoop showed a slight advantage in favour of *P. pungens* (+3%) but a definite depression in growth of grafts on *P. sitchensis* (−21%). Other grafters very much favour *P. abies* over *P. pungens*, stating that after one season, stem girth of scion varieties grafted to *P. abies* is twice that of those on *P. pungens*; furthermore, *P. abies* root system is more tolerant of high temperatures in containers than *P. pungens*[‡]. The ready availability of *P. abies* in Europe and the USA, and comparative lack of availability of *P. pungens* in Europe, is another reason for its popularity.

Pot grown rootstocks are invariably used; specifications for these vary from 1+1P to 2+1P. Dutch authorities specify 4–6mm diameter, but many rootstocks seen in nurseries are substantially larger. If the recommended 2+1P grade rootstocks are used, it is hard not to exceed the optimum size.

There can be little doubt that use of oversized rootstocks has a negative effect on results. The 'juvenility effect' of smaller, younger rootstocks and retention of buds and needles close to the graft union has a very positive influence on securing a successful take. It is important that grafters make clear to rootstock suppliers that young, well-grown seedlings, normally 1+0, are required for potting to make rootstocks of grade 1+1 4–6mm maximum, at grafting time. There will be some resistance to this request because 1-year seedlings grade out badly and price returns are poor. An alternative is for grafters to raise their own seedlings, preferably in plugs.

Comparisons between well-established plants after one season in the pot before grafting and those potted in the autumn proceeding winter/spring grafting produced results of 75–80% for well-established plants and 45–50% for those autumn potted[§].

One-year plug grown seedlings in fairly small plugs and potted on into 9 × 9 × 12.5cm pots make ideal rootstocks after a further season's growth. It is possible to grow graftable plants in plugs throughout, and discussions with suppliers would be worthwhile. An accelerated plug-grown seedling of suitable girth, clothed almost to ground level with needles and some buds present spaced evenly along the stem, makes an ideal rootstock for grafting.

Spruce seedlings are a magnet for the Green Spruce aphid (*Elatobium abietinum*). This is a year-round pest but often overlooked as the adults are very small and effectively merge in with the foliage during the summer months. Damage becomes very obvious over winter, particularly in the spring, and infested rootstocks brought into warmer conditions in the poly-tents quickly defoliate with serious loss of grafts. Routine monitoring coupled with preventative sprays during the growing season is necessary. Red spider mite (*Tetranychus urticae*) is occasionally also a problem and manifests itself as a bronzy or yellowish cast to the foliage. It is most often present during the mid-summer/

---

* Mahlstede (1962).
† Pellekooren (1974).
‡ Buchholtz (2017).
§ Wells (1955) pp.261–264.

early autumn period and causes significant damage to plants under protection. It is rarely a problem in overhead, irrigated beds but routine miticide spraying is a wise precaution.

SCION WOOD

Dutch grafters take the view that scions should not be collected in wet weather as the waxy coating of the needles tends to be removed; tests to support this contention would be worthwhile. Despite suggestions to the alternative, scion collection in cold, frosty weather is standard procedure for grafters in Scandinavia where there is often no option[*].

Views on length of time between collection and use vary. General view at an IPPS discussion group in 1979[†] was that only sufficient material for 1 day's work should be collected. Some take the view that storage of scions packed in polythene bags and stored at 4°C for limited periods is successful[‡]. Dutch recommendations for cold storage are that it should be for a limited period as long storage can cause an irregular take[§]. Others hold a different view, and scions collected in November–December are held in plastic bags in cold store at 4°C until March[¶]. For a hardy genus such as *Picea* it seems likely that scion material stored at a sufficiently low temperature will remain viable for long periods. From experience with other genera, 0.5 to 1°C would seem appropriate, and, in this temperature range, storage without deterioration for up to 6 weeks should be satisfactory.

Where possible, scions for the major commercial types are best obtained from stock mother plants pruned to provide optimum scion wood; procedures for achieving this are discussed in Chapter 10: Scions – Conifers. In Boskoop it is common practice to collect scion material from 1-year-old plants growing in lining-out beds.

Species of botanical interest, or dwarf, compact types in botanic gardens, specialist collections or private gardens require different treatment. In these situations when mother plants are growing without any pruning, to achieve sufficient girth and length a basal portion of 2- or even 3-year-old wood must be included. For dwarf and compact forms, branched scions are favoured (Figure 44.2). Upright formal types gathered from these sources must be selected to provide as much vigorous, well-budded shoot length as possible, but often a branching scion, with a basal portion of 2-year-old wood is unavoidable, and will provide better results than weak, short, 1-year material.

Comparisons between 1-year-old scions and those with a 1-year base have been made. The most extreme example with *P. orientalis* 'Atrovirens' provided results of 81% for scions of only 1-year-old wood, compared with 14% for those taken from lateral shoots with a 2-years-old basal portion[**]. At the Boskoop Trial station comparisons between 1-year scions of *P. omorika* 'Nana' and those with a basal portion of 2-year wood produced 95% for the former and 75% for those with the older base[††]. Trials at the same station in 1965[‡‡] using scions of *P. breweriana* compared 1-year scion wood with those including a 2-year basal portion. Takes were not significantly affected, but subsequent growth from the 1-year scion was more vigorous and shapely after one season's growth.

Some see no difference in takes between 1- or 2-year basal portion[§§]; others routinely collect scions with a 2-year-old base[¶¶]. Results here have been in favour of the older base, since it avoids any possibility of insufficiently ripe wood. For varieties such as *P. orientalis* 'Aurea' or *P. orientalis*

---

[*]   Thomsen (1978).
[†]   Macdonald (1979).
[‡]   Macdonald (1989).
[§]   Alkemade & van Elk (1989) and van Elk (1965).
[¶]   Thomsen (1978).
[**] Okken (1998).
[††] van Elk (1968).
[‡‡] Cave (1982) and van Elk (1965).
[§§] Willard (1968).
[¶¶] Thomsen (1978).

FIGURE 44.2   *Picea* scions. Left, *Picea pungens* 'Iseli Fastigiate', 2-year basal portion with a whorl of shoots above to encourage early development of low branch structure. In the centre, *P. pungens* 'Hoopsii', a conventional scion of a 1-year shoot, including the basal portion and terminal bud, well-furnished with lateral buds. On the right, a branched scion of the dwarf form *P. sitchensis* 'Papoose', with a 2-year basal portion and whorl of 1-year shoots.

'Skylands' with somewhat restricted vigour compared with green forms of the same species, the use of larger scions enables faster and quicker development to a saleable stage. The 2-year base also allows inclusion of a basal whorl of shoots which, if carefully selected, are able to form the basis of a shapely plant more quickly than when reliance is placed on uniform and well-spaced growths from scion buds (Figure 44.2). Possibly union development is slightly slower when a two-year basal portion is included, but the difference appears insignificant.

An important aspect of scion selection is that wherever possible the terminal bud of the formal upright types should be surrounded by at least 3 lateral buds, preferably more. Other buds evenly distributed along the length of the stem are particularly important for less formal species, such as *P. breweriana*, but are always helpful in aiding the production of a well branched, shapely speci-men. *Picea breweriana* responds well to inclusion of a basal portion furnished with well-placed shoots with a two-year-old portion below (Figure 44.3). Shoot placement to provide symmetry may be further aided by the use of 'spacer canes' to be fully described in Growing on.

Preparation of the scion involves cutting to length, always retaining the terminal bud and includ-ing firm, ripened, 1- or 2-year wood. Removal of the needles for most types is demanding and described in Chapter 10 and Figure 10.11: Conifer Scion Preparation.

SYSTEMS

Two main systems are in use; (i) the conventional warm callusing system, where grafts are placed in poly-tents and subject to normal, enclosed environment management, this system can be used for winter or summer grafts; or (ii) the supported warm callusing system used during the winter graft-ing period, when grafts are placed on an open bench and subjected to frequent sprinkling or misting with water, normally carried out manually. Dry fog would be a good and labour saving alternative.

The supported warm callusing system has had significant use in the USA[*] and also in Scandinavia[†]. In some cases, grafts are stood on open benches with the union buried in a substrate

---

[*] Carville (1970), Mahlstede (1962), Wells (1955) p.261 and Dirr & Heuser Jr. (2006) pp.262–267.
[†] Thomsen (1978).

FIGURE 44.3 *Picea breweriana* scions. On the left, a conventional scion of 1-year wood, including the basal portion and terminal bud. Notice the shoot is well-furnished with lateral buds throughout. On the right a branched scion comprising a 2-year basal portion with a whorl of well-placed, 1-year shoots above. Experience is that this material makes a well-balanced plant more quickly than the unbranched scion.

chosen from peat, peat/bark or peat/perlite mixtures; one grafter used sand. Some stand grafts on open benches with the graft union sealed. Despite protecting the union from desiccation, it is surprising, as often suggested, that sufficient humidity can be constantly and reliably maintained on an open bench by the use of manual sprinkling only. In many instances, structures housing such grafts are very low profiled 'sash houses' or 'pits' with no side walls, more akin to a walk-in cold frame, and this may be a factor in maintenance of suitable conditions. Despite these considerations, this system cannot be recommended as an efficient and secure method of providing optimum conditions for callusing and union formation of spruce grafts.

Conventional warm callusing systems are to be preferred for winter or summer grafts using polytents, which may be raised to 90cm above the standing down bed (Figure 44.4). Frequent ventilation may be appropriate, and an internal lining of horticultural fleece will prevent unwanted drips. Polytents housed in modern glasshouses, with appropriate overhead shade and ventilation, can also give reasonable control of excessive temperatures.

Walk-in grafting polytunnels could also be considered, but there would be an additional cost of heating a larger space during the winter. Summer/autumn grafting was traditionally carried out under double-glass housed in cold frames. Excessively high temperatures were mitigated by wooden lath shades and skilful ventilation of the cold frame lights.

PHYSIOLOGICAL AND ENVIRONMENTAL FACTORS

Root pressure problems are a consideration with *Picea*. At an IPPS discussion group of UK grafters in 1979, consensus opinion was that *Picea* rootstocks require some level of drying to control excessive bleeding before grafting*. For winter grafts, this is achieved by withholding water and bringing

---

* Macdonald (1979).

FIGURE 44.4   *Picea pungens* cvs short tongue side grafted to *P. abies* plugs, bedded up in peat filled Boskoop trays, and placed in a poly-tent (Golden Grove Nursery).

the rootstocks into an unheated structure maintained at 10°C for 2–3 weeks. This procedure also has the effect of initiating root extension, most grafters feeling that grafting should not commence until at least a small number of white extension roots are visible. Judgements are made on compost water content and, if necessary, the pots are removed to aid drying-off. In Holland, summer grafts made in August are also dried for several days; Dutch grafters favour quite severe drying but ensure the standing-down medium (normally peat) is thoroughly wetted, resulting in high humidity and gradual water uptake through the base of the pot. There is obviously a fine line between too wet and too dry, experience must be the guide (see also Chapter 9: Drying-Off).

Not all grafters agree drying-off is necessary and some also recommend soaking the rootstocks before grafting commences*. Neither is it universally agreed that root extension before grafting is required[†]. In trials of *P. breweriana*, grafted in late February at the Boskoop trial station, best results were obtained from those grafted immediately the rootstocks were brought in (88%) compared with those with consequent root development after some drying (66%)[‡].

Experience with winter/early spring grafts indicates that some level of rootstock drying-off is sensible. Until the importance or otherwise of active root extension is established by scientifically based investigation, it is recommended to use the 'cold drying down' system outlined in Chapter 9 but placed to receive natural light. Under these conditions root tips will be active but with little or no extension at grafting time. After some days under warm callusing conditions, root extension commences along with initial callus development.

Summer grafts may be made on rootstocks 'on the dry side' but careful watch must be made to ensure the root-ball does not become excessively dry or the graft will be lost.

Picea grafts, in common with most others in Pinaceae, require low temperatures. Using a type of modified hot-pipe system[§], detailed investigation into the effect of heat on Sitka spruce (*P. sitchensis*) clones grafted to Sitka spruce rootstocks, showed no advantage of applied heat at the grafting stage

* Carville (1970), Hatch (1982) and Wells (1955) pp.261–264.
† Decker (1998).
‡ van Doesburg (1960).
§ Barnett & Miller (1994).

to the grafting area of the rootstock, or to the graft union. Most agree that for the winter/early spring period, maximum temperature should not exceed 18°C. Allowing for temperature rise, due to periods of sunshine, this means that, in practice, bottom heat settings should be between 10 and 15°C with ±12°C as optimum. Higher temperatures can be tolerated during the summer but, if possible, should not exceed 22°C for significant periods; heavy shading is required to achieve this.

After the union is complete, a few grafters head-back the rootstock after about 12 weeks*. Most advocate a three-stage heading-back: the first once union formation is apparent and the others for convenience at handling stages later. Retention of the rootstock stem for a period allows those types with a soft leading shoot such as *P. breweriana* to be tied-in to the rootstock for a period, until time for caning-up is available (see Figure 20.5 in *Abies*, Chapter 20 for example).

TOP-WORKED PLANTS

A number of the dwarf, compact forms of *P. abies* 'Doone Valley', 'Little Gem', etc; *P. glauca* 'Humpty Dumpty', 'Jalako Gold'; *P. omorika* 'Pimoko'; *P. pungens* 'Globosa'; *P. sitchensis* 'Papoose'; and others are suitable for top-working at heights between 60 and 90cm (Figure 44.5).

Taller stems (180–200cm) may be used for some of the weeping forms such as *P. abies* 'Inversa'. (For a full range of suggestions, see *RHS Encyclopaedia of Conifers*.)[†]

P. abies is normally used as the rootstock stem, often achieved by tying to a cane and training in the field; although, short stems may be grown throughout in a 2- or 3-litre-deep pattern pot. Field grown stems are normally potted up as the grafting procedure overall is lengthy and root-balled plants would be inconvenient to handle. Short stem grafts may be housed in tall poly-tents, but grafts on tall stems are often individually enclosed in poly-bags which may be gradually cut away as the union develops.

FIGURE 44.5   Top-worked *Picea*. LEFT, young *Picea pungens* 'Globosa' grafted to *P. abies* stems at 45cm high (Kilworth Conifers). RIGHT, *Picea sitchensis* 'Papoose', top-worked plants at saleable stage, grafted to stems of *P. abies* at 60cm high, growing in final 10-litre pots (Yorkshire Plants).

---

* Hatch (1982).
† Auders & Spicer (2014).

For compact forms, one scion per stem provides the best long-term effect. To avoid wastage, two scions may be grafted to the stem close above each other and, if the lower is successful, the rootstock is headed back to this point. Weeping types are usually more balanced if two or three scions are multi-grafted to the top of the stem.

GROWING ON

Spruce grows well in either containers or in the field. For upright formal shaped types, such as the Colorado blue spruce cultivars and less formal types such as *P. breweriana*, the main concern is to achieve a symmetrical, well-balanced, branched structure as quickly as possible after grafting. This is dependent upon choice of cultivar and also upon good choice of scion material. An important additional factor is a grafting procedure which ensures the least stress on the scion, and retention of a full complement of lateral buds, all of which are able to grow out at the appropriate stage after grafting.

These requirements are more easily stated than achieved. A further period of growth and some level of formative pruning will normally be necessary before the plants achieve top quality. Cultivars of blue spruce claimed to quickly provide shapely plants include *P. pungens* 'Edit' ('Edith'), 'Fat Albert' and 'Hoopsii'. Other claimants have appeared, but experience is that few match those listed. Doubtless, further searching will reveal other reliable types, but it is a slow procedure.

Use of 'spacer canes' as a method of improving shape and symmetry deficiencies has been mentioned in Chapter 20: *Abies*. These are inserted into the area around the young graft in such a way that the existing branch structure is deflected into the desired position. It is sometimes necessary to attach the branch to the cane by hand-tying or a machine tie (Max Tapener®), but often sufficient to rely on the pressure of the cane pushed into the substrate. Left in position for one or two growing seasons, branches become fixed into the new position and the canes may be removed then or at some convenient time later. This technique speeds up branch symmetry without loss of bulk more quickly than pruning, and in some instances can speed up plant development to saleability by a growing season (Figure 44.6).

FIGURE 44.6 *Picea pungens* 'Oldenburg', 2-year plants with 'spacer canes' in position to aid uniform distribution of branches. *P. orientalis* 'Skylands' (golden foliage), seen to the left, is more shapely and, apart from the odd branch, requires this treatment less often.

45 Pinus (Pinaceae) – Pines

INTRODUCTION

A large genus, the majority of species found throughout the northern hemisphere, with a small incursion into the southern. They are most numerous in temperate areas but are found from sub-alpine to sub-tropical locations. Many are major timber producers, a number are grown for their ornamental value. Some species produce edible seeds, sought after by gourmets and in past times an important food source for certain Native American tribes.

As Pines are difficult to induce to root from cuttings, grafting procedures are necessary to produce most cultivars and those species for which seed is rarely available. Grafting can also be used to provide the basis of elite seed mother plants for forestry purposes.

COMPATIBILITY

Pine compatibility appears to have never been fully resolved and varies from the view that virtually all types are compatible on the rootstocks *Pinus mugo*, *P. mugo* subsp. *uncinata*, *P. contorta* or *P. contorta* var. *latifolia**; to a contention that pines should always be grafted to species with the same number of leaves per fascicle, i.e. three-needled pines to three-needled, two to two, five to five[†]. In view of the size and complexity of the genus, the first would seem too simplistic and experience has shown the second is not certain for all combinations.

Taxonomy of this large genus is complex, subject to disagreement and amendment between botanical authorities. The genus is split into two, or arguably three subgenera, which in turn are split into sections and subsections. As always, it is a sensible strategy, when possible, to make links between botanical affinity and graft combinations; with this in mind, a simplified 'Taxonomic Tree' is set-out below in Table 45.1.

TABLE 45.1

Pinus – Taxonomic 'Tree'*

Subgenus	Section	Subsection	Selected Species
Pinus	Trifoliae	Australes (three needled)	*P. muricata, P. patula, P. radiata*
		Ponderosae (five or three needled)	*P. coulteri, P. durangensis, P. engelmannii, P. jeffreyi, P. montezumae, P. ponderosa, P. pseudostrobus*
Pinus	Trifoliae	Contortae (two needled)	*P. banksiana, P. contorta, P. virginiana*
Pinus	Pinus	Pinus (two needled)	*P. densiflora, P. mugo, P. mugo subsp. uncinata, P. nigra, P. resinosa, P. sylvestris, P. tabuliformis, P. thunbergii*
		Pinaster (two needled)	*P. halapensis, P. heldreichii, P. pinaster, P. pinea, P. roxburgii*
Strobus	Quinquefolia	Strobus (five needled)	*P. armandii, P. ayachuite, P. cembra, P. koraiensis, P. lambertiana, P. monticola, P. parviflora, P. pumila, P. strobus, P. wallichiana*
		Gerardianae (three needled)	*P. bungeana, P. gerardiana*
	Parrya	Cembroides (variable one to five needled)	*P. cembroides, P. culmonicola, P. edulis, P. monophylla*
		Balfourianae (five needled)	*P. aristata, P. balfouriana, P. longaeva*

*Adapted from Phylogeny and classification of Pines in Gernandt et al. (2005).

---

* Briggs (1977), Thorburn (1973) and Thomsen (1978).
[†] Dirr & Heuser Jr. (2006) p.270 and Hess (1953) p.35.

It is likely that any individual species within the same subsection can be considered as a candidate for best, most compatible rootstock for others in the group. Second choice would be species selected from the same section regardless of subsection. Some suggest that selecting a 'major' species such as *P. strobus* or *P. mugo* subsp. *uncinata* (*mugo* var. *rostrata*) from the relevant subgenus may suffice for all species within that subgenus. Personal experience is that this is not sufficient and further refinement of choice is necessary to achieve a permanent, reliable union. Rootstock availability often determines what is used, but this expediency can result in eventual graft failure, which has a negative effect on the perception of grafted plants.

Arguably, subgenus Strobus, the so called soft pines containing the largest number of species, may present a less complex problem than subgenus Pinus (hard pines). Most within subsections Strobus and Balfourianae appear compatible on *P. strobus* or, for reasons to be discussed later, the preferred alternative *P. wallichiana*. Subsections Gerardianae and Cembroides are less easily accommodated. *Pinus bungeana*, *P. cembroides*, *P. culmonicola*, *P. gerardiana* and *P. monophylla* have not succeeded well here on *P. strobus* or *P. wallichiana*. Personal experience is that an ideal rootstock for these has yet to be identified, a view supported by others*.

A review of suggested graft combinations reveals some interesting queries and conclusions. Bonsai production in Japan embraces a number of pine species, especially the Japanese White Pine (*Pinus parviflora*). This is in the subgenus Strobus; subsection Strobus, along with many other five needled pines, all of which appear compatible on other five needled species in the same subsection. It is surprising therefore that a number of ancient examples of *P. parviflora*, grown as Bonsai, are grafted to *P. thunbergii* from within the subgenus Pinus. Reasons for this choice are to take advantage of the thick trunk and picturesque bark of *P. thunbergii*, which creates the impression of age more effectively than the relatively smooth bark of *P. parviflora*†. With this combination, leaf development of the scion variety is reduced and in keeping with the miniature trees, an effect which may be due to partial incompatibility. A speculation supported by the fact that some clones of *P. parviflora* are incompatible on *P. thunbergii*.

Rootstock/scion combinations recorded at The Arnold Arboretum mostly confirm the previous suggestions, but exceptions occur‡. For example in the subgenus Pinus, *P. sylvestris* and *P. thunbergii* are combined with species in the same subgenera but in subsection Trifoliae, i.e. *P. banksiana*, *P. ponderosa* and *contorta* var. *latifolia*. Perhaps the most surprising combination is *P. resinosa* in subgenus Pinus, with *P. flexilis* in subgenus Strobus, which is said to form a satisfactory union (longevity not stated).

In the Pacific Northwest, some suggest *P. sylvestris* as a suitable rootstock for *P. coulteri* and *P. gerardiana*. Personal experience has been that these combinations do not result in a permanent union.

In a personal communication, an interesting suggestion is made for a range of two needled pine species, *banksiana*, *densiflora*, *mugo*, *sylvestris*, *thunbergii*, etc. The preferred rootstocks for these are five needled pines which, it is stated, exhibit good takes and long-term compatibility§.

Work at the Trial Station in Boskoop, Holland over some years¶ has come to the following conclusions: for most of the five needled pines (Subgenus Strobus), use of two needled rootstocks (subgenus Pinus) produces poor results and shortened lifespan. Exceptions exist, *P. parviflora* 'Glauca' and 'Negishi' survive to a level of 60–80% after many years. *Pinus thunbergii* as rootstock for five needled pines produced good results with *P. pumila*, *P. parviflora* 'Negishi' and *P. monticola* 'Skyline', but most others in the Strobus subgenus performed badly on this rootstock. Surprisingly, *P. sylvestris* is suggested as a better rootstock than *P. mugo* for many *P. mugo* cultivars, a conclusion likely to be contested by many conifer grafters.

---

* Buchholz (2016) and Buchholz (2017).
† Baker (1976) and Valavanis (1982).
‡ Alexander III (1998).
§ W. Barnes (pers. comm.).
¶ Joustra, Ruesink & Verhoeven (1990).

It is hard to summarise the above results and conclusions. Often, long-term incompatibility is masked using combinations inherently incompatible, but which survive and grow with varying degrees of success for periods of time. Unfortunately nurserymen may be guilty, as apparently successful combinations survive nursery life only to fail at some point in the future. Graft compatibility in Pines presents problems which may take many years of patient investigation to overcome.

TIMING

In the UK and USA, grafting takes place in winter (late December–February); in Holland, many species within the five needle group and cultivars of *P. sylvestris* may also be grafted in October. Reasons for this are not clear but may be linked to the fact that good results are achievable at this time, and it enables grafters to spread the work load.

GRAFTING METHODS

The majority of grafts used are side veneers. Choice between short tongue and long tongue is dependent upon rootstock quality. Young rootstocks permit use of the preferred long tongue veneer (Figure 45.1). In Japan and Korea, the apical wedge (cleft) or a type of saddle graft (inlay) appears standard procedure. Splice grafts have been used occasionally in the Netherlands, but the majority agree it is preferable to retain the rootstock top by use of a side graft.

Most consider sealing is not required because of the highly resinous content of pine wood and the resultant exudation. Use of non-enclosing rubber ties is normal and as for some other genera in Pinaceae (*Abies,* etc.), recommendations are to leave gaps between the coils to allow excess exudates to escape (Figure 45.1).

In Pines, leaf arrangement in bunches or fascicules often means leaves may be pulled-off rather than cut-off, consequently easing and speeding scion and rootstock preparation. Older rootstocks will still need secateurs to remove obstructing shoots.

Heading-back the rootstock takes place after the union is well established; many favour this being done in two stages: about half being removed in the first and complete removal in the second.

FIGURE 45.1   Short tongue (left) and long tongue (right) side veneer grafts on *Pinus wallichiana* rootstocks. Note: resin exudate running down the right-hand side of the stem, below the tie on the left-hand graft.

ROOTSTOCKS

Trials at the Boskoop Trial Station demonstrated the advantage of young rootstocks over older ones*. Grafters should emphasise to rootstock growers the importance of supplying relatively light grade seedlings for potting up to produce 1+1P 3–5 or 5–7mm potted rootstocks. Some species with heavier wood will require thicker rootstocks and a 3–5mm 2+0 seedling for potting-up will be required to produce 2+1P 6–8mm after one growing season (Figure 45.2).

The large majority of pines require well-drained conditions, and therefore 9 × 9 × 12.5cm pots are preferable to the commonly used 9 × 9 × 9cm pots. Compost should also reflect the drainage requirement and the addition of bark to 30% or more of the total will ensure production of a healthy, disease-free root system for species in the subgenus Pinus.

The most commonly used species in subgenus Strobus, *P. strobus*, is particularly sensitive to root disease problems, despite using bark rates as high as 45%. As a result, many grafters in Europe have switched to *Pinus wallichiana* as the best choice of rootstock for species in this subgenus (Figure 45.3).

Plug grown plants with the advantages of vigour and juvenility are a very good choice of rootstock. Well grown, 1-year seedlings in large plugs may reach graftable size after one growing season; small plug seedlings require moving-on into 9cm pots or larger plugs for a further season.

SCION WOOD

Scion collection follows procedures described in Chapter 10. After collection in the October grafting period, scions are recommended to be grafted within 48 hours. In December–February, much more latitude is possible and cold storage at 0.5–1°C is safe for 3–4 weeks, probably longer.

FIGURE 45.2   Two needled Pine rootstocks (*Pinus mugo* subsp. *uncinata*) prepared ready for grafting. Note the handling system, the plants in large plug trays stacked on Danish trolleys (Golden Grove Nursery).

* Joustra, Ruesink & Verhoeven (1989).

FIGURE 45.3 *Pinus wallichiana*, a grafted rootstock with pot removed, showing excellent quality root system, extending white roots are visible as a result of higher temperatures during warm callusing.

One-year-old wood is usually the preferred choice for most medium- and tall-growing species. Retention of the terminal bud, with surrounding lateral buds, is important to ensure good, reasonably straight extension growth and balanced axillary branching (Figure 45.4). Selection of suitable shoots is influenced by the ideal length of 100–125mm and the need to include the basal portion of the shoot as well as the terminal bud. If good, well-ripened and adequately sized 1-year wood is not available, use of a basal portion of 2-year wood is perfectly acceptable. Compact growing species and dwarf cultivars invariably require scion wood which includes a two- or even three-year basal portion.

SYSTEMS

Warm callusing or supported warm callusing systems are generally considered essential for grafting pines. Field grafting is said to be possible in some situations on the Pacific Northwest's coast of the USA. No figures for the percentage of successful grafts using this system are available, but the fact that it is recorded indicates that pine grafts are capable of tolerating far more ventilation than most evergreen genera.

FIGURE 45.4   Selection of prepared *Pinus* scions. Left, *Pinus sylvestris* 'Nisbets Gem', a dwarf compact form, including basal portion of 2-year wood. Centre, *P. sylvestris* 'Aurea', 1-year wood only. Right, *P. sylvestris* 'Hillside Creeper' a pendulous form with basal 2-year wood.

Conventional warm callusing is used by most grafters. This is normally provided by polythene-covered poly-tents (Figure 45.5). It is advisable for these to be designed with hoops at least 90cm high to give good clearance above the grafts and mitigate very high humidity and drips. Under these conditions, plunging the graft union, a labour-consuming job, is not required. Use of one layer of internal fleece will not unduly shade the grafts and has a beneficial effect in keeping dryer conditions within the poly-tent.

Some shade may be required for October grafts; the level recommended on the chart in Chapter 13: Shading for cuttings (40% on bright days) should be sufficient. Grafts during the December–February period should also be given the shading regime recommended for cuttings, which for much of the period will mean none required.

Careful watch should be kept for any sign of fungal disease, and routine spraying of a broad spectrum fungicide is considered a sensible precaution, especially for the subgenus Strobus or 'soft pines'. Ventilation can be applied more liberally than for many genera; during periods of dull overcast weather in polytunnels, and in modern glasshouses with low air exchange, ventilation is possible for some hours morning and evening.

Several grafters recommend that grafts should be placed on open benches in greenhouses with the unsealed graft union plunged in a free-draining, open mulch, often comprising a 1:1 v/v mixture of peat and perlite, or peat and Styrofoam. The grafts are kept in high humidity conditions (supporting warm callusing)*. This is achieved by hand misting at regular intervals or by an overhead mist system. An ideal labour-saving solution would be use of dry fog which could be adjusted to provide uniform, predetermined humidity levels. Optimum humidity has not been recorded but ±90% RH is suggested, being lowered as graft union development progresses.

---

* Decker (1998) and Gregg (2013).

FIGURE 45.5   Two needled pine grafts; rootstocks are from large plugs placed in Boskoop trays, and laid-in peat to leave the graft exposed, before standing down under a poly-tent (Golden Grove Nursery).

PHYSIOLOGICAL AND ENVIRONMENTAL FACTORS

As discussed in Chapter 18, Pines fall into the low temperature requirement group of Pinaceae. Opinions differ on the optimum temperature; some suggest a minimum as low as 1°C, but at that temperature few if any woody species, will produce callus; a minimum of 6°C is best, with an average maintained at or above 10–12°C. Tests in Holland with *P. strobus* cultivars and *P. pumila* 'Glauca' grafted to both *P. strobus* and *P. wallichiana* surprisingly produced better results with air temperatures of 18°C rather than 11°C. Personal experience is that good results have been obtained with a bottom heat setting of 10°C, which means on bright days in the October- or December–February periods, with little shading, air temperatures inside the poly-tent may increase to 16–18°C for short periods. In low temperatures, callus production is slow, 40 days or more will be required for visible callus development, and a further minimum of 30 days for graft unions to be considered united.

Virtually all grafters suggest that pine rootstocks should not be grafted until the roots are active and extending with 5–6mm of white root visible. The arguments for and against this have been discussed at length previously (see Chapters 11 and 18).

Despite general agreement that pines respond to high light levels, Japanese methods using apical rather than side grafts, are said to involve placing completed grafts into darkness for 6–8 weeks. Perceived wisdom would suggest this is guaranteed to negate good results; presumably this cannot be the case. Whether the technique is still in use today has not been possible to establish, but the results of such an enquiry would certainly be interesting.

PERFORMANCE COMPARISONS

Most disappointing results are linked to compatibility issues, which result in immediate or eventual graft failure. Aside from this, golden foliaged forms usually represent more of a challenge than other types. Maintenance of disease-free scion wood is important in the success of grafting procedures.

TOP-WORKED PLANTS

Dwarf, slow growing and pendulous forms may be top-worked onto suitable stems, occasionally at 1.8 or 2 metres high, but more usually from 60–80cm, to produce novelty or Patio plants. Often side veneer multi-grafts are used to accelerate development to saleable size. Completed grafts are housed in purpose made, tall poly-tents or individually enclosed in polythene bags. In the latter method, the bag may have one corner cut off to permit ventilation, closed off as required by use of a peg. Once the grafts have fully united, the bags are removed and the plants grown-on for eventual sale (Figure 45.6).

GROWING ON

Well-drained soils or growing media are essential for virtually all species. Although tolerant of some shade, pines thrive best in full sun.

FIGURE 45.6   *Pinus sylvestris* 'Watereri' (left) and *Pinus densiflora* 'Alice Verkade' (right) grafted to *P. sylvestris* stems and grown as top-worked Patio plants (Yorkshire Plants).

46 Prunus (Rosaceae) – Ornamental Prunus

INTRODUCTION

Mostly deciduous, small-to-medium-, occasionally large-sized trees, but that include a few dwarf and evergreen species. Restricted mainly to temperate regions in the northern hemisphere, this large genus of over 200 species contains some of the most beautiful flowering trees and shrubs. For centuries certain Chinese and Japanese species have been hybridised and selected in Japan to constitute what have become known as the Japanese flowering cherries or Sato Zakura.

Prunus mostly show typical Rosaceae characteristic of being relatively easily grafted and many are produced in the field using summer budding or winter grafting techniques. Bench grafting has its place, especially for the production of limited numbers of rare species – those with thin, light wood not well adapted to field grafting, and where severe weather conditions prevail during the winter.

COMPATIBILITY

As so often, taxonomy plays a significant role in the combinations which can be considered compatible. Table 46.1 (see below) lists the subgenera, sections and associated species, and Table 46.2 (see below) the hybrids etc. providing guidance for grafters on combinations likely to succeed for the species. Prunus hybrids with their parentage and suggested rootstocks are listed in Table 46.2. Some clonal rootstocks are listed but others developed for use with Almonds and Stone fruit are not included; for details of potential dwarfing rootstocks developed for fruit crops but suitable for ornamentals see Chapter 47.

TABLE 46.1

Prunus Taxonomy – Suggested Rootstocks*

Subgenus	Section	Species (Cultivars Not Listed)	Rootstock (Suggested)
Prunophora	Euprunus (European/ Asian Plums)	*cerasifera, consociiflora, divaricata, domestica, institia, salicina, simonii, sogdiana, spinosa*	Brompton, *cerasifera* seedling, Myrobalan B
	Prunocerasus (American Plums)	*alleghaniensis, americana, angustifolia, maritima, munsoniana, nigra*	Brompton, *cerasifera* seedling, Myrobalan B
	Armeniaca (Apricots)	*armeniaca, mandschurica, mume*	*armeniaca* seedling, *cerasifera* seedling, Myrobalan B, *persica* seedling, St Julian A
Amydalus	Almonds and Peaches	*davidiana, dulcis, fenzliana, kansuensis, persica, x persicoides (amygdalopersica), scoparia, tangutica, tenella, triloba*	*dulcis* seedling, *persica* seedling, Myrobalan B, St. Julian A, Brompton
Cerasus	Microcerasus (Lithocerasus)	*besseyi, glandulosa, incana, jacquemontii, japonica, prostrata, pumila, tomentosa*	*avium* seedling, F 12/1, Colt
	Pseudocerasus	*campanulata, canescens, cerasoides, concinna, himalaica, hirtipes (conradinae), incisa, jamasakura, lannesiana, latidentata, mugus, nipponica, pendula, pseudocerasus, rufa, sargentii, serrula, serrulata, sieboldii, speciosa, x subhirtella, verecunda, x yedoensis*	*avium* seedling, F 12/1, Colt
	Eucerasus	*avium, cerasus, fruiticosa*	*avium* seedling, F 12/1, Colt
	Loropetalum	*cantabrigensis, cyclamina, dielsiana*	*avium* seedling, F 12/1, Colt
	Phyllomahaleb	*macradenia, maximowiczii*	*avium* seedling, F 12 /1, *mahaleb*
	Mahaleb	*emarginata, mahaleb, pensylvanica*	*avium* seedling, F 12 /1, *mahaleb*
	Phyllocerasus	*litigiosa, pilosiuscula*	*avium* seedling, F 12 /1, *mahaleb*
Padus	(Bird cherries)	*cornuta, grayana, maackii, obtusata, padus, serotina, virginiana*	*padus* seedling, *avium* suggested for *maackii. P. serotina* has failed on *virginiana*.
Laurocerasus	(Cherry laurels)	*caroliniana, ilicifolia, ilicifolia subsp. lyonii*	*lusitanica* seedling or ex cuttings

*Adapted from Bean (1976) vol.3 pp.348–349 and Rehder (1940) pp.452–455.

TABLE 46.2
Prunus Hybrids, Parentage and Suggested Rootstocks*

Name	Parentage (Where Known)	Rootstock (Suggested)
'Accolade'	*sargentii* x *P. x subhirtella*	*avium* seedling, F 12/1, Colt
x *blireana*	*cerasifera* 'Pissardii' x *mume*' Alphandii'	*cerasifera* seedling, Myrobalan B
x *cistena* ('Crimson Dwarf')	*besseyi* x *cerasifera* 'Pissardii'	*cerasifera* seedling, Myrobalan B
'Collingwood Ingram'	'Kursar' x ?	*avium* seedling, F12/1, Colt
x *dasycarpa*	*armeniaca* x *cerasifera*	*cerasifera* seedling
x *dawyckensis*	*? canescens* x *dielsiana*	*avium* seedling, F 12/1
x *dunbarii*	*americana* x *maritima*	*avium* seedling, F 12/1
x *gondouinii* (x *effusa*)	*avium* x *cerasus*	*avium* seedling, F 12/1
'Hally Jolivette'	(x *subhirtella* x *P. x yedoensis*) x *P. x subhirtella*	*avium* seedling, F 12/1, Colt
'Hillieri'	*sargentii* x ?	*avium* seedling, F 12/1, Colt
'Hilling's Weeping'	? weeping form of *P. incisa*	*avium* seedling, F 12/1, Colt
x *incam* 'Okame'	*campanulata* x *incisa*	Best on own roots. *P. avium* seedling, F 12/1
x *incam* 'Shosar'	*campanulata* x *incisa*	Best on own roots. *P. avium* seedling, F 12/1
x *juddii*	*sargentii* x *P. x yedoensis*	*avium* seedling, F 12/1, Colt
'Kursar'	*campanulata* x *nipponica* var. *kurilensis*	*avium* seedling
'Moerheimii'	*? incisa* form	*avium* seedling, F 12/1, Colt
'Pandora'	*pendula f. ascendens* 'Rosea' x *P. x yedoensis*	*avium* seedling, F 12/1, Colt
'Pollardii'	*dulcis* x *persica*	*dulcis* seedling, *persica* seedling, *cerasifera* seedling, Myrobalan B
'Pink Shell'	? seedling x *yedoensis*	*avium* seedling, F 12/1, Colt
x *schmidtii*	*avium* x *canescens*	*avium* seedling, F 12/1, Colt
'Snow Goose'	*incisa* x *speciosa*	*avium* seedling, F 12/1, Colt
'Spire' (*P.* 'Hillieri Spire')	*sargentii* x *? incisa*	*avium* seedling, F 12/1, Colt
'Sunset Boulevard'	? hybrid or seedling of *P. serrulata* var. *spontanea*	*avium* seedling, F 12/1, Colt
'The Bride'	seedling of 'Pandora'	*avium* seedling, F 12/1, Colt
'Trailblazer'	*cerasifera* 'Pissardii' x 'Shiro' (? *salicina* x *cerasifera*)	*cerasifera* seedling, Myrobalan B
'Umeniko'	*incisa* x *speciosa*	*avium* seedling, F 12/1, Colt
'Wadae'	*pseudocerasus* x *P. x subhirtella*	*avium* seedling, F 12/1
Sato Zakura – Japanese Flowering Cherries	Hybrids and selections from *P. serrulata* and forms, *P. pendula*, *P. incisa*, *P. sargentii* and doubtless others.	*avium* seedling, F 12/1, Colt. Dwarfing rootstocks developed for dessert cherries may also be considered for Patio trees/container production etc.

*Derived from the following sources: La Rue & Johnson (1989) pp.9–11, Dirr & Heuser Jr. (2006) pp.283–294, Sheat (1965) pp.318–322, P.C.R. Dummer (pers. comm.), A. Wright (pers. comm.), Farmer Jr. & Hatmaker (1970), Alexander III (1998) and Duncan, Connell & Edstrom (2010).

As always with compatibility issues, nothing is straightforward, and Tables 46.1 and 46.2 must be taken as guidelines within which a number of variables occur. When deciding combinations, the following should be taken into account:

- *P. dulcis* seedling rootstocks are not compatible with all *P. persica* varieties, European and Japanese plum species.
- *P. persica* seedling rootstocks are compatible with Peach, Apricot, Almond, but not with most Japanese and many European plums.
- *P. cerasifera* and Myrobalan B show incompatibility symptoms with some Almond, Peach and Apricots.
- *P. maackii* clones ('Amber Beauty' etc.) are best grafted to *P. maackii* seedlings, but *P. avium* is thought to be suitable. Growth is slow on *P. padus*, and *P. avium* induces good extension growth but long term compatibility may be in doubt. In one report, *P. maackii* grafted to *P.* Colt failed after one season, snapping-off at the union while being lifted; another grafter reports *P.* Colt as being a suitable rootstock for *P. maackii*.

The Sato Zakura Japanese Cherry group, and a number of other hybrid flowering cherries, show some inconsistences in performance when grafted to various rootstocks. *P. avium* seedlings were the rootstock of choice for many years until the development of the clonal selection F12/1, and subsequently P. Colt, and the now apparently defunct P. Cob. The latter two were developed as slightly dwarfing rootstocks for fruiting cherries. Both these selections, but particularly Cob, promoted early growth and development of the Japanese scion varieties. Cob was particularly successful as a rootstock for the important cultivars 'Ukon' and 'Kanzan'. Not all hybrids and Sato Zakura flourish on *P. avium* and F12/1. Colt rootstocks are reported to be suitable across a wide range of Japanese Flowering cherry cultivars.

TIMING

Mid-winter/early spring (late January to mid-March) is the favoured period. Cold storage of scions and/or rootstocks permits extension of the grafting period until the end of April.

GRAFTING METHODS

Apical grafts are normal for all *Prunus*. Splice, whip and tongue or, less often, short tongue veneers are favoured. Good cambial matching is more important for *Prunus* than most other Rosaceae, and the use of machines or tools needs careful consideration as it often results in some wood shredding and uneven matching, leading to poor results. Short tongue side veneer is recommended for multi-grafting top-worked stems; sometimes the less accurate side wedge is used.

ROOTSTOCKS

In the taxonomic groupings above, if material is available, a number listed as scion varieties can also act as rootstocks. For example, *P. americana* is used for apricots and plums in parts of the USA because it is available and much hardier than alternatives[*]. *Prunus tomentosa* has been suggested as suitable rootstock for some apricots, cherries, nectarines, peaches and plums[†]. However, long-term compatibility had not been established beyond one season, and some combinations may display delayed incompatibility.

---

[*] Davis II (1976).
[†] Mezitt (1973).

Prunus avium seedlings may show up to 25% infected with prune dwarf virus* and consequently affect subsequent takes, growth and performance.

Bare-root seedlings of the species listed are available from suppliers as 1+0 6–10mm or if heavier grades are favoured 1+1 10–12mm. Those propagated vegetatively (Brompton, Colt, F12/1, Myrobalan B and St. Julien A) are available as rooted layers or hardwood cuttings as 0+1 5–7mm, 7–9mm or 9–12mm (the latter grade often 0+1+1).

Pot grown rootstocks are rarely used for deciduous species although rare specimens may warrant the extra trouble; 1-litre-deep pots will normally suffice. Rootstock stems for top-working may be grown in pots, 2-litre being suitable for stems up to 90cm; for larger stems, root-balled plants are often favoured. *Prunus lusitanica* is available as a pot grown item in 9 × 9 × 9cm pot or as 1+2 6–8mm bare-root seedling.

Comparisons of growth rates for bench grafts of *Prunus* 'Amanogawa' and 'Kanzan' between the rootstocks *P. avium*, F12/1 and Colt were made in Holland at the Proefstation voor de Boomteelt[†]. After planting out and one season's growth, Colt produced a 34% increase in growth over F12/1 for *P.* 'Amanogawa' and 5% for *P.* 'Kanzan'. The one-year-old trees were cut back at the end of the first year, and after re-growth, subsequent overall height by the following season was equal for all trees on different rootstocks. The tallest first year trees of 'Amanogawa' on Colt produced more branched and better quality trees than those on *avium* and F12/1.

SCION WOOD

Well-ripened, current-year wood is the usual requirement. For dwarf, slow-growing species or sources from arboreta or gardens where preparatory pruning has not been carried out, inclusion of a portion of 2-year wood is unlikely to adversely influence results. Collection should take place when the material is thoroughly dormant and can then be stored until required, preferably in a cold store, at −2 to 0.5°C. Scions normally include three to five buds and strong shoots may provide two or three scions before terminal portions become too soft and thin.

SYSTEMS

Compared with fruiting cherries, the ornamental types are generally required in much greater diversity and smaller numbers. However if numbers are sufficient to warrant the cost of installing suitable equipment, sub-cold callusing is worth consideration. For the main varieties, bare-root rootstocks are used while for rare types and those with thin wood, required only in small numbers, potted rootstocks may be more convenient.

Grafts produced on bare-root rootstocks are normally sealed, plunged in suitable, well-drained medium in boxes, and placed in unheated structures as discussed previously (Chapter 18). Successful grafts may be potted off and placed in frost-free growing structures. The alternative system of planting directly in the field is usually delayed until soil conditions and ambient temperatures are suitable. In northern areas, grafts normally require a period in cold store until these conditions prevail. A good strategy is to delay grafting to a later date by cold storing scions and rootstocks; this reduces the required period in cold store between graft union completion and planting.

Warm callusing can rarely be justified or required; however, particularly scarce or rare material may warrant the additional effort. Temperatures between 18°C and 20°C will result in rapid callus production and formation of a union. Early bud-break and premature extension growth is always a potential problem and will require careful management of the poly-tented enclosure to avoid problems with damping-off.

Hot-pipe systems might be considered worth the extra input because of the certainty of a near 100% take. Use of large rootstocks 10–12mm diameter, established in 2-litre pots, grafted with

---

* Sweet, Goodall & Campbell (1978).
† Proefstation voor de Boomkwekerij (1984).

large scion material, possibly including two-year basal wood, placed in a hot-pipe system, followed by growing on in a frost-free structure, should guarantee high takes and rapid development to saleable size.

PHYSIOLOGICAL AND ENVIRONMENTAL FACTORS

Good cambial matching is important for this genus. Root pressure, particularly with bare-root rootstocks, is not a problem. Optimum temperature for callus production and union formation is 18–21°C.

TOP-WORKED PLANTS

The strongly weeping forms such as 'Hilling's Weeping', *incisa* 'Pendula', the pendulous forms of *P. pendula*, *P.* x *yedoensis*, and the weeping Sato Zakura ('Kiku-Shidare-Zakura', etc.), are all commonly top-worked to stems of *P. avium* or preferably F12/1; Colt is not suitable for this purpose as it is liable to produce unsightly adventitious roots at various positions. Use of stems with attractive bark such as *P. serrula* has been discussed previously (see Chapter 4). Top-working may be carried out in the field, but more reliable results are obtained by bench grafting and placing the grafts in structures heated sufficiently to provide frost protection. The rootstocks are either lifted from the field and root-balled or potted for more or less immediate grafting. Less often, they may be established in the pot for a growing season before potting.

For the types mentioned previously, grafting height is normally 180–200cm. Another group of smaller, lighter wooded types, such as *P. glandulosa* forms, *P. incisa* 'Kojo-no-mai' and *P. tenella* forms such as 'Fire Hill', may be grafted to shorter stems at 45–80cm high, creating a Patio tree. In this case, it may often be more convenient to grow the rootstock as a potted item in a 2–4-litre-deep pattern pot.

GROWING ON

Container production is now a common strategy for nurserymen producing ornamental cherries. Larger specimens may be grown in the field for a period before containerisation for sale. Cherries are not demanding regarding soil conditions, tolerating most types except poorly drained or thin, dry, acid conditions.

47 Prunus (Rosaceae) – Almonds and Stone Fruits

INTRODUCTION

Worldwide production of almonds is considerable at ±1.25 million tons, of which the USA produces about 68%, Spain 16%, Italy and Greece about 4% each and the other major producers approximately 1% each. Overall, production of stone fruits is massively greater at ±43 million tonnes, of which peaches and nectarines account for over half at ±25 million tonnes, half of this being produced in China. Others include cherries (Sweet and Sour), plums and apricots.

Two other stone fruits within the *Prunus* genera are apriums, hybrids between apricot and plum with apricot dominating and pluots, hybrids between plum and apricot with plum dominating. At present, production of these does not appear to be significant.

COMPATIBILITY

Broad issues of compatibility have been dealt with in the Ornamental Prunus, Chapter 46, but will be further discussed on an individual basis in the section Rootstocks below.

TIMING

Basic principles of timing will follow those discussed for ornamental *Prunus*. In view of the large numbers involved, cold storage of rootstocks and scion wood will often be utilised to provide more flexibility in grafting and planting operations. Use of these facilities can mean that cultural procedures and storage are extended over a period from scion collection in mid-/late December to planting grafts as late as early/mid-July.

GRAFTING METHODS

Good matching of the cambia of rootstock and scion is important for *Prunus*, and for this reason most adopt manual grafting methods despite a requirement for very large numbers. Knifesmanship skills can be concentrated by use of whip and tongue grafts, allowing assistants to take over tying and sealing procedures. This allows skilled individuals to achieve very high outputs while maintaining accurately matched grafts.

Rootstocks derived from micropropagation, and subsequently potted, are sometimes grafted with a chip bud or short tongue side veneer graft.

ROOTSTOCKS

In many countries the health and quality of clonal rootstocks for top fruit and nut crops is subject to regular monitoring and appraisal by licenced agencies, often linked to government departments (see Chapters 8 and 9 for further discussion).

Modern production of almonds and stone fruit crops demands rootstocks with the ability to form a long-term compatible union, able to control overall growth and cropping characteristics, and be resistant to various soil conditions, diseases and pests. In general, vegetatively produced clonal rootstocks are the only guarantee that these desirable characteristics will be expressed in the rootstock and transmitted to the scion. However, certain selections come reasonably true from seed and, because of economic considerations, some grafters make use of these.

Rootstocks for almonds and stone fruits are the subject of constant improvement and an ongoing search for new types through breeding and selection. Consequently, recommendations will be likely to become redundant in the years ahead due to the development of superior varieties. Producers should be prepared to review any information presented before making final decisions on choice of rootstock.

Rootstocks are required to grades 1+0 or 1+1 8–10mm for seedlings, or from a clonal source, 0+1 or 0+1+1 8–10mm. Micropropagated clonal rootstocks are increasingly being produced, particularly for peaches and almonds; these are normally sold as containerised items in plugs or deep pots from 0.5 to 1-litre capacity.

Almonds

At present, in the major almond producing area of USA, highest cropping per planted acre is still achieved by the largest trees, which, as far as possible, fill the orchard space. Harvesting is carried out by machines which shake the trees to release the nuts; these are then collected from the ground by purpose built equipment, often relying on vacuum systems.

New developments involve close planting of trees on clonal rootstocks which provide a growth dwarfing effect. This enables harvesting to take place by straddling the rows with high-clearance machines which gather the nuts directly from the trees. Harvesting becomes more efficient, and this strategy seems likely to gain increasing importance as greater numbers of dwarfing and semi-dwarfing rootstocks become available. See Table 47.1 below for a description of vigorous and dwarfing seedling and clonal rootstocks; their major characteristics with regard to growing requirements and disease resistance are also shown.

Apricots

As with almond producers, those producing apricots have demonstrated less interest to date in semi-dwarf and dwarf trees than for many other fruit crops. It seems likely that the trend towards dwarf trees and intensive planting will extend to include apricots. Suitable seedling and clonal rootstocks together with their major characteristics are listed below in Table 47.2.

Peaches and Nectarines

Seedling and clonal rootstocks listed for almonds and apricots contain many types also suitable for peaches and nectarines; in most cases their influence on tree size, form, behaviour and disease resistance is unchanged from those previously listed. Additional choices suggested specifically as rootstocks for peaches are listed below in Table 47.3.

TABLE 47.1

Rootstocks for Almonds*

Name	Origin	Vigour	Soil	Bacterial Canker	Phytophthora	Armillaria	Crown Gall	Comments
Nemared	Seedling	Vigorous	Susc. to high pH	Susc.	Susc.	Susc.	Susc.	–
Nemaguard	ditto	Vigorous	ditto	Susc.	Susc.	Susc.	Susc.	–
Lovell	Seedling	Vigorous	Susc. to high pH	Susc.	Susc.	Susc.	Susc.	–
P. dulcis (Almond)	Seedling	Very vigorous	Well drained	Susc.	Susc.	Susc.	Susc.	Too vigorous delayed cropping
P. davidiana (Chinese wild Peach)	Seedling	Vigorous	–	Susc.	Susc.	Susc.	Very susc.	Unreliable
P. cerasifera (Myrobalan)	Seedling	Vigorous	Suitable for heavy soils	Mod. Susc.	Resistant	Resistant	Mod. Susc.	Now usually replaced by Marianna 2624
P. armeniaca (Apricot)	Seedling	Semi vigorous	Well drained	Susc.	Susc.	Susc.	Susc.	Eventually incompatible
Hansen's 536; Nickels; Bright's Hybrid; Cornerstone and Titan	Clonal	Vigorous large to very large trees	Tolerates drought	Very susc.	Very susc.	Very susc.	Very susc.	Better yields than Lovell, etc.
Marianna 2624; Brompton and St Julien A	Plum and plum derivatives	Dwarfing	Suitable heavy soils/poor drainage	Very susc.	Tolerant	Resistant	Tolerant	Sucker profusely. Not compatible all vars. Delayed incompatibility
Krymsk 86	ditto	Vigorous	ditto	Very susc.	Tolerant	Resistant	Tolerant	ditto
Viking; Atlas and Mirobac (Rootpac 20)	Interspecific hybrids peach x almond x apricot x plum	Vigorous	Tolerate high pH	Tolerant	–	–	–	High yielding. Good anchorage. Disadvantages yet to be identified

*Sources of information for Tables 47.1, 47.2 and 47.3 have been derived from those listed in the footnote below Table 47.3: Rootstocks for Peaches and Nectarines.

848

TABLE 47.2
Rootstocks for Apricots*

Name	Origin	Vigour	Soil	Bacterial Canker	Phytophthora	Armillaria	Crown Gall	Comments
Prunus armeniaca (Apricot) and Apricor	Seedling	Vigorous	Well drained	Susc.	Susc.	Susc.	Susc.	V. good compatibility some sucker production
Nemaguard; Nemared and Lovell	Seedling	Vigorous	Susc. To high pH	Susc.	Susc.	Susc.	Susc.	Good compatibility with most varieties
P. cerasifera (Myrobalan)	Seedling	Vigorous large trees	Tolerates heavy wet soils	Mod. susc.	Resistant	Resistant	Mod. susc.	Large, productive trees cropping delayed. Some varieties weak union
Myrobalan B	Clonal	Very vigorous	Tolerates wet soil	Mod. susc.	Resistant	Resistant	Mod. susc.	More reliable and consistent than *P. cerasifera*
Brompton	Clonal	Very vigorous	Tolerates wet soils	Mod. susc.	Resistant	Resistant	Mod. susc.	Better compatibility than Myrobalan
Krymsk 86	Clonal	Large vigorous	Ditto	Very susc.	Tolerant	Resistant	Tolerant	Little sucker production
Penta	Clonal	90% of most vigorous	Ditto	–	–	–	–	Very compatible some suckers
Jaspi and Tetra	Clonal	Semi vigorous	Ditto	–	–	–	–	Some suckering
St Julien A and Torinel	Clonal	Medium/large	Well drained	–	–	–	–	Good cropping some suckers
Plumina and Pixy	Clonal	Semi dwarf	Well drained	–	–	–	–	Good croppers. Pixy needs permanent staking on most soils

*Sources of information for Tables 47.1, 47.2 and 47.3 have been derived from those listed in the footnote below Table 47.3: Rootstocks for Peaches and Nectarines.

TABLE 47.3

Rootstocks for Peach and Nectarine*

Name	Origin	Vigour	Soil	Bacterial Canker	Phytophthora	Armillaria	Crown Gall	Comments
Seedling rootstocks as listed above except *P. dulcis*, which has delayed incompatibility	Seed	Very vigorous	Dislike high pH soils	Susc.	Susc.	Susc.	Susc.	Very good compatibility. Large vigorous trees sometimes with delayed cropping
Guardian (peach selection)	Seed	Very vigorous	Dislike high pH soils	Susc.	Susc.	Susc.	Susc.	Good long-term compatibility and heavy cropper popular in S.E. USA
Atlas and Bright's Hybrids	Clonal	Vigorous	Tolerant to high pH and drought	Tolerant	V. susc.	V. susc.	V. susc.	Large trees heavy croppers
Controller 9	Clonal	90% size of Lovell	–	–	–	–	–	Good cropper
Viking	Clonal	Vigorous	Tolerant to high pH	Tolerant	–	–	–	High yield good anchorage
Penta	Clonal	80–90% size of Lovell	Tolerant to high pH and heavy wet soil	–	Resistant	Resistant	–	High yield good fruit size
Citation	Clonal	Slight vigour reduction	Tolerates wet soils	–	–	–	Susc.	Very compatible with apricot and plum varieties
Mirobac (Rootpak R)	Clonal (*P. cerasifera x P. dulcis*)	Vigorous	Tolerant to heavy, wet soil	–	–	–	–	Good compatibility Good anchorage
Controller 8 (HBOK 10)	Clonal	Semi dwarf	Well-drained soil. Poor tolerance to high pH	–	–	–	–	Good compatibility. Good anchorage
Controller 7 (HBOK 32)	Clonal	Medium	Poor tolerance to high pH	–	–	–	–	Good compatibility. Good anchorage
Controller 5	Clonal (*P. salicina x P. dulcis*)	50% vigour of Lovell	Well drained soil	Susc.	–	–	–	Good compatibility Good anchorage
Krymsk 1 (VVA-1)	Clonal (*P. tomentosa x P. cerasifera*)	Dwarfing	Tolerates wet, high pH soils	Susc.	–	–	Susc.	Poor compatibility with many peach varieties. Very good with plum varieties

*Sources of information on Tables 47.1–47.3 above of rootstocks for Almonds, Apricot, Peach and Nectarine have been derived from the following sources: Sierra Gold Nurseries (2017), Yoshikawa, Ramming & LaRue (1989), Duncan, Connell & Edstrom (2010), Micke (1996), Johnson & DeJong (2009–2012), Burchell Nursery (2017), Fowler Nurseries Inc. (2017), Orange Pippin Fruit Trees (2018) and Frank P Matthews Ltd. (2018).

PLUMS

The rootstocks listed above, particularly hybrids derived from plum hybrids, will suit most plum varieties and have similar effect on tree size, characteristics and cropping.

Not listed is a recent rootstock, Myruni, a selection from *P. institia* (Bullace) said to be non-suckering and produce a semi-vigorous tree with good compatibility to plum cultivars and good cropping potential.

APRIUMS AND PLUOTS

Developed in the USA during the 1980s, these fruits are hybrids of apricot and plum. Aprium shows more dominant apricot features and pluot more plum. Both are compatible on the rootstocks previously suggested for apricots and plums.

SWEET AND SOUR CHERRIES

Rootstocks for cherries follow a different choice to those above. Interest in dwarfing rootstocks has led to a revolution in cherry growing, the previously large and difficult to harvest trees being replaced by smaller more manageable ones. Some dwarfing rootstocks have made it possible to produce cherry fruit within structures, mostly polytunnels, enabling heavier, earlier cropping, protection from bird depredation, easier harvesting and culture in areas previously considered too cold or unsuitable for cherries.

Rootstocks in Table 47.4 are currently in use although constantly being appraised and amended with additions and improvements. They are arranged in descending order from the largest to the most dwarfing.

TABLE 47.4

Rootstocks for Sweet and Sour Cherry*

Name	Size % of P. avium (Mazzard)	Precocity/Productivity	Compatibility	Suckering Suckers	Support/Anchorage	Comments
avium (Mazzard)	100%	Poor/ quite good	Very good also for ornamentals	Slight	V. good	Harvesting v. difficult
avium F12/1	100%	ditto	ditto	ditto	ditto	Tolerant to bacterial canker
Maxma 14 (*mahaleb* x *avium*)	90–95%	Good	Good	Little	good	Good in high pH soils. Tolerant to bacterial canker. Some have experienced heavy unexplained losses after 10 years.
Mahaleb (*P. mahaleb*)	90%	–	Not all varieties	–	good	Good in very dry high pH soils
Krymsk 5 (*fruiticosa x lannesiana*)	85%	Not overly precocious or productive	Good	–	Good no support	Good fruit quality Tolerates heavy soils. Very hardy
Gisela 12 (*cerasus x canescens*)	80–85%	Precocious and productive	Good	–	Good no support	–
Colt (*avium x pseudocerasus*)	80%	–	Good with ornamentals also	–	Good no support	Tolerates wide soil range. Resistant to Phytophthora and bacterial canker. Productive in warm temperate climates
Gisela 6 (*cerasus x canescens*)	75–80%	Highly productive	–	–	Normally no support	Tolerates wider range of soils than G5
Gisela 5 (*cerasus x canescens*)	50%	Precocious and v. productive	–	–	Needs staking	Most widely planted of Gisela series.
Clare (*cerasus* selection ex. Michigan State Univ. – MSU series)	50%	–	–	–	Needs staking	Promising, trials only commenced 2015
Gisela 3 (*cerasus* Schattenmorelle x *canescens*)	35–40%	Precocious, fruit can be small	–	None	Needs staking	Requires deep fertile soils and irrigation

*Sources of information for Table 47.4 are: N. Dunn (pers. comm.), Washington State University (2018), Long, Brewer, & Kaiser (2014), Sierra Gold Nurseries (2017), Orange Pippin Fruit Trees (2018) and Frank P Matthews Ltd. (2018).

SCION WOOD

Scion material for all almonds and stone fruits must comprise suitable, well-ripened material from virus-free mother trees. In many areas, such material is obtainable from certified stock.

After collection, the wood is normally trimmed into 3–5 bud lengths and immediately grafted or placed into a cool shed or cold store until required. Storage at −2 to 0.5°C permits prolonged delay of several weeks before use.

Scions for micropropagated rootstocks will normally require thinner grade material, achieved by additional pruning of the mother plants, possibly incorporating some level of deep shade or etiolation in the early stages.

SYSTEMS

Major cherry tree production is still accomplished by field budding in the summer; however, as previously discussed, sub-cold callusing is gaining popularity, especially in northern Europe. Cold callusing using bare-root rootstocks is sometimes favoured. Hot-pipe systems can rarely be justified but guarantee very high takes.

In some areas, notably California, there is increasing interest in container-grown, planting-out stock. The rootstocks are normally produced by micropropagation and grafted in the container (usually a deep plug or cell), placed in cold or warm callusing conditions before potting-on into a pot of 0.5- to 2.0-litre capacity, for eventual field planting. In the future use of techniques involving rootstock propagation from soft, semi-ripe cuttings placed under ventilated fog and treated with appropriate growth substances might challenge those produced from micropropagation.

PHYSIOLOGICAL AND ENVIRONMENTAL FACTORS

Provided cambial matching is reasonably good, cherries follow the typical Rosaceae pattern of being easily grafted and takes in excess of 90% can be expected.

48 Pyrus (Rosaceae) – Ornamental and Fruiting Pears

INTRODUCTION

The genus is split between those grown for their ornamental value and those for fruit production.

The ornamentals include about 15 species, found only in the Old World. Mostly small- to medium-sized trees, they display white flowers and attractive glossy green or grey foliage, sometimes having attractive autumn colours. Tolerant of a range of soil types, low temperatures, hard pruning and sometimes having an upright, compact growth habit, they provide excellent subjects for street planting as well as use in garden and parkland situations.

Fruiting pears provide a range of fruit suitable for dessert and culinary purposes. World-wide production in 2013 was ±23 million tonnes, of which 15 million were produced in China. The USA, Italy and Argentina each produced ±700,000 tonnes. Turkey, Spain, South Africa, India, Netherlands and Belgium each between 3–400,000 tonnes; the UK produced 21,000 tonnes. No figures are available for Asian pear production.

COMPATIBILITY

ORNAMENTAL PEARS

Taxonomic treatment has so far distinguished some subspecies but no sections appear to have been identified. Recent comprehensive investigations based on phylogeny of the genus have not been discovered; however, one made by Challice and Westwood in 1973 included an initial investigation*. This concluded that the genus was derived from a single source (monophyletic), always an encouraging result, suggesting that incompatibility is less likely to be a problem than when the origins are from several sources.

The genus can be split into clusters of species thought to be most closely related as follows:

- *P. calleryana*, possibly the most primitive species.
- *P. dimorphophylla* and *faurei* closely related to, or considered forms of, *calleryana*.
- *P. cossonii* (*longipes*), *pashia* and *ussuriensis*, to these perhaps could be added *boissieriana*.
- *P. pyrifolia*, a group apparently on its own, though *phaeocarpa* may belong here.
- *P. cordata* and *betulifolia*.
- *P. communis*, *nivalis*, *amydaliformis*, *salicifolia* and *elaeagnifolia* constitute possibly the most advanced species.

Recommendations are that species within these groups should be treated as potentially the most compatible combinations.

Personal experience has demonstrated that a wide range of compatibilities are possible.

P. cossonii, P. amydaliformis, P. pashia, P. betulifolia, P. nivalis and *P. elaeagnifolia* have succeeded well using *P. communis* as the understock.

Some grafters have suggested that *P. calleryana* grafted to *P. communis* does not make a compatible union[†]. In the UK, varieties of *calleryana* budded to *P. communis* produced reasonable,

---

* Chalice & Westwood (1973).
† Mathies (1985), McMillan Browse (1979) and Dirr & Heuser Jr. (2006) p.299.

but not good takes, and subsequently a proportion (5–10%) failed to grow vigorously and exhibited signs of incompatibility. A significant number grew satisfactorily and made saleable and eventually long-lived plants.

Pyrus calleryana 'Autumn Blaze', 'Chanticleer' and 'Trinity' grafted to *P. communis* rootstocks have grown here for 30 years and made trees 6–7.5 metres high; however, the rootstock suckers heavily, often a sign of some level of incompatibility. *Pyrus faurei* and *dimorphophylla*, both closely related to *calleryana*, have made a union on *P. communis*, but growth is poor, as is *P. ussuriensis* on the same rootstock. Trials of quince rootstocks as alternatives to *Pyrus* have failed to produce acceptable levels of compatibility.

It would seem *P. calleryana* is the appropriate rootstock for its cultivars, but the characteristic of this species for late seasonal growth, especially in maritime climates, means that it is subject to damage by early frosts; to avoid this, *P.* ussuriensis has been recommended as a suitable alternative[*]. However, trials in the Netherlands reported in 1988[†] comparing *P. ussuriensis* with *P. communis* as rootstocks for *P. calleryana* 'Red Spire', *communis* 'Beech Hill', *communis* subsp. *caucasica* and *salicifolia* 'Pendula' resulted in lower takes and less growth for all types on *P. ussuriensis*.

FRUITING PEARS

As with other important fruit- and nut-producing Rosaceae, modern practice is to find rootstocks which will control growth and encourage early fruiting. Unlike *Malus* and *Prunus*, the *Pyrus* are partially or totally compatible with a range of other genera in Rosaceae. This is important because pears grafted to the traditional rootstock, *Pyrus communis,* produce medium- to large-sized trees, slow to reach fruiting stage, and difficult and costly to harvest.

To overcome this problem, Quince (*Cydonia oblonga*), having a significantly dwarfing effect, has been used as a rootstock for many years in areas with suitable winter temperatures. Compatibility is not complete with all varieties, and for a number a mutually compatible inter-stem such as 'Beurre Hardy' or 'Fertility' must be used to allow double-working and provide a viable long-term union. Unfortunately, *Cydonia* is not sufficiently hardy to withstand the low temperatures encountered in northerly production areas of America, Europe and Asia, and there has been a search for alternatives over many years.

Four other genera have been investigated for their potential as dwarfing rootstocks to provide trees 40% below the size of those on *P. communis*. So far, possibilities have been discovered in the *Amelanchiers, Crataegus, Malus and Sorbus*. The species *Pyrus nivalis* has also been investigated in what are known as the Brossier series from France. Other genera such as *Aronia, Docynia* and *Pyronia* have also been tested but so far results are inconclusive.

Amelanchier hybrids of *A.* x *grandiflora, A. alnifolia* and *A. spicata* have been developed by Michael Neumüller in Germany at the Bavarian Fruit Centre, with potential as very dwarfing rootstocks combined with extreme hardiness. Trials some years ago in Oregon showed possibilities for significantly improved field performance in precocity and cropping, using *Amelanchier* instead of the conventional pear rootstock OH (Old Home) x F87 (Figure 48.1). Many additional trials and development will undoubtedly be required to establish the long-term performance and reliability of this breakthrough in dwarfing rootstocks for pears.

Crataegus oxycantha has shown good compatibility with the pear variety 'Old Home'. When this variety was used as an inter-stem, a number of other significant pear varieties produced compatible combinations with thorn.

It has been stated that pears are not compatible on apple (*Malus*). This is generally the case but one variety from Indiana, USA, *Malus* 'Winter Banana', has proved to be an exception. Grafted to a dwarfing apple rootstock such as M27 or M9 and used as an inter-stem for a pear variety, scion

---

[*] Mathies (1985).
[†] Driver (1988).

FIGURE 48.1 Pear 'd'Anjou': LEFT, grafted to an American semi-dwarfing pear rootstock selection, OH x F87. RIGHT, grafted to *Amelanchier* hybrid rootstock. Both rows pictured were planted on the same date in adjoining rows. Very little, if any, fruit has been produced on the semi-dwarfing pear rootstock. (photos: Dr. Todd Einhorn, Michigan State University).

control of growth might be achieved. Other apple varieties and rootstocks such as M16 have shown some level of compatibility with pears, the pear variety 'Fertility' being particularly amenable*.

Sorbus aucuparia has also been investigated as a potential rootstock for pears. A number of varieties have proved compatible but not 'Old Home', a crucially important Fire blight (*Erwinia amylovora*)-resistant variety for warm climates, doubtless further investigation will take place in the future.

Dependent upon overcoming issues of incompatibility, a number of exciting possibilities for the development of truly dwarf fruiting pear rootstocks remain to be discovered.

TIMING

Scion material is collected during mid-winter and stored until required under cold conditions, preferably in a cold store at 0–0.5°C, where they can remain for many weeks if necessary. Grafting takes place between late January and early March; cold storage permits this to extend to late April/early May if required to be delayed for field planting.

During the summer, micropropagated rootstocks grown in pots, possibly under protection, may be chip budded on open benches, or short tongue veneer side grafted in poly-tents. Winter grafting can take place on potted rootstocks using side or apical methods.

* Garner (2013) p.46.

GRAFTING METHODS

Splice and whip and tongue are usually employed for the ornamentals; the culinary pears may be produced similarly. Grafting machines may be considered where very large numbers are involved.

Bare-root rootstocks destined for field planting may be grafted sufficiently low at the basal end of the grafting cuts to include a portion of the hypocotyl, normally seen as a blanched portion of stem just above commencement of the root system. This should enhance the chance of early scion rooting; a desirable outcome for ornamental pears where the rootstock is not required to control growth*.

ROOTSTOCKS FOR FRUITING PEARS

Bare-root rootstocks are favoured for both ornamentals and fruiting pears. Seedling rootstocks are normally 1+0 or 1+1 8–10mm. Unfortunately, *Pyrus communis* seedlings are notable for producing a coarse, 'fangy' root system, not well-suited to transplanting procedures. Clonal types, particularly quince, are produced by stooling, layering or hardwood cuttings as 0+1 or 0+1+1 8–10mm. There is limited micropropagation of the newer introductions because of difficulties in rooting certain pear rootstocks.

For many years the pear variety 'Old Home' (OH), identified as resistant to Fire Blight (*Erwinia amylovora*), has provided a reliable genetic pool for inducing resistance in a number of pear rootstock hybrids. In areas such as southern Europe, and parts of the USA, with significantly higher spring and summer temperatures than the UK, this has proved an important feature for this very Fire Blight susceptible genus.

The following list of rootstocks* is arranged in groups according to their effect on vigour. These are shown as percentages of the size of the most vigorous types on *P. communis*. Where possible, the origin of the rootstock is shown and, where known, cultural comments are added:

- **<40%** – *Amelanchier*; *P. nivalis* Brossier series; *Crataegus*; *Malus*; *Sorbus aucuparia*; these have already been discussed in the section on compatibility.
- **40–60%** – Eline quince; QR517/9; QR719-3; Quince EMA; Quince BA29; Quince EML; Quince C132; Quince EMC; Quince C132; Quince EMH; Sydo Quince. <u>Origin</u> – Quince selections BA 29; EMC; and EMH; these are selections from a commercial nursery. <u>Comments</u> – Quince rootstocks allow high-density planting, but some varieties require double-working. Quince rootstocks are particularly well adapted for the Comice pear varieties. Staking or some support is generally required.
- **61–70%** – Fox 9; Fox 11; Fox 16; Horner 10; OH x F40; OH x F51; OH x F69; OH x F87; OH x F333; Pyro 2–33; Pyro dwarf. <u>Origin</u> – Fox rootstocks are *P. communis* selections from Italy. Horner is a clonal series from Oregon State University. OH x F 40, is an 'Old Home' x 'Farmingdale' hybrid. Pyro and Pyrodwarf are *P. communis* selections from Germany. <u>Comments</u> – Fox 16 is considered the best cropper. The OH x F series has good resistance to fire blight and number 69 gave very good yields in trials in the USA.
- **71–90%** – BM 2000; OH x F97.
- **91–100%** – Bartlett seedling; Horner 4; Winter Nelis seedling; *P. calleryana* D6; *P. calleryana* seedling. <u>Origin</u> – Winter Nelis seedling is 'Winter Nelis' x 'Bartlett', the seed is collected from 'Winter Nelis' growing in a *P.* 'Bartlett' orchard. <u>Comments</u> – It produces consistent size and productivity and has a wide soil tolerance.
- **100% +** *Pyrus betulifolia* seedling; *P. communis*. <u>Comments</u> – *Pyrus betulifolia* is a surprising recommendation from the USA as in the UK *P. betulifolia* generally makes a smaller tree than *P. communis*. This might be because it is normally seen in British

---

* Davis II (1982).

collections as a tree grafted to *P. communis* rootstock. It is said to provide an ideal root-stock for Asian pears but at present appears unavailable in any quantity from rootstock suppliers in northern Europe.

Less interest is evident in dwarfing rootstocks for Asian pears and current practice is to use *Pyrus betulifolia*; *P. communis* is said to reduce overall size by 50% at maturity.
See footnote on the sources of information for the listed rootstocks:

SCION WOOD

Well-ripened, current year's wood is required. Two or three sufficiently ripe scions should be obtain-able from a single shoot provided by well-pruned mother plants. Collection is best carried out in mid-winter, and cold storage at 0–0.5°C allows storage for many weeks until required.

SYSTEMS

The ornamental pears are normally grafted bare-root and cold callused. Rootstocks are laid in a well- drained medium leaving the sealed union exposed. After the union has formed, the grafts may be cold stored for a period until direct planting in the field takes place. For the ornamental types, some grafters favour planting so that the union is buried. Potting up the completed grafts before or after union formation is a further option.

Fruiting pears may be sub-cold callused using methods described previously. When clonal root-stocks are used, the graft union must be planted well above soil level to prevent any possibility of scion rooting.

Cold callusing is an alternative for fruiting pears but, unless cold storage is available, provides less flexibility in planting dates. This cannot safely be long delayed because pears break bud early and growth commences at quite low temperatures.

PHYSIOLOGICAL AND ENVIRONMENTAL FACTORS

Pears are able to produce callus at low temperatures, though below ±6°C little development takes place. Root pressure does not appear to be a problem when bare-root rootstocks are used.

TOP-WORKED PLANTS

Pyrus salicifolia 'Pendula' is occasionally top-worked; this is normally in the field but, to be certain of a high take, *P. communis* stems may be run-up as single stems, lifted when dormant and grafted under cover. The root system may be laid-in during callusing or the completed grafts potted-up into suitable containers. Multi-grafting using two scions can provide saleable plants in one season.

GROWING ON

Pears grown in the field thrive in most conditions tolerating dry or wet soils. Growth in dry, acid, peaty soils is slow. For commercial purposes, many ornamental types are now grown in containers.

---

* N. Dunn (pers. comm.), Weinstock (2016), Washington State University (2017), Stebbins (1995), Elkins (2012), Orange Pippin Fruit Trees (2018) and Frank P. Matthews Ltd.

49 Quercus (Fagaceae) – Oaks

INTRODUCTION

A large genus of over 500 species but less than this number is in cultivation. They are widely dispersed throughout the northern hemisphere occurring from cold to warm temperate areas and entering the southern hemisphere in Indonesia. In the wild, due to habitat loss and excessive harvesting for timber, several species are vulnerable, endangered or critically endangered.

Oaks may be propagated readily from seed, but if this is collected from plants in cultivation, they are very likely to be hybrids. True-to-name plants can only be certain from wild collections and even here hybridisation can sometimes occur. Most species are difficult to root from cuttings and to date micropropagation has not provided a reliable means of producing a wide range of types. Grafting remains the major method of producing forms, cultivars and species, when true to type seed is unobtainable.

COMPATIBILITY

This genus possibly represents the most challenging of temperate genera to achieve long-lasting compatible graft unions. An example is seen over many seasons of grafting *Quercus robur* 'Fastigiata Koster' to *Q. robur* rootstocks. This combination regularly provides good initial takes (80%+), but subsequently, after two growing seasons, 20% or more of apparently successful grafts can show poor extension growth accompanied by premature flowering, classic signs of delayed incompatibility. The response varies from crop to crop, presumably influenced by rootstock seed source.

The strategy of double-working, discussed in Chapter 12, Compatibility: Future Progress, has particular relevance for *Quercus*. The *Quercus lamellosa* grafted to *Q. myrsinifolia*, shown in Figure 49.5, is one of only very few successes in a trial batch which comprised many grafts. The sucker shoot from the successful graft pictured could be encouraged to grow, and when appropriate, re-grafted to a *Q. myrsinifolia* rootstock. This may then form the foundation of a mother plant to produce inter-stems showing compatibility between *Q. lamellosa* scions and *Q. myrsinifolia* rootstocks.

Many genera have some low-level incompatibility, but with oaks, after a period of two to five years, rarely are any batches of grafts free of the problem. It seems that more than purely taxonomic and phylogenic factors influence the outcome. Nevertheless, until the underlying causes are better understood, taxonomic and phylogenic characters must guide the choice of rootstock/scion combinations. To ignore them usually leads to very poor results (Figure 49.1).

The size and complexity of the genus has led to numerous taxonomic proposals still ongoing and subject to disagreement and revision. The more recent science of phylogeny does not yet appear to provide published data giving a comprehensive and categorical overview of the oaks but, where possible, use has been made of the information available. In the future further information on this aspect may aid better decisions on suitable rootstock/scion combinations, bearing in mind that "kinship is no guarantee of compatibility".

From a taxonomic standpoint, the conventional view is that the genus is split into two main subgenera, Quercus, comprising the majority of species and Cyclobalanopsis (ring-cup oaks), mostly evergreen species found in Asia from the western Himalayas to eastern and south eastern Asia, including some sub-tropical and tropical areas*. A more recent proposal based on combined taxonomic and phylogenic factors suggests a different view, the genus again being split into two, the

---

* Bean (1976) pp.454–459 and Amory (2009) pp.32–33.

FIGURE 49.1   LEFT, *Quercus* x *hispanica* 'Fulhamensis' (section Cerris); left, grafted to *Q. cerris,* and on the right, to *Q. robur* (section Quercus), an unsuitable rootstock. RIGHT, *Quercus variabilis* (section Cerris) grafted to rootstocks of *Q. robur.* These can fail after the first season (left) or be successful (right), with likely long-term compatibility.

main subgenera becoming Cerris and Quercus*. In both proposals, the subgenera are split into a number of sections.

Table 49.1 largely retains the conventional layout but also combines aspects of phylogeny to provide proposals for suitable rootstock/scion combinations. As always, species within the same sections are likely to demonstrate the most compatible partnering; however, it can be assumed with many of the suggestions there will always be a certain level of delayed incompatibility.

Further Discussion on Compatibility

Subgenus Cyclobalanopsis

Little information is available regarding suitable grafting combinations for the subgenus Cyclobalanopsis. Two reasonably hardy species for southern/midlands regions of the UK are *Quercus myrsinifolia* and *Q. schottkyana*. Both have proved compatible for *Q. lamellosa*, but with a low percentage of success, and so far on a short timescale of 4 season's growth. A third species, *Q. glauca*, shows promise with good early callus development and retention of a healthy 'tongue' on long tongue side veneers. It too is said to be hardy, but subject to late frost damage on young growths in the spring. This disadvantage would not apply in its role as a rootstock. Of the three species tested, *Q. myrsinifolia* has less leaf drop in the poly-tent and may callus earlier than *Q. schottkyana*. Eventually *Q. schottkyana* produces plentiful callus and stimulant buds develop rapidly.

---

* Denk & Grimm (2017).

TABLE 49.1

Quercus, Subgenera and Sections – with Suggested Rootstocks

Subgenus Cyclobalanopsis	Subgenus Quercus				
	Section Cerris	Section Mesobalanus	Section Quercus	Section Protobalanus	Section Lobatae (Erythrobalanus)
Suggested rootstocks: Q. myrsinifolia Q. schottkyana Q. glauca	**Suggested rootstocks:** Q. cerris •alternative Q. ilex •see comments below	**Suggested rootstocks:** Q. robur see comment below	**Suggested rootstocks:** Q. robur	**Suggested rootstocks:** Q. ilex? Q. robur?	**Suggested rootstocks:** Q. palustris Q. coccinea •see comments below
acuta	acutissima	canariensis	alba & cvs	chrysolepis	acerifolia•
delavayi	alnifolia•	dentata & cvs	aliena & cvs	tomentella	buckleyi
gilva	aquifoliodes•	frainetto & cvs	arizonica	vaccinifolia	canbyi•
glauca	aucheri•	Macon	bicolor	candicans	
lamellosa					
multinervis					
myrsinifolia	baloot•	macranthera	x bimundorum Crimschmidt (Crimson Spire)		coccinea & cvs
oxyodon	castaneifolia cvs•	Pondaim	x hickelii		crassifolia
salicina	cerris cvs	pontica	macrocarpa		ellipsoidalis & cvs
schottkyana	coccifera•	pyrenaica & cvs	mongolica		emoryi•
	x hispanica cvs	sadleriana	muehlenbergii		gravesii•
	ilex cvs•		petraea cvs		imbricaria
	x kewensis•		prinus•		incana
	x libanerris		robur•		kellogii•
	libani•		x rosacea		x ludoviciana•
	macrolepis & cvs		rugosa		marilandica•
	monimotricha•		X turneri & cvs.		Mauri•
	phillyreoides•		x warburgii		mexicana•
	rotundifolia•		X warei cvs.		nigra & cvs
	semecarpifolia				palustris cvs
	suber•				phellos cvs•
	trojana				rubra cvs
	tungmaiensis•				rysophylla•
	variabilis•				x schochiana
					shumardii•
					texana & cvs
					velutina & cvs
					wislizeni

Subgenus Quercus: Section Cerris

Quercus cerris provides a good, reliable rootstock for many in this section (Figure 49.2). Those with a green bullet may be more successful when *Q. ilex* is used as the rootstock. However, *Q. baloot* is thriving on *Q. cerris* after nine years. *Quercus alnifolia* can grow well when grafted to *Q. ilex*. A tree on this rootstock had to be cut back after 30 years because it was endangering overhead cables*. To highlight the inconsistencies of oaks, the same combination here has failed after five years. *Quercus ilex*, *alnifolia*, *aucheri* and *coccifera* are closely linked phylogenically into a monophyletic (same origin) group. *Quercus monimotricha* has failed at three attempts to graft it to *Q. cerris*; *Q. ilex* may be worth a try.

Subgenus Quercus: Sections Mesobalanus and Quercus

A view shared by others is that the most reliable of oak combinations are between *Q. robur* and species in the Mesobalanus and Quercus section[†]. As noted previously, there are some problems with delayed incompatibility but less than those in Cerris and considerably less than those within Lobatae.

Where *Q. robur* is unavailable, or unsuited to growing conditions in a given area, other species in the two sections prove satisfactory substitutes. In many parts of the USA, *Q. bicolor* or *Q. alba* are commonly used as rootstocks for species in these sections[‡]. In Oregon, *Q. macrocarpa* from a North Dakota provenance is used because of its hardiness and tolerance of high pH[§]. Closely resembling a diminutive *Q. pontica* , the charming little oak *Q. sadleriana* , in the Mesobalanus section, is the only species which has so far registered total failure after three attempts on *Q. robur*. *Quercus pontica* used as an inter-stem on a *Q. robur* rootstock may solve the problem. Alternatively, *Quercus* 'Bill George' or *Q.* x *hickelii* are possible candidates for the same purpose.

FIGURE 49.2   *Quercus* x *hispanica* 'Lucombeana' (Cerris section), a clear line of the graft union with the Q. cerris rootstock showing just above soil level. The graft was made in 1895 and one of an avenue of trees adjoining the University of Cambridge (photo: John Dyter).

---

* I. Dickings (pers. comm.).
† Coggeshall (1996).
‡ Brotzman (2012).
§ M. Krautmann (pers. comm.).

Subgenus Quercus: Section Protobalanus

This small section has not been investigated and no references for graft combinations between the species have been discovered.

Subgenus Quercus: Section Lobatae

Species in Lobatae consistently show unpredictable long-term compatibility. The suggested root-stocks *Q. palustris* and *Q. rubra* are suitable for a range of species but by no means all. In the list, those with red bullet are unreliable on one or other, sometimes both of these species.

Work by Santamour[*] has suggested that because of enzyme linkages, *Q. rubra* is an unsuitable rootstock for most species in Lobatae, other than *Q. rubra* cultivars, the preferred option being *Q. palustris*. In nurseries, Knap Hill Scarlet Oak (*Quercus coccinea* 'Splendens') was notorious for delayed incompatibility due to the original practice of grafting it to *Q. rubra* rootstocks. A switch to *Q. palustris* appears to provide a more satisfactory combination, but time is required to confirm this beyond doubt. *Q. rubra* rootstock generally appears reliable for *Q. rubra* cultivars, though not totally confirmed because some assert *Q. rubra* cultivars are not compatible on *Q. rubra* rootstocks[†].

In practice, Santamour's well-researched hypothesis concerning peroxidase enzymes linked to compatibility does not appear to have stood the test of time, and instances of graft failure in oaks point to more than this as being the underlying cause. Suggestions that *Q. rubra* is an unsatisfactory rootstock for all Lobatae, other than *Q. rubra* and its varieties, are not supported by experience. A report by Coombes relates to a trial in Italy comparing results of *Q. rysophylla* grafted to *Q. rubra*, *Q. palustris*, *Q. cerris* and *Q. robur*. Best results were obtained with those on *Q. rubra*, less good with *Q. palustris*, poor on *Q. cerris* and a failure on *Q. robur*[‡]. *Quercus rysophylla* grafted to *Q. palustris* failed here after 8 years whereas a plant grafted to *Q. rubra* has survived 15 years.

Among the others with an asterisk listed in Lobatae, *Q. acerifolia, emoryi, gravesii, kellogii, marilandica* and *shumardii* have all shown dubious results when grafted to *Q. palustris*. Most initially appear to take well, but over time, sometimes as little as one season, a varying percentage fail or become stunted and show typical incompatibility symptoms. To reduce eventual losses, it is reported that one American nurseryman is grafting selections and cultivars only to seedling-raised rootstocks of the same species[§]. This enlightened approach reduces but does not eliminate long-term incompatibility. There is also the problem of obtaining seedlings of the desired species.

Quercus phellos hybrids, *Q.* x *heterophylla* (*phellos* x *rubra*) and *Q.* x *ludoviciana* (*pagoda* x *phellos*) present a challenge. *Q.* x *ludoviciana* is said to be compatible on *Q. phellos* rootstocks[¶] but a repeated attempt in 2010 failed to achieve a result, the tree failing after four years of proceeding poor growth. A retry using *Q. palustris* rootstocks also failed. Probably a few among a number of grafts will succeed on a long-term basis, reflecting the influence of seedling variation.

Quercus mexicana, marked with a red bullet, has performed well here on *Q. palustris* rootstock but is said to also succeed well on *Q. robur*.

Inter-Sectional Combinations – Potential Inter-Stems

Three species in the Cerris section marked with a blue bullet *Q. castaneifolia, libani* and *variabilis* can succeed for a time period when grafted to *Q. robur* in the Quercus section[**]. Subsequent recent personal experience of *Q. variabilis* grafted to *Q. robur* confirms the finding reported earlier; individual grafts within small batches have failed to show delayed incompatibility at any stage after five season's growth. All three species succeed well on *Q. cerris*, and their apparent ability to unite

---

[*]   Santamour Jr. (1988) and Santamour Jr. (1996).
[†]   Upchurch (2009).
[‡]   Coombes (2015).
[§]   Brotzman (2012).
[¶]   P. Dummer (pers. comm.).
[**] Humphrey & Dummer (1967).

with species from other sections suggests their potential use as an inter-stem. A further example of inter-sectional compatibility is seen in a link between sections Cerris and Lobatae with the hybrid *Q.* x *kewensis* (*cerris* x *wislizeni*). It regularly succeeds well on *Q. cerris* rootstock and it is therefore conceivable that, because of its alternative parent *Q. wislizeni*, there is the possibility of using it as an inter-stem in combinations using species from Lobatae.

Effects of Particular Combinations

The influence of rootstock choice on an unusual oak, *Quercus dentata* 'Pinnatifida' (section Mesobalanus) was reported in the 1999 Japanese region meeting of the IPPS by Watanabe*. This concerned the influence of different rootstocks on this particular form, distinguished by narrow, elegantly lobed leaves which make it quite unlike any other oak. When grafted to seedlings of the parent species, *Q. dentata,* it produces the most vigorous growth; on *Q. mongolica*, in the closely related Quercus section, growth is more compact and leaf lobes become curled at their tips. Species from Cerris section, *Q. acutissima* or *Q. variabilis*, both from Japan, appear to form a compatible union (perhaps not so surprising in view of the known ability of *Q. variabilis* to unite with *Q. robur*). What is surprising is that the foliage colour of the resultant graft is reputedly an attractive golden colour. This phenomenon links with attempts over several decades by Sir Harold Hillier to locate a golden form of 'Pinnatifida' he had heard existed in Japan. Scion material of what was purported to be this form was repeatedly sent from Japan to his nursery, and duly grafted to *Q. robur*, the normal choice for this species. Invariably the resultant growth was the usual green, and after several attempts, it was considered the Japanese nurserymen and botanic gardens wished to retain the golden form uniquely to Japan. Sadly, Sir Harold, unaware of the role of *Q. variabilis,* never did find his golden 'Pinnatifida'.

TIMING

Winter/spring grafting is popular with many, while others favour summer/autumn grafting. *Q. robur* 'Fastigiata' is not recommended for summer grafting as heavy losses can occur over winter.

Opinions on precise timing for the early period vary; some recommend a start as late as mid-April,[†] presumably based on the late date of the emergence of extension growth of many oaks. If bare-root rootstocks are used most agree an earlier start in January or February is preferable. An early start is also required for root grafts: delaying until April resulted in failure[‡]. Work in Holland has shown that grafting after March produces lower results[§]. Later grafting should work well if cold stored scion material is used.

During the summer/autumn period, grafting does not usually commence before mid-August, potted rootstocks are normally used at this time, but by delaying until early September, use of newly lifted bare-root rootstocks is possible.

GRAFTING METHODS

Good knifesmanship is required. Accurate cambial matching is important for a genus which is slow to form a strong graft union and, unless well attached, is likely to fail within the first or second season. A further problem is that oak wood is among the hardest encountered by grafters, particularly when as often recommended, older wood is used for scions. The scion wood of many species, other than in the Cyclobalanopsis subgenus, may be rather large in diameter, often more than 6mm. All

---

* Watanabe (1999).
† Krüssmann (1964) pp.494–497.
‡ Leiss (1984).
§ van Doesburg & van Elk (1963).

these factors indicate the need to use a substantial well sharpened knife; a Tina 600A, sharpened to razor sharpness on a Tormek® sharpening machine, provides the ideal tool.

Splice grafting is frequently favoured; although, the importance of obtaining a good match suggests that veneer grafts are a better choice. Side veneer grafts are to be preferred over apical and probably essential for the evergreen species. To facilitate tying-in, the rootstock top may be reduced in height when grafting in the dormant period*.

Use of the long tongue side veneer takes advantage of the fact that oak callus development in the veneer tongue/scion interface occurs more copiously, and earlier, often by some weeks, than the rootstock stem/scion interface. Work in Holland compared the influence of short or long tongue grafts on takes of *Q. coccinea* Splendens, short tongue grafts gave 36% and long tongue 64%[†]. This result is supported by personal experience which continually highlights the superiority of long tongue side veneers over other methods. In the subgenus Cyclobalanopsis, a stimulant bud and, if possible, associated leaf should be present on the tongue. For deciduous species, the value of a stimulant bud is evident at all times (Figure 49.3).

FIGURE 49.3   *Quercus* 'Pondaim' grafted to *Q. robur* using a long tongue side veneer graft. Note the presence of a stimulant bud just beginning to extend at the end of the tongue. This has maintained viability of the tongue, and stimulated callus formation beneath the bud. The graft was not sealed but placed in warm callusing conditions. After removing buds and side shoots from the rootstock stem above the graft, the scion can be tied-in to this, together with any extending growth, to provide early support and training.

---

* Flemer III (1962).
† Kloosterhuis (1975).

ROOTSTOCKS

Pot and Plug Grown Rootstocks

Most agree pot grown rootstocks are likely to produce better takes and significantly better subsequent growth after grafting. Unless seedlings are raised in plug growing systems incorporating air root pruning, field grown, 1-year-old seedlings can be expected to produce a main root 200–250cm long. In the field, the undercutting technique will encourage root branching but the blade must be set to a reasonable depth (±175mm) to avoid a severe check to top growth. Seedlings of 1+0, 1u1 and 1+1 grade will produce root systems significantly longer than can be accommodated by an average rootstock pot. Use of extra deep pots will be required (Figure 49.4) and has been fully discussed in Chapter 9: Pot Sizes and Types.

Pot grown rootstocks may be 1+1P, 1u1+1P or 1+1+1P, 6–10mm. The first comprising top grade 1-year seedlings followed by one year's culture in the pot. 1u1 and 1+1 seedlings may be used if only second grade field grown seedlings are available or if summer grafting is involved with a shorter growing season. Seedlings grown in a small pot take a considerable time to develop into vigorous plants.

The additional juvenility of 1+0 and plug grown seedlings direct sown into deep plugs or pots and grown on under protection makes it likely they will produce the best takes (Figure 49.4). The

FIGURE 49.4 LEFT, *Quercus palustris* rootstocks showing, on the left, use of extra deep pattern pot for a 1+0 field grown seedling; compared with (right), standing on an upturned pot, a plug sown seedling in a 9 × 9 × 9cm pot (grafted with *Q. rysophylla* 'Maja'). RIGHT, *Quercus palustris* rootstock potted-up as a small plug grown seedling into a 0.8 litre 'Long Tom' pattern plastic pot. This will be suitable for grafting the following spring. Plants in the small pots but will grow only slowly for an initial period.

young juvenile stems well furnished with buds resulting from this technique will produce first quality long tongue veneer grafts containing a stimulant bud. These are able to regenerate callus and quickly develop a graft union.

Some species, such as *Q. phellos*, and a number of the Lobatae, transferred from the warmer summers of North America are unlikely in northern Europe to make sufficient size as 1+0 seedlings for immediate potting and must be potted as 1+1, 1u1, or plug grown.

Dependent upon scion material, most rootstocks are required with a diameter of 6–10mm.

BARE-ROOT ROOTSTOCKS

Bare-root rootstocks, 6–10mm diameter, may be top grade 1+0 or 1u1 or 1+1 seedlings. They are convenient for use in hot-pipe systems or grafted in early September and placed in warm callusing conditions. Hot-pipe facilities for oaks must take into account the length of the roots of many species, the 'Economy Design' being particularly well suited for this type of material (see Chapter 16: Hot-Pipe – Economy Design and Figure 16.13).

ROOT GRAFTS

Pieces of root from successful graft combinations previously planted in the field may be obtained when the plants are lifted. This is the best chance of guaranteeing compatibility. Sizes should be a minimum of 10–15mm in diameter and 100mm long. Some suggest that branched roots are best[*].

SCION WOOD

For deciduous species well-ripened, 'robust' scion wood is recommended by one authority. Collection follows the normal pattern and is best in late December to mid-January. Scion wood may be stored packed in just moistened well sealed polythene bags and can be kept in a cold store at just above freezing for up to 3 months or in a domestic refrigerator for 6 weeks[†].

A number prefer older scion wood and inclusion of a basal portion of 2- or even 3-year-wood is recommended[‡]. Personal experience would support the need for this view for some species. However, work in Holland has suggested that 1-year wood from young trees produces better results than scions with a 2-year basal portion taken from older trees[§]. Given good, well-ripened wood, especially from well-managed stock plants, the basal portion of current year wood, 3–5 buds long, is often the best choice. Grafters must use their judgement, bearing in mind that 1-year wood produced in the UK will be less well-ripened than in continental countries.

Scions of deciduous species collected for summer grafting raise the question of removal or retention of leaves; personal experience with side grafts would favour removal.

A trial at the Boskoop Trial Station of *Q. coccinea* 'Splendens' grafted to *Q. palustris* compared de-leaved scions with leafy scions; results were in favour of leafless scions at 97% compared with leafy at 81%[¶]. Early September bare-root grafts of oaks normally involve apical grafts and here some scion leaves must be retained to avoid the adverse effects on takes of total defoliation.

Evergreen oaks are best with some leaf retention. Species such as *Q. lamellosa* taken in the winter or early autumn following summer dormancy, usually involve a scion 125–175mm long, including a basal portion, often of 2-year wood, if possible including the apical bud. Leaves are best limited to 3 or at most 4, each being reduced in area by up to a half or more (Figure 49.5). A small trial of *Q. multinervis* (Cyclobalanopsis) with scion leaves removed and placed in a hot-pipe system at

---

[*] Leiss (1995).
[†] Coggeshall (1996).
[‡] Krüssmann (1964) pp.494–497.
[§] Alkemade & van Elk (1989).
[¶] van Elk (1978).

FIGURE 49.5   LEFT, *Quercus lamellosa* (subgenus Cyclobalanopsis), a scion comprising current season's growth prepared for grafting with some leaves removed, and those remaining reduced. In this example, 1-year wood only is used; a basal portion of 2-year wood may improve results. RIGHT, a 2-year graft of *Q. lamellosa* showing good extension growth. The rootstock *Q. myrsinifolia* has produced a sucker (red leaved shoot), not always a good sign, but the graft has subsequently survived and grown well (see Chapter 12 and section Compatibility for further discussion).

±22°C produced good callus formation but subsequent failure almost certainly due to incompatibility. Further investigation using this technique with other Cyclobalanopsis species is warranted.

Many oak species regularly produce Lammas shoots from mid-July onwards. This material should not be used as scion wood because takes are invariably poor. Occasionally, it may be appropriate to include a piece of this wood, but it is important that the main proportion of the scion should constitute the basal part of the 'early wood'.

SYSTEMS

WARM CALLUSING

Warm callusing using potted rootstocks was the favoured method for deciduous species until the advent of the hot-pipe system. Evergreen oak species must always be grafted using warm callusing unless a suitable means of handling evergreens in hot-pipe systems can be devised.

The system is successful during winter/early spring or late summer/early autumn following the methods described previously (Chapters 13 and 18). Personal experience is that summer grafting takes of *Q. robur* 'Fastigiata Koster' are normally high, but losses over winter can be severe with up to 50 or 60% failing to grow away the following spring. Curiously, the problem seems to be restricted to this cultivar *Q. castaneifolia*, and many others invariably come through unscathed.

One grafter describes a system of potting rootstocks in late January/early February and grafting 3–4 weeks later[*]. It is hard to imagine that such a short period for establishment could replicate the advantages of an established pot grown rootstock; delaying potting until it is known whether the graft has been successful seems a more efficient strategy.

Although it is generally agreed that pot grown rootstocks produce the best results, bare-root rootstocks may be used for grafting the deciduous species in the winter/spring period. Rootstocks

---

[*] Coggeshall (1996).

are plunged into suitable, open, well-drained media. To avoid sealing, the union can be buried, but this is not recommended for oaks; dipping in hot wax or using non-porous enclosing ties and waxing exposed cut surfaces is preferable

The early September bare-root grafting system developed in Holland at the Leinden trial station for tree crops and described for a number of other species, (*Acer, Carpinus, Fagus,* etc.) works well for *Q. coccinea* 'Splendens' and *Q. frainetto.* Methods follow those described previously, the unsealed grafts being placed in poly-tents maintained at high humidity and provided with 18°C bottom heat. Sunlight would almost certainly raise ambient temperatures significantly higher, a positive advantage for oaks. Results were claimed to be as good as those achieved using pot grown rootstocks.

Hot-Pipe

Hot-pipe systems have proved extremely successful for the deciduous species. Pot grown rootstocks provide convenient handling, but when using extra-deep pots fitting them into hot-pipe runs may present problems. The 'Economy' hot-pipe is well-suited for use with bare-root oak grafts (Chapter 16, Figure 16.13).

Personal experience is that grafts require longer in the system than is sometimes recommended. Temperatures of ±22°C for at least 3 weeks, and often 4, are required to produce adequate callus to support graft union development. If cold storage after hot-pipe is planned, a period of 7–10 days' acclimatisation at lower temperatures is advisable before placement in the store at 2°C. Potting-up successful bare-root grafts using appropriate deep pots is a better option.

Root Grafting

Root grafting is rarely used but has potential advantages in tackling the problem of incompatibility. For this reason, it has been championed by grafters in North America[*]. Procedures follow those described in appropriate sections in Chapters 12 and 18. Root grafting takes place in mid-February, the roots collected at lifting time. Most maintain roots should be stored at relatively high temperatures, 20–25°C, for three weeks before grafting takes place. A storage temperature of 12–15°C for 25 days was not so successful[†]. Apical veneer grafts are recommended using scions 3–6 buds long, they are then sealed or laid-in with the graft union buried in well drained media and maintained at 20–24°C for 4–6 weeks, by which time the graft union has formed. Takes are not high (±50%), but rootstock costs are minimal, and for notoriously incompatible combinations, subsequent losses due to delayed incompatibility may be mitigated by choosing only roots from known compatible combinations.

Nurse-Seed Grafting

Large seeds produced by oaks lend themselves to nurse-seed grafting. Generally, the system used falls into the modified procedure described for *Aesculus* in Chapter 23, the acorn being germinated, and once sufficient stem diameter is achieved, a light-grade scion is grafted to the hypocotyl, normally using in inverted wedge graft. This appears more successful than the conventional nurse-seed graft described in Chapter 18. Growth after this procedure is likely to be very slow and consequently has a dubious role in commercial production.

---

[*] Coggeshall (1996), Leiss (1988) and Leiss (1995).
[†] Leiss (1988).

PHYSIOLOGICAL AND ENVIRONMENTAL FACTORS

Root pressure does not seem to be a significant problem with oaks and, with the exception of the Cyclobalanopsis species, no mention of adverse effects on grafting takes has been discovered or experienced. The usual rule of 'on the dry side' for methods involving warm callusing in the dormant period is safest policy. Anatomical studies have indicated that the Cyclobalanopsis subgenus may lack tyloses in the wood and are therefore potentially most likely to show evidence of bleeding. Personal experience would support the need for a level of drying-off for this group before grafting.

Response to temperature and callus formation is most active in the range 22–24°C. At lower temperatures, callus formation is slower and union formation may be delayed. Evergreen species in subgenus Cyclobalanopsis are better with basal temperatures at 18–20°C, ambient temperatures will often be higher; in the summer bottom heat should not be set above 16°C to reduce the possibility of excessive daytime temperatures and consequent loss of humidity.

PERFORMANCE COMPARISONS

The Robur section grafts most reliably, with plentiful callus production and subsequent graft union development, and much the same can be said for those in Cerris. Lobatae are less satisfactory and union formation can be delayed. Evergreen species, particularly those in the Cyclobalanopsis section, are the most demanding, with less callus production evident than with others. The final consideration must always be whether or not the chosen combination proves to be compatible.

GROWING ON

Growth is promoted by high temperatures and good nutritional levels, especially nitrogen. Good soil moisture levels permit best nutrient uptake. In cool, maritime climates such as the UK, oaks respond to higher ambient temperatures provided by protected structures. The Lobatae and Cyclobalanopsis thrive best in soils with a pH below 6.5.

Trials work in Holland has demonstrated that the growth of *Q. robur* 'Fastigiata' was significantly enhanced by two applications of gibberellic acid (GA4+7) made to the plants in May*.

---

* Boomteeltproeftuin Midden-Nederland (Lienden) (1985).

50 Rhododendron (Ericaceae)

INTRODUCTION

Horticulturally, the Rhododendrons are divided into what might be regarded as 'true' Rhododendrons (subgenus Hymenanthes and Rhododendron) and the Azaleas (subgenus Pentanthera and Tsutsusi). Apart from a few specialist producers, most Azaleas are now produced by cuttings. Techniques of grafting are somewhat different to those in subgenus Hymenanthes and Rhododendron, which are the subject of this chapter. For information on grafting Azaleas, see Part Eight.

Rhododendrons must be considered as one of the most important groups of ornamental woody plants. They range in size from low prostrate shrubs with tiny leaves to trees up to 25 metres high; some have huge leaves and enormous flower trusses. The genus is very large with possibly over 1000 species and, including the Azaleas, 28,000 registered hybrids, with many more unregistered. They occur in the northern and southern hemispheres in regions from tundra-like mountaintops to tropical rainforest.

In many temperate areas Rhododendrons are a major nursery crop. Plants hybridise freely and clonal methods of propagation are essential. Propagation by cuttings must be considered the most important. Wild collected and self-pollinated seed is significant for the species and to a lesser extent for the production of rootstocks. Since the mid-1970s, micropropagation has assumed an increasingly important role. Grafting was a major method of propagation before the introduction of mist systems and widespread use of growth substances to promote root initiation. The value of rootstocks tolerant of less than ideal soils has been a compelling reason to retain grafting as an important technique for Rhododendron production.

COMPATIBILITY

Because of the size and significance of the genus, taxonomic and, more recently, phylogenic investigations into Rhododendron taxonomy has been extensive, and is still ongoing, with constantly changing interpretations and revisions. From the grafter's perspective, two major divisions are of importance: these are the subgenus Rhododendron containing the so called scaly or lepidote types, mostly propagated by cuttings, and the subgenus Hymenanthes, those without scales (elipidote), constituting the vast majority which may be grafted.

Grafts between subgenus Rhododendron and Hymenanthes fail, usually immediately, but sometimes grafts linger for a while. Attempts at grafting *R. excellens* (subgenus Rhododendron) to *R. calophytum* (subgenus Hymenanthes) initially appeared successful. For three growing seasons, subsequent growth of the scion variety was slow but positive. In the fourth, flower bud production was prolific and it was apparent that the union was tenuous; the scion wood failing to unite properly with the rootstock and seen to be breaking away. It seems likely that after five to six years the graft will collapse.

The subgenus Hymenanthes contains only one section (Ponticum) but numerous subsections, and it is between these that incompatibility can and does appear to occur. The popular rootstocks *R. ponticum*, *R.* 'Cunningham's White', *R. catawbiense* 'Bourshalt', and others less commonly seen, are in subsection Pontica. The new hybrid group INKARO®-containing rootstock selections, some also with ornamental potential, is a complex of species and hybrids; subsections Pontica and Fortunea are mainly involved.

There can be little doubt that subsections Falconera and Grandia, containing the so-called 'big leaved or large leaved' species, when grafted to other subsections, show incompatibility symptoms and eventual failure. Rootstocks of *R. calophytum* in subsection Fortunea have been tested for suitability, using as the scion variety *R. magnificum* (subsection Grandia). Grafts may achieve a union, but subsequent growth is very depressed, and leaves are less than full size. Doubtless profuse bud formation will occur before eventual failure (Figure 50.1). However, with this combination, scions do remain alive for some time (at least two seasons) and are able to be re-grafted when a suitable rootstock becomes available. Attempts to graft *R. macabeanum* to rootstocks of *R. ponticum* (subsection Pontica) resulted in death of the scion after a few seasons*.

A possible link between the large leaved types, in this case subsection Falconera, and the normal leaved subsections Arborea and Pontica, respectively, is found in the hybrids *Rhododendron* 'Colonel Rogers' and *Rhododendron* 'Great Dane'. These hybrids, respectively *falconeri x niveum*, and *yakushimanum x rex*, form a compatible and long-lasting union with *R*. 'Cunningham's White' rootstock. It is possible that these, and possibly others with a similar parentage, would provide an inter-stem between the usual rootstock of *R*. 'Cunningham's White', *R. ponticum* or *R*.INKARO®, and scions from the large leaved Falconera or Grandia subsections. The suggestion has not been tested but if successful (a likely outcome) would avoid the need to raise the large leaved group as rootstocks.

Species provide the best guidelines for identifying any possible examples of incompatibility between subsections. Hybrids with a complex of parents are difficult to link with a pattern in this respect and must rely on individual examples based on past results. Experience of grafts of a wide range of species and varieties, combined with the major rootstocks, has produced very few verified examples of incompatibility. Some report difficulties in achieving good results with *R. lacteum* in the Taliensia subsection[†], a result supported by personal experience. Others in Taliensia include *R. faberi*, which, after initially forming an apparently successful union, has failed to produce extension growth and is alive only because of the use of side grafts. This possible pattern of incompatibility

FIGURE 50.1   *Rhododendron magnificum* (subsection Grandia). Left, grafted to *R. calophytum* (subsection Fortunea). Right, grafted to *R. sinogrande* (subsection Grandia). Both grafts after two season's growth. Note significantly more extension growth on *R. sinogrande* rootstock.

---

* M. Foster (pers. comm.).
† Miller (1979).

is negated by positive results for others in Taliensia: *R. bureauvii, roxieanum, traillianum* and *wiltonii* all successfully grafted to *R.* 'Cunningham's White'.

The Glischra subsection may also contain species which are incompatible with Pontica. *Rhododendron glischrum*, after forming an apparently successful union with 'Cunningham's White', has consistently failed to survive beyond two seasons. *Rhododendron crinigerum* and *recurvoides* in the same subsection have not been tested. Some subsections represent more of a challenge than others; Barbata containing *R. barbatum* and *R. exasperatum* (aptly named from the grafter's viewpoint) have consistently failed to produce very reliable results on 'Cunningham's White'.

Subsection Neriiflora contains several species, *beanianum, mallotum, parmulatum* and *sperabile*, which always presents a challenge. This difficulty is probably not due to incompatibility, as some successful grafts are normally achieved, even when only rather poor scion material is obtainable.

Little information is available on possible incompatibility problems within subgenus Rhododendron because they are so rarely grafted. Species from subsections Maddenia, Boothia, Lapponica and Edgworthia may be grafted to whatever is available within the subgenus. Results are usually rather poor, but this is probably more due to the difficulty of providing suitable environmental conditions than factors of incompatibility. The foliage and even stems of this group are liable to damping-off in moist conditions, but if this is not provided, desiccation is probable. They test to the limit the grafter's skill in controlling enclosed environments. Dry fog with a fairly low RH setting (90%?) may provide the best chance of success but this suggestion has not been tested.

TIMING

Rhododendrons may be grafted over a long period, from late July until early May. Most favour more conventional timing in December/early January until mid-February, but cuttings-grafts, gaining in importance for *Rhododendron* production, are normally carried out from late August/September to November/early December. One amateur enthusiast recommended the period of June 20 until July 10 as the best for what are described as 'Green Grafts'*.

Grafting to pot grown or bare-root rootstocks is best from early January to late February, when accurate temperature may be achieved largely by controlled heat sources. The danger of excessive temperatures caused by sunlight is reduced and the need for heavy shading and unwanted ventilation is minimised

GRAFTING METHODS

Despite side grafting being considered for most evergreens, methods for Rhododendrons have mainly focused on apical grafts. The form of apical inlay known as saddle graft is still used by a few traditional propagators, but has mainly given way to the long or short tongue veneer or wedge, all easier and quicker to execute. Some grafters, particularly in Germany, favour a simple splice graft.

For large scale production, because of its convenience and speed, there is a strong case to be made for an apical graft when using top quality propagation material and varieties relatively amenable to grafting. However, there can be little doubt that the use of side grafts will result in substantially higher takes for the more demanding species and cultivars. Production of Rhododendrons by cuttings-grafts, now popular in Germany, involves only the use of side grafts.

The long tongue side veneer including a stimulant bud, where possible with an accompanying trimmed leaf, provides the best choice of grafting method for *Rhododendron*. It affords opportunity for maximum cambial contact and copious callus production, in part induced by the stimulant bud. These features together aid the formation of a strong graft union, resistant to physical damage in subsequent handling. It also maintains active root growth and a nutrient level within the young graft until it is able to fully re-establish itself (Figure 50.2 & 50.3).

---

* Leach (1960).

FIGURE 50.2   LEFT, *Rhododendron* 'Nimbus'. RIGHT, *Rhododendron beanianum*. In both cases, note the use of long tongue side veneer, retaining both stimulant bud and leaf. The stem of the rootstock of both grafts is still at the green stage, with no evident development of a corky periderm. Note the tongue of the veneer graft is viable to the very tip.

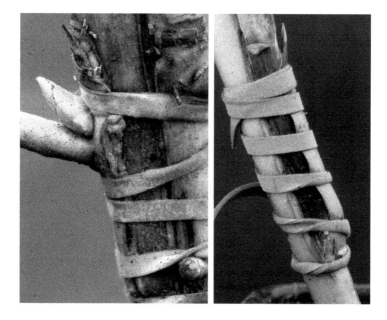

FIGURE 50.3   The influence of stimulant bud on the graft. LEFT, *Rhododendron traillianum* showing extensive callus development and viable tongue, virtually to the tip, supported by the presence of a stimulant bud and leaf. RIGHT, two aspects of *Rhododendron* grafting are illustrated: (i) the scion is of *R.* 'Naomi Hope', characterised by wood which quickly develops a dark brown corky periderm, giving the impression of dead tissue, but is totally viable beneath; and (ii) no stimulant bud on the tongue, resulting in noticeably less callus development at the distal end. Copious callus formation is seen at the base of the tongue, resulting in a successful graft.

Tying-in is best done using non-enclosing rubber ties left unsealed.

Foliage retention on the rootstock, by means of side grafting until the scion variety is able to make regrowth, is essential for the recovery of species and cultivars in poor health, often suffering from some disease such as Rhododendron mildew (*Erysiphe (Microsphaera) azalea*). Personal experience of grafting many demanding species has demonstrated beyond doubt the value of side grafting. Sometimes all foliage on the scion is lost within a relatively short period after the grafts are placed in high humidity and warm ambient temperatures. Despite this, the rootstock is often able to sustain the scion to allow union formation and subsequent extension growth (Figure 50.4).

ROOTSTOCKS

ROOTSTOCK SELECTION

Species in subgenus Hymenanthes were traditionally grafted to seedling *R. ponticum*, with a specification of 2+1P 6–8mm. The exception was for large leaved species in subsection Falconeri and Grandia. Improved clones of *R. ponticum* are now produced from cuttings, but general agreement is that *R. ponticum* is not ideal for three main reasons:

- It is susceptible to root diseases particularly *Phytophthora sp.*
- It freely produces 'true' suckers with shoots arising from the root system, and also basal shoots from the original rootstock stem.
- The stems, when quite young, quickly produce a 'woody' periderm which becomes suberized and forms the typical bright brown bark protective layer of dead tissue*. This has a negative effect on graft takes especially when the long tongue side veneer is used.

FIGURE 50.4   Side grafted scions which became defoliated after grafting, maintained by the rootstock of the side graft. LEFT, *Rhododendron subansiriense,* which lost all its leaves 10 days after placement in the poly-tent. The apical bud was retained, because of concern over the health of the foliage, and is seen to be commencing regrowth, made possible by sustenance from the rootstock. RIGHT, *Rhododendron* 'Michael Hall', foliage was badly infected with Rhododendron mildew, and the leaves dehisced a few days after grafting. The terminal was a flower bud that could not be retained. Fortunately, axillary buds were present, and can be seen to be starting regrowth. Both grafts, and others in the batch similarly infected, were successfully established.

---

* Humphrey & Dummer (1967).

Choice of suitable rootstocks has now largely transferred from *R. ponticum* to forms of *R. caucasicum*, most particularly a hybrid between this species and a white form of *R. ponticum*: R. 'Cunningham's White'. Another *caucasicum* hybrid, occasionally recommended is 'Christmas Cheer'*. Forms and hybrids of an American species *R. catawbiense*, such as 'Bourshault', 'Grandiflorum', and 'Roseum Elegans' are favoured by some; other varieties are suggested by individual grafters but are not in general use. Of the rootstocks used by far the most popular is R. 'Cunningham's White'[†]. Reasons for this relate to:

- Ease of rooting from cuttings.
- Almost total freedom from suckering.
- Tolerance of poor soil conditions, even a relatively high pH up to neutral.
- Retention of young, green wood for longer than many alternatives.

A new hybrid race, the INKARO® Rhododendrons, has been developed in Germany as a result of a cooperative effort between growers and a German government trial station. For a full account of these see an article by H. Shepker in the Rhododendron, Camellia and Magnolia RHS Group Year Book for 2016[‡]. These plants, originally intended as rootstocks for tolerance to a high pH, have been patented and are on restricted distribution to licenced growers. A broad view of their performance is not yet available. Some clones will also have potential as ornamental plants and this aspect appears to be under development by the group. To date, of the various clones tested, No. 37 seems to be performing best in the US where it is known as 'Lakeview Pink'[§]; it has also performed well in Europe[¶].

When planted in soil with a pH close to neutral, foliage colour of INKARO® is not as dark green as 'Cunningham's White'. Neither variety has received supplementary feed, but it is understood that INKARO® requires good levels of nutrition. Trials at RHS grounds Wisley showed no improved tolerance to elevated pH levels of INKARO® compared with 'Cunningham's White'. Rooting success from cuttings is less good than for 'Cunningham's White' but it is possible that higher levels of IBA are required to show its full potential in this respect. A favourable aspect is retention of young, green wood for a significant period of time and the stems are vigorous and straight, making them ideal for producing a long tongue side veneer graft.

Rootstocks for Large Leaved Types

No clonal rootstocks for the species and hybrids in Falconera and Grandia exist, and therefore, seedlings from any species within these subsections are potential rootstocks. Choice is often governed by what is available, but where possible the hardier species should be chosen to widen the scope for planting in areas where climatic conditions are not ideal: *R. macabeanum*, *R. rex*, *R. falconeri* and *R. sinofalconeri* are among the most suitable. Seedlings are potted-up from the seed trays into small 7×7cm pots and grown on for a year before being potted-on into a rootstock pot. In view of the leaf size and bulky nature of the plants, a 1 litre deep pot is most suitable. Every attempt should be made to obtain rootstocks with some weight to match the heavy scion wood produced by species and hybrids within this group. It is advantageous if, at the same time, the ideal young stem with an absence of corky periderm can be retained. The next growth flush after potting-on from 7cm to 1 litre often provides best material to meet this requirement. Experience is that at this stage the plants should be discouraged from further extension growth by removal of the terminal bud. Plants which

---

* Peek (1986).
[†] Humphrey & Dummer (1967), Eichelser (1967),Peek (1986), McGuire, Johnson & Dawson (1987), Miller (1979), Millais (1996) and Megre, Kondratovis & Dokane (2007).
[‡] Schepker (2016).
[§] Decker (2013).
[¶] Schepker (2016).

develop less rapidly can be left untouched (Figure 50.5). Left for a further season, the stems inevitably develop an undesirable bark covering and will be less than ideal when grafted.

ROOTSTOCK STEM

The significant advantage of active green wood on the stem of *Rhododendron* rootstocks at the point of graft placement cannot be over emphasised. Speedy development of callus tissue and absence of any hint of necrotic patches on rootstock or scion tissues are characteristics of grafts placed on young green stems, compared with those which have commenced stages of bark development (Figure 50.6). The situation worsens when, older 'heavy' rootstocks with quite a thick covering of bark (periderm) are used, as is sometimes recommended for saddle grafts. However, it would not be correct to always suggest that only young stems are suitable. Stems which have only recently developed a brown periderm, particularly if they are well furnished with lateral buds, even better if leaves are present at the point of grafting, can produce good results especially with the more easily grafted types (see Figure 50.9).

An alternative method of providing suitably active wood has been described by Leach*. Potted rootstocks, normally *R. ponticum*, are cut down to within 10mm of the compost level in early spring and resultant sucker shoots are reduced to one only. By late June/early July it is suggested growth has developed sufficiently to permit grafting using an apical wedge graft. The method has been described as 'Green Grafting'.

There seems little point in increasing the likelihood of loss by grafting as early as is suggested. Delay until August/early September permits further stem development and use of a more suitable grafting technique such as apical long tongue veneer or better long tongue side veneer. The stem is unlikely to have developed a noticeable degree of corky periderm by the following winter/spring, and it should be possible to graft the selected shoot using normal procedures described previously. The technique is rather time consuming and likely to predispose *R. ponticum* rootstocks to quite rapid further sucker development. How other rootstocks such as 'Cunningham's White' or INKARO® will respond to the treatment is not certain. The technique does, however, offer a potential means of returning rootstocks, deemed too old and woody, back to a much more suitable condition.

FIGURE 50.5 Rootstocks for subsections Falconeri and Grandia. LEFT, *Rhododendron sinofalconeri*, showing excellent extension growth after potting-on into 1-litre deep pot. It was considered that growth was optimum, and the terminal bud was removed to prevent further extension. RIGHT, *R. sinofalconeri*, a slower growing seedling. Removal of the apical bud was not considered necessary at the pre-grafting stage. Note the prominent axillary buds at the node between previous and current growth extension. These will aid callus development at the base of the tongue of the long tongue side veneer graft which is to be used. Neither rootstock has developed corky periderm at the site of proposed graft placement, above the bottom whirl of leaves.

* Leach (1961) pp.340–345.

FIGURE 50.6   *Rhododendron* 'Cunningham's White' rootstocks produced from cuttings in plugs, and potted into 9 × 9 × 12.5cm pots. LEFT, after one growing season in the pot, an 'ideal', excellent, green stem for grafting at first or second node is produced. RIGHT, held for a further year the stem has developed a corky periderm, and is less suitable. Grafts placed on an older rootstock stem, carrying plentiful lateral (stimulant) buds, or a leaf with an axillary bud, will still usually produce satisfactory results. See *Rhododendron* 'Loderi Venus' in Figure 50.9 as an example.

POT GROWN VERSUS BARE-ROOT

Personal experience is that pot grown rootstocks are more convenient and often produce marginally better results than those which are bare-root, the latter comprising a small unwrapped root-ball as lifted from the field or open-air production beds. Work in Boskoop has not shown any advantage in using potted rootstocks over bare-root; however, scions of relatively non-demanding Hardy Hybrid varieties were being investigated*.

RAISING CLONAL ROOTSTOCKS

Clonal rootstocks are best raised from cuttings inserted into 75mm deep plug trays using a rooting medium of 50–75% fine pine bark and 25–50% peat or 25–33% perlite to peat. The higher bark or perlite content is best for mist systems producing high precipitation in the August early September period; the higher peat content is better for enclosed mist systems, dry fog or contact polythene, especially if used later in the season when drying-out is a possibility. Cuttings should be prepared

---

* van Elk (1967).

from current year's growth to provide an overall length of approxinately 125mm, including the basal and terminal portion of the stem, plus the additional length of the leaves. Cutting size will normally influence the finished size of the rootstock; small cuttings provide shorter plants with thinner stems, sometimes an advantage for species such as *R. roxieanum* or hybrids such as *R.* 'Moonstone'. Cuttings made from flowering shoots will often produce branched rootstocks useful for multi-grafted Rhododendrons (Figure 50.7). Some growers using *R. ponticum* recommend removal of all lateral buds to prevent suckering*; this is unlikely to provide a long-term solution and has not found to be necessary with 'Cunningham's White' or INKARO®.

Growth substance treatment and bottom heat are essential to induce adventitious root production. If cuttings are wounded, 8–10,000ppm IBA in talc has proved satisfactory. Non-wounded cuttings root well if treated with 2,500ppm IBA applied as a 50% alcohol and water, quick dip. INKARO® cuttings will require a higher concentration of 5–10,000ppm IBA quick dip, or wounding plus 20,000ppm IBA (2%) formulated as a dust in talc. Bottom heat is best set at 20–21°C. In most seasons a high proportion of cuttings of 'Cunningham's White' inserted in early/mid-August will be well-rooted by early January, INKARO® may take longer.

SPECIFICATIONS

Specifications for seedling *Rhododendron* rootstocks are 2+1P 6–8mm; large leaved species require 8–12mm. Plug production of seedlings under protection could potentially produce 1+1P 6mm with a much higher proportion of green stems. For cuttings-raised rootstocks 0+1P 6–8mm is the norm.

Rootstocks for side grafting should not be excessively large; 30cm is an absolute maximum height, 20cm being preferable. To avoid overcrowding at grafting time in the enclosed conditions of the poly-tent, overlarge rootstocks will need heading-back and some leaf removal, especially around the scion placement area.

SCION WOOD

For grafts made in the winter/spring period, scion material is normally collected as required and used shortly after collection. Cold storage at −2 to +1°C permits Hardy Hybrid scions to be stored for up to 2 months[†]. More tender types can also be stored for a few weeks at 1 to 2°C. In the late

FIGURE 50.7   *Rhododendron* 'Cunningham's White' rootstocks. LEFT, influence of cutting size on final plant size after one growing season in the rootstock pot. On the left, normal size cutting; on the right, small cutting. RIGHT, plant produced in the same timescale from a cutting selected from a flowering shoot, showing typical multi-stem characteristics.

---

* Peek (1986).
[†] van Elk (1967).

summer/early autumn period grafts and cuttings-graft material will be satisfactory at temperatures of 2 to 5°C for up to 10 days. In all cases, leafy material must be packed in slightly moistened, tightly sealed, polythene bags.

One-year old wood is always the preferred choice for *Rhododendron* scions. When material is taken from old or ailing mother plants, a basal portion of 2-year-old wood may be unavoidable (Figure 50.8). This often results in lower takes and slower union formation, reflecting the effect discussed previously of using rootstocks with older 'woody' stems. Ideally, the scion should include the apex of the shoot with or without the apical bud, and where possible, the basal portion where it joins the previous year's wood. For vigorous species and varieties this may mean avoiding strong leading shoots which are overlarge. Best material is often provided by well-balanced laterals, smaller in stem diameter than the leading shoots but of sufficient girth to provide a stem ±6–8mm in diameter and 125–150mm long.

Where it is impossible to avoid over-long scions, it is preferable to retain the basal portion of the shoot and cut to length to a well-developed lateral bud. Cutting higher above the base risks using insufficiently ripened wood and, if possible, should be avoided. Sometimes viable buds exist only at the terminal portion of the shoot and use of the upper portion with less than ideal scion material is necessary. Removal of the vegetative apical bud is recommended for vigorous scion material with plentiful healthy lateral bud development (Figure 50.8). This immediately encourages a well-branched habit and has the effect of delaying extension growth, providing a longer period for union development before the requirement to support soft new growth. The overall quality of the finished plant is improved, particularly for varieties which tend to produce a straggly growth habit.

After trimming to length, scions require reduction of the leaf area to prevent excessive moisture loss and avoid overcrowding in the poly-tent enclosures. This is normally achieved by removal of all but three or four of the leaves and reduction of remaining leaf area by one-third to one-half (Figure 50.9).

No comparison between the influence on results of flowering and non-flowering shoots is available but it is essential that flower buds are removed. If this is not done, flower development in the enclosed conditions of the warm callusing system normally results in fungal infection.

FIGURE 50.8   LEFT, use of a basal portion of 2-years-old wood was unavoidable for this large Grandia hybrid 'Sandringham'. RIGHT, removing the apical bud from the scion will improve branching of the hybrid *Rhododendron* 'Snow Queen'.

FIGURE 50.9 LEFT, *Rhododendron* 'Loderi Venus' long tongue veneer side grafted to 'Cunningham's White' rootstock. The scion has been reduced to three leaves which have each been reduced in size by about one-third to one-half. The rootstock stem was overly mature but the presence of numerous lateral buds (seen growing out) has promoted good callus production. RIGHT, prepared *Rhododendron* 'Fantastica' scion, comprising four leaves reduced in area by one-third to one-half. Note basal portion of the scion shoot is included, and in this case the terminal bud was not removed.

Cuttings-grafts require scion material following the same pattern as conventional grafts. Because of size limitations of the cuttings-graft rootstock, there may be a need to select lighter grade shoots, but soft unripe material must be avoided. Where material is in short supply or to quickly bulk-up numbers of a new variety, the use of chip bud scions has been investigated*. Comparisons between chip buds and conventional scions showed no difference in takes between the two types of scion wood. In these investigations, the chip bud had the attached leaf retained and completed cuttings-buds were placed under misting jets. Results for both conventional cuttings-grafts and cuttings-buds were poor but improved techniques might change this. Time taken to produce a saleable plant from a chip bud is likely to be substantially longer than for a conventional scion.

The large leaved species and hybrids in Falconera and Grandia present a particular challenge with regard to size of scion wood. Mother plants should be disbudded and pruned, encouraging the production of numerous shoots substantially thinner and shorter than normal (Figure 50.10). This will ensure a good match with the ideal rootstocks having stems at an optimal stage for grafting (see Figure 50.5 previously). Except in favoured areas, such as the extreme southwest, scion mother plants are best housed under protection during the winter. To ensure the wood is sufficiently ripened, they must be moved into the open-air with only light shading during the summer months.

SYSTEMS

Warm callusing is the main system used for Rhododendrons. It normally takes the form of conventional enclosed conditions using poly-tents; a few grafters, particularly for summer grafting, favour supported warm callusing systems using overhead mist lines or, preferably, dry fog. Some recommend burying the union, stressing that the medium covering the grafts must be well drained and not over-watered†.

Poly-tents should be drip-free and supported well above the grafts. Horticultural fleece within and over the poly-tent should provide sufficient shade from December until mid-/late February,

---

* McGuire, Johnson & Dawson (1987).
† Miller (1979).

FIGURE 50.10 *Rhododendron macabeanum* KW 7724 grafted from the original collection by Kingdon Ward, now growing at Trewithen estate garden, UK. Resultant stock mother plant grown under protection during the winter, with apical buds removed to encourage a cluster of lighter grade shoots more suited as scions for young rootstocks. To ensure the wood is thoroughly ripened, mother plants are best grown in the open-air during the summer.

after which time shading levels should follow those suggested in Chapter 13. For the first two weeks, ventilation is not usually required on a regular basis. Inspection every two or three days will indicate if more ventilation is necessary to lower excessive moisture. Once plentiful callus formation is evident, daily ventilation is required, being gradually extended from half an hour as experience dictates.

Union formation is very dependent upon temperature. Poor quality scion wood at temperatures in the 12–15°C range may take up to two months, occasionally more, to form a union. Good quality scions of relatively easily grafted types, at temperatures of 18°C, may become united in five weeks or less.

Management of Rootstock – Heading-back

Management of the rootstock portion of side grafts will be determined by the ease or difficulty of the species being grafted and the quality of the scion wood. Healthy, well-united grafts may be partially or completely headed back at the first handling during removal from the poly-tents. At the same time, any rootstock extension growth and breaking buds should be pinched out.

Weak, poor-quality scions require additional support and, unless there is danger of excessive crowding, only rootstock extension growth and breaking buds should be removed. Careful monitoring is required to maintain supportive rootstock growth until scion viability reaches a stage where it is considered the graft is sufficiently established to allow partial rootstock removal. When scion growth is well established, complete heading-back can take place, ideally at a convenient stage such as potting-on or planting out (Figure 50.11).

Summer Grafting

Summer grafting during August or September is possible; procedures follow those outlined for the winter/spring period. Plant material involved may well be in an excellent state for ensuring good

FIGURE 50.11   *Rhododendron* 'Golden Torch', successful graft just headed back.

takes and tender species will not have sustained winter damage. More shade will be required to control potential excessive temperatures and maintain high humidity levels. At this period, the need for bottom heat is minimal, consequently there is no danger of root damage as a result of excessive compost temperature.

CUTTINGS-GRAFTS

A number of genera have been suggested for propagation by leafy cuttings-grafts; of these, Rhododendrons have achieved the most success and are produced in the largest quantity. One company in Germany is reputed to provide one million plants per year by this method*.

As discussed in Chapters 8 and 18, the rootstock has no root system and is reliant on this being produced more or less simultaneously with formation of the graft union. Megre and others[†], following a close study of the developmental phases, reported that wound vascular cambium (cambial strand) was formed by day 30, and root initials simultaneously formed in the phloem rays.

---

* Adlam (2007).
† Megre, Kondratovičs & Dokāne (2007).

'Cunningham's White' appears to be almost exclusively used as the rootstock, probably reflecting the reliable rooting ability of this variety*. Cuttings are selected and prepared as described for conventional rootstocks; they are then grafted with the chosen scion variety; the base of the scion being placed 35–45mm above the base of the cutting. A long tongue side veneer is used, including, if possible, a stimulant bud. Tying-in is achieved with a non-enclosing rubber tie which will degrade after some months (Figure 50.12).

When the graft is completed, cutting bases are treated with appropriate growth substance (rooting hormone) as discussed for conventional cuttings. Indolyl butyric acid (IBA) is invariably the choice for this purpose. Quick dip formulations comprising IBA dissolved in 50% alcohol and water are to be preferred, as wounding is then unnecessary and only the basal 5–7mm requires treatment.

After preparation and treatment, the cuttings-grafts are placed in suitable receptacles, ideally plugs but trays are an alternative. These are filled with a cutting's rooting medium and the cuttings-graft inserted so that the union remains just above the surface. Provision of bottom heat at 18–20°C is important. After cuttings' insertion, filled receptacles are placed in low poly-tents or under contact poly. Both should be lined with one or preferably two layers of pre-moistened fleece to reduce drips. Some suggest placement under mist (Figure 50.13). With this arrangement there is always the danger of damage due to water droplets running down the stem and into the interface of the graft. To avoid this, grafts will require sealing, at least at the top of the rootstock/scion interface. Others suggest the graft union should be buried; however, the risk of rotting-off is increased. Cuttings-grafts made in late August/early September are usually united and rooted by January/February and may be lifted when convenient for potting-up or bedding out in beds under protection. They must be kept frost free at this stage.

FIGURE 50.12   Cuttings-graft of *R.* 'Halfdan Lem' to *R.* 'Cunningham's White'. The basal portion should be wounded if powder formulation root hormones are applied. Quick dip and solution formulations can be used without a need for wounding.

---

* Eichelser (1967), Peek (1986) and McGuire, Johnson & Dawson (1987).

FIGURE 50.13   A trial batch of cuttings-grafts on a grower's nursery. These have been inserted into plugs in peat rooting medium under an enclosed mist system. With this arrangement, unless the graft is sealed, water is likely to make some ingress into the union (Millais Nursery).

Good results of 75–80% can be expected from cuttings-grafts (Figure 50.14).

MODIFIED CUTTINGS-GRAFTS

To avoid the possibility of lost grafts through a failure of the rootstock to root, a modified method can be used, really converting the system into conventional grafting at an earlier stage than normal.

Cuttings may be rooted as described for rootstock production. Once rooted in January/early February, they should be graded according to how well they have rooted and only the best grafted; the others being returned for potting-up and grafting as normal the following season. Rooting individual cuttings in plugs is particularly advantageous as the stem and root system is presented as a 'package', lending itself to convenient handling (Figure 50.15).

FIGURE 50.14   *Rhododendron yakushimanum* 'Koichiro Wada', produced from a cuttings-graft, one season after potting-on.

FIGURE 50.15   Modified cuttings-grafts; *Rhododendron* 'Cunningham's White' cuttings. LEFT, in plug trays. RIGHT, rooted, removed and the best grade are waiting to be grafted.

After completion, grafts are carefully bedded into a very free-draining, open medium such as composted bracken peat or fibrous peat/perlite mixtures and placed in a poly-tent to permit formation of the union (Figure 50.16). The excellent quality of the rootstock stem usually ensures very high takes, even for some of the more demanding species. Once the graft has united, usually by March/April, it is best potted-up, and subsequently, after root growth and aerial extension, the rootstock may be headed-down to the union.

PHYSIOLOGICAL AND ENVIRONMENTAL FACTORS

Root pressure does not seem to be an issue with Rhododendrons, and, although it is not advisable to have the pot or root-ball very wet, it is best to avoid any danger of drying-out.

Light levels can be low; therefore, heavy shading can be used to reduce damaging high temperatures. Rhododendrons perform best in moderate temperatures not below 10°C or above 21°C.

Having evolved to grow in light, open structured soils Rhododendron roots have a high oxygen requirement and they respond very badly to poor drainage. This may account for their sensitivity when subjected to high substrate temperature, often leading to root infection, or in worst cases to Phytophthora and eventual root death.

Consequently, when bottom heat is used, particularly at reasonably high settings of ±18°C, there is a case for standing rootstock pots on an elevated base (upturned shallow trays are convenient) to avoid direct contact with the warm bed surface (Figure 50.17).

FIGURE 50.16   Modified cuttings-grafts. LEFT, *Rhododendron yakushimanum* successfully grafted and potted-off, waiting for sucker removal and heading-back. Note, in centre of photo, large extension growth on stimulant bud. RIGHT, *Rhododendron roxieanum* var *oreonastes* successfully grafted and headed-back before potting-off.

FIGURE 50.17   *Rhododendron* grafts standing on upturned Boskoop trays to raise the pot root-balls above the bottom heated floor of the poly-tent. Shallower trays could be used with equal success (Penwood Nursery).

To avoid this potential problem, some grafters in Germany advocate the use of only very low basal heat settings (2–3°C), which in their view should take priority over optimum temperature for callus formation.

In the winter/early spring, choice of appropriate ambient temperature should be influenced by the type of material being grafted. Given a highly humid environment coupled with ambient temperatures above ±15°C, non-vigorous, old, possibly even somewhat diseased scion wood will quickly develop necrotic areas which spread to the union causing loss of grafts. With this type of material, it is necessary to set bottom heat in the 8–10°C range. Coupled with this is a need for careful ventilation to control temperature rise and ensure that drips and excess moisture are kept to a minimum. Callus formation is slow under these conditions and prolonged, careful management is required to achieve success.

For healthy, vigorous material, optimum temperature for callus formation and union development is in the 18–20°C range; the only proviso for this is that consideration should be given to avoid excessive root temperature by elevating the pots or root-balls above the warm base. Many grafters stand the pots directly on the heated base with no apparent problem. However, it is noticeable that in such situations, if the pot root-ball is knocked out and examined, roots at the base of the pot are often dead.

With some grafts, particularly the species, bud extension from the apical bud and/or lateral buds after grafting is much delayed or absent; a similar response to that noted for *Aralia* in Chapter 25. The same procedure has been suggested, i.e. to treat the buds with gibberellic acid (GA 4+7)*. Unfortunately, treatment with GA at 500ppm on a range of species has failed to achieve bud break; possibly, a higher dose is required. In most (except *R. faberi*), bud break occurred naturally in the

* Cox (1979).

spring of the following year. This would seem to indicate that to avoid the problem, a cold period is necessary. But the response is very inconsistent; for a number of individuals in the same batch, extension growth the first season after grafting can occur normally.

TOP-WORKED PLANTS

There is limited demand for top-worked plants on 80–120cm high stems. 'Cunningham's White' or a vigorous INKARO® selection may be trained to a stem for grafting at appropriate height. In the case of 'Cunningham's White', this is a slow process, taking more than one season and adding to costs.

Where hardiness is of less concern, a good choice for stem production is *R*. 'Anna Rose Whitney'. This is a very vigorous cultivar capable of making a straight stem in two seasons with stem girth sufficient to enable two scions to be grafted at the desired height of 80–120cm. Grafting usually takes place in high poly-tents; alternatively, although less satisfactory, poly bags can be fitted over individual grafts. Takes are usually very high as might be expected from vigorous rootstocks and healthy scion wood grafted at height. Scion varieties are best chosen for an ability to make a reasonably compact symmetrical head, some of the more vigorous but shapely *R. yakushimanum* hybrids such as 'Kalinka' or 'Rendezvous' are ideal.

GROWING ON

Container culture of Rhododendrons is an important aspect of production especially of smaller younger plants destined for Garden Centre outlets. Correct compost, good drainage, irrigation and shade levels are crucial for success. Budded plants are always in demand and stressing the plants with applications of phosphate in various forms is a means of encouraging early flower bud formation.

51 Sorbus (Rosaceae) – Mountain Ash, Whitebeam

INTRODUCTION

A large genus found in northern temperate regions, in some instances as a major constituent of high-altitude forests, where they occur as smallish compact trees. Horticulturally important, they are grown mostly for their beauty of foliage, autumn colour and fruit. The majority are small-to-medium in size with a few forming what might be considered large trees (*S. croceocarpa* ('Threophrasta'), *S. sargentiana*, *S. torminalis*, 'Wilfred Fox', etc.).

Many of the most ornamental types from within the Mountain Ash (Rowan) and Whitebeams are relatively easily propagated in the field by summer budding or spring grafting. A percentage of these and especially those from other groups, Micromeles, Cormus and Torminaria, are more demanding and require bench grafting.

COMPATIBILITY

Here traditional use of the name *Sorbus* to cover all types is retained; although, many botanical authorities hold the view that the genus should be broken up into a number of separate genera, *Sorbus*, *Aria*, *Chamaemespilus*, *Torminalis* and *Cormus*. It is not surprising, therefore, that differing views on taxonomic linkages and categorisation within the genera, subgenera and sections make it almost impossible to provide a presentation agreed by all authorities. From the grafter's perspective the important issue is identification of close family connections, most likely to produce compatible graft unions. Table 51.1 is an attempt to achieve this by grouping together those thought to be most closely related. It does not follow strict rules of taxonomy.

Within Main Group A (Sorbus) on Table 51.1, the issues are relatively simple, as all species subspecies, varieties, cultivars and hybrids within Category numbers 1–8 appear compatible on *S. aucuparia*. Using combinations of species within each Category may or may not yield better results; however, the ready availability of *Sorbus aucuparia* means that it has become the primary choice of rootstock in Europe. Use of *S. aucuparia* for group 8 (Reducta) is the one exception. Despite no obvious incompatibility, excessive sucker production is normal, and if not continuously controlled, it can lead to poor performance or death of the scion. The best strategy if seed of any one species within the category can be obtained is to use the resultant seedlings to provide rootstocks for the others.

An indication of the complex inter-relationships within Sorbus is highlighted by the use of *S. intermedia* as a preferred rootstock for some in Category 1 (Aucuparia). The following are able to take advantage of the uniformity, vigour and excellent root system of this apomictic species: *S.* 'Apricot Lady', *S. aucuparia* forms – 'Beissneri', 'Cardinal Royal', var. *edulis*, 'Red Marbles', 'Sheerwater Seedling', var. *xanthocarpa*, *S.* 'Ethel's Gold', *S. pseudohupehensis* 'Pink Pagoda' (*hupehensis*), *S.* 'Pearly King', *S.* 'Signalman', *S.* 'White Wax', and undoubtedly others will be revealed.

Problems of compatibility arise within Main Groups B & C, and further discussion may be helpful. Where possible, those in Category 9 are best grafted to *S. intermedia* as this rootstock produces best growth, and provides a well-anchored, easily transplanted root system, when compared with the alternative *S. aria*. Category 10 presents a more complex picture: only *S. wardii* and *S. pallescens* seem totally happy on *S. aria*. *S. wardii* is also successful on *S. intermedia* and may possibly be

TABLE 51.1

Sorbus: Closely Related Categories and Groups with Suggested Rootstocks*

	Category Numbers	Representative Species and Selected Cultivars	Suggested Rootstock
		Main Group A	
Sorbus	**1** (Aucuparia)	*S. aucuparia* – var. 'Asplenifolia', var. *edulis*, 'Fastigiata' (*scopulina*), subsp. *maderensis*, and 'Sheerwater Seedling', *S.* x *kewensis*, *S. esserteauana, S. pohuashanensis.*	*S. aucuparia*, but some selected forms better grafted to *S. intermedia* (see text after table).
	2 (Commixta)	*S. americana, S. commixta* and its following cvs and vars. – 'Embley', 'Jermyns', var. *rufoferruginea* and 'Serotina', *S. randaiensis, S. sargentiana, S. ulleungensis* 'Olympic Flame', *S. wilsoniana.*	*S. aucuparia*, experience is that *S. intermedia* is not successful with this group.
	3 (Separate, not closely related to others)	*S. californica, S. gracilis, S. sambucifolia, S. tianshanica.* Probably *S. fansipanensis*	*S. aucuparia*
	4 (Insignis)	*S.* 'Ghose', *S. harrowiana,* *S. insignis.* Probably *S. helenae*	*S. aucuparia*
	5 (Vilmorinii)	*S.* 'Joseph Rock', *S.* 'Pearly King', *S. pseudovilmorinii, S. rehderiana,* *S. vilmorinii.*	*S. aucuparia*
	6 (Koehneana etc.)	*S. cashmeriana, S. fruiticosa, S. koehneana,* *S. microphylla, S. prattii.*	*S. aucuparia*
	7 (Hupehensis)	*S. discolor, S. forrestii, S. glabriuscula,* *S. pseudohupehensis* 'Pink Pagoda' (*S. hupehensis* misapplied).	*S. aucuparia*, and some grafted to *S. intermedia*
	8 (Reducta)	*S. monbeigii, S. poterifolia, S. reducta.*	*S. aucuparia*? A better choice would be to produce from seed one of those listed and use it as rootstock for the others.
		Main Group B	
Aria	**9** (Aria)	*S. aria* 'Lutescens', *S. aria* 'Magestica', *S. intermedia* 'Brouwers', *S. leyana,* *S. umbellata, S. umbellata* var. *cretica.*	*S. intermedia* for *aria* forms and *S. leyana*, not tested for *umbellata.*
	10 (Thibeticae)	*S. hedlundii, S. hemsleyi, S. pallescens,* *S. thibetica, S. vestita, S. wardii.*	*S. aria* for most but growth or success can be poor for some, especially *S. hedlundii.* *S. latifolia* has been suggested as an alternative but further investigation is necessary.
	11 (Micromeles)	*S. caloneura, S. eleonorae, S. keissleri,* *S. megalocarpa, S. meliosmaefolia,* *S. subulata* (*verrucosa* var. *subulata*).	*S. aria* for *S. caloneura* and *S. subulata*, less satisfactory for *S. eleonorae* and *S. meliosmaefolia.*
	12 (Ferrugineae)	*S. epidendron, S. rhamnoides.*	*S. meliosmaefolia* and *S. alnifolia* suggested, possibly as inter-stems on *S. aria* rootstock.

(Continued)

TABLE 51.1 (CONTINUED)

Sorbus: Closely Related Categories and Groups with Suggested Rootstocks*

	Category Numbers	Representative Species and Selected Cultivars	Suggested Rootstock
	13 (Alnifoliae)	*S. alnifolia, S. folgneri, S. japonica, S. yuana, S. zahlbruckneri.*	*S. folgneri* satisfactory on *intermedia*, other species on *S. aria.*
	14 (Chamaemespilus)	*S. chamaemespilus, S. sudetica.*	*Crataegus* has been suggested but grafts are short lived. *S. aria* or *S. meliosmaefolia* preferred choice.
	15 (Torminaria)	*S. croceocarpa* ('Theophrasta'), *S. torminalis.*	*S. latifolia*, but may not be compatible with *S. torminalis.*
		Main Group C	
Cormus	**16** (Cormus)	*S. domestica* f. *pomifera*, f. *pyrifera*	Ideally *S. domestica* seedling but *Pyrus communis* can be used for f. *pomifera* and f. *pyrifera.*

*P. Dummer (pers. comm.), C.Lane (pers. comm.), Alexander III (1998), McMillan Browse (1979), Krüssmann (1964) Nelson (1968) and C. Sanders (pers. comm.)'.

---

considered as an inter-stem for double-working some of the more intractable, but highly desirable, members of this group, such as *S. hedlundii*. Experience here is that some clones of *S. hedlundii* fail after one or two seasons when grafted to *S. aria*, and fail almost immediately with *S. intermedia*. Individual examples of *Sorbus vestita* grafted to *S. aria* have shown delayed incompatibility and failed after a period of some years*. As *Sorbus aria* is not apomictic and seedlings show typical variation, some graft combinations in this category show good health and vigour while some do not.

Categories 11–14 represent species within what was originally known as the Micromeles section, now split across other sections. *Sorbus folgneri* appears a very amenable species and may prove a possible candidate for use as an inter-stem for certain combinations yet to be identified. The same comments made for *S. aria* regarding Category 10 apply here.

Category 15 contains the very distinctive Chequer tree *S. torminalis*, which is normally only propagated by seed because there appear to be no cultivars of this species. *Sorbus latifolia* is suggested as a possible rootstock (untested). This would be a useful development as *S. torminalis* is notorious for a 'fangy' unbranched root system, difficult to transplant. Category 16 with one species, *S. domestica*, the Service tree, is particularly interesting because the rootstock of a separate genus, *Pyrus*, appears to show good short- and long-term compatibility.

TIMING

Although budding in the field during the summer is an important method for *Sorbus*, grafting takes place only during the winter/early spring period (January to March). Because of their relative ease of grafting, timing can often be adjusted to fit in with work programmes. Some species, such as *S. caloneura*, are among the earliest of all temperate species to commence bud break in the spring, and for these a careful watch is necessary; cold storage of early collected scion wood may be required.

---

* D. Jewell (pers. comm.).

GRAFTING METHODS

Splice grafts are normal for potted rootstocks. The majority of bare-root grafts may also be whip and tongue grafted to permit the use of separate tying and sealing by grafter's assistants.

ROOTSTOCKS

Sorbus aria seedlings for bare-root grafting are suitable as 1+0 6–8mm or preferably, because of the unbranched 'fangy' root structure, 1u1/1+1 6–10mm. Rootstocks of this species potted as 1+0 seedlings at 4–6mm should make good quality 1+1P 6–8mm for grafting after one season.

Sorbus aucuparia is more vigorous and for bare-root grafting 1+0 is normally available at 6–10mm girth. For those who require a heavier grade with the hope of producing larger plants more rapidly, the next grade, 1+1, will produce 8–12mm rootstocks. Potted *Sorbus aucuparia* grows rapidly with good compost and irrigation, and lighter seedlings 4–6mm will make 1+1P 6–8mm grade for grafting.

Sorbus intermedia is available as 1+0 6–8mm or 1+1 slightly larger at 6–10mm. Light grade seedlings 4–6mm should provide 6–8mm after one season's growth in the pot.

Similar grades are available for *Sorbus domestica* and *S. torminalis* for bare-root use, and are often available to purchase as pot grown plants.

Species in categories 11, 12 and 14 are rarely available, and in-house production requires finding a seed source and producing the seedlings. Most are finished as pot grown rootstocks.

SCION WOOD

Current year wood is normally preferred, but for a genus so easily grafted, a basal portion of 2-year wood is perfectly acceptable. Scions, ideally 100 to 200mm long, if possible, are gathered to include the basal portion of the shoot and, for upright growing types, retention of the apical bud.

SYSTEMS

Cold callusing is suitable for a genus able to produce callus in temperatures above 6°C. Material which is rare, valuable or in poor condition will respond well to the use of a hot-pipe system, with the heating chamber set at a temperature of ±18°C.

PHYSIOLOGICAL AND ENVIRONMENTAL FACTORS

Sorbus appear to be unaffected by root pressure problems and, although it is wise to apply the rule 'on the dry side' for pot grown rootstocks, drying-off is not required.

GROWING ON

Most species are amenable to container culture and thrive well in virtually all soil types, provided they are reasonably well drained. Some from the cooler, moister temperate maritime areas do not thrive in regions with a warm dry continental climate. When making decisions on what to plant, it is best to carry out a pre-planting search of their geographical origins.

52 Syringa (Oleaceae) – Lilac, French Lilac

INTRODUCTION

A genus of about twelve species which includes small- and medium-sized shrubs and some which eventually become tree-like. The majority may be described as strong, often suckering, shrubs. Distributed in the northern hemisphere, they occur from S.E. Europe to China and Japan. They are virtually all showy flowering shrubs, many also having strong fragrance.

Syringa vulgaris has dominated the types in cultivation and possibly over 2000 cultivars, otherwise known as French Lilacs, have been selected and named. It is mainly this group which is discussed in the following account.

Because of the suckering habit of lilacs, special consideration must be given to which grafting methods are followed. *Syringa vulgaris* has commonly been used as a rootstock for the French Lilacs. Unfortunately, its use has diminished the perceived value of grafted plants, too often plants purchased as a given variety disappear in a forest of sucker growth from the rootstock of an inferior type.

Grafting remains an important method, because cuttings propagation of French Lilacs has proved challenging. Micropropagation has been adopted for a limited range of varieties, but unfortunately plants produced by this system take a long time to reach the flowering stage. Root cuttings have been suggested and described as a viable alternative, but so far this technique has not established any significance.

COMPATIBILITY

The most compatible combination for *Syringa vulgaris* cultivars is to use *S. vulgaris* seedlings as rootstocks; unfortunately, this immediately initiates the potential for a long-term rootstock suckering problem. To eliminate this and allow suckers from the scion variety to develop, the objective of many grafting procedures is to encourage scion rooting. This may be achieved by choosing a rootstock acting as a 'nurse graft', the rootstock surviving sufficiently long for scion root development to take place, and later dying out due to delayed incompatibility. Once on its own roots, the suckering habit of the scion variety presents no problem. For further discussion, see Chapter 18: Root Grafting.

In trials at the Dutch Boskoop Trial Station, other *Syringa* species, *S. reticulata (S. japonica)*, *S. josikaea*, *S. komarowii* subsp. *reflexa (S. reflexa)* and *S. tomentella*, have been tested for their suitability as rootstocks. All appeared to unite well with *S. vulgaris,* and early indications were that compatibility was satisfactory. An average take of 91% was obtained after two years, but longevity of the combinations was not totally established in this short-term investigation. Final conclusions were that none were better choices than *S. vulgaris* for French Lilacs[*].

Syringa tomentella has been used by others who report that French Lilacs grafted to it apparently do not produce scion rooting but retain the *S. tomentella* root system. Unfortunately, this often proves insufficient to properly support a mature French Lilac bush, and the plant rocks in strong winds causing root damage and early death[†].

---

[*] Caron (1959).
[†] C. Lane (pers. comm.).

A review of rootstock combinations in the 1968 IPPS Proceedings reported that *Syringa chinensis* as a rootstock for *S. oblata* subsp. *dilatata* survived only one year, and *S. reticulata* subsp. *amurensis* (*S. amurensis*) gave a low percentage of takes and produced dwarf, short lived plants*. These results indicate that there are potential incompatibilities within the genus.

For the French Lilacs, choices of suitable rootstocks with delayed incompatibility have been from within the same family Oleaceae, and include the following: *Fraxinus pennsylvanica, F. americana, F. excelsior, Ligustrum ovalifolium* (called California Privet in the US), *L. vulgare, Forsythia* x *intermedia, Osmanthus* x *burkwoodii* and *Jasminum nudiflorum*. Of these, the last two failed to achieve a union; *Ligustrum vulgare* produces copious root suckers, *Fraxinus Americana* is considered less compatible than *F. pennsylvanica*[†] and *Fraxinus excelsior* does not appear to have ever been further investigated, possibly because it too is insufficiently compatible. Although *Forsythia* x *intermedia* was promising in early investigations[‡], no further progress has been reported.

TIMING

The major period for grafting is January to March. Where scion rooting is required early grafting is recommended to allow as long as possible for this to occur. Grafts involving the use of *S. vulgaris* rootstocks, when scion rooting is not a consideration, can occupy the later period. The *Syringa vulgaris*/French Lilac combination may be summer grafted using apical methods during August and September. Another strategy, less often adopted, is using the technique of cuttings-grafts, employing *Ligustrum ovalifolium* as the unrooted rootstock. To suit the rooting requirements of this, timing is best in August/September.

GRAFTING METHODS

Splice, apical veneer and wedge grafts may be used. Side veneer grafts are employed for top-worked plants and cuttings-grafts. To achieve good results, very tight tying-in and good cambial matching is considered to be important. Even with the relatively easily grafted genus *Syringa*, matching cambia on one side only can reduce takes by 20%.

Combinations involving *S. vulgaris* and the French Lilacs mostly utilise a simple splice. Positioning this is open to debate; some favour grafting as low as possible, others higher at 150–200mm above soil level. Reasons for the latter choice are that any suckering shoots may be quickly recognised and removed; others argue that low grafting reduces the possibility of sucker development. It seems likely that sucker production is a complex of rootstock and scion interactions, always exacerbated by the fact that *Syringa vulgaris* is a species naturally producing 'true' suckers from the root system. With stronger growing cultivars, if suckers are diligently removed in the early years, sucker production is often diminished once the scion variety has reached substantial size.

Nurse grafts and nurse root grafts may be achieved using a simple splice. In one method, grafts are tied using what is described as 'grafting thread' rather than a grafting tie. This is preferred because it is possible to tie more tightly; it does not break-down and cuts into the tissues encouraging adventitious root formation. The thread is assumed to have a nylon content to provide strength and rot resistant properties. Grafts are sealed by dipping into melted wax or shellac. They are then bundled and placed into polythene bags, stored in warm conditions (21–26°C) for 12 days to encourage callus development before cold storage at 0.5 to 1.0°C until required for field planting[§].

A number of methods involve the use of *Ligustrum ovalifolium*. Rooted stems of this species are selected at the appropriate grade and headed-down to within ±50mm of the root system. The scion

---

* Nelson (1968).
† Wedge (1977).
‡ Hunt (1971).
§ Wedge (1977).

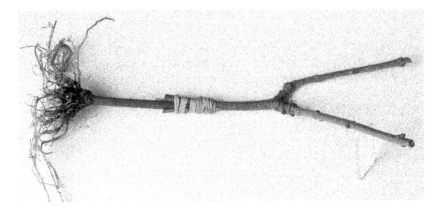

FIGURE 52.1   *Syringa* 'Madame Lemoine' grafted to bare-root *Ligustrum ovalifolium* rootstock using an inverted wedge graft. The 'Y' shaped scion has the basal stem split to above the tie (just visible above top of tie, centre of picture). Using an artist's paint brush facilitates IBA quick dip application to only this portion. Without the added stimulation of IBA, adventitious roots may also form between the tie loops on the arms of the wedge.

is often attached to the rootstock by a splice. Others adopt a similar procedure with variations in the use of an apical veneer rather than splice[*] or an inverted wedge graft[†].

Use of an inverted wedge is particularly relevant, as it has been shown that for many species a split cut on the basal portion of hardwood cuttings, combined with an IBA growth substance treatment, enhances adventitious root formation[‡]. This effect may be replicated if the rootstock is prepared by cutting the apex of the *Ligustrum* stem to a wedge shape, the *Syringa* scion split at the base being pushed onto this wedge and tied (Figure 52.1). A sensible modification to preparation of the scion would be to increase the depth of the split by ±10mm; then, using an artist's paint brush to allow accurate application, IBA in quick dip formulation may be applied to the split area above the graft. This should stimulate adventitious root production and need not add an unreasonable amount of time to the overall procedure. It would avoid the previously identified possible adverse effect of IBA applied directly to the graft.

Sealing or omitting sealing warrants further discussion. Most appear to seal the graft, but one report states no discernible difference in takes between sealed and unsealed[§]. Since the scion is meant to emulate a cutting, it can be assumed that unsealed grafts should produce adventitious roots more readily than those sealed. Propagators would not consider the possibility of dipping the basal portion of a cutting into grafting wax before insertion into the rooting medium. Unsealed grafts will require extra care in handling as field planting or bedding-out can risk desiccation. To avoid the union drying-out while awaiting the next stage, strategies for dealing with this must be considered, such as containment in polythene bags until the last moment.

ROOTSTOCKS

Syringa vulgaris seedlings are grafted while bare-root 1+0 or 1+1 6–8mm, or 8–10mm seedlings, or as potted 1+1P 6–8mm seedlings. Before grafting, all terminal and adventitious buds on the root-stock are removed to reduce the number of suckers produced. To enable early recognition of sucker

[*] deWilde (1964).
[†] Kern (1956) and Sheat (1965) pp.374–377.
[‡] Howard, Harrison-Murray & Mackenzie (1984).
[§] deWilde (1964).

growth, some suggest that seedlings with dark buds should be used for white flowering French Lilac varieties, and green budded types for dark flowered forms.

Root pieces of *Syringa* may be used, but it is a somewhat more demanding technique and, provided due diligence is given to disbudding the conventional rootstock, suckering is not reduced by using roots only. Reversing root polarity results in increased sucker production*.

Ligustrum ovalifolium is a popular choice for nurse grafting in the milder, temperate areas. It is normally used as a 0+1 6–8mm or 8–10mm bare-root plant. This is produced from hardwood cuttings inserted into light sandy soil in late October to December. Long cuttings are planted deeply and root at a number of nodes. On lifting the following late autumn, they are cut to produce separate rooted portions comprising a ±50mm long piece of stem with attached roots. All buds and unwanted roots are removed. The prepared portions constitute individual rootstocks and are stored until required in polythene bags in a cold area or cold store.

Fraxinus pennsylvanica (Green Ash) provides a hardier rootstock for cold areas. Root pieces are taken from 1+0 seedlings to provide 8–10mm diameter × 100mm long portions. Those close to the top of the hypocotyl (crown) produce slightly higher takes but tend to sucker more freely. Roots from older plants are useable but slower to tie-in because of obstructing lateral growth†. As with *Ligustrum*, *Fraxinus pennsylvanica* displays delayed incompatibility and dies out after a period. Trials to encourage rootstock death and scion rooting by grafting the ash rootstock in an inverted position (inverted root grafting), produced only a 13% take‡.

SCION WOOD

Ideally, scions should be gathered only from stock mother plants regularly pruned to keep them in a vegetative state with minimal flower bud production. Scions normally comprise 1-year-old wood cut to lengths between 100 and 200mm with the upper portion, potentially carrying flower buds, discarded§. Shoots collected can normally provide two scions, the upper portion cut to include 3 pairs of buds, the lower 2 pairs.

An interesting variant is to use a 'Y' shaped scion incorporating a basal portion of 2-year-old wood. This can be expected to provide an excellent quality graft with low branching characteristics, but additional stock plants will be required. There is no reason to suppose graft takes will be jeopardised by use of 2-year-old wood, but it is possible that scion rooting will be delayed. This type of scion could be readily grafted using the inverted wedge method (see Figure 52.1).

SYSTEMS

Winter/early spring grafting using *S. vulgaris* rootstocks may be accomplished by cold callusing.

Nurse and nurse root grafts are claimed to be more successful if heat is applied for a short period after grafting to enhance callus production; 21–22°C for 12 days has been suggested¶. The duration and amount of heat needs to be judged to promote callus but avoid bud-break and extension growth. Choice of 10–15°C for a slightly longer period may be preferable.

Sub-cold callusing systems with grafts packed in polythene bags or polythene lined boxes held in cold storage at ±0.5°C allow planting to be delayed until later in the season, when higher soil temperatures promote callus formation.

Hot-pipe systems appear not to have been investigated. They should guarantee 100% take and scion root promoting procedures such as wounding the scion stem and applying IBA, may be

---

* Hunt (1971).
† Wedge (1956).
‡ Wedge (1956).
§ Wells (1955) pp.291–294.
¶ Wedge (1977).

conveniently incorporated during subsequent handling after removal from the heating chamber at planting or potting-off.

FIELD PLANTING

Traditional methods involve deep planting the grafts into beds leaving only one or two scion buds exposed; here they remain for one growing season. Careful handling, such as retaining in polybags until the last possible moment, is essential for unsealed grafts, or losses due to desiccation will be considerable. Some protection after planting, such as a polytunnel, cold frame or polythene covered cloche, should be provided. If this is not possible, the sub-cold callusing system may be a better option.

Soils chosen for this system are important; they must be well-drained, light-textured, and irrigation should be available. Ideally, the land should have been sterilised before use.

POTTING UP/ON

If nursery procedures are not geared to field plant nurse grafts, they may be potted deeply into suitable containers. Deep pattern pots are necessary and decisions on which type will be influenced by subsequent treatment. Where continuous container production to sale is planned, quite large containers, up to 4 litres deep, may be suitable. A smaller, extra-deep pattern pot, or poly-bag, such as used for rootstocks, will give more flexibility and according to requirements permit potting-on or field planting. The application of root promoting treatments (IBA) to the scion to encourage early adventitious root production may be conveniently incorporated into the routine at the potting-up stage.

SUMMER GRAFTING

Pot grown rootstocks may be summer grafted from August to early September. French lilacs grafted to *S. vulgaris* normally produce very high takes. If suitable rootstocks can be identified, the system may also prove a safe option for grafting certain species such as *S. meyeri* 'Palibin' and cultivars which are difficult to root or develop slowly from cuttings.

After completion, grafts are placed in a poly-tent with a bottom heat setting of 18–20°C and quickly produce callus and unite in temperatures elevated by sunlight to give ±24°C. Shading is necessary to prevent excessive temperatures. Side grafts using de-leaved scions or apical grafts, such as splice or veneer, are equally successful. For apical grafts, approximately 50% of scion leaves should be retained.

CUTTINGS-GRAFTS

Cuttings-grafts have been suggested as a further option for French Lilac propagation*. As with cuttings or micropropagation-raised plants, they will take longer than other systems to reach saleability. This method occupies far less space in propagation area than the alternatives and may offer a well-integrated and efficient system for container plant production. 75–100mm long unrooted cuttings of *L. ovalifolium* are either grafted using an inverted wedge as described previously, or side grafted using a short tongue veneer placed ±25mm above the base of the cutting. The former method results in total loss of foliage of the cuttings-graft rootstock and even for a species as easily rooted as privet must increase the possibility of poor results. The latter method enables retention of leaves on the rootstock and should promote quick and reliable rooting.

---

* Kern (1956).

After preparation, the graft union is inserted into a suitable rooting medium, ideally comprising a fine bark or peat/bark/perlite mixture. Deep trays will be required, as cuttings-grafts must be inserted deeply, the graft union buried. These are then placed into a poly-tent, which may have mist lines installed, and given bottom heat, 20–21°C, to promote rapid rooting. Successful grafts may be potted-off or planted-out the following spring at which time side grafts can be headed-off to just above the graft union. At the same time, it is possible to wound the base of the lilac scion and apply IBA treatment to promote scion rooting.

PHYSIOLOGICAL AND ENVIRONMENTAL FACTORS

Syringa grafts readily produce callus even at low temperatures (10°C) and can be considered among the easier genera to graft. No particular problems with root pressure have been recorded, and although the usual rule of grafting potted rootstocks 'on the dry side' applies, drying-off is not required.

Syringa vulgaris varieties are susceptible to the bacterial disease *Pseudomonas syringae*, and it is best to avoid growing the most disease prone types. Little can be done to control the disease in moist maritime areas such as the UK, but it is less prevalent on the dryer eastern seaboard of that country and in continental climates.

TOP-WORKED PLANTS

Novelty top-worked plants of a few types have been suggested for grafting. *Syringa reticulata* subsp. *amurensis* 'Ivory Silk' is grown in stem form in a pot or root-balled and top-worked with various types. During late winter, top-working with multi-grafts at the chosen height is carried out in a poly-tunnel. After grafting, these are sealed by dipping or brushing with grafting wax and maintained at a minimum of 10°C, which may rise to 20°C+ in sunny weather; rapid callus and union development occurs at these temperatures*.

GROWING ON

Field production is best in medium-textured, fertile soils, with a pH above 7. Container production has become increasingly important in recent years. Lilacs grow well in standard container composts, but use of overhead irrigation, combined with high nutrient levels, may increase the incidence of Lilac Blight (Pseudomonas syringae).

---

* Mezitt (1973) and Bakker (1995).

53 Tilia (Malvaceae) – Lime-trees, Lindens, Basswoods

INTRODUCTION

Some authorities consider the earlier taxonomic treatment which placed *Tilia* in its own family, Tiliaceae, should be retained. This genus comprises about 23 species of deciduous trees having some subspecies and hybrids rarely seen in cultivation. Restricted to the northern hemisphere, all become trees of various sizes, from small and shapely to tall and majestic. Many of the major species are tolerant of hard pruning, making them suitable for roadside planting. This feature also allows for 'pleaching', a system of trained horizontal main branches carrying shoots which are pruned back annually to control growth, adding structure to formal landscape plantings.

A few, propagated vegetatively, are among the most important trees produced by nurserymen and are grown in very large numbers, mostly from budding in the field during the summer. A more limited range of rare species and hybrids are bench grafted, as they are required only in modest numbers and are often more demanding in their requirements.

COMPATIBILITY

Apart from well-established combinations involving the major species, *Tilia* represents a genus with many unresolved problems of incompatibility. This may be due in part to the inherent variability within the genus, further emphasised by the propensity of some species to produce natural hybrids.

Taxonomic investigation has split the genus into four sections. They do not represent separate evolutionary development, unlike groupings in some other genera; therefore, decisions on combinations to achieve graft compatibility may be less reliable than those established following phylogenic (evolutionary) principles. This is immediately underlined by inclusion of *Tilia cordata* and *T. platyphyllos* in the same taxonomic section, as it is known when cultivars of either are grafted to the other, an incompatible union results. Additionally, other species compatible on the one are not usually successful on the other (Figure 53.1). To confuse the issue still further, the two species cross-pollinate readily to form the ubiquitous natural hybrid *Tilia* x *europea*.

Initial but unpublished reports of DNA analysis have shown that *T. platyphyllos* and *T. cordata* are not closely related. The former constitutes what might be considered a rather primitive form more akin to *T. endochrysea* than *T. cordata*, which itself is closely allied to the Asian species *T. amurensis, japonica* and *mongolica**, and for which it provides a reliable rootstock.

The four sections[†] with representative species from each are as follows:

- Anastraea: *T. americana*: *T. cordata, T. dasystyla, T. kiusiana, T. mongolica* and *T. platyphyllos.*
- Astrophilyra: *T. callidonta, T. caroliniana, T. chinensis, T. maximowicziana,T. nobilis, T. oliveri, T. tomentosa;* and *T. tuan.*
- Endochrysea: *T. endochrysea.*
- Henryana: *T. henryana.*

---

* D. Pigott (pers. comm.).
† Pigott (2012) pp.45–48.

FIGURE 53.1   LEFT, *Tilia maximowicziana*. A graft to *T. cordata* rootstock on the left; on the right, a graft to *T. platyphyllos* rootstock. RIGHT, *Tilia oliveri*. A graft to *T. cordata* rootstock on the left; on the right, a graft to *T. platyphyllos* rootstock. In both instances, the larger plant in the photograph continued normal growth, with a compatible union. The smaller plant developed more severe incompatibility symptoms and subsequently failed.

An interesting feature of Table 53.1 below, giving a review of stock/scion combinations, is the potential advantage of using *T. tomentosa* as a rootstock. It has not featured as an important species for this purpose in the UK, partly because seedlings are not so readily available, and it has not performed as well as *T. cordata* and *platyphyllos* in trials in Holland*. *T. tomentosa* may prove a suitable rootstock for *T. nobilis*, which is weak growing when grafted to *T. cordata* and has failed on *T. platyphyllos*. In the USA, *T. tomentosa* and *T. cordata* are generally the two major species used for comparative trials of compatibility.

The possibility of double-working to overcome incompatibility should not be overlooked. A good example is seen with *Tilia callidonta*, which has been successfully grafted using this technique (Figure 53.2), fully described in Chapter 12 in the section Future Progress – Strategies to Overcome Incompatibility.

TIMING

Grafting normally takes place in the winter/early spring period (January to March). Summer grafting in August-early September is also possible and has the advantage of reducing root pressure problems.

GRAFTING METHODS

Splice or whip and tongue grafting methods are normally employed. In view of the propensity for *Tilia* to bleed badly when headed back, it is surprising that side grafts are rarely suggested.

Sucker production (not true suckers but basal, often underground stems) is a potential problem when *T. cordata* or *T. platyphyllos* are used as rootstocks. Use of root grafts should prevent this problem because *Tilia* does not produce stem buds on its root system[†]. Roots may be obtained from established plants, or by heading-back seedlings to just below the hypocotyl and grafting the root

---

* Proefstation voor de Boomkwekerij (1982).
[†] Pigott (2012) pp.366–368.

TABLE 53.1

Suggested Combinations for Tilia Rootstocks and Scions*

Rootstock	Scion
T. cordata	T. amurensis
	T. chinensis
	T. cordata and cvs
	T. x euchlora
	T. henryana
	T. japonica (japonica 'Ernest Wilson')
	T. kiusiana
	T. mongolica
	T. oliveri
	T. paucicostata
T. platyphyllos	T. americana and cvs
	T. endochrysea
	T. x euchlora
	T. x europea and cvs (but not 'Zwarte Linde')
	T. maximowicziana
	T. x moltkei
	T. platyphyllos cvs
	T. x stellata (michauxii)
	T. tomentosa and tomentosa 'Petiolaris'
T. cordata or platyphyllos	T. x euchlora
	T. x europea cvs
	T. x flavescens 'Glenleven'
	T. tuan var. chenmoui
T. tomentosa	T. americana cvs
	T. caroliniana subsp. heterophylla
	T. henryana
	T. oliveri
	T. tomentosa cvs
	T. tomentosa 'Petiolaris'
T. americana	T. americana 'Redmond'

Possible Compatible Combinations (not all proven)	
T. cordata	T. 'Harold Hillier' (japonica 'Ernest Wilson' x mongolica)
	T. nobilis
	T. tuan
T. platyphyllos	T. caroliniana and subsp. incl. mexicana
	T. chingiana
	T. dasystyla
	T. mandschurica
	T. miqueliana
	T. tuan

*Alkemade & van Elk (1989), Pigott (2012), Okken (1998), Flemer III (1980), Nelson (1968), Alexander III (1998), T. Brotzman (pers. comm.) and D.Harris (pers. comm.).

FIGURE 53.2   *Tilia callidonta*, a double-worked graft. Rootstock is *T. cordata*, with *T. chinensis* inter-stem grafted to it. Grafted to *T. chinensis* is *T. callidonta* (showing extension growth). The *T. chinensis* stem is headed back to the *T. chinensis/callidonta* union after a period to promote scion growth. This apparently successful graft combination has continued to develop, providing a lasting compatible union, to date, for five years.

beneath. Wedge grafting is often recommended for this; a short tongue veneer may enable more accurate matching of the cambia of root and shoot.

If potted rootstocks are chosen, the plants are partially knocked out to gain access. The rootstock is headed back to just below the hypocotyl and the root exposed and grafted using a wedge or short tongue apical veneer. Sealing is not required for enclosed, warm callusing conditions.

ROOTSTOCKS

Use of *Tilia* seedling rootstocks may cause problems because batches of seedlings labelled *T. cordata* or *T. platyphyllos*, particularly the latter, may contain a substantial percentage of the natural hybrid *Tilia* x *europea* (x *vulgaris*). It is difficult to estimate what effect this has on compatibility, but it must confuse the issue and add further to uncertainties. There is no easy solution beyond suggesting investigation into the seed source of the rootstock supplier and, if possible, an inspection of the growing crop before delivery.

Use of bare-root rootstocks substantially slows development of the plant after grafting, compared with those budded in the field. Hot-pipe systems will partially reduce the imbalance; however, potted rootstocks promote extension growth more effectively, especially if combined with hot-pipe*.

---

* Flemer III (1980).

Some Dutch growers claim one-year-old seedlings (1+0) of *T. cordata* can be produced to give 7–10mm grade for *T. platyphyllos* and 8–12mm grade for *T. tomentosa*. In the cool maritime climate of the UK, transplants (1+1) are usually required to achieve similar grades for both species.

Potted, best-grade seedlings should produce 1+1P 6–10mm grade suitable for grafting during the winter/spring period. The shorter growing period available for summer grafted rootstocks will require transplanted seedlings, 1+1+1P, to produce an equivalent grade. Considering the large root system of *Tilia* seedlings, there is a case for potting rootstocks into 1-litre-deep pots rather than the more usual 9×9×12cm.

Recognition of the value of clonal rootstocks in producing uniform grades of high-quality trees prompted work in the mid-1980s on this aspect at East Malling Research Station. Early candidates for this initiative were chosen from seedling populations of *T. cordata*, *T. platyphyllos* and *T.* x *europea*. The selected clones needed to be relatively easily propagated from hardwood cuttings or by stooling methods. They should also produce good extension growth to permit easy working by budding or grafting and produce vigorous, highly uniform trees.

Several selections were made and given code numbers, of which *T. cordata* clone 6 and *T. platyphyllos* clone 231 appeared the most promising, being relatively easily propagated by hardwood cuttings and producing trees superior in growth and uniformity to seedling rootstocks*. The chosen scion variety was *Tilia* x *euchlora* but it seems likely that these rootstocks would produce superior trees for a number of other species and cultivars. Due to funding constraints, the work was not continued to its ultimate conclusion; the original selections are no longer at East Malling and may now be lost.

SCION WOOD

Best material is provided by well-ripened, current year wood trimmed to between 120 and 180mm including the basal portion. Compact growing species such as *T. kiusiana* will require inclusion of 2-year wood at the base. Similar material is chosen for summer grafting; side grafted scions are best de-leaved before grafting, leaves are retained for apical grafts.

SYSTEMS

In the field, failed buds from summer budding are 'grafted over' the following spring in an attempt to redress losses; in the UK takes are often rather poor. Due to higher temperatures compared with the open, cold callused bench grafts produce more reliable results. Warm callusing with bottom heat set at 18–20°C is to be preferred, particularly for rare and light wooded species.

Hot-pipe systems are very effective and, provided temperatures external to the heating chamber are low (<5°C), root pressure is significantly reduced and all that is required is to ensure potted rootstocks are 'dryish' when placed in position. Optimum temperature for the heating chamber is ±20°C.

Investigations at the Dutch Trials Station into late summer/early September grafting of avenue trees were reported in 1995[†]. Bare-root rootstocks grafted with apical grafts left unsealed, and with scion leaves retained, were placed into poly-tents set at 18°C and maintained at 95%+ humidity until union formation was evident. Results for *T.* x *moltkei*, *T. tomentosa* 'Brabant' and 'Nijmegen' were slightly superior to those achieved by conventional winter/spring grafting in warm callusing conditions.

Summer grafting earlier in August involves the use of potted rootstocks on which side grafts are placed. Unlike the apical grafts described, the scion is de-leaved, otherwise techniques are identical. Summer grafting systems have a number of advantages because, while comparable to field

---

* Howard (1995).
[†] Ravesloot (1995).

budding in many respects, environmental conditions are under careful control, and high takes can be expected.

PHYSIOLOGICAL AND ENVIRONMENTAL FACTORS

During the late winter/early spring, root pressure problems are significant with *Tilia*, particularly *T. platyphyllos* are significant (Figure 53.3). To prevent excessive and damaging bleeding when warm callusing systems are used, severe drying-off is required. Those side grafted in August can still 'bleed' despite the presence of leafy rootstock tops, and at this time pot grown rootstock should be 'on the dry side' when grafted. The Dutch system using bare-root rootstocks grafted rather later in early September largely avoids this problem.

For hot-pipe systems, pot grown rootstocks should be fairly well-dried and are best de-potted to allow inspection of the root-ball and permit some light spraying over or watering, if needed.

Tilia callus readily, even in fairly low temperatures but respond best in the region of 18–21°C. Callus formation is extensive and can become excessive. Well-judged ventilation of poly-tents is important to avoid development of 'fluffy callus', likely to become infected and kill the developing tissues of the graft union. Hot-pipe may be set at the normal 20–22°C.

GROWING ON

Limes respond to fertile soils or rich compost with adequate moisture. If these requirements are met, full sun is preferable for most but some golden forms; notably *Tilia* x *europea* 'Wratislaviensis', can be scorched in full exposure and light shade should be provided.

FIGURE 53.3 *Tilia* x *stellata* (*T. michauxii*); multi-grafted to *Tilia platyphyllos* in a warm callusing system during winter/spring. Clearly visible are the results of excessive callus production: root pressure causing bleeding and rotting callus. The callus has also become infested with fruit fly larvae (Drosophila); all grafts failed.

54 Viburnum (Adoxaceae)

INTRODUCTION

A large genus of deciduous and evergreen, predominantly shrubby species, with some becoming small trees. They are found in northern, temperate areas of the northern hemisphere, with incursions into the southern hemisphere via southeast Malaysia and South America. Many have outstanding ornamental value in their flowers, fragrance, foliage, fruit, autumn colour or mixtures of these.

Most Viburnums are propagated by seed or cuttings. Many root relatively easily from semi-ripe cuttings but some subsequently grow very slowly and, for commercial reasons, are grafted to accelerate growth. A few species are difficult to induce to root, and along with types top-worked on a stem to produce standards or novelty Patio plants, are always propagated by grafting.

COMPATIBILITY

To explore the potential for suitable combinations, Table 54.1 below is based on phylogenic (evolutionary) factors and shows ten clades containing related species grouped together. These might provide the basis for good decisions on rootstock/scion combinations. Unfortunately, not all significant species are included, and, among others, *V. harryanum*, *V. grandiflorum* and *V. henryi* are missing; the last two are expected to fall in the Solenotinus section although *V. henryi* may well be more closely allied to the Tinus group. Those proceeded by '?' are the author's speculation.

Viewed with a plantsman's eye, it will be seen that potential anomalies in the groupings occur, which highlights the problem and complexity of graft compatibility in woody plants. Inclusion of *V. odoratissima* with *V. farreri* seems an unlikely combination, and *V. lantana* with *V. rhytidophyllum* may not succeed. Most other groups do suggest logical categories, and it is interesting to note that *V. plicatum* is the sole member of its own section and may indicate a possible reason for difficulties with this species.

Incompatibility can occur in the main commercial types, and comparatively few investigations into suitable combinations of rootstock and scion are available for less well-known species and cultivars.

The Lantana group, containing some important commercial types such as *V. x carlcephalum* and *V. carlesii*, when grafted to *V. lantana* appear compatible for prolonged periods, most probably indefinitely. Grafting *V. davidii* 'Angustifolium' to *V. lantana* quickly produced incompatibility symptoms and far less extension growth than a cutting of the same type rooted at the same time (Figure 54.1). It is now clear that *V. tinus* may have been a better choice of rootstock for this species.

Dutch grafters state that *V. plicatum* forms show delayed compatibility when grafted to *V. lantana** and favour *V. opulus* as the rootstock. Plants grafted on *V. opulus* have shown no signs of incompatibility for the two or three years they remain on the nursery before sale[†]. It is also claimed growth is faster on *V. opulus* than *V. lantana*. Whether the combination will remain compatible indefinitely is much less certain; a wise precaution might be to plant deeply to encourage scion rooting. Trials of *V. plicatum* cultivars grafted to *V. opulus* at the Royal Horticultural Society's trial ground at Wisley (UK) have demonstrated incompatibility symptoms on several plants after two or three season's growth. There is a noticeable improvement in vigour and foliage health where grafted plants have shoots which have self-layered and produced branches growing on their own root systems (Figure 54.1).

---

[*] Alkemade & van Elk (1989).
[†] K. Verboom (pers. comm.).

TABLE 54.1

Phylogenic Grouping for Viburnum*

Group (Clade)	Species
Pseudotinus	*furcatum*
	lantanoides
Lentago	*cassinoides*
	lentago
	prunifolium
Lantana	*macrocephalum*
	carlesii
	rhytidophyllum
	lantana
Tomentosa	*plicatum*
Solenotinus	*chingii*
	erubescens
	farreri
	?grandiflorum
	?henryi
	odoratissima
Tinus	*tinus*
	cinnamonifolium
	davidii
	?harryanum
	?henryi
Oreinodontotinus	*dentatum*
	jamesonii
Opulus	*opulus*
	sargentii
Coriacea	*cylindricum*
	?harryanum
Succodontotinus	*foetidum*
	betulifolium
	wrightii
	setigerum

*Amended from Clement & Donoghue (2012).

Possibly *V. lantana* as a rootstock for *V. plicatum* should not be dismissed too readily; personal experience is that *V. plicatum* 'Rosace' and *V. plicatum* 'Molly Schroeder' on this rootstock have survived without signs of incompatibility for some time - *'Rosace'* for nine years.

TIMING

Viburnum grafting is dominated by summer grafting (July to early September). Winter grafting is possible in warm callusing systems, and hot-pipe is always an option although rarely considered necessary.

FIGURE 54.1   LEFT, *Viburnum davidii* 'Angustifolium'. On the left, showing very little extension growth when grafted to *V. lantana*, compared with the plant on the right, on its own roots, produced from a cutting. The graft subsequently died. RIGHT, *Viburnum plicatum* cultivar grafted to *V. opulus* rootstock. The original graft (upper half of picture) shows early senescence and restricted growth, compared with one lateral branch, which has self-rooted and growing with health and vigour (photo by Maurice Foster).

GRAFTING METHODS

Because of the relative ease of grafting Viburnums, summer grafting often involves the use of apical grafts. Apical methods using a splice, or short tongue veneer, are suggested after potted rootstocks are headed back to 25mm above compost level*. Such low grafting is intended to reduce possible future suckering. Access for the knife will be hampered when working at this height and use of a curved (hooked) bladed knife to aid preparation of the splice may of assistance. Others suggest use of apical long tongue veneer graft with the base of the rootstock cut finishing ±12mm above the compost level of the potted rootstock[†].

Side veneer grafts may be used and can achieve equal, or more, numbers of grafts completed per hour if rootstock tops are cut back to a suitable height, permitting rapid tying-in. The graft is more easily accomplished because the rootstock stem can be pulled back to provide more exposure, better access and provide more choice of exact, low placement. Takes may well be slightly improved because any problems caused by root pressure are mitigated, and the rootstock root system is not compromised by removal of all top growth when the plant is in full growth. Extra time taken to head-back rootstocks may be delayed until early the next season, most conveniently when preparing grafts for potting-on.

ROOTSTOCKS

The most important rootstock species used in Europe are *V. lantana* and *V. opulus*. The distinctive foliage of *V. opulus* makes it a popular choice for those who see the advantage of being able to quickly recognise sucker shoots, a common problem with this naturally suckering genus.

In parts of the USA, *V. lantana* is subject to attack by a leaf spot (unspecified but possibly *Monochaetia sp.*), and some recommend substituting it with a native species, *V. dentatum*. This is claimed to produce better results than the *V. lantana* alternative[‡]. There are reports of *V. dentatum* failing to make a good union with *V. x carlcephalum*[§].

---

* McMillan Browse (1970).
† Hoogendoorn (1988).
‡ Hoogendoorn (1988).
§ Nelson (1968).

1+0 4–6mm or 5–7mm seedlings of *V. lantana, opulus* or *dentatum* are potted to produce 1+1P 6–8mm grades by mid-summer; grade 6–10mm is achieved for grafting the following winter.

An alternative rootstock may be produced from *V. lantana* propagated by firm semi-ripe cuttings taken during August to early September. *Viburnum opulus* has not been identified for this procedure but should work equally well. The lower lateral buds of the cutting are removed to give ±75mm of disbudded stem before rooting substance treatment (IBA) and insertion into the rooting medium. Rooting into trays using a loose (bark + perlite) mix is recommended to allow easy removal of the medium at potting-up time, so that further checks can be made to ensure no sucker development has taken place during the rooting period.

Rooted, disbudded cuttings provide 0+1+1P 6–8mm grade, which should substantially reduce the risk of suckering, provided grafts can be made low enough to avoid any risk of viable rootstock buds remaining.

For top-working 1+2P *Viburnum lantana* are required with a stem girth of 8–10mm or 10–12mm at 80cm high. Root-balled plants can be used if grafting is delayed until September, but potted rootstocks in 2- or 3-litre deep pots provide more flexibility on timing and easier handling at the grafting and potting-on stage.

SCION WOOD

Stock mother plants are required to provide strong, reasonably heavy, and mostly non-flowering scion material for the main commercial types, which are usually grafted. These include: *V. bitchiuense, V.* x *burkwoodii, V.* 'Anne Russell', *V.* 'Mohawk', *V.* 'Park Farm Hybrid', *V.* x *carlcephalum, V. carlesii* 'Aurora', *V.* 'Diana', *V.* 'Chesapeake', *V.* 'Eskimo', *V.* x *juddii* and *V. macrocephalum*. Where possible, current year's wood is selected, but this easily grafted genus unites equally well when a basal portion of 2-year-old wood is included. For apical summer grafts, normally two sets of leaves are retained, often being trimmed back by half or more; scions for side grafts may be de-leaved.

SYSTEMS

Warm callusing is required during the summer and preferable for winter grafting. During the summer, ambient temperatures are normally sufficient and no supplementary heat should be necessary. Warm callusing during the winter is best with bottom heat settings of 15–18°C, which result in higher air temperature during bright periods. Grafts are normally callused and united 4–6 weeks after grafting.

PHYSIOLOGICAL AND ENVIRONMENTAL FACTORS

Root pressure does not appear to be a problem. The woolly leaves of *V. lantana* and *V. carlesii* types can encourage fungal infection. Adequate ventilation and the avoidance of drips while the grafts are in poly-tents should avoid the problem.

TOP-WORKED PLANTS

Viburnum carlesii cvs *V.* x *carlcephalum* and *V.* 'Eskimo' can make attractive Patio plants when grafted at 60–90cm high to stems of *V. lantana*. Normally two scions per stem are side grafted in the summer (August) and placed in a tall poly-tent or walk-in grafting polytunnel to callus and unite. *V.* 'Eskimo' requires two growing seasons following grafting to make a full, balanced head (Figure 54.2).

FIGURE 54.2   LEFT, Boskoop trays containing top-worked *Viburnum lantana*, and awaiting standing down in an overwintering polytunnel. Rootstocks are 1+2P 8–10mm at 80cm, potted into deep, 2-litre pots. These were side veneer grafted with *Viburnum* 'Eskimo' at 80cm high, in a walk-in grafting polytunnel. RIGHT, top-worked *V.* 'Eskimo', with a one-year head of branches, awaiting potting-on into 5-litre pots to be grown on to sale.

GROWING ON

Easily grown in a range of soils, most tolerating or relishing chalky soils; very few, including *V. furcatum*, demand lime-free conditions. Commercially, most plants are now grown in containers. The woodland species, *V. furcatum*, etc., thrive better under some shade.

55 Vitis (Vitaceae) – Vines, Grapevines

INTRODUCTION

The vines comprise a genus of woody climbers with 60–80 species found predominately in the northern hemisphere. Within the genus are those grown for ornament and those for producing fruit (grapes) for dessert or wine. The ornamental types are almost all readily propagated by cuttings. The situation is different for the Grapevine (*V. vinifera*) which is highly susceptible to a soil borne infection, Phylloxera, caused by a diminutive root aphid (*Daktulosphaira vitifoliae*), responsible for decimating the wine growing areas of Europe in the late 19th century until a control strategy was devised. This involved using rootstocks of species and hybrids from America to provide a Phylloxera-resistant root system.

Commercial vines also suffer from other soil-borne pests, notably nematodes (Dagger and Root-knot eelworm). To provide planting material with rootstocks also having a level of resistance to these problems, bench grafting is used on a very large scale in wine producing areas. Rootstocks may also be selected to influence growth and vigour and adapt to particular soil conditions.

COMPATIBILITY

Genetic/phylogenic studies have shown two major groups are identifiable: the American and European cultivar group, and the Asiatic wild grape group. Many of the ornamental types are from the wild Asiatic group and most are successfully propagated by stem cuttings. *Vitis davidii*, a decorative climber from China is an exception and rooting response is poor; tests on suitable rootstocks have not been reported. In view of the comments above, compatibility issues may arise for this species and other related Asiatics may prove best.

Fortunately, the American and European cultivar group contains both the Grapevine (*Vitis vinifera*) and the disease resistant species from America. Consequently, few examples of incompatibility between the two types have been encountered.

TIMING

The bulk of grafting is carried out in the winter, early spring months, December to March. A less often used technique, known as Green Grafting, can take place throughout the year if stock mother plants providing rootstock and scion stems are given appropriate growing conditions in protected structures.

GRAFTING METHODS

Grapevines are grafted using cuttings-grafts techniques. They are particularly suited to grafting by machines or tools because both rootstock and scion comprise exceptionally hard, straight and smooth wood. Methods in which the components fit tightly together and do not require tying-in are favoured. This approach highlights the need before grafting commences to grade rootstock and scion stems to achieve diameters matching as closely as possible.

The Omega graft made by the appropriate grafting machine or tool with an 'Ω'-shaped cut is popular (Figure 55.1); also popular is the 'V notch' inlay graft, otherwise described as inverted

FIGURE 55.1   Omega grafts of Grapevines (photo by Omega Star H+L Wahler Weinstadt–Schnait Germany).

saddle, which requires tying-in (Figure 2.10, Chapter 2). Both of these methods produce frayed wood edges as the cutting tool emerges on the exit side of the cut (see Chapter 2: Machine and Tool-Made Grafts). Some grafters stress the importance of matching the frayed edges together[*], presumably to ensure the relatively undamaged tissues of the entry side are well aligned, allowing tissue repair, callus development and formation of the cambial strand.

Manual grafting methods may be used for smaller numbers if knifesmen with the necessary skills are available. Splice grafts are suitable but must be tied-in or held together by pegs or staples. A whip and tongue graft, using a larger than normal tongue and sealed with grafting wax, provides sufficient strength to make tying-in unnecessary[†].

Green Grafting, using lighter, thinner, semi-ripe material, is carried out by manual grafting methods such as an inverted apical wedge, splice, or apical short tongue veneers. Hand tying is required using rubber ties, leaving gaps between the coils; alternatively, clothes pegs, clips or staples can be used.

For some grafts, which are not tied-in, using a strong adhesive wax (high resin content) aids strengthening of graft unions.

The importance of oxygen availability for grafts of this genus has been discussed previously, consequently, several grafters omit or delay sealing until after completion of union formation when protection from outside conditions following planting is highly desirable[‡].

ROOTSTOCKS

Phylloxera resistance is a basic requirement for vine rootstock selection. All those listed below in Table 55.1 of species and species hybrids show this resistance; also, additional characteristics where known are listed, including nematode resistance, type of root system, drought resistance, vigour, length of growing season and comments on hardiness and suitability for various soil types. Varieties are not included as they are subject to change, with new selections regularly appearing. These may be raised in government-supported and commercial plant breeding establishments.

For winter/spring grafting, stems of current year wood are collected in the December to February period when it is considered the ratio of starch to sugar is optimum[§]. The stems may then be cut or roughly trimmed to length and stored in a cold store at 1°C until required. Material should be stored surface-dry to prevent fungal infection and packed in polythene bags or poly-lined cases. Some

---

[*] Villanueva & Maniaci (1996).
[†] C. Foss (pers. comm.).
[‡] Lentz (1999), Hartmann, Kester, Davies Jr. et al. (2014) pp.791–2 and Waite, Whitelaw-Weckert & Torlay (2015).
[§] Ziraldo (1974).

TABLE 55.1

Characteristics of Phylloxera-Resistant Species and Hybrids*

Name	Nematode Resistance	Root System	Vigour	Growing Season	Comments
rupestris	Low/medium	Deep drought resistant	–	Long	Good for areas with long growing season.
berlandieri	Medium	Deep	Vigorous	–	Hardy.
riparia	Low	Shallow	Low	–	Non-calcareous moist soils.
x champini (candicans x rupestris)	Medium/low	–	–	–	Has been used as a parent for other rootstocks and cultivars.
riparia x rupestris	Moderate	–	Low/medium	–	Requires deep fertile soils. May hasten ripening.
berlandieri x riparia	Medium	–	–	–	Requires moist soil.
berlandieri x rupestris	–	–	Moderate/high	–	Drought tolerant. Best for hot dry soil.

*Adapted from Goldammer (2018), Grape Grower's Handbook.

recommend these bags should be perforated with well-spaced, 7mm holes to allow some air circulation. Preparation before grafting involves cutting the stems into suitable lengths, normally between 25 and 36cm, and 10–12mm in diameter; the basal cut is made just below a node. Rootstock stem lengths may be varied according to customer's requirements, up to 90cm may be specified*.

Rootstocks for Green Grafts comprise the same varieties, but handling procedures are different. Stock plants are placed in protected structures and maintained in a state of active growth by a combination of heat and, if required, artificial illumination. They are grown in deep pallet boxes with a 2.4–3 metres high trellis above to support the shoots which are trained on strings attached to the trellis and fixed to the plants at ground level. The mother plants need 'resting' on a routine basis; this being achieved by lowering temperatures and day length within the structure, or possibly use of a cold store. Material is selected with a minimum diameter of 6mm to provide material for rootstock or scion. Rootstock shoots are cut 30cms long and disbudded, leaving one leaf at the top with the lamina reduced by two thirds.

SCION WOOD

Current year wood for scion material comprising stems 25–30cm long and 8–12mm in diameter, carrying 5 buds, is collected from stock mother plants. After collection, any tendrils are removed; it is then packed in polythene bags and cold stored at 1–2°C. Scion preparation just before grafting involves cutting the collected shoots to length with 20mm of wood above a single bud and 50–60mm below.

Green Graft scions are prepared similarly, but one leaf is retained, and the lamina of this is reduced by two-thirds. They are normally lighter-grade material, 6–10mm in diameter.

SYSTEMS

Most commercial grapevine production revolves around the cuttings-graft system. Disbudding rootstock stems is essential and normally takes place at the pre-grafting stage following a period of storage. Before bud removal, some recommend pre-warming the stems at 21°C and 85% RH "to increase cell activity"[*]. This procedure raises the previously posed question of how quickly woody tissue responds to applied heat. It is possible that this treatment, especially when applied to scion wood carrying buds, may result in a reduction of carbohydrate reserves and have a negative rather than positive effect.

Following disbudding, procedures again differ; many soak the vines before grafting, the reason given is to make the wood less brittle. Soaking times vary from between 3 to 48 hours. Some take the view that soaking significantly increases the likelihood of disease infection[†]. Sterilisation in some form is required in this pre-grafting phase to reduce the incidence of Phylloxera, nematodes and Mycoplasma (grapevine yellows, Pierce's disease). Chemical treatments were favoured and the practice of spraying or dipping in Copper sulphate (0.3gm/litre of water) or 8 hydroxy–quinoline sulphate (Chinosol®) is still in use. Modern practice is to replace chemicals with hot water treatment. Standard timing is 30 minutes at 48°C for plants in cool climates, 50°C for those in moderate climates, such as the UK, and 53°C for those in warm areas such as southern Europe and southern US states.

Hot water treatment shortens storage life of the rootstock stems, so it is preferable to treat after storage and before grafting. If this is not possible, cuttings treated before storage should be allowed 24 hours to return to a surface dry state before re-packaging. For those to be grafted shortly after treatment, 12 hours should be allowed for the material to return and remain at ambient temperature. Hot water equipment needs careful design and use; water temperature should regain specified temperature within 2 minutes of the baskets of cuttings being inserted. Tank capacity should be approximately 1/3 to 1/2 litre per cutting, and the water should be circulated to ensure even heating.

Just prior to grafting, the top of the rootstock stem is re-cut to leave fresh tissue. Grafting methods have been discussed, and once assembled, the grafts are either immediately packed in callusing boxes or sealed by dipping in melted wax. Because the cut end of the scion is exposed, this will require sealing if the rest is not enclosed in wax.

Callusing boxes may be purpose made from ply-wood, plastic, polystyrene, stainless steel or adapted from existing materials such as modified pallet-based potato boxes. They may be constructed as described in Chapter 39: Juglans, with an open top and one removable side. Dimensions are fixed to accommodate the chosen rootstock cutting length plus an extra 100mm to accommodate the scion. If a removable side is incorporated, this is replaced after packing and held in position by webbing straps or a hinge between the adjoining sides.

Scrupulous cleanliness regarding the callusing boxes and surrounding areas is highly desirable to prevent reinfection from what is normally material of carefully monitored, very high-health status. Boxes are often replaced or fully sterilised between crops, stainless steel is costly but provides the most easily cleaned container.

Vine grafts are packed into the box horizontally in layers while it is laid on its side. Some pack the grafts solid with no infill between, but a more satisfactory method is to infill 10–40mm deep between each layer, placed so that the extreme tip of the scion is just protruding from the top of the medium. Mixtures of peat and perlite, 4 to 1, have been popular for this purpose but many use pure, fine-grade perlite or mixtures of perlite and pumice. Whether or not the grafts have an infill, they are placed on a basal layer (15–25mm deep) of well-drained substrate comprising peat/perlite mixtures or pure perlite.

An entirely mineral material, such as Perlite, is less likely to harbour harmful microbes and fungal spores than organic materials such as peat. Whichever material is used it must be carefully moistened to maintain the tissues of the graft components in a fully turgid condition. If grafts are not infilled between the layers, they must be maintained in high humidity conditions (90%+) to ensure that they remain sufficiently hydrated. Consequently, a fully supportive environment for the

---

* Lentz (1999).
† Waite, Grainaje, Whitelaw-Weckert et al. (2013).

grafts is maintained and still equates to warm callusing although very different to conventional systems.

Once packed, the side cover is replaced, the box rotated so that the grafts are in a vertical position. It is then transported to a warm, climate-controlled storage room, or placed on a bench fitted with heating cables. The most sophisticated installations have full climate control with heat, illumination to PAR levels (photosynthetically active range) and, if necessary, humidification equipment. Grafts are left in these conditions until callusing of the union and root initiation, possibly with some root emergence. The time taken for this is dependent upon the chosen temperature regime to be discussed in Physiological and Environmental Factors.

When development is sufficient, the callus boxes are taken from the heating area and the grafts removed. Treatment following varies according to the next stage of handling. In suitable areas, the grafts can be directly field planted, any extension growth from the scion bud during the callusing stage is pinched back and this is re-sealed along with the graft union normally by dipping quickly into melted wax followed by a cold water dip. Any adventitious roots which have extended beyond a few millimetres are also trimmed back. The grafts may then be immediately planted in the field or cold stored at 2–3°C for a period until climatic and soil conditions are ideal for planting.

At the planting stage, the rootstock stem is pushed into the soil, leaving a minimum of 150mm plus the grafted scion above soil level. Weed control and soil moisture levels are more easily maintained if the grafts are pushed through polythene film mulch, ideally with punched holes spaced to accommodate the grafts at the correct distance. This procedure requires appropriate well textured soil, very good climatic conditions and normally involves soil amelioration, soil sterilisation and irrigation facilities.

Some producers want a pot-grown item or are not able to provide suitable conditions for immediate planting in the field. In this situation, callused grafts are pushed into suitable containers held in a glasshouse or polytunnel. Planting tubes $5 \times 5 \times 25$cm are a popular choice[*]; others favour paper tubes or rigid pots. Conditions are maintained at high humidity and shaded until the rootstocks have developed a supportive root system and extension growth has commenced.

For Green Grafts, once grafting is completed, the basal end of the rootstock stem is treated with 1500ppm IBA as a 5 second dip to promote adventitious root formation[†]. They are then inserted into deep cells containing a mixture of peat and perlite (4 to 1), and the whole placed in a poly-tent to provide a supportive warm callusing environment. Shading is necessary to maintain the high humidity required.

PHYSIOLOGICAL AND ENVIRONMENTAL FACTORS

Vitis fall into the group of genera requiring high temperature to achieve a union, as high as 32°C has been suggested. It is now generally recognised that temperatures above 29°C can have a deleterious effect[‡]. The interaction between temperature and callus formation discussed previously applies here, and lower temperature settings require longer in the callusing boxes. 21 days at 26.5°C is probably optimal. Shorter periods at 30°C followed by a period at 26°C are favoured by some. Exposing grafts to periods of high temperatures or prolonged treatment can result in excessive callus production. Callus should not protrude beyond the graft union more than 2–3mm. Excessive callus is vulnerable to infection, and also presents further distance for the cambial strand to achieve linkage between the cambia of rootstock and scion, resulting in delayed formation of a graft union (see Chapter 11: Graft Union Formation for further discussion).

---

[*] Hartmann, Kester, Davies Jr. et al. (2014) pp.791–2.
[†] Lentz (1999).
[‡] Waite, Whitelaw-Weckert & Torlay (2015).

GROWING ON

Numerous publications, investigative work in the USA and other important wine producing areas, together with educational and training courses in the UK, USA and elsewhere, cover all details of Viticulture.

Ornamental vines are invariably grown as container plants in nurseries, where the main requirement is to ensure trimming and tying-in to support canes is regularly maintained until plants reach saleable size.

Part Eight

Genera Grafting Guide Tables

56 Grafting Table List

INTRODUCTION

See the introduction to Part Seven for a description of climatic conditions in *S.* England, which will influence some of the recommendations given in the following table.

Genera are arranged in alphabetical order. A number of genera are dealt with in greater detail in Part Seven; where this applies, it is indicated in the Comments column by 'see also Chapter', with the relevant chapter number shown.

- Unless otherwise stated, pot grown rootstocks are used. Specifications, 1+1P, etc., are generally indicated and the neck collar diameter Ø, when shown (mm), is at grafting time; allowance must be made for size increments in pot culture to establish the required potting-up specification.
- Where more than one method or system is shown, possible choices and preferences are indicated by coloured stars and/or comments.
- Unless otherwise indicated, the hot-pipe temperature is set at 20–22°C.
- Where long-term compatibility is uncertain, it is always worth planting the graft union deeply to attempt to induce scion rooting.
- Preferred rootstock is listed first.
- '?' Is used for a rootstock that is suggested but not proven.
- To conserve space, text used in the comments is often abbreviated.
- Star colours meanings are shown below:

●= preferred choice   = also possible ●= least successful ●= suggestion, not proven

Name	Rootstock	Method		Systems				Comments
		Apical	Side	Cold callus	Warm callus - winter	Warm callus - summer	Hot-pipe	
Abies sp. & cvs	*A. alba, A. nordmanniana* or *A. koreana* 2+1P 5–8mm or 2+1P 8–10mm for thick wooded types.	●			●	●		Require cool conditions. Grafted late August–April most in Jan–Feb. Some favour September to December. **See also Chapter 20.**
Acacia sp. & cvs	*A. dealbata* 1+1P 4–6mm	●			●			Temp requirement 18–20°C. Slow callus former. Difficult.
Acer sp. with non-milky sap	Strong growing species (*A. pseudoplatanus, A. rubrum* etc.) 1+1P 7–10mm. Smaller (*A. palmatum*) 5–7mm	●	●		●	●	●	Rootstocks listed are suitable for a large range but by no means all. Most Maples are best grafted in the summer using side grafts placed in an enclosed environment. Drying–off is important for many species. Stronger growing species may be bare-root grafted in warm callus conditions during late August-early September. **See also Chapter 21.**
Acer sp. with milky sap	*A. cappadocicum, A. platanoides, A. campestre.* 1+1P 5–7mm or 7–10mm							
Acradenia franklinie	? *Choisya ternata*		●		●			An unproven combination.
Actinidia sp. and *A. deliciosa* (Kiwi fruit) cv. 'Hayward' is main variety	A. 'Bruno' 1+0 seedlings or 0+1 cuttings 7–10mm A. 'Bounty 71' clone. Both clones resistant to PsaV root disease	●		●				Only Kiwi fruit normally grafted. Rootstock 'Bruno' requires light, well-drained soil. 'Bounty' is suitable for wetter heavy soils. Temperature during cold callusing not below 6°C. Often machine or tool grafted using 'V' notch. **See also Chapter 22.**
Adenocarpus species	? *Laburnum anagyroides*	●			●			Very difficult from cuttings. Rootstock suggestion only.
Aesculus sp. and cvs	*A. hippocastanum* 1+0 or 1+1P 8–10 to 10–12mm or *A. flava, A. pavia* or *A. x carnea* 1+1P 6–8 to 8–10 for lighter wooded types	●		●			●	Cold callusing suitable for warmer and maritime temperate areas. Hot-pipe superior to warm callusing. Optimum temperature 20–22°C. Nurse seed grafting possible. Top-working good for *A. neglecta* 'Erythroblastos'. **See also Chapter 23.**
Afrocarpus: (*Podocarpus*) *falcatus, A. gracilior*	*Prumnopitys* (*Podocarpus*) *andina* 1+1P 6–8mm	●			●			Other rootstock species in this group could be used.

(Continued)

Name	Rootstock	Method		Systems				Comments
		Apical	Side	Cold callus	Warm callus - winter	Warm callus - summer	Hot-pipe	
Agathis australis	*Araucaria araucana*		●		●			Rootstock compatibility not confirmed, considered likely.
Ailanthus: altissima 'Pendulifolia', 'Purple Dragon', *A. vilmoriniana*	*A. altissima* 2+0 8–10mm or root grafting possible	○		○				Bare-root rootstocks.
Alangium species and varieties	*A. platanifolium* 0+1+1P 5–8mm	○			○		○	Readily produces callus.
Albizia julibrissin cvs	*Albizia julibrissin* ? *Acacia dealbata* 1+1P 6–8mm	○			○			An uncertain combination. Cuttings are difficult.
Alnus: glutinosa cvs, *incana* cvs, x *cordinca*, other species and hybrids	*A. glutinosa* or *A. incana* 1+1 6–8mm bare-root or 1+1P 6–8mm (can pot just before grafting to reduce bleeding)			○	○		○	No problems identified with compatibility. *A. cordata* is a potential rootstock and more amenable to the required drying-off. However, species in sub-genera Alnobetula may demonstrate long-term incompatibility.
Amelanchier: alnifolia 'Obelisk', x *grandiflora* 'Ballerina', 'Robin Hill', 'Rubescens', *laevis* 'R.J.Hilton', 'La Paloma'	*A.* x *lamarckii* 1+0 6–8mm 1+1 8–10 or 1+1P 6–10. Or *Cotoneaster acutifolius*, *C. bullatus*. Also suggested: *Crataegus monogyna*, *Sorbus aucuparia*, *S. intermedia*, *Pyrus communis*	○		○			○	Grafting takes tend to be lower on *Amelanchier* than other rootstocks. *Cotoneaster* can be considered as alternative rootstock. Long-term growth and compatibility best on *Amelanchier*. Hot-pipe at 18–20°C produces best take. Some cultivars: *A* x *grandiflora* 'Ballerina', 'Robin Hill', 'Rubescens', *A. laevis* 'R.J. Hilton' best for producing small trees. **See also Chapter 24.**
x *Amelosorbus jackii*	*Amelanchier lamarckii* 1+1P 6–8mm or 2+0 8–10mm bare-root	○		○				*Crataegus phaenopyrum* has been suggested as an alternative rootstock. *Sorbus aucuparia* may be another possibility.
Amenotaxus argotaenia	*Torreya californica* 6–8mm		●		●			Rootstock ex. cuttings or seed.
Ampelopsis								See *Parthenocissus*.
Aralia elata: 'Aureo-variegata', 'Silver Umbrella', 'Variegata'	*A. elata* bare-root (often root-balled) 15–25mm or seed raised (often pot grown) 1+2P 12–20mm	○	○		○			Often apical graft using a bud plus shield as a scion. Short, lateral branches also possible. A complicated procedure described fully in Chapter 25. **See also Chapter 25.**

(Continued)

Name	Rootstock	Method		Systems				Comments
		Apical	Side	Cold callus	Warm callus - winter	Warm callus - summer	Hot-pipe	
Araucaria sp.	*Araucaria araucana* 1+1P or 2 + 1P 8–10mm	◐		◐				Warm temperate species successfully grafted at RBG Kew.
Arbutus: andrachne, x *andrachnoides*, *menziesii*, x *reyorum* 'Marina'	*A. unedo* 1+1P or 2+1P 6–8mm or ? *A. x andrachnoides* 0+1+1P 6–8mm	◐		◐	◐			Doubts about compatibility of *A. andrachne* & *A. menziesii* grafted to *A. unedo*. *Arbutus* x *andrachnoides* a possible inter-stem for *A. andrachne*.
Arctostaphyllos tall sp.	? seedlings of tall sp. *A.manzanita*	●		●	●			Speculative recommendations.
Argyrocytisus battandieri (*Cytisus battandieri*)	*Laburnum anagyroides* 1+1P 6–10mm	◐		◐				Easily grafted, ambient temperatures should not stay long below 6°C.
Aristolochia macrophylla	Roots of *A. macrophylla* 10–12mm Ø x 80mm	◐			◐			Root grafts potted into individual plug or pot. 18–20°C bottom heat.
Arthrotaxis sp.	*Cryptomeria japonica* 0+1+1P 6–8mm	◐				◐		*Cupressus glabra* may also be a suitable rootstock.
Austrocedrus chilensis	*Thuja occidentalis* 1+1P 6–8mm or ? *Platycladus orientalis*	◐				◐		*Calocedrus decurrens* is an alternative rootstock.
Berberis deciduous sp: *temolaica*, evergreen sp: x *lologensis, trigona* (*linearifolia*)	*B. thunbergii* or *B. vulgaris* 1+1P 6–8mm or 2+0 bare-root	◐		◐	◐			Apical graft, Nov–Feb or also possible for evergreens, Aug–Oct. Deciduous species can be cold callused, not long below 6°C; or, as for evergreen sp., warm callus at 15–18°C. Scion wood with 2yr basal portion if necessary. **See also Chapter 26.**
Betula sp. and cvs	*B. pendula* or *B. nigra* 1+1P 6–8mm or 2+0 6–10mm bare-root. *B. lenta* or similar for some in Asperae subgenus	◐	◐	◐	◐		◐	Rootstock must be well dried before grafting, especially those warm callused. Cold callus not long below 6°C; warm callus 18–21°C; hot-pipe 20–22°C. Late summer grafting possible. Incompatibility a problem with some in subgenus Asperae. **See also Chapter 27.**
Bignonia capreolata cvs	*Campsis radicans* 1+1P 6–8mm or roots 8–10mm Ø x 80–100mm	◐			◐		◐	Hot-pipe/potted rootstocks most successful. Root grafting/warm callus possible.
Broussonetia: kazinoki, papyrifera	*Broussonetia* roots 10–12mm Ø x 80–100mm	◐			◐			*Morus alba* or *M. nigra* roots are alternatives for root grafting.

(Continued)

Name	Rootstock	Method		Systems				Comments
		Apical	Side	Cold callus	Warm callus - winter	Warm callus - summer	Hot-pipe	
Callitris sp.	? *Cupressus glabra*		●		●			Unproven combination.
Calocedrus decurrens cvs	*Calocedrus decurrens* 1+1P 5–8mm or mm or *Platycladus* (*Thuja*) *orientalis* ?*Thuja occidentalis*		●			●		Graft March–May or preferably August–October. Procedures follow those for *Chamaecyparis* given below and also in Chapter 32.
Calophaca: *grandiflora, wolgarica*	*Caragana arborescens* 1+1P 6–8mm	●		●				Grafts grow more quickly than seedlings.
Camellia sp and cvs especially *C. reticulata* cvs and some *reticulata* hybrids	*C. japonica* strong growing cvs (e.g. 'R. L. Wheeler') 0+1+2P 6–8mm. If available, strong seedlings are to be preferred	●	●		●	●		Timing Jan–March or Aug–early September. Rootstocks dryish for winter grafting. Warm callus temp. 18–20°C. Heavy shading in summer to keep grafts cool. Cuttings-grafts possible in late July–August. A difficult genus. **See also Chapter 28.**
Campsis: *grandiflora*, x *tagliabuana* cvs	*Campsis radicans* 1+1P 6–8mm or *C. radicans* roots 8–10mm Ø x 80–100mm	●			●		●	Hot-pipe or warm callus using potted rootstocks is very successful. Warm callus for root grafts. *Catalpa bignoniodes* root pieces said to be a possible alternative rootstock.
Caragana: *arborescens* cvs 'Lorbergii', 'Pendula', 'Walker', *Caragana* species	*C. arborescens* Bottom-worked 1+1P 5–8mm or stems 1+2P potted in 2-litre deep depth	●		●				Bottom-worked species should be successful in hot-pipe. Top-worked weeping cvs on 120+cm 3yr stems, root-balled or potted. Scion 4–5 buds minimum. Cold callused temp. 6°C minimum.
Carpinus betulus cvs and others in Carpinus section. *Carpinus* in Distegocarpus section: *cordata fangiana* cvs, *rankinensis*	*C. betulus* for those in Carpinus section 1+1P 6–8mm or 1+2P 8–10mm or bare-root 1+1 8–10mm. If possible, use *C. japonica* for Distegocarpus section	●	●		●	●	●	Most in Distegocarpus section are compatible on *C. betulus* for a long period if not permanently. Late summer/early autumn apical grafting to bare-root rootstocks successful for *C. betulus* cvs. Pot grown generally favoured for others. **See also Chapter 29.**

(Continued)

Name	Rootstock	Method		Systems				Comments
		Apical	Side	Cold callus	Warm callus - winter	Warm callus - summer	Hot-pipe	
Carya sp. and cultivars	C. cordiformis or C. ovata 1+1P 8–12mm or root pieces 10–12mm Ø × 100–120mm C. illinoinensis seedlings best used for Pecan nuts	○			●		○	Hot-pipe, or for root grafts, warm callus. Temperature ±24–27°C. Pecan (C. illinoinensis) cultivars use potted rootstocks in very deep pots. See Juglans (Chapter 39) for more information. Juglans nigra and J. regia roots are said to be compatible with Carya.
Castanea sativa cvs and species in variety	C. sativa 1+1P 6–10mm or 1+1 bare-root 8–10mm	○			○		○	Hot-pipe the best system. Scions of some species could be incompatible on C. sativa. Nurse seed grafting possible.
Catalpa bignoniodes cultivars and other species. The following are best top-worked: bignoniodes 'Aurea', x erubescens 'J. C. Teas' ('Purpurea'), 'Pulverulenta', speciosa	C. bignoniodes 1+1P 8–10mm or 1+1 or 2+0 8–10mm bare-root. Alternatively, root pieces 10–12mm Ø x 100–120mm For top-working stems 1+2 potted 9–12 litre pots or 2+1 root-balled 10–12mm at 150–180cm high	○		●	○		○	Hot-pipe for bottom-worked species and cultivars. Use warm callusing for root grafts. Catalpa bignoniodes 'Aurea' and other foliage forms can be top-worked at 150–200cm high on stems which are root-balled or potted. Grafts placed in cold or warm callusing conditions.
Cathaya argyrophylla	? Pseudotsuga menziesii		●		●			Unproven combination but some grafts have survived 4 years. Larix decidua a possible alternative rootstock.
Cedrus sp and cvs	C. deodara 1+1P 6–8mm		○		○	○		Timing is either late July to November or February to March. In the winter period, grafts kept cool as possible, not above 18°C. **See also Chapter 30.**
Celtis species in variety	C. australis or C. occidentalis 1+1P 5–8mm	○			○		○	Easily grafted in late winter. Hot–pipe very successful. No compatibility issues discovered.
Cephalotaxus: fortunei cvs, harringtonia cvs	Taxus baccata 2+1P 6–8mm		○		○			Grafting follows system recommended for Taxaceae. See Chapter 18: Conifer Grafting.
Cercidiphyllum japonicum 'Amazing Grace', 'Heronswood', 'Globe', 'Morioka Weeping', f. pendulum	C. japonicum 1+1P 6–8mm for bottom-working. Stems for top-working in 2-litre deep pots or root-balled		○		○		○	Bottom-worked grafts succeed well in hot-pipe, 20–22°C. Globular and weeping forms may be top-worked at 120–180cm high. Summer grafts are also possible.

(Continued)

Name	Rootstock	Method		Systems				Comments
		Apical	Side	Cold callus	Warm callus - winter	Warm callus - summer	Hot-pipe	
Cercis: siliquastrum cvs, canadensis cvs, other sp.	C. siliquastrum or C. canadensis 1+1P 6–10mm. Bare-root is occasionally used but not recommended		○		○	●	○	Hot-pipe gives by far best results in maritime climates. High temperature requirement of 25–26°C. Good quality rootstocks very important. Use only well-ripened, current year wood for scions. Long tongue side veneer favoured. **See also Chapter 31.**
Chaenomeles species, hybrids and cultivars	C. japonica or C. speciosa 1+1 or 2+0 6–8mm bare-root	○		○				Grafting rarely required but possible. Easily grafted but suckering a problem.
Chamaecyparis: formosensis, lawsoniana cvs, obtusa cvs, pisifera cvs, thyoides cvs	C. lawsoniana 1+1P 6–8mm or 1+1 or 2+0 bare-root. Thuja occidentalis is proposed as alternative for C. obtusa	●	○		○	○		C. obtusa eventually makes an ugly union on C. lawsoniana. Long-term compatibility of C. lawsoniana on C. thyoides uncertain. Bare-root apical grafting possible autumn and late Feb. Long tongue side grafts to potted rootstocks favoured. Temperature requirement 10–12°C bottom heat. **See also Chapter 32.**
Chimonanthus praecox cvs: 'Concolor', 'Grandiflora', 'Luteus', etc.	Chimonanthus praecox 1+1P 6–8mm. ?Calycanthus floridus	○	○		○		○	Graft early January. Remove flower buds from scion wood to avoid Botrytis in enclosed environment. Calycanthus floridus is a possible rootstock.
Chionanthus retusus	Chionanthus virginica 1+1P 5–8mm or Fraxinus ornus or own root graft	○			○	●	○	Hot-pipe gives better results than warm callusing. Summer grafting also possible. Experience is that after 1–2yrs grafted to F. ornus the graft often fails.
x Chitalpa tashkentensis cvs	Catalpa bignoniodes (see Catalpa)	○			○		○	Hot-pipe successful for bottom-worked grafts. Top-working 180cm high on stems of Catalpa bignoniodes is successful.
Chosenia arbutifolia (bracteosa)	Salix viminalis or S. x smithiana 0+1P or 0+1	○		○				Long-term compatibility of this combination is not certain.
Chrysolepis: (Castanopsis) chrysophylla, sempervirens	Roots from same genus or? seed of Quercus from various sections to be trialled	●				●		Root grafting in poly-tent. Temp: ?18–20°C. C. sempervirens has failed on Quercus ilex rootstock. ? Nurse seed graft using Q. robur acorns or those from Cyclobalanopsis section.

(Continued)

Name	Rootstock	Method		Systems				Comments
		Apical	Side	Cold callus	Warm callus - winter	Warm callus - summer	Hot-pipe	
Citrus: x *aurantium, cavaleriei* (*ichangensis*), x *insitorum*, x *limon* 'Meyer', *japonica* (Fortunella)	*Citrus trifoliata* (*Poncirus trifoliata*) 1+1P 5–8mm. Alternatives are not generally hardy in cool temperate areas such as UK		●		●	●		Other rootstocks are also used. These have mostly been developed in California for resistance to soil pathogens. In some southern areas grafts are achieved using cuttings-grafts.
Cladrastris: kentukea 'Perkins Pink' & other cvs., *sinensis*	*Cladrastris kentukea* 1 or 2+1P 5–8mm		●		●		●	Hot-pipe best. Winter apical or summer side grafting in poly-tent successful. Temperature 20°C +.
Clematis: armandii cvs, *finetiana*	*Clematis vitalba* or *C. flammula* 1+0 5–8mm Headed back to hypocotyl or just below at grafting time	●			●			Early Dec. to Feb. Scion 5–8mm Ø stem, one inter-node with 2 buds. Grafted to root of *C. vitalba* using apical short tongue veneer. After graft completed potted union buried, buds just above compost. Temp. 17–21°C. Objective to promote scion rooting.
Clematis: alpina, florida, macropetala, Large-flowered garden *Clematis*	*Clematis vitalba* or *C. flammula* 1+0 4–6mm Headed back to hypocotyl or just below at grafting time	●			●			Mother plants forced in heat from January to produce ripened but actively growing shoots for scion wood. Grafted March–April. Scion stem split into two, with bud on each piece. Splice graft method. Tied-in with 5mm of splice projecting at base to encourage scion rooting. Temp. 20–21°C. Potted after preparing graft is completed.
Cleyera japonica 'Fortunei'	*C. japonica* ex. cuttings. 0+1+1P 5–7mm		●		●			Cuttings of 'Fortunei' possible but difficult.
Colutea: *arborescens* 'Bullata', 'Bushei', *orientalis*	*C. arborescens* 1+1P 6–8mm	●		●	●			Cuttings possible but 'Bullata' very slow growing and best grafted.
Cornus: alternifolia cvs, *controversa* cvs	*C. alternifolia* or *controversa* or *C. alba* or *C. sanguinea* 1+1P 6–8mm or *C. alba* cvs 0+1+1P ex cuttings		●		●	●	●	Rootstocks *of C. alba* cvs ex cuttings favoured over *C. controversa* or *alternifolia*. Hot-pipe may change the recommendation. Winter warm callus 15–18°C, summer <±25°C. Root pressure a potential problem. **See also Chapter 33.**

| Name | Rootstock | Method | | Systems | | | | Comments |
		Apical	Side	Cold callus	Warm callus - winter	Warm callus - summer	Hot-pipe	
Cornus: mas cvs, officinalis cvs	C. mas 1+1P 6–8mm	○	○		○		○	Generally grafted Jan –March. Other recommendations as above. **See also Chapter 33.**
Cornus: florida cvs, nuttallii and cvs	C. florida 1+1P 6–8mm		○		●	○	○	Rutgers and other hybrids may respond slightly differently to C. florida and C. kousa. Further trials required. Warm callus, winter 15–18°C, summer <±25°C. Hot-pipe 20–22°C. Heavy scions with 2yr wood basal portion are a successful strategy, also use of heavy rootstocks. **See also Chapter 33.**
C. hybrids, 'Eddies White Wonder', Rutgers Hybrids (C.x rutgersensis)	C. florida or C. kousa 1+1P 6–8mm							
C. kousa cvs, capitata, hongkongensis	C. kousa 1+1P 6–8mm							
Corylus: avellana cvs, chinensis, colurna cvs, maxima cvs, 'Te Terra Red'	C. avellana 1+0 or 1+1 6–8 or 8–10mm; 1+1P 6–8mm or 8–10mm. C. colurna as above and top-worked 60–120cm high stems	○	○	●		○	○	Doubts about long-term compatibility of some Hazelnuts on C. colurna. Summer grafting a good strategy, especially for top-worked methods. Hot-pipe temperature 22–24°C. **See also Chapter 34.**
Cotinus obovatus & hybrids	C. coggygria 1+1P 6–8mm	○			○		○	Hot-pipe very successful.
Cotoneaster - bottom-worked forms: 'Cornubia', 'Exburiensis'. Top-worked forms: 'Hybridus Pendulus', radicans 'Eichholz', x suecicus 'Coral Beauty'	C. bullatus, C. frigidus, C. x watereri 1+1P or 1+0 or 0+1. Also can be used as stems for grafting at 120–150cm. Crataegus phaenopyrum recommended as a stem in very cold areas	○		○	○			Mostly propagation by cuttings so bottom-working not normally required. For top-working, usually 2 scions grafted approximately opposite at chosen height. May be grafted at any time between August and March. Choice largely dependent upon working schedule.
+Crataegomespilus: 'Darda rii', 'Jules d'Asnieres'	C. laevigata (oxycantha) or monogyna 1+1P or 1+1	○		○				May be grafted at any time between August and March. Careful selection of scion wood to maintain true characteristics.

(Continued)

Name	Rootstock	Method		Systems				Comments
		Apical	Side	Cold callus	Warm callus - winter	Warm callus - summer	Hot-pipe	
Crataegus Eurasian species: *laevigata* 'Paul's Scarlet', 'Punicea', 'Rosea Flore Pleno *monogyna* 'Biflora', 'Flexuosa', *orientalis, pinnatifida* var. *major*	*C. laevigata* (*oxycantha*) or *monogyna* 1+1P or 1+1 6–8mm. In the Netherlands *Sorbus intermedia* has produced high takes from field budding. No long-term investigations on compatibility have been discovered	◦		◦				*Crataegus* appear to be easily grafted; apical grafts using cold callus systems produce consistently high takes. Forty Series of *Crataegus* have been recognised. The Old World (Eurasian) species and New World are not all closely related botanically, which may be an indication of potential compatibility problems. Some grafters maintain incompatibility does exist between the Old and New world species. A view supported by personal experience.
Crataegus Hybrids between Eurasian and New World species and *Crataegus* New World species and hybrids	For these *C. persimilis* (*prunifolia*), *C. crus-galli, C. mollis* have been suggested as potentially superior to *C. laevigata* or *C. monogyna*	◦		◦				In field-budded stands, the Eurasian species, hybrids between each other, and Old and New world hybrids exhibit low level incompatibility (±2%) when field budded on *C. laevigata* or *C. monogyna*. Growth of some species (e.g. *C. marshallii*) is poor when grafted to *C. monogyna* or *C. laevigata*.
x *Crataemespilus grandiflora*	*Crataegus monogyna* 1+1P or 1+1 bare-root 6–10mm	◦		◦			◦	Treatment as for *Crataegus*.
Cryptomeria japonica: tall forms; those with irregular foliage and dwarf forms	*Cryptomeria japonica* 1+1P 5–8mm or *Cryptomeria japonica* Elegans 0+1+1P 5–8mm		◦			◦		Timing April/early May or August to early November. Procedures follow those for *Chamaecyparis lawsoniana* and in Chapter 32.
Cunninghamia: *lanceolata* 'Glauca', *konishii*.	*Cunninghamia lanceolata* 1+1P	●				◦		Best temperature ±'15°C. No waxing necessary but avoid drips.
Cupressus: *arizonica* cvs, *lusitanica, macrocarpa* cvs, *sempervirens*	*C. macrocarpa* or *C. arizonica* var *glabra* 1+1P 5–8mm. Stems 60–90cm high for top-working		◦			◦		Winter grafts best in temperatures above 15°C. Summer grafts, mid-July to mid-September <25°C. Susceptible to Botrytis some enclose grafts in fleece rather than polythene. **See also Chapter 35.**

(Continued)

Name	Rootstock	Method		Systems				Comments
		Apical	Side	Cold callus	Warm callus - winter	Warm callus - summer	Hot-pipe	
x *Cuprocyparis*: *notabilis*, *ovensii*	x *C. leylandii* 0+1+1P 6–8mm	●				●		Treatment as for *Chamaecyparis*, Chapter 32.
Cydonia oblonga (Quince), 'Meech's Prolific', 'Serbian Gold', 'Vranja'.	Quince A 0+1+1 or *Cydonia oblonga* 1+1 8–10mm bare-root	●		●				Cold callusing very successful. Hot-pipe would work well but not essential.
Cytisus battandieri								See *Argyrocytisus battandieri*
Cytisus: x *kewensis*, *praecox*, *praecox* 'Allgold', other varieties possible.	*Laburnum anagyroides* 1+1P, 1+1 or 2+0 stem (splice) 8–10mm at 120–150cm	●			●			Only top-worked. *Laburnum* stems either potted or lifted autumn and root-balled before early spring grafting.
Dacrydocarpus dacrydiodes (*Podocarpus dacrydiodes*)	? *Prumnopitys* (*Podocarpus*) *andina* 1+1P or *Dacrydium cupressinum* 0+1+1P	●			●			? Best temperature 15–18 °C.
Dacrydium								See *Halocarpus*, *Lagerostrobus* and *Lepidothamnus*.
Daphne: *bholua* cvs, x *burkwoodii*, *mezereum* cvs, *odora* cvs, x *transatlantica*	*D. acutiloba* or *D. longilobata* 1+1P or 1+2P 5–7mm or *D. odora* 0+1+1 5–7mm. For some *D. mezereum* bare-root or 1+1P 6–8mm	●			●	●		*D. mezereum* rootstock for *D. mezereum* cvs and *D. jezoensis*. Dec–Feb grafts, temperature 15–18°C, early July–mid-August temp. <25°C. Summer grafting avoids winter damaged scion wood. Side grafts maintain rootstock health. **See also Chapter 36.**
Daphniphyllum: *macropodum* var. *humile*, *macropodum* 'White Margin'	*D. macropodum* 1+1P 6–10mm	●			●	●		Rootstocks difficult to obtain. Easily grafted. Other variegated forms exist.
Davidia involucrata 'Sonoma', and other cultivars	*D. involucrata* or *D. vilmoriniana* 1+1P or 1+2P 6–8mm. *Nyssa sylvatica* has been suggested but is eventually incompatible	●	●		●			Relatively easily grafted. Difficult to cultivate in a pot due to root infections. If material is in limited availability, the scion may be a single bud. Hot-pipe should be successful. Temperature 20–22°C.
Diospyros: *kaki* and cvs, *lotus*, *virginiana*	*D. kaki* or *D. virginiana* 1+1P or 1+2P 6–8mm for *D. kaki*. ? *D. virginiana* for *D. lotus*. Dwarfing rootstocks for *D. kaki* being developed	●					●	Hot-pipe ± 22°C likely to produce best results. Summer grafting in August should also be successful. *D. lotus* has shown long-term incompatibility as a rootstock for *D. kaki* cultivars. Compatibility of *D. lotus* as scion variety on other species unknown.

(Continued)

Name	Rootstock	Method		Systems				Comments
		Apical	Side	Cold callus	Warm callus - winter	Warm callus - summer	Hot-pipe	
Dirca: occidentalis, palustris	? *Daphne longilobata*		●			●		Very difficult from cuttings, possibility of grafting to *Daphne* not investigated.
Docynia: delavayi, indica	*Cydonia oblonga* or possibly Quince A, bare-root or potted *Pyrus communis* 1+1 6–8mm bare-root or 1+1P 6–8mm	●		●				Pear rootstock sometimes recommended. *Cydonia* or Quince A may show better long-term compatibility.
Edgeworthia: chrysantha 'Grandiflora', 'Red Dragon'	*E. chrysantha* 1+1P or 2+1P 5–7mm		●		●		●	Relatively easily grafted but acquisition of rootstocks can be difficult. Hot-pipe worth investigation and might work well.
Elaeagnus: x *ebbingei* cvs, pungens cvs, macrophylla	*E. umbellata* 0+1 (hardwood cuttings) or 1+0 Bare-root or 1+1P 6–10mm. *E. multiflora* 1+0 has also been suggested. *E. pungens* 0+1+1P 6–8mm produces fewer suckers than *umbellata*	●	●		●	●		*E. macrophylla* more demanding than *E.* x *ebbingei* and *E. pungens* and best late summer side grafted onto pot grown *E. pungens* rootstocks. Bare-root grafting in spring for easier types. Lower buds should be removed from hardwood cuttings of *E. umbellata* to prevent suckering.
Erica Cape Heath species	? *E. arborea alpina* 0+1+1P 5–7mm. Other species may be necessary for compatibility		●		●	●		A potential method. Temperature 18–20°C in early spring. 20–22°C in late August. To avoid Botrytis careful ventilation and no drips.
Eriobotrya deflexa	*E. japonica* 1+1P 6–8mm. *Photinia beauverdiana* 1+1P 6–8mm or *P.* x *fraseri* cvs 0+1+1P		●		●	●		Relatively easily grafted late winter or late summer. Long-term compatibility on *Photinia* uncertain.
Eucalyptus: (*Corymbia*) ficifolia, megalocarpa	? *E. gunnii*, ideally 1yr plug grown 6–8mm							Very little experience in Europe with this difficult genus. Juvenile rootstocks considered essential.
Eucryphia sp. and cultivars as required.	*E.* x *nymansensis* 0+1+1P 5–7mm, *E.* x *intermedia* 1+1P 5–7mm					●		Rarely grafted and difficult. Side grafts in late summer have most potential for success.

(Continued)

Name	Rootstock	Method		Systems				Comments
		Apical	Side	Cold callus	Warm callus - winter	Warm callus - summer	Hot-pipe	
Euonymus species: *europeus* cvs, *fortunei* cvs (as top-worked Patio plants)	*E. europeus* 1+0/1+1 6–8/8–10mm 1+1P 6–8mm. Stems 2+0 or 1+2P T/W @ 80cm	○		○	○			Rarer species, e.g. *E. clivicola, E. cornutus*, grafted to pot grown rootstocks. Many others bare-root grafted winter or late summer August/September. Rootstock can sucker badly.
Eurya: *emarginata* 'Microphylla', *japonica* 'Variegata',	*E. japonica* 0+1+1P or 1+1P 6–8mm							Seedling rootstocks rarely available.
Evodia (Euodia)								See *Tetradium*
Exbucklandia populnea	? *Distylium racemosum* 1+1P. Plus an inter-stem of ? *Rhodoleia championii* or *R. henryi*		●		●	●		An unproven suggestion to meet compatibility requirements. *Hamamelis* rootstock has proved incompatible; a different rootstock combination will be required.
Fagus: *engleriana*, *orientalis* cvs, *sylvatica* cvs,	*F. sylvatica* 1+1P 6–8mm. Bare-root 1u1 or 1+1 6–8 or 8–10mm	○	○	○	○	○	○	Winter grafting Jan–March cold callus or warm callus 15–18°C or late July–early Sept. Bare-root apical grafts early Sept also possible. Hot-pipe 18–20°C. **See also Chapter 37.**
Firmiana simplex 'Variegata'	*F. simplex* 1+1P 6–8mm		○				○	Hot-pipe 22–24°C should produce good results.
Fitzroyia cupressoides	*Cupressus macrocarpa* 1+1P 5–7mm		○		○			Normally propagated by cuttings.
Fokienia hodginsii	*Platycladus (Thuja) orientalis* 1+1P 5–7mm. *Thuja plicata* a possible alternative		○			○		Treatment as for *Chamaecyparis*. Chapter 32.
Fontanesia: *fortunei*, *phillyreoides*	*F. ornus* 1+1P 6–8mm or *F. excelsior* 1+1P 6–8mm	○			○			Both *Fraxinus* species worth trying as a rootstock. Some doubt over long-term compatibility on either.
Fortuneria sinensis	? *Corylopsis sinensis* or *C. spicata* 1+1P 5–7mm		●		●		●	A potential method. Long-term compatibility uncertain.
Fortunella								See *Citrus*
Fothergilla: x *intermedia, major*	*Parrotia persica* 1+1P 6–8mm or stem 8–10mm Ø at 60–120cm high		○		○			Top-worked stems of *Parrotia persica* have proved compatible for 30 years. Bottom-worked grafts should be successful.

(Continued)

Name	Rootstock	Method		Systems				Comments
		Apical	Side	Cold callus	Warm callus - winter	Warm callus - summer	Hot-pipe	
Frangula: (formerly *Rhamnus) alnus* 'Asplenifolia', *californica* cvs, *purshiana*	*Rhamnus cathartica* 1+1P 6–8mm	●	○		○		○	De-leaved *F. californica* for hot-pipe system.
Fraxinus sections: Fraxinus Melioides and Sciadanthus: *americana* cvs, *angustifolia* cvs, *excelsior* cvs, *pennsylvanica* cvs	*F. excelsior* 1+1P 6–8mm or 1+0/1+1 6–8mm; 8–10mm bare-root. *F. americana* or *F. pennsylvanica* also suitable. Stems 8–10mm Ø at 60–80cm high for *F. excelsior* Nana as a Patio plant	○			○			Grafting frequently takes place in the field. Cold callusing normally satisfactory. Warm callusing saves the need to seal the grafts.
Fraxinus sections: Dipetalae and Ornus: *bungeana, chinensis, dipetala, griffithii, sieboldiana (mariesii)* cvs	*F. ornus* 1+1P 6–8mm	○			○		○	The Ornus and Dipetalae sections are more demanding and warm callusing or hot-pipe can be considered. All types, but especially Ornus section, can be summer grafted from late July to early September.
Garrya elliptica cvs, Other species and hybrids between them.	*Garrya elliptica* 0+1+1P 6–6mm		○			●	○	Normally propagated by cuttings. Late summer grafting gives good results.
Genista: lydia, pilosa 'Vancouver Gold',	*Laburnum anagyroides* 1+1P or 1+1 stem 8–10mm Ø at 60–120cm		○		○			Same method as for *Cytisus*. Suitability and long-term performance of this combination open to question.
Ginkgo biloba cvs	*Ginkgo biloba* 1+1P or 1+2P 6–8mm. or stems 2+ 2+1P or 2+ 2/3 root-balled 8–10mm Ø at 80–150cm	○			○			Some dwarf cultivars and 'Pendula' side grafted with two scions to provide a top-worked Patio tree. Strong growing and upright forms 'Autumn Gold'; 'Tremonia', bottom-worked to potted rootstocks, using splice or apical veneer. Cold callusing conditions suitable, top-worked grafts often placed in warm callus conditions.

(Continued)

Name	Rootstock	Method		Systems				Comments
		Apical	Side	Cold callus	Warm callus - winter	Warm callus - summer	Hot-pipe	
Gleditsia triacanthos cvs and *G.* species	*Gleditsia triacanthos* f. *inermis* (spineless) 1+1P or 1+0/1+1 bare-root. For top-working 1+2P or 1+2 8–10mm Ø at 150–180cm	○		○			○	Rootstock f. *inermis* preferred over *G. triacanthos.* No evidence of incompatibility to date. Compact or weeping cultivars top-worked using 2 scions placed in warm callusing conditions. Hot-pipe for pot grown rootstocks very successful, especially for the species.
Glyptostrobus pensilis	*Taxodium distichum* 1+1P 6–8mm	○		○				Procedures follow those described for other deciduous conifer species.
Grevillea hardier species and varieties: 'Canberra Gem', *juniperina* 'Olympic Flame', *victoriae*	Any available species. *G. lanigera* 0+1+1P 4–7mm a good choice. Seedlings of any hardier species would be ideal		●		●	●		Normally propagated by cuttings but always difficult. Late summer side veneer grafts offer best chance. Requirements will be high humidity but no drips and as little shade as possible. Dry fog may be ideal. Grafts said to withstand 40°C.
Gymnocladus dioica 'Variegata'	*G. dioica* 2+2P 8–10mm. For top-working 2+2/3+1P or root-balled stem 10–12mm at 180cm high		○		○			Very slow growing if bottom-worked. Top-worked at 180cm high, using side veneer and 2–3 scions makes a larger plant more quickly, extension growth is still slow. Scion selection to avoid green reverted shoots is very important. Scion wood for subsequent collection should be marked while in leaf.
Halesia: Monticola Group, *H. carolina* Vestita Group and cvs within the Groups.	*H. carolina* 1+1P or 1+2P 6–8mm	●	○		○		○	Some have had success cold callusing. Side grafts and warm callus at 18–21°C is recommended. Hot-pipe, preferably using side grafts, gives best results. *H. diptera* compatible on *H. carolina* only in the short/medium term.
Halesia diptera Magniflora Group	? *Pterostyrax hispida* or *P. corymbosa*		●		●		●	No investigation of this suggested combination has taken place.
Halimodendron halodendron	*Caragana arborescens* 1+1P 5–7mm	○		○				Summer cuttings very liable to foliage rotting. Grafts grow faster than cutting raised plants.

(Continued)

Name	Rootstock	Method		Systems				Comments
		Apical	Side	Cold callus	Warm callus - winter	Warm callus - summer	Hot-pipe	
Halocarpus biformis (*Dacrydium biformae*)	? *Prumnopitys* (*Podocarpus*) *andina* 1+1P		●		●			A suggested method not tested. Best temperature likely to be 15–18°C.
Hamamelis: x *intermedia* cvs, *japonica* cvs, *mollis* cvs, *vernalis* cvs	*H. virginiana* 1+1P 6–8mm or *H. virginiana* 1+1 or 2+1 bare-root		○	●		○		Side grafts late July to mid-September. Winter grafting cold (>6°C) or warm callusing (bottom heat 15–18°C) Jan-March also possible but less successful. Bare-root grafting using apical graft possible, takes not high. **See also Chapter 38.**
Hedera colchica 'Dendroides' (Arborescens), *H. helix* 'Arborescens'	*H. helix* 1+1P or bare-root. x *Fatshedera lizei* has been suggested. Grown as a 1+1P 6–8mm, it would be more convenient in use than *Hedera helix*	○	○			○		September/October is recommended as best time for grafting but any period during dormancy should be successful for this relatively easily grafted species. Grafting on a short stem using x *Fatshedera lizei* to create Patio plants is a possibility.
Hemiptelea davidii	? *Zelkova serrata* 1+1P 6–8mm		●		●		●	Compatibility unknown. Hot-pipe likely to be most successful.
Heteromeles salicifolia	? *Photinia davidiana* 1+1P		●		●			Should graft easily. Some doubt about compatibility
Hibiscus: paramutabilis, sinosyriacus cvs, *syriacus* cvs	*H. syriacus* 1+0 6–9mm; 1+1 8–10mm; 2+0 8–12mm all bare-root grades or 1+1P pot grown. Root pieces 10–12mm Ø from older plants also possible. For top-working *H. syriacus* potted stems 8–10mm Ø at 120–150cm high are used	○		○				Bare-root grafted at point of hypocotyl using splice or wedge (inverted wedge preferred). A late breaking genus means enclosed environment not essential, bottom heat 12–14°C is suitable. Graft may be buried to avoid need for sealing. Potted after grafting or when callus evident. Pot grown rootstocks rarely used except when top-working. September grafting using complete root system or root pieces. Warm callusing is required at this period.
Huodendron sp.	? Try either inter-stem of *Rehderodendron* or *Melliodendron* grafted to *Pterostyrax hispida*		●		●	●		Has proved to be incompatible with *Pterostyrax hispida*. If a suitable rootstock can be identified, de-leafing and hot-pipe a possibility.

(Continued)

Name	Rootstock	Method		Systems				Comments
		Apical	Side	Cold callus	Warm callus - winter	Warm callus - summer	Hot-pipe	
Idesia: polycarpa male & female; *I. polycarpa* var. *vestita*.	*I. polycarpa* 1+1P 6–8mm	◦	◦		◦			Hot-pipe not investigated but likely to produce best results.
Ilex: x *altaclerensis* cvs, *aquifolium* cvs, *cornuta* cvs, *kingiana*, x *koehneana* cvs, *latifolia*, *opaca* cvs, other sp. and cvs	*I. aquifolium* or *I.* x *altaclerensis* 1+1P or 0+1+1P 5–7mm. *I. cornuta* 'Burfordii' or I. 'Nellie Stevens' suggested as an alternative. The latter is reputedly very resistant to poor drainage		◦			◦		A large family and compatibility issues have not been resolved for all. Some e.g. *I. pernyi* survive but do not grow freely on *I. aquifolium* rootstock. Deciduous species may require deciduous rootstocks. Timing is best late summer. Temperature for winter 18–20°C, for summer <24°C.
Illicium species	*I. floridanum* 0+1+1P 6–8mm		◦			◦		Normally propagated by cuttings but grafts reliably in late summer.
Juglans: x *intermedia*, *nigra* cvs, *regia* cvs including vars. grown for nut production	*J. regia* or *J. nigra* or *J.* Paradox hybrids, 1+0, 1u1, 1+1 8–12mm or 2+0 12–22mm; all above bare-root. 1+1P 10–12mm or ex. Microprop 6–10mm pot grown	◦	◦		◦		◦	Warm callusing, callusing box, hot-pipe all successful. In N. Europe hot-pipe best. Optimum temp. 25–27.5°C. Drying-off crucial to success. Scion wood must be non-pithy and 2yr. basal portion may be included. Root grafts are possible. **See also Chapter 39.**
Juniperus: chinensis cvs, *deppeana*, *horizontalis* cvs, *sabina*, *scopulorum* cvs, *virginiana* cvs	*J.* x *pfitzeriana* 'Hetzii' (*J.* x *media* 'Hetzii') 0+1+1P 5–7mm. For top-worked stems *J. scopulorum* 'Skyrocket' 0+1+2P 7–10mm		◦		◦			Winter/spring grafting (Dec-April) most favoured; temp. 18–22°C. Summer/Autumn (late Aug-Oct) less often used. Warm callus or supported warm callus. Possible from cuttings-grafts to unrooted *J.* x *pfitz.* 'Hetzii'. Also, top-worked stems for weeping and dwarf types to produce Patio plants. **See also Chapter 40.**
Kalmia latifolia cvs	*Kalmia latifolia* 1+2P 6–8mm		◦		●	◦		More successfully grafted in mid-August/early September than late winter, as normally recommended. If available graft on portion of rootstock stem with green bark. Heavy shading required.

(Continued)

Name	Rootstock	Method		Systems				Comments
		Apical	Side	Cold callus	Warm callus - winter	Warm callus - summer	Hot-pipe	
Keteleeria davidiana and other species.	*Abies alba* or *A. nordmanniana* 1+1P 6–8mm		●			●		Treatment as for *Abies* (Chapter 20). Late summer grafting preferred.
Koelreutaria: paniculata 'Coral Sun', 'Fastigiata', 'Rose Lantern'.	*K. paniculata* 1+1P 6–8 / 8–10mm or root pieces 8–10/10–12mm Stems 2+1P or 2+0 root-balled 8–10mm at 150–180cm high	●		●	●		●	'Coral Sun' and 'Rose Lantern' may be top-worked to provide instant impact. Top-working best achieved using two side grafted scions and placed in a tall poly-tent or bagged individually.
+ *Laburnocytisus* 'Adami'	*Laburnum anagyroides* 1+0, 1+1 8–10/10–12mm	●		●				Scions should be selected to ensure typical growth is maintained.
Laburnum: anagyroides cvs, x *watereri* cvs	*L. anagyroides* 1+0; 1+1 8–10/10–12mm. Stems 2+1P or root-balled 180cm high	●		●				Weeping forms and 'Quercifolium' may be top-worked and cold callused or a poly-tent can be used to house the grafts. Hot-pipe an option.
Lagarostrobus (*Dacrydium*) *franklinii*	*Dacrydium cupressinum* 0+1+1P 6–8mm		●			●	●	An untested combination which should be compatible
Larix species and cultivars	*L. decidua* 1+1P 6–8mm or Stems 2+1P or root-balled 8–10mm at 60–150cm high	●			●	●		Weeping forms grafted at 120–150cm high, some dwarf forms may be grafted at 60cm to create small novelty tees.
Lepidothamnus: (*Dacrydium*) *intermedius, laxifolius*	? *Prumnopitys andina* 1+1P or 1+2P trained into a small stem 4–6mm at 15cm high		●			●		A suggested method. *L. laxifolius* can be grafted to a 15cm stem to provide some substance to the young plant.
Libocedrus: bidwillii, plumosa	*Calocedrus decurrens Platycladus* (*Thuja*) *orientalis* 1+1P 5–7mm		●			●		Procedures as for *Calocedrus* and *Chamaecyparis* (Chapter 32).
Ligustrum: compactum, confusum, japonicum cvs, *lucidum* cvs, *obtusifolium* var. *regelianum, quihoui, sinense* cvs	*Ligustrum ovalifolium* 0+1+1P or *L. vulgare* 1+1P 6–10mm Bare-root rootstocks can be used for *L. compactum* and *L. lucidum*		●			●		*L. ovalifolium* produced from partially disbudded hardwood cuttings is thought to sucker less than seedling *L. vulgare*. Growth from grafted *Ligustrum* stronger than when rooted from cuttings.
Liquidambar: acalycina, formosana Monticola Group, *orientalis, styraciflua* cvs	*L. styraciflua* 1+1P 6–8mm or stems 2+1 or 2+2P grown in suitable pots (2–4 litre deep) for grafting at 60–150cm	●		●	●	●	●	Many options possible in this easily grafted genus. *L. styraciflua* 'Golden Treasure' and 'Naree' benefit from grafting to a low stem 60cm, for added vigour. *L. sty.* 'Gumball' and 'Oconee' best at 80–150cm.

(Continued)

Name	Rootstock	Method		Systems				Comments
		Apical	Side	Cold callus	Warm callus - winter	Warm callus - summer	Hot-pipe	
Liriodendron: chinensis, chinensis x *tulipifera* hybrids, *tulipifera* cvs	*L. tulipifera* 1+1P or 2+1P 6–10mm, or bare-root 2u1, 2+1. 6–10mm	●	●		●	●	●	Good results grafted August/ early Sept. on bare-root or potted rootstocks placed in a poly-tent. Scion wood must be mature and a 2yr basal portion can improve results. Late winter grafting also successful, poly-tent or hot-pipe set at 22–24°C.
Lithocarpus: henryi, variolosus, other sp. and cvs	*L. edulis* 2+1P 5–8mm or root pieces from parent species 100mm x 10–12mm Ø	●						*L. henryi* successfully grafted using root section from parent species, a difficult procedure. Compatibility of all species on *L. edulis* not established and probably unlikely.
Lonicera shrubs: *elisae, pyrenaica, setifera.* climbers: *alseuosmoides, hildebrandtiana, implexa*	*L. xylosteum* 1+1P 5–7mm for shrubby sp. For climbers: evergreen or semi-evergreen species such as *L. japonica* 0+1+1P	●				●		Listed shrubby sp. can be propagated by cuttings but this is always difficult, and they are easily grafted. Hot-pipe can be used for deciduous sp.
Maackia: amurensis subsp. *buergeri, fauriei, hupehensis*	*M. amurensis* 1+1P 6–8mm	●				●		No compatibility issues identified. Papilionaceae normally fairly amenable even to intergeneric graft compatibility.
Maclura: hybrida, tricuspidata	*M. pomifera* 1+1P 5–7mm	●		●	●		●	*Morus alba* could be a possible alternative rootstock, not tested.
Magnolia sp. and cvs Includes those previously *Michelia, Manglietia,* etc.	*M. kobus* or *M. obovata* 1+1P 6–8mm or 1+2P 8–10mm. From cuttings, *M. stellata M.* x *loebneri* 'Leonard Messel'; *M.* deVos or Kosar hybrids or *M. liliiflora* 0+1P 6–8mm or 0+1+1P 8–10	●	●		●	●	●	Most, if not all, within the genus are compatible. For deciduous types, hot-pipe produces best results, temp. 22–24°C. Warm callus summer also v. successful and essential for evergreens, winter or summer. Winter warm callus for deciduous less successful than summer. Drying-off required for winter grafts. Apical or side veneer favoured over splice. **See also Chapter 41.**
Mahonia: fremontii, nevinii, trifoliata var *glauca*	*Berberis thunbergii* or *B. vulgaris* 1+1P 6–8mm		●		●	●		Graft Nov-early March or Aug-early Sept. Warm callus 18–22°C. Side veneer grafts best. Scions can have 2–3yr basal portion. **See also Berberis, Chapter 26.**

(Continued)

Name	Rootstock	Method		Systems				Comments
		Apical	Side	Cold callus	Warm callus - winter	Warm callus - summer	Hot-pipe	
Malus sp. & flowering crabs	*M.* Antonovka, *M.* Bittenfelder, *M. domestica, M. pumila,* or *M. sylvestris* 1+0 or 1+1 8–12mm. Or clonal rootstocks 106mm; 111mm; M 25, Budagovsky, Geneva 0+1 or 0+1+1 8–15mm	○		○				Root grafting is possible. Clonal rootstocks must be virus free. M 27 rootstock can be used for Patio plants. Top-working at 1.5–1.8m high to lifted stems possible. Little callus production below 6°C. Hot-pipe unnecessary, sometimes used for container production. Early autumn grafting to newly lifted bare-root rootstocks is possible. **See also Chapter 42.**
Malus: Orchard Apples	Modern orchard production demands high health status, vegetatively produced clonal rootstocks. A comprehensive list is found in Chapter 42	○		○				Machine or tool-made grafts sometimes used (see Appendix K). Sub-cold callusing can be considered. Grafting height may be tailored to specific customer requirements. Bench grafting/bare-root grafting followed by field planting often adopted. **See also Chapter 42.**
Melliodendron xylocarpum	*Pterostyrax hispida* 1+1P or 0+1+1P 5–7mm. *Halesia carolina,* suggested alternative, but eventually shows incompatibility		○			●	○	Hot-pipe produces best results. Growth on *Pterostyrax* rootstock excellent (<2 metres/yr) little or no suckering. Long-term compatibility seems very likely. 8yr old plants still very vigorous. No suckers. Grafts on *Styrax japonica* failed.
Meliosma: alba (*beaniana*), *pinnata* var. *oldhamii, veitchiorum*	One of the pinnate leaved species. 1+1P or 1+2P 6–10mm. Grafting on parent roots possible		●			●	●	Trials using simple leaved rootstock (*M. dillenifolia*) failed due to incompatibility after 1 year. Pinnate leaved combinations probably essential. In hot-pipe, simple leaved species easily grafted to simple leaved rootstock.
Mespilus: canescens (x *Crataemespilus canescens*), *germanica* cvs (Medlar)	*Crataegus monogyna* 1+1P 6–8mm or bare-root 1+1 6–10mm. For Medlar varieties: Quince C, Quince A or *Pyrus communis*	○		○			○	Use of Quince or *Pyrus* rootstocks may result in delayed incompatibility. For grafts on *Crataegus* rootstocks plant deeply to encourage scion rooting. *M. canescens* grafts satisfactorily on *Crataegus*.

(Continued)

Name	Rootstock	Method		Systems				Comments
		Apical	Side	Cold callus	Warm callus - winter	Warm callus - summer	Hot-pipe	
Metasequoia glyptostroboides cvs	*M. glyptostroboides* 0+1+1P 8–10mm	○		○				Easily rooted from soft or hardwood cuttings. Compact and weaker selections more vigorous when grafted.
Morus: alba cvs, *cathayana, nigra* cvs, *rubra* cvs	*Morus alba* 1+1P or 1+1 bare-root 8–10mm. *Morus alba* 1+2P or 2+1 stem grafted at 150–180cm	○		○	○		○	Hardwood cuttings are a standard method of propagation. Weeping and dwarf forms may be top-worked. A suggestion that *M. nigra* is incompatible when grafted to *M. alba* has not been substantiated.
Nothofagus deciduous species	? *N. obliqua*	●			●			Hot-pipe should provide best takes.
Nothofagus evergreen species	? *N. dombeyi*				●			Late summer grafting August to September may produce better results than late winter.
Nyssa: sinensis cvs, *sylvatica* cvs	*N. sylvatica* 1+1P 6–8mm	○		○	○		○	Relatively easily grafted. Hot-pipe provides most reliable results.
Olea: europea cvs, *europea* subsp. *africana*	*Olea europea* 2+1P 6–8mm		○		○	○		Relatively easily grafted. Late summer grafting avoids risk of frost damaged wood
Osmanthus species and cvs	*O. x burkwoodii* 0+1+1P 6–8mm		○		○	○		As for *Olea*.
Osteomeles subrotunda	?*Pyracantha coccinea* (or other species) 0+1P 6–8mm		●		●			Normally propagated by cuttings but grafting to *Pyracantha* should provide medium-/long-term compatibility.
Ostrya: japonica, virginiana	*Ostrya carpinifolia* or *Carpinus betulus* 1+1P 6–8mm or bare-root		○	○	○		○	Hot-pipe most reliable. *Carpinus* may not provide long-term compatibility.
Ostryopsis davidiana	? *Corylus avellana* or *Carpinus betulus* 1+1P 6–8mm	●		●	●		●	May be problems of long-term compatibility on suggested rootstocks.
Paeonia: delavayi cvs, Gansu group, Itoh cvs, *lemoinei* cvs, *suffruiticosa* cvs	*P.* Gansu seedlings 1+1P 7–9mm. Or root pieces of Gansu, Itoh or herbaceous *P. lactiflora* vars. 'Early Windflower', 'Red Charm', 'Krinkled White', 100–150mm x 15–20mm Ø	○		●	○	○		Conventional side graft to suitable pot grown seedling Gansu type produces very reliable results. Root grafting summer or winter, cold callused but with initial 3 weeks warm period ±20°C. Following summer grafting, many field plant autumn or spring. **See also Chapter 43.**

(Continued)

Name	Rootstock	Method		Systems				Comments
		Apical	Side	Cold callus	Warm callus - winter	Warm callus - summer	Hot-pipe	
Parrotia: *persica* cvs, *subaequalis*	*Parrotia persica* 0+1+1P or *Hamamelis virginiana* 1+1P 6–8mm. *Parrotia* Stems 0+1+2P or root-balled 8–10mm at 150–180cm		○	●	○	○		See *Hamamelis* (Chapter 38) for full description. Stems can be grown for *P. persica* 'Pendula', which is usually summer grafted and placed in tall poly-tents or enclosures. *Parrotia* makes a more reliable rootstock than *Hamamelis*.
Parrotiopsis jacquemontana	*Hamamelis virginiana* 1+1P 6–8mm		○			○		See *Hamamelis* (Chapter 38) for full details
Parthenocissus tricuspidata cvs,	*P. quinquefolia* 1+0 or 2+0 4–6mm or 0+1 5–7mm ex hardwood cuttings	○			○			Select scion wood with vegetative buds. Splice, wedge or veneer graft. Potted after grafting and placed in poly-tent at 18–20°C. Easily grafted.
Paulownia: *fargesii*, *fortunei* cvs *kawakamii*, *tomentosa*, 'Lilacina'	*P. tomentosa* or other available species seedlings 1+1P. 8–10mm or root pieces 12–20mm Ø × 100–120mm	○		●	○		○	Potted seedlings warm callused or hot-pipe at 20–22°C. For root grafting - graft to root pieces, pot up after grafting and place in poly-tent, bottom heat 18–20°C.
Petteria ramentacea	*Laburnum anagyroides* 1+1P 6–8mm	○		○	○			Easily grafted should succeed well in hot-pipe. Long-term compatibility not certain.
Phellodendron: *amurense*, *chinense*, *japonicum*	Any available *Phellodendron* seedlings 1+1P 6–8mm. Root pieces 100mm x 8–10mm Ø	○			○			Potted seedlings warm callused or hot-pipe at 20–22°C or graft to root pieces, pot up after grafting. Warm callus in poly-tent, bottom heat 18–20°C.
Phillyrea latifolia	*P. angustifolia*, or ?*Osmanthus x burkwoodii* 0+2P 6–8mm		○		○	○		As for *Olea*.
Photinia: *beauverdiana*, *integrifolia*, *nussia*, *prionophylla*, *serratifolia*	*P. davidiana* 0+2P 6–10mm		○		○	○		Generally propagated by cuttings but some of the rare forms listed may only be available as scions. Summer grafting avoids frost damaged scion wood.

(Continued)

Name	Rootstock	Method		Systems				Comments
		Apical	Side	Cold callus	Warm callus - winter	Warm callus - summer	Hot-pipe	
Picea: abies cvs, *breweriana,* omorika cvs, *pungens* cvs	*P. abies* 1+1P 4–6mm or plug grown well-furnished to soil level with needles and buds. Stems for top-working, *P. abies* run-up in field to 60–180cm		○		○	○		Warm callus is usual in Europe. Some in USA favour supported warm callus when grafts stood on bench and frequently hand syringed or sprayed. Optimum temp for winter ±12°C. In late summer <22°C. Scions 1yr shoot with apical and good lateral bud formation, or 1yr branched shoot with a 2yr basal portion. **See also Chapter 44.**
Picrasma quassioides	? *Ailanthus altissima* roots 12–15mm Ø x 120–150mm	●			●			Very difficult from cuttings, seed rarely available. Compatibility of this combination uncertain.
Pieris sp. and cultivars	*Pieris japonica* or *P. formosa* 0+1+1P 6–8mm		○		○			If required, easily grafted; conditions as for *Rhododendron* Hymenanthes subgenus.
Pinus species with 2 needles per bundle	*P. sylvestris, P. mugo uncinata, P. contorta, P. contorta latifolia.* 1+1P 4–6mm or 2+1P 5–7mm, 6–8mm for heavy wooded types	●	○		○			Most grafted Dec–Feb but October possible. Warm callusing or supported warm callusing when grafts stood on open bench and frequently syringed or hand sprayed. Plenty of ventilation of enclosed structures required. Minimum temperature of 6°C. Bottom heat set at 10–12°C. Scions well ripened 1yr wood. Branched 1yr with 2yr basal portion is often required for dwarf, compact forms. Top-working possible to produce Patio plants normally on stems 60–80cm high of *P. sylvestris* or *P. wallichiana/strobus.* **See also Chapter 45.**
3 needles per bundle	*P. strobus?, P. radiata?, P. ponderosa?, P. mugo uncinata?* 1+1P 4–6mm or 2+1P 5–7mm, 6–8mm for heavy wooded types							
5 needles per bundle	*P. wallichiana, P. strobus.* 1+1P 4–6mm or 2+1P 5–7mm, 6–8mm for heavy wooded types							
Piptanthus nepalensis	*Laburnum anagyroides* 1+1P 6–8mm		○		○			Warm callusing with basal temp of 18–20°C. Long-term compatibility of this combination uncertain.

(Continued)

| Name | Rootstock | Method | | Systems | | | | Comments |
		Apical	Side	Cold callus	Warm callus - winter	Warm callus - summer	Hot-pipe	
Pistachia: vera, terebinthus, other species hybrids	*P. terebinthus* 1+1P (deep patterned pot) 6–8mm. *P. chinensis*, the obvious candidate is said to be incompatible				●			Field grafted in New Zealand but said to be difficult. Should respond well to hot-pipe at 22–26°C. Seedlings must be raised in deep plugs as the plant has a very long tap root.
Pittosporum: dallii, divaricatum, turneri	*P. tenuifolium* 0+1+1P 6–8mm		●		●	●		Summer grafting for these rather tender evergreens will avoid winter damaged scion wood.
Planera aquatica	*Ulmus glabra* 1+0 6–8mm	●	●	●	●			Normally propagated by semi-ripe cuttings. Compatibility with *U. glabra* unproven.
Platanus: 'Augustine Henry', x *hispanica* cvs, *orientalis* cvs, *racemosa*	*P.* x *hispanica* clone (probably *P.* x *hispanica* 'Pyramidalis') 0+1+1P or 0+1+1 bare-root	●		●	●		●	Hot-pipe at temperature settings of 20–22°C will produce best results. Warm and cold callusing is also successful. One report stated that *P. occidentalis* rootstocks for *P. orientalis* produced delayed incompatibility. Surprising considering numbers of grafts on *P.* x *hispanica* rootstocks have survived for decades.
Platycladus (Thuja) orientalis cultivars	*P. orientalis* 1+1P		●			●		See *Chamaecyparis* for details of methods and requirements - Chapter 32.
Podocarpus: cunninghamii, elatus, ferruginea, macrophyllus, salignus, taxifolia	*Prumnopitys andina* 1+1P or *Podocarpus nivalis* 0+2P 6–8mm		●		●			Poly-tent enclosed conditions and temperatures in the 12–18°C range are best. Callus and union formation are quite slow. Subsequent growth is better than from cuttings propagation.
Populus Leucoides section: *glauca, heterophylla, wilsonii*	*P. lasiocarpa* 1+1P or 1+1 bare-root 8–10mm	●		●	●			Hot-pipe should not be necessary. Warm callusing at low temperatures 12–15°C avoids the need to seal grafts.
Populus White Poplars: *alba* 'Richardii', x *canescens* cvs	*P. alba* 0+1 8–10mm bare-root	●		●	●			Treatment as for Leucoides section. *P. alba* suckers very freely, sometimes 20 metres from parent rootstock.

(Continued)

Name	Rootstock	Method		Systems				Comments
		Apical	Side	Cold callus	Warm callus - winter	Warm callus - summer	Hot-pipe	
Populus Aspens: *grandidentata*, 'Hiltingbury Weeping', *tremula* cvs, *tremuloides* cvs	*P. tremula* 1+0 or 1+1P 8–10mm. Stems 2+1P or 2+1 root-balled 10–12mm Ø at 150–180cm high	●		●	●			Treatment as for Leucoides section. Pendulous forms top-worked using one or two scions. Rootstocks from other sections said to be compatible for at least 2 or more years.
Prumnopitys: ferruginea, taxifolia								See *Podocarpus ferruginea* and *P. taxifolia*
Prunus Plum sp: *cerasifera* cvs, *institia* cvs, *sogdiana*	*P.* Brompton; *cerasifera*; Myrobalan B; 1+0 or 1+1 6–12mm or clonal 0+1 5–7mm to 7–9mm or 0+1+1 9–12mm	●		●				Cold callusing bare-root rootstocks is favoured method, temperature not long below 6°C. If sufficient numbers warrant investment sub-cold callusing can be considered. Hot-pipe is rarely justified. Side grafts may be required for top-worked stems. These are normally for weeping cherries grafted to stems of *Prunus* F 12/1 at 1.8–2.0 metres high. Accurate cambial matching is important for *Prunus* and short tongue apical veneer may replace splice grafts. Successful grafts are field planted or potted. Field planting may need to be delayed in northern areas. Cold storage of successful grafts may be necessary or cold storage of scions and rootstocks allows grafting and planting to be delayed. **See also Chapter 46.**
Apricots: *armeniaca* cvs, *mume* cvs	*P. armeniaca P. cerasifera*. 1+0 or 1+1 6–12mm or clonal 0+1 5–7mm to 7–9mm or 0+1+1 9–12mm							
Prunus: Ornamental Almonds and Peaches	*P. dulcis P. persica P.* Myrobalan B; St Julien; Brompton 1+0 or 1+1 6–12mm or clonal 0+1 5–7mm to 7–9mm or 0+1+1 9–12mm							
Prunus: Ornamental Cherries (Sato Zakura etc.).	*P. avium P.* F12/1; *P.* Colt; *P. mahaleb*. 1+0 or 1+1 6–12mm or clonal 0+1 5–7mm to 7–9mm or 0+1+1 9–12mm							
Prunus: Bird Cherries	*P. padus, P. avium* 1+0 or 1+1 6–12mm							
Prunus: Almonds and Stone Fruit	A range of clonal rootstocks is being continually developed to meet the requirements for production of Nut and Stone fruits within this group. These are extensively reviewed in Chapter 47	●		●				Sub-cold callusing is gaining in importance for the production of the very large numbers of plants required to meet fruit grower's demands. Development of dwarfing rootstocks for cherries has enabled production under protection. As a result, cherry fruit can now be economically produced in northerly areas. For further details on all aspects of these **see also Chapter 47.**

(Continued)

Name	Rootstock	Method		Systems				Comments
		Apical	Side	Cold callus	Warm callus - winter	Warm callus - summer	Hot-pipe	
Pseudocydonia sinensis	*Cydonia oblonga* or Quince A 1+1 bare-root	○		○				Hot-pipe should not be necessary. Cold callusing should provide good results.
Pseudolarix amabilis	? *Abies nordmanniana*		●		●			Incompatible on *Larix*. *Abies* may be worth trying.
Pseudotsuga menziesii cvs	*Pseudotsuga menziesii* 1+1P 6–8mm. Stem 1+2P 6–10mm at 60–120mm high			○	○			Treatment as for Pinus (Chapter 45) but more shade tolerant. *P. menziesii* 'Glauca Pendula' makes an attractive top-worked Patio specimen.
Ptelea: crenulata nitens, trifoliata cvs	*P. trifoliata* 1+1P 5–7mm	○			○			Hot-pipe may not be justified for this easily grafted genus.
Pterocarya: fraxinifolia var. *dumosa, macroptera* var. *inignis, stenoptera* 'Fern Leaf'	*P. fraxinifolia* 1+1P 8–12mm or ? root pieces 20–25mm Ø × 100–125mm for root grafting (speculative)	○			○		○	Not easy warm callused. Hot-pipe should provide best results for stem grafts. As the genus is a member of Juglandaceae, ±25°C may be optimum temperature. Root grafts best in warm callus conditions at 20–22°C?
Pterostyrax: corymbosa, psilophyllus, psilophyllus var. *leveillei*	*P. hispida* 1+1P 6–8mm or 0+1+1 P also possible	●	○				○	Hot-pipe using side graft preferred method. Surprisingly *P. psilophyllus* var. *leveillei* does not seem entirely compatible on *P. hispida*, frequently producing suckers and little free growth.
x *Pyrocomeles vilmorinii*	*Pyracantha coccinea* 1+1P 6–8mm	○			○			Normally propagated by cuttings but grafting a reliable option.
+*Pyrocydonia* 'Daniellii'	*Pyrus communis* 1+1P 8–10mm or 1+1 bare-root	○		○				Diligent choice of scion wood to ensure true to type of material selected. Use of Quince rootstock may produce long-term incompatibility.
x *Pyronia veitchii* cvs	*Pyrus communis* 1+1P 8–10mm or 1+1 8–10mm bare-root	○		○				Same remarks as for x *Pyrocydonia* apply.
Pyrus: calleryana cvs, *communis* cvs, *salicifolia* 'Pendula'.	*P. communis* 1+0 or 1+1 8–10mm; *P. calleryana* can be considered for *P. calleryana* cvs	○		○				Cold callusing is the usual method adopted. For *P. salicifolia* 'Pendula' top-working is usually carried out in the field, but *P. communis* stems can be lifted and bench grafted at 1.8–2.0m high. **See also Chapter 48.**

(Continued)

Name	Rootstock	Method		Systems				Comments
		Apical	Side	Cold callus	Warm callus - winter	Warm callus - summer	Hot-pipe	
Pyrus Fruiting pears.	A detailed account of pear rootstocks is given in Chapter 48	○		○				Some fruiting pear rootstocks are now micropropagated and the resulting pot grown plants may be chip budded under protection in the summer. **See also Chapter 48.**
Quercus: Quercus & Mesobalanus sections: *canariensis*, *frainetto* cvs, *petraea* cvs, *robur* cvs, x *warei*	*Q. robur* 1+0 or 1u1 or 1+1 6–10mm. 1+1P or 1u1P or 1+1+1P 6–10mm. Plug grown 1yr or 2yr seedlings	●	○		○	○	○	Evergreen species within these sections are not suitable for hot-pipe. For deciduous species hot-pipe produces the most reliable results. Temperature 22–24°C. **See also Chapter 49.**
Cerris section: *castaneifolia* cvs, *cerris* cvs, x *hispanica* cvs	*Q. cerris* Same specification as above or *Q. ilex* 1+1P for some evergreen sp.							
Lobatae section: *coccinea* cvs, *palustris* cvs, *rubra* cvs, *rysophylla*, *texana* cvs	*Q. palustris* or *Q. rubra* or other seedling Lobatae. 1+0 or 1u1 or 1+1 6–10mm. 1+1P or 1u1P or 1+1+1P 6–10mm. Plug grown 1yr or 2yr seedlings							
Cyclobalanopsis section: *delavayi*, *lamellosa*, *oxyodon*	*Q. myrsinifolia* or *Q. glauca* or *Q. schottkyana* 1+1P or 2yr pot grown	●	○		○	○		Warm callus 18–20°C winter and bottom heat of 16°C summer while ambient will frequently be well above this. **See also Chapter 49.**
Rehderodendron macrocarpum	*Pterostyrax hispida* 1+1P or 0+1+1P 5–7mm		○		●		○	Best in hot-pipe at 20–22°C. Good takes and short-term compatibility, but long-term compatibility is in doubt.
Rhamnus imeretina (*Rhamnus frangula* see *Frangula alnus* 'Asplenfolia')	*R. cathartica* 1+1P 6–8mm	●	●		●		●	Compatibility of this combination not confirmed.

(Continued)

Name	Rootstock	Method		Systems				Comments
		Apical	Side	Cold callus	Warm callus - winter	Warm callus - summer	Hot-pipe	
Rhododendron subgenus Pentanthera - the deciduous Azaleas: Ghent, Knap Hill, Mollis, Occidentale, Rustica, & some sp.	*R. luteum* 1+1P 3–5mm (summer) or winter 2+0 bare-root or potted 2+1P 5–7mm		○		○	○	●	Summer grafted - apical wedge or side graft with leafy semi-mature scions. Winter dormant scions, apical or side possible. *R. luteum* may not be compatible with some sections within Pentanthera, i.e. *R. schlippenbachii*, etc. See also Chapter 18: Grafting Systems.
Rhododendron subgenus Tsutsusi - evergreen species - Indian azaleas, *nakahari, oldhamii, weyrichii, yedoense,* etc.	Azalea *phoenicea* 'Concinna' – an unresolved name for southern *indica* azalea used by specialist growers for florist Indica hybrid and simsii Azaleas. 0+1P ±4mm		○			○		Grafted late summer to cutting raised rootstock produced 8 weeks earlier. Some use vertical wedge graft; the scion is cut on one side into long wedge shape and pressed into 75mm long vertical slit made along one side of rootstock. Then tied-in or fixed with pegs. Side veneer may be better.
Rhododendron subgenus Hymenanthes - the smaller leaved species and hybrids Hymenanthes - large leaved (subsection Falconeri & Grandia)	*R.* 'Cunningham's White'; *R.* INKARO® selections; *R.* clonal selections of *R. ponticum*; 0+1P 6–8mm *R.* species from either subsection 1+1or2P 8–12mm		○		○			Possible to graft between late July and early May. Best late Dec–mid-February. Summer grafting Aug–Sept. Cuttings-grafts Sept.–Dec. Apical long or short tongue veneer, splice or preferably side veneer. Long tongue with stimulant buds ideal. If possible, rootstock stems still green wood. Temperature best between 10 and 20°C, dependent upon quality of scion wood. Cuttings-grafts bottom heat 18–20°C. Top-worked stems of INKARO® or 'Anna Rose Whitney' grafted at 80–100cm. **See also Chapter 50.**
Subgenus *Rhododendron*	Seedlings or cuttings from subsections Maddenia; Boothia or Edgworthia. 0+2P or 2+1P 4–7mm		○		○			Environmental requirements difficult to provide. Risk of leaf and stem damping in high humidity, desiccation in low. Dry fog set at ? ±90% RH may be best. **See also Chapter 50.**
Rhodoleia championii	? *Distylium racemosum* 0+1+1P	●			●			Compatibility of this combination not confirmed.

Name	Rootstock	Method		Systems				Comments
		Apical	Side	Cold callus	Warm callus - winter	Warm callus - summer	Hot-pipe	
Ribes uva-crispa (*grossularia*) Gooseberry. Grown as standards	*R. aureum* (best clones 'Brecht' or 'Fritsche') or *R.* x *nidigrolana* (Jostaberry) 2+0 or 2+1P 8–10mm at 100cm		○	○				Top-worked Gooseberries. Grafted to stems in the field or preferably under protection. Rootstocks root-balled or pot grown in 2-lt.-deep pots. Cold callused in protection or grafts covered with polythene bags. Best multi-grafted at 100cm.
Robinia: species cvs hybrids	*R. pseudoacacia* 1+0; 1+1 8–10mm bare-root	○		○				Late to grow out in spring, so March–April possible grafting dates. Easily grafted, hot-pipe unnecessary.
Rosa selected species and hybrids such as: *bracteata, ecae* cvs, 'Golden Chersonese', 'Mermaid'	*R. canina* or *R. corymbifera* (*dumetorum*) 'Laxa' 1+1P or 1+0 bare-root 5–8mm. Unrooted of above or *R. rugosa R. chinensis*, etc.	○		○		○		Cold callusing of potted or bare-root rootstocks is satisfactory. *R. bracteata* and 'Mermaid' grafted in late June to avoid frost damaged scion wood will require warm callus conditions. Cuttings-graft is possible onto unrooted rootstocks which form a graft union and roots.
Salix: caprea 'Curly Locks', 'Kilmarnock', *integra* 'Hakuro Nishiki'	*S.* x *smithiana* or *S. viminalis*. Unrooted or rooted stem 80–120cm long 12–14mm or 14–20mm Ø at grafting height	○	○	○				Vigorous clones of *S.* x *smithiana* or *S. viminalis* available from specialist rootstock producers or selected 'in house'. Often 2 or more scions grafted per stem using splice and side wedge.
Sambucus: racemosa 'Plumosa Aurea', 'Sutherland Gold'	*S. nigra* 1+0 6–10mm bare-root or *S. nigra* root pieces 100mm × 10–12mm Ø	○		○	○			Root grafting standard practice, grafts cold callused or better warm callused at 15–18°C. *S. nigra* roots are better than *S. racemosa*. Conventional stem graft possible.
Saxegothaea conspicua	? *Prumnopitys andina* 1+1P 6–8mm or ? *Araucaria araucana*		●		●			Compatibility of this combination has not been confirmed.
Schefflera sp.	*Schefflera* sp. 1+1P or 2+1P as available		●		●			No accounts of grafting available. As this genus is in Araliaceae grafting should be relatively simple. Side veneer graft of chosen scion variety. See *Aralia* Chapter 25.

(Continued)

Name	Rootstock	Method		Systems				Comments
		Apical	Side	Cold callus	Warm callus - winter	Warm callus - summer	Hot-pipe	
Schima argentea	? *Stewartia* sp. 1+1P 5–7mm or *S. pteropetiolata* 0+1+1P		●		●		●	Long-term compatibility unlikely. De-leafing the scion may permit hot-pipe conditions to be used with advantage.
Sciadopitys verticillata cvs	*S. verticillata* <2/3+2P or 0+1+2P 5–8mm Root pieces of *S. verticillata*		●		●			Rootstock production by seed or cuttings very slow. Warm callusing in cool temp. 10–15°C. Root grafting Feb–March reputedly successful.
Schizophragma: hyrangeoides 'Roseum', integrifolium	*S. hydrangeoides* 2+1P 6–8mm		●		●			When available cuttings preferred method. Hot-pipe may work well.
Sequoia sempervirens cvs	*S. sempervirens* 1+1P or 1+2P 6–8mm or *Sequoiadendron giganteum* 1+1P		●		●			Summer grafting less often used but possible. *Sequoiadendron giganteum* rootstock said to impart extra vigour.
Sequoiadendron giganteum cvs	*S. giganteum* 1+1P or 1+2P 6–8mm or *Sequoia sempervirens*		●		●			Early training of weeping cultivars required.
Sinojackia: rehderiana, xylocarpa	Either species 0+1+1P or 1+1P 5–7mm *Pterostyrax hispida?*		●				●	A possible means of grafting one or other of the species. Relatively easily rooted from cuttings. *Pterostyrax* rootstock has not been tested here.
Sinowilsonia henryi	*Hamamelis virginiana* 1+1P 5–8mm		●				●	Grows well initially but eventually fails. However, a possible method for obtaining cuttings material.
Skimmia species forms and varieties	*Skimmia* seedling or cuttings raised from vigorous clone (0)1+1P or (0)1+2P 6–8mm		●		●	●		Easily grafted if necessary. Temperature 15–20°C. No sealing, graft in warm callus conditions.
Sophora species and cvs	*S. microphylla* or *S. tetraptera* 2+1P or 0+1+1P 6–8mm ? *Styphnolobium japonicum* may prove suitable		●		●			Not easily grafted. Temperature 18–21°C. Unsealed grafts well ventilated and no drips. Long-term compatibility grafted to *Styphnolobium* is uncertain and trials here with scions of *S. cassioides* on *Styphnolobium* have failed.
Sophora japonica								See *Styphnolobium*
x *Sorbaronia:* alpina, dippelii	*Sorbus aria* 1+1P 6–8mm or 1+1 6–8mm bare-root	●		●				Hot-pipe should not be necessary.

(Continued)

Name	Rootstock	Method		Systems				Comments
		Apical	Side	Cold callus	Warm callus - winter	Warm callus - summer	Hot-pipe	
x *Sorbaronia fallax, hybrida, sorbifolia,*	*Sorbus aucuparia* 1+1P 6–8mm or 1+1 bare-root	◦		◦				Hot-pipe should not be necessary.
x *Sorbocotoneaster pozdnjakovii*	*Sorbus aucuparia* 1+1P 6–8mm or 1+1 6–8mm bare-root	◦		◦				Hot-pipe should not be necessary.
x *Sorbopyrus: auricularis, auricularis* 'Malifolia'	*Pyrus communis* 1+1P 6–8mm or 1+1 6–8mm bare-root	◦		◦				Hot-pipe should not be necessary
Sorbus sp. and cultivars.	Main rootstock species are: *S. aria, S. aucuparia*; and *S. intermedia*. 1+0 or 1+1 6–8, 8–10mm. A number of other species may be involved in this complex genus. Readers are referred to the compatibility table in Chapter 51 for full account	◦		◦				Generally easily grafted, requiring only cold callusing conditions. As with others, little callus formation occurs below 6°C. Splice or whip and tongue grafts mainly used. The genus falls into three main groups: Aria (Whitebeams) grafted to *S. aria* or *S. intermedia*; the Aucuparia (Mountain Ash) grafted to *S. aucuparia* or some to *S. intermedia* and the Micromeles with a more complex stock/scion requirement. **See also Chapter 51.**
Staphylea: bumalda, colchica cvs, *emodii, holocarpa* cvs	*S. colchica* or *S. pinnata* 1+1P 6–8mm		◦		●	◦	◦	Winter hot-pipe or summer side graft the most successful.
Stewartia species and cultivars	*S. pseudocamellia* or other available species seedlings 1+1P or 1+2P 5–7mm		◦		●		◦	Use of hot-pipe system has significantly improved results for this difficult-to-graft genus. Temperature of 20–22°C and use of side graft combined with accurate knifesmanship boosts takes.
Styphnolobium japonicum cvs (*Sophora japonica* cvs)	*S. japonicum* 1+1P 6–8mm	◦					◦	Successful in hot-pipe at ±20°C. Much easier than Sophora.
Styrax species and cultivars	*S. japonicus* or *S. obassia* 1+1P or 1+2P 5–7mm. *Styrax* grafts have failed on *Pterostyrax hispida*		◦		●		◦	Callus production is slow and limited. Rapid extension growth occurs once heat is applied to dormant material. Hot-pipe offers best chance of success. Maintaining health of root system in pot or container always difficult. Follow same recommendations as for *Stewartia*. A difficult genus.

(Continued)

Name	Rootstock	Method		Systems				Comments
		Apical	Side	Cold callus	Warm callus - winter	Warm callus - summer	Hot-pipe	
x *Sycoparrotia semidecidua* cvs	*Hamamelis virginiana* 1+1P 5–8mm		○		●	○		De-leafed for hot-pipe. Also, summer grafting possible.
Syringa sp. and cvs *S. vulgaris* cvs (French Lilacs)	*S. vulgaris* 1+0, 1+1 6–8/8–10mm or 1+1P 6–8mm; *Ligustrum ovalifolium* 0+1 6–8/8–10mm; or *F. pennsylvanica* root pieces 8–10mm Ø x 80–100mm long, including hypocotyl. Other *Syringa* sp. suggested as rootstocks often not successful	○	○	○				Other than use of *Syringa* rootstock the objective is to encourage scion rooting due to delayed incompatibility. Best achieved with inverted wedge graft in winter. Temperatures not long below 6°C. Cuttings-grafts also possible, using side graft to *Ligustrum ovalifolium* in Aug.-Sept. with mist and bottom heat 18–20°C. French Lilacs can be grafted to potted *S. vulgaris* in the late winter or Aug.-Sept. using splice graft. **See also Chapter 52.**
Taiwania cryptomerioides	*Cryptomeria japonica* 1+1P 5–8mm or *C. japonica* 'Elegans' 0+1+1P		○			○		Relatively easy in temperatures of 10–15°C.
Taxodium: distichum cvs, *mucronatum*	*T. distichum* 1+1P or 2+1P 6–8mm or *Cryptomeria japonica* 'Elegans' 0+1+1P is suggested	○	○	○				*T. mucronatum* semi-evergreen, requires warm callusing 12–15°C or defoliation and cold callusing. Long-term compatibility of *Cryptomeria* rootstock uncertain, deep planting for scion rooting recommended. Summer grafting (August) also possible.
Taxus: baccata cvs, *brevifolia, chinensis, floridana,* x *media* cvs	*T. baccata* 2+1–2P or T x *media* 'Hicksii' 0+1+2P 5–8mm. *T.* x *media* 'Hicksii' does not transplant as well as *T. baccata*		○			○		Cool temperature 10–15°C is suitable. Best without bottom heat as rootstock very susceptible to root disease. Careful watering required. Some favour late summer (Sept/Oct.).
Tetraclinis articulata	? *Platycladus orientalis* 2+1P 5–7mm. *Calocedrus decurrens* a possible alternative		●		●	●		Similar conditions to *Chamaecyparis*. Late summer grafting avoids possibility of winter damage and should work well. Compatibility unproven.

(Continued)

Name	Rootstock	Method		Systems				Comments
		Apical	Side	Cold callus	Warm callus - winter	Warm callus - summer	Hot-pipe	
Tetradium (Euodia): *daniellii*, *ruticarpum*,	? Root graft; roots 10–12mm Ø x 100mm from the parent plant. Conventional graft ? *Phellodendron amurense* 1+1P 6–8mm	○	●		○		●	Root grafts offer best chance of success. Methods are described in Chapter 18. Arnold Arboretum quotes *Phellodendron* as a suitable rootstock. Stem grafts may be best in hot-pipe at a temperature of ±20–22°C.
Thuja occidentalis cvs	*T. occidentalis* 1+1P 5–8mm or *T. occ.* 'Fastigiata' 0+1+1P		○			○		Same treatment as for *Chamaecyparis*, Chapter 32.
Thuja: *plicata* cvs, *standishii*.	*T. plicata* 1+1P 5–8mm or *T. plicata* 'Atrovirens' 0+1+1P 5–8mm		○			○		Same treatment as for *Chamaecyparis*, Chapter 32.
Thujopsis dolobrata cvs.	*Platycladus orientalis* 1+1P 5–8mm		○			○		Same treatment as for *Chamaecyparis*, Chapter 32.
Tilia cordata cvs, x *euchlora*, *henryana*, *oliveri*,	*T. cordata* 1+0 6–8mm or 1+1 6–10mm or 1+1P 6–10mm	○	○		○	○	○	Cold callusing with temperatures not long below 6°C. Warm callusing Jan–March 18–20°C or August-early September 20°C provides better results. Hot-pipe ±20°C better still. Root pressure significant at all times drying-off essential. For full details including further information on compatibility issues, **see also Chapter 53.**
americana cvs, x *euchlora* x *europea* cvs, *platyphyllos* cvs, *tomentosa* cvs	*T. platyphyllos* 1+0 6–8mm or 1+1 6–10mm or 1+1P 6–10mm							
Torreya species and cultivars.	Any *Torreya* rooted cuttings suitable for potting as rootstocks 0+1+1P 5–7mm. ? *Taxus baccata* 2+1P or 1+2P 5–7mm		●		●			Compatibility of *Torreya* on *T. baccata* not established.
Tsuga: *canadensis* selected cvs species, except *mertensiana*	*Tsuga canadensis* 1+1P 4–6mm. *T. mertensiana* for *T. mertensiana* cvs		○			○		Graft in early spring or late summer. Temperatures best in the 10–18°C region.

(Continued)

Name	Rootstock	Method		Systems				Comments
		Apical	Side	Cold callus	Warm callus - winter	Warm callus - summer	Hot-pipe	
Ulmus species and cultivars	*Ulmus glabra* 1+1 8–10mm bare-root or potted. This rootstock appears to demonstrate good compatibility with a wide range of *Ulmus* species and cultivars	●		●				*U. laevis*, said to be DED resistant, unfortunately is thought to show incompatibility symptoms with many types. *U. glabra* does not sucker and shows some resistance to DED. Compatibility issues with *Ulmus* have not been determined. The highly complex taxonomy of the genus means these are unlikely to be quickly resolved.
Vaccineum corymbosum cvs	*V. arboreum* 2+2P or 0+1+2P 5–7mm *V. virgatum*, same grade, also possible		●		●			Grafting Blueberry on a tall growing rootstock is a recent development and facilitates mechanical harvesting and widens the soil types for successful cultivation. Bench grafting using side grafts is replacing grafting in the field. Hot-pipe has not been tested but has potential.
Viburnum species and cultivars including: 'Anne Russell', x *carlecephalum*, *carlesii*, 'Eskimo', *macrocephalum*	*V. lantana* or *V. opulus* 1+0 4–6 or 5–7mm or 1+1P 6–10mm or disbudded cuttings 0+1+1P 6–8mm. Stems 1+1+1P or 1+2P 8–10/10–12mm at 80cm high	●	●	●	●	●		Ease of grafting means that even summer grafts may be apical splice or apical veneer. Side grafts for some demanding species. Main timing July–Sept but winter grafting 15–18°C possible. Top-worked stems grafted in the summer warm callus or individual polythene bags. **See also Chapter 54.**
Vitis vinifera Grapevine cultivars.	Cuttings-graft; stem 250–300mm x 6–12mm Ø or for 'Green Grafts' 300mm × >6mm Ø Rootstocks of disease resistant hybrids from American species	●			●	●		Grafts for grapevines normally produced by machine or grafting tool; Omega or 'V' notch popular. Scion normally a single bud. Grafts placed in callusing box to simultaneously unite and produce roots on the rootstock stem. Normally followed by cold storage and field planting when conditions are suitable. **See also Chapter 55.**

(Continued)

Name	Rootstock	Method		Systems				Comments
		Apical	Side	Cold callus	Warm callus - winter	Warm callus - summer	Hot-pipe	
Widdringtonia species	? *Platycladus orientalis* or *Cupressus macrocarpa* 1+1P 5–7mm		●		●	●		Compatibility on suggested rootstocks not confirmed. Grafting in late winter or summer should be possible. Conditions as for *Cupressus*.
Wisteria species cultivars and hybrids	*W. sinensis* bare-root and roots. 1+0 6–8mm Ø or 1+1 8–10mm Ø Root pieces 6–8mm Ø x 75mm–100mm. Conventional stem graft to 1+0P 6–8mm rootstocks also possible	●		●	●		●	Rootstock: basal portion of seedling rootstock stem with root attached or root only severed at hypocotyl, also root pieces below this and from established plants. Scion wood: current or mature wood (2yr+) from established plants. The latter may induce earlier flowering. Optimum temperature in poly-tent 15–18°C. For hot-pipe optimum temperature is 18–20°C. Use a wedge, long tongue or short tongue apical veneer.
Wollemia nobilis	*Araucaria araucana* 1 or 2 +1P 8–10mm		●		●			Combination not tested but may well succeed.
Xanthocyparis nootkatensis	*Xanthocyparis nootkatensis* 1+1P 6–8mm *Platycladus* (*Thuja*) *orientalis* 1+1P 6–8mm X *Cuprocyparis leylandii* 0+1+1P 6–8mm	●				●		Late summer grafting produces good results. Most graft Feb–March. Temperature best on cool side, bottom heat 10–12°C. Other details with *Chamaecyparis*. **See also *Chamaecyparis*, Chapter 32.**
Zelkova species and cultivars	*Zelkova serrata* 1+1P 6–8mm. *Ulmus glabra* 1+1P 6–8mm	●	●		●		●	Long-term compatibility of *Ulmus glabra* not confirmed. Hot-pipe should provide best results. Temperature 20–22°C likely to be optimum.

Appendices

Appendix A: List of Families and Genera Field Grafted

List of families and genera grafted in the field in cool, temperate zones in Northern and Southern Hemispheres:

In these regions, the range of species which can be reliably and consistently produced by grafting or budding in the field is essentially restricted to the following families and genera:

- Betulaceae - *Betula*
- Caprifoliaceae - *Viburnum*
- Caesalpiniaceae - *Gleditsia*
- Fagaceae - *Fagus*
- Grossulariaceae - Gooseberry top grafted onto stems of *Ribes aureum*
- Hamamelidaceae - *Hamamelis*
- Oleaceae - *Fraxinus, Syringa*
- Papilionaceae - *Caragana, +Laburnocytisus, Laburnum, Robinia*
- Rosaceae - *Amelanchier, Cotoneaster, Crataegus, Cydonia* (Quince), *Malus* (including culinary and dessert Apple), *Mespilus* (Medlar), *Prunus* (Almonds, Apricots, Bird Cherries, ornamental and fruiting Cherries, Peaches including ornamental, a few hybrid *Prunus* within these groups), *Pyrus* (including culinary Pears), *Rosa, Sorbus* (Whitebeam, Mountain ash)
- Salicaceae - *Populus, Salix*
- Sapindaceae - *Acer, Aesculus*
- Tiliaceae (Malvaceae) - *Tilia*
- Ulmaceae - *Ulmus, Zelkova*

Appendix B: List of Families Bench Grafted and with Potential to Graft

Listed are the Families grafted, and with potential to graft by bench grafting. The possible number of candidate genera (almost 300) is shown in parenthesis after the family name (two family names are shown where botanical authorities differ on classification):

Actinidiaceae (2), Alangiaceae (1), Anacardiaceae (3), Annonaceae(1), Aquifoliaceae (1), Araliaceae (7), Aristolochiaceae (1), Aurucariaceae (3), Berberidaceae(3), Betulaceae (6), Bignoniaceae (3), Boraginaceae (1), Buddlejaceae (Loganiaceae) (1), Buxaceae (2), Cactaceae (4), Caesalpiniaceae (6), Calycanthaceae (2), Caprifoliaceae (5), Celastraceae (3), Cercidiphyllaceae (1), Cistaceae (3), Clethraceae (1), Cornaceae (Nyssaceae) (5), Cupressaceae (25), Daphniphyllaceae (Euphorbiaceae) (1), Ebenaceae (1), Elaeagnaceae (2), Ericaceae (12), Eucryphiaceae (1), Fagaceae (6), Flacourtiaceae (3), Ginkgoaceae (1), Grossulariaceae (2), Guttiferae (1), Hamamelidaceae (9), Hydrangeaceae (8), Juglandaceae (4), Lauraceae (5), Lythraceae (1), Magnoliaceae (6), Malvaceae (Tiliaceae) (4), Meliaceae (2), Meliosmaceae (1), Mimosaceae (2), Moraceae (4), Myrsinaceae (1), Nyssaceae (2), Oleaceae (10), Onagraceae (1), Papilionaceae (18), Paulowniaceae (Scrophulariaceae) (1), Pinaceae (10), Platanaceae (1), Podocarpaceae (11), Punicaceae (1), Ranunculaceae (2), Rhamnaceae (3), Rosaceae (38), Rubiaceae (2), Rutaceae (6), Salicaceae (3), Sapindaceae (includes Aceraceae and Hippocastanaceae some authorities) (5), Sciadopityaceae (1), Simaroubaceae (2), Stachyuraceae (1), Staphyleaceae (1), Sterculiaceae (2), Styracaceae (6), Symplocaceae (1), Tamaricaceae (1), Taxaceae (3), Theaceae (2), Thymelaeceae (2), Trochodendraceae (2), Ulmaceae (7), Verbenaceae (2), Vitaceae (3)

Appendix C: Practice Holders

This aid to training is pictured in Figures 4.5 and 4.6 in Chapter 4.

Details of construction are shown in the diagrams (Figure C.1). Dimensions chosen should represent the most commonly used pot(s) in a given situation. Pictured is a design based on a 9 × 9 × 12.5cm pot, commonly used for rootstocks. Minimal skill is required to form the appropriate shapes in wood; although, round pot simulation may represent more of a challenge. Hinges can be recessed into the base so that the pots stand steadily upright. The two halves may be held together by one of two methods:

1. A spring claw toggle catch fixed on opposite sides provides a relatively quick and convenient closure system. The spring is essential to allow for variations in stem diameter. Not shown on the diagram is a slight depression, which may be cut on the inside facing surface of each half, to provide anchorage for the stem.
2. The through-bolt design is slower in use but is more suited for variable stem diameters, especially when large diameter stems are used. The wing nut shown can be supplemented by a longer steel strip 'paddle winder' welded on to provide quicker closure (see Figure 4.5 in Chapter 4).

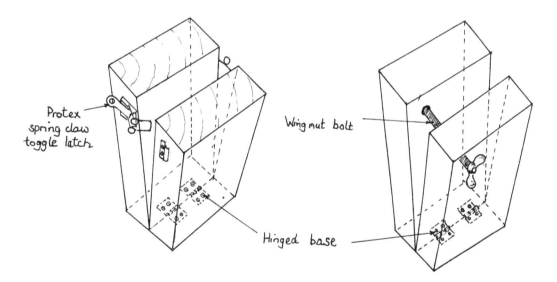

FIGURE C.1   Practice holders showing latch (left) and bolt (right) closing designs.

Appendix D: Outputs for Bench Grafting

TABLE D.1

Output for Various Types of Graft and Method**

Type of Graft	Output /hr	Country	Comments	Source
On Potted Rootstocks				
Whip and tongue	100–120	USA	Assume sealing by assistants	Légaré (2007)
Side veneer (*Cupressus macrocarpa* var. Lutea)	69. 60. 53.*	UK	Assume unsealed	Gaggini (1971)
Side veneer (*Picea pungens* Koster)	48. 42. 37.*	UK	Includes removal of needles. Assume unsealed	Gaggini (1971)
Long tongue side veneer (*Fagus sylvatica* cv.)	51. 44. 39.*	UK	Includes grafter sealing	Gaggini (1971)
Side veneer (long or short tongue)	50	UK	Non-specialist trained nursery staff	Mathews (1982)
Side veneer (long or short tongue)	100–130	UK	Specialist Dutch grafting team. Sealing by assistants	Mathews (1982)
Long tongue side veneer	100–120	USA	Assume sealing by assistants	Légaré (2007)
On Bare-root Rootstocks				
Splice	50	UK	Grafter seals own grafts	Bradley (1982)
Splice	63	USA	Assume grafter seals own	Légaré (2007)
Whip & tongue	120–150	USA	Assume sealing by assistants	Légaré (2007)
Whip & tongue	63. 55. 48.*	UK	Grafter seals own grafts	Gaggini (1971))
Apical veneer (*Hibiscus*)	94. 82. 72*	UK	Assume sealing by assistants	Gaggini (1971)
Apical veneer (single bud *Clematis* to root piece)	97. 84. 74*	UK	Assume sealing by assistants	Gaggini (1971)
Grafting tool (Raggett 'V' Notch)	<125	Europe	'Normal output'. Sealing by assistants	Légaré (2007)
Pneumatic-assisted Raggett 'V' notch	375	Europe	'Ultra-fast' worker. Sealing by assistants	Légaré (2007)

*First figure represents output assuming use of rubber ties. Second figure includes allowance of 14% relaxation time. Third figure is using raffia ties + 14% relaxation time.
**Bradley (1982), Gaggini (1971), Légaré (2007) and Mathews (1982).

Appendix E: Knife Types Other than Tina

Single-angle blades with one flat side are available from the Swiss manufacturer Victorinox but possibly only for right handed grafters. Most models on offer from this company have plastic composition handles. Blades are very resistant to corrosion, being a type of stainless steel. Disadvantages are that the edge is not retained as well as harder steels.

A French knife made by Opinel has been suggested as a possible grafting knife*. It has a partially straight-edged blade which curves back at the tip and is double angled so suitable for right- or left-handed grafters. Steel quality is quite good and holds an edge fairly well. The balance is reasonably good with a comfortable wooden handle but, after use, the blade soon develops sideways movement.

The Italian company SIFF make a range described as grafting knives, but the blades are all curved (hooked), and therefore more suitable for field grafting.

Of the North American companies, Spyderco, based in Oregon, appears among the biggest, offering a huge range of knives of all types. No specific grafting knives are listed.

---

* Garner (2013) pp.101–102.

Appendix F: Knife Sharpening – Cutting Edge Grind Types

Grind types relevant to grafting knives are briefly discussed and shown in the following cross-section diagrams of knife blade grinds and cutting edge grinds (Figure F.1).

CUTTING EDGE GRIND TYPES

SABRE GRIND ON UNIFORM THICKNESS BLADE (A)

The Sabre grind is the most commonly used for blades of uniform thickness. Disposable blades with uniform blade thickness normally have a cutting edge angle of 15 deg. on each side, together producing 30 deg., which is rather wide for extreme sharpness.

SABRE GRIND ON BLADE WITH INCLINED ANGLE AND A SPINE (B)

The presence of a spine and inclined angle permits use of narrower edge angles (<25 deg.) without compromising strength.

CHISEL GRIND ON SINGLE-ANGLE BLADE WITH A SPINE (C)

Single-angle knives with a spine are able to maintain an edge angle of 17.5 deg., producing extreme sharpness. Unless soft, woody material or herbaceous species are being handled, steel blades with an edge angle less than 17.5 deg. usually become blunted after a few hours use. Many grafters who hand-sharpen often produce a cutting edge angle of approximately 20 deg., providing an excellent compromise between edge retention and extreme sharpness.

CONCAVE CUTTING EDGE GRIND (D)

Some new grafting knives have a concave cutting edge resulting in a sharper cutting angle with less edge restriction than alternatives when making cuts. Disadvantages can be a reduction in blade edge strength and rapid loss of sharpness. Using a large diameter grindstone produces a shallow concave profile, providing the best possible compromise between sharpness and strength. Specialised sharpening equipment and procedures are required to maintain this grind and are discussed in Appendix J.

CONVEX CUTTING EDGE GRINDS (E)

This is invariably the result of hand-sharpening procedures. The edge retains sharpness longer than the alternatives but tends to be steeper, with the result that the knife is less sharp than concave and flat edges. Use of hand strops eventually always leads to the formation of a convex cutting edge.

MULTI-BEVEL CUTTING EDGE GRIND (F)

Almost identical to the convex edge but is a series of flat faces produced by applying the cutting edge to a sharpening stone at slightly different angles, which together form a convex profile. When

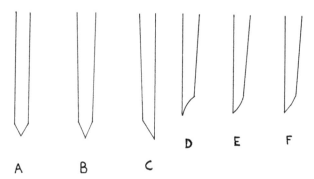

FIGURE F.1   Cutting Edge Grinds - A. Sabre grind on uniform thickness knife blade. B. Sabre grind on knife blade with an inclined angle on each side. C to F. Cutting edge grinds on a single angle, flat-sided blade having a spine and inclined angle on one side only. C. Chisel. D. Concave cutting edge. E. Convex cutting edge. F. Multi-bevel cutting edge.

two angles are involved (dual cutting edge), the first is steeper than the inclined angle (±12–17 deg.) and the second, providing the cutting edge, is steeper still, normally 20–22 deg. This edge grind is produced as a result of using only fine honing stones rather than a strop to maintain sharpness.

Appendix G: Types of Sharpening Stones

Sharpening stones may be natural or synthetic:

NATURAL STONES

These are less commonly used than previously because they are more expensive than synthetics and less reliable regarding consistent quality. The natural stone surface has two advantages. Powder produced in the sharpening process simultaneously sharpens and polishes the blade. In some types of stone, loss of surface prevents blockage by ground-off steel, and these do not require lubrication and are used dry.

Lubrication for most stones prevents surface blockage and overheating. Fine grade natural stones are mostly lubricated by water when they are known as water or wet-stones. Oil may occasionally be used as a lubricant, i.e. Arkansas stones. Natural stones are still available from the USA, Washita and Arkansas being the most popular. In Japan, some ultra-fine wet-stones up to 30,000 grit are available and marketed in the UK (Shapton Pro Stone®/Murasaki Ko 711®). A Belgian finishing stone (Belgian Coticule® or Belgium Brocken®) is available throughout Europe.

SYNTHETIC STONES

Normally a single or mixture of precisely graded abrading materials, suspended in a bonding compound, often a clay-like material. When wetted, the stone releases the abrasive materials rapidly and provides quick sharpening. Manufacturing improvements provide a product which has better uniformity of grit size, consistency and availability than its natural stone counterpart. For sharpening grafting knives, stones are available between 1000 and 8000 grit. Some stones incorporate industrial grade diamonds* which may be imbedded in hardened plastic or steel laminated to a plastic base. Advantages claimed are quick sharpening and retention of a perfectly flat surface. To date, no finishing grade diamond stones are on offer. Diamond paste or spray applied as a dressing to a leather strop provides an alternative.

---

* Holden (2001).

Appendix H: Honing - Strops and Honing Machines - Corrosion Prevention

LEATHER DRESSINGS

The favoured material for honing by hand is leather. To help eliminate any burr, the surface is normally dressed with an abrasive paste. These can vary from coarse, fine or ultra-fine, the choice between them being dependent upon the stones used previously, and the desired level of sharpness. Honing pastes, such as valve grinding polishing grade, can be mixtures of heavy grease incorporating appropriate abrasives. Others are Jewellers Rouge, Dovo Solingen® razor paste, Veritas® Honing compound, Flexicut® (Gold) polishing compound, Diamond paste (spray formulation) and Tormek® honing paste.

STROPS

Hand strops are an undressed or dressed leather strip between 50 and 80mm wide and 250 and 350mm long. These may either be fixed to a rectangular backing board, forming a Bench Strop or, when provided with a handle, they then become a Paddle Strop. Use of a strop without a backing board will produce a convex cutting edge more quickly with resultant reduction in edge sharpness.

Bench and Paddle strops may easily be homemade by purchasing suitable grade leather and attaching to a backing board, either using adhesive or tape (Figure 5.7, Chapter 5). Strops of suitable size, i.e. sufficient width, for the knives in use can be homemade, not always the case with ready-made items.

HONING MACHINES

Honing machines fall into two main categories: belt honers and wheel honers:

Belt Honers

Belt honers normally involve a 50- to 75mm-wide fabric belt, dressed with appropriate paste, held under tension between motor driven revolving pulleys. The knife blade is offered up to this, either free-hand or placed in an angle holder. Unless the belt has a suitable back-plate, applying the blade causes some distortion and results in a convex cutting edge.

Wheel Honers

Wheel honers have a leather or leather substitute glued to a power-driven wheel, which must rotate away from the cutting edge. The blade is offered up to this free-hand or in an angle holder, and as the hone has a rigid backing, the desired concave cutting edge is produced. See Appendix J, Knife Sharpening Machines: Figure J.1.

CORROSION PROTECTION

After cleaning off plant sap and any deposits, oil should be smeared over the blade at the end of each working day. If knives are stored for prolonged periods, there will be a need for corrosion protection. This is especially the case with high quality knives, with blades made of the type of steel to provide the sharpest edge but which is rather susceptible to corrosion. Rust inhibitors, such as the Protec® products or similar, can be considered.

Appendix I: Sharpening Aids

These take two forms: sharpening guides and rod-guided systems.

SHARPENING GUIDES (HONING GUIDES)

1. A simple system involves attaching a suitable diameter cylindrical guide to the blade spine thereby raising it to the cutting edge angle against the stone. Guides available are designed for large kitchen knives with a broad blade and thick spine, unsuitable for use with a grafting knifes. The concept is simple, and using tubular plastic mouldings or other materials, such as a wooden dowel, tubes of appropriate diameter can be acquired and it should be possible to fabricate a simple and suitable guide for single-angled grafting knives.
2. Blade guides have blade gripping mechanisms which have been developed for woodworking tools such as chisels and smoothing plane blades. The mechanism must be supplemented by a suitable knife blade holder obtained as a spare part from one of the knife sharpening machines, such as the Tormek Super Grind® or Robert Sorby Knife sharpening jig. Once fixed in position, the grinding angle against the stone can be obtained by adjusting the length of projection of the cutting edge from the mobile guide.

Sharpening is achieved by pushing the blade up and down the sharpening stone, holding the blade guide firmly in position on the stone surface. The result is a flat rather than convex cutting edge, superior in accuracy and consistency to that achieved by hand.

To work well and effectively, stone width needs to be matched to blade length.

ROD GUIDED SYSTEMS

Known as rod guided, several sharpening systems involving jigs and the use of special, edge-forming tools are available, in the USA and Europe. The knife blade is clamped in position and sharpened with a type of hand file, with a working face of abrasive material. This has a handle modified by having a long rod attached. The rod is passed through a hole in a partner jig, adjustable to hold the required angle of the file to the knife. This simple, flexible design can produce a reasonably accurate, flat cutting edge provided care and judgement are used. Shortcomings are the narrowness of the sharpening tool and the relatively limited range of grit sizes. One model (Lansky®) does offer a leather honing tool claiming to provide a sharp finished edge.

Appendix J: Knife Sharpening Machines

These take two main forms:

BELT SHARPENERS

Belt sharpeners depend on driving a continuous belt between two or more pulley wheels. Advantages are that belt widths matching the length of many knife blades are available. The disadvantages are the relatively poor choice of grit sizes, often restricted to coarse and medium. There is a need to support the belt from behind to avoid distortion and poor edge angles. Most of the belt sharpeners available are manufactured in the USA and increasingly in the Far East. One UK manufacturer makes a model which can be adapted for knife sharpening (Robert Sorby Pro-edge®). It does reportedly have good edge angle adjustment by the use of knife holder and guide.

GRINDSTONE SHARPENERS

A number of purpose built sharpening machines are available. These are primarily for woodturning and carpentry tools but offer modifications for knife sharpening. Some are manufactured in the USA (Grizzly Wet Grinder®, Dakota Wetstone® Sharpening System, etc.) and are also available in the UK.

A machine of this type is offered in the UK by the Scandinavian company Tormek. A centrally placed, slow speed electric motor drives spindles with a grindstone mounted on one side and honing cylinder on the other. A number of attachments are available, including an edge angle setting system involving an adjustable guide bar and adjustable knife holder. To produce the required angle, the blade is held in position against the grindstone or hone. An angle measuring guide is used to make the necessary adjustments to bar and holder to produce the desired cutting edge angle. The setting can be locked in position providing accurate repetition of the sharpening procedure time after time.

In practice knives, can be maintained in razor sharp condition quickly and relatively easily (Figure J.1).

Normally, it is necessary to only use the hone dressed with ultra-fine paste to retain the best edge. It is also used to polish the inclined angle and flat face to a mirror-like finish. Sole use of the ultra-fine hone ensures little steel is removed, giving the blade prolonged serviceability. Although Tormek offer curved grindstones and honing wheels as accessories, they are not suited to sharpening curved knives, and it is best to consider the machine as satisfactory only for straight edge blades.

The Tormek sharpening machine provides the opportunity to produce an accurate, polished, concave cutting edge, unrivalled by hand sharpening.

FIGURE J.1 Tormek Super Grind® sharpening machine, set-up for knife sharpening. The grindstone has been removed and replaced by a second honing wheel. The two wheels are each treated separately with medium or ultra-fine grade honing paste. The grafting knife (Tina 600) is shown mounted in the knife sharpening holder and applied to the honing wheel, which is revolving anti-clockwise away from the cutting edge. The precise angle of blade engagement with the wheel is set by the angle guide, pictured beneath the holder to the left of the machine. On this can be seen graduated settings and two black adjustable pointers. Once set this angle is constantly maintained. The knife holder is moved along the guide bar to engage the blade on the wheel for its entire length. The width of the wheel equivalent to the length of the blade would be ideal but is unavailable. Some discolouration (light corrosion) is seen on the flat facing side of the blade. This should be removed and both sides polished to a mirror finish by holding each surface flat on the honing wheel. The machine pictured is still working reliably after 20 years and approximately 2000 hours of use.

Appendix K: Grafting Machines and Grafting Tools

GRAFTING MACHINES

Machines which give matching cuts for both scion and rootstock normally produce these by cutting the wood with a knife or saw blade. Despite damage to the wood, especially on the exit side of the blade, some easily grafted species still produce successful results. In less easy types, the result can lead to failure or very poor results. An apple grower, preparing 25,000 grafts achieved a 25% take using an Omega machine and 90% with hand-made grafts*. Another grower using a similar machine for 33,500 apple tree grafts obtained a 50–51% take compared with 68% for handmade grafts. The machine grafts took several days longer to callus†. It is most probably important to place together 'inlet' and 'exit' sides of the cut so that mutually undamaged surfaces allow unhindered wound healing and callus production. Omega blades in particular require careful sharpening and timely replacement to ensure cuts are as clean and undamaged as possible.

GRAFTING MACHINES USING SAW BLADE SYSTEMS

Some saw blade machines produce a single mortise and tenon joint at the apex of the scion and rootstock. It is important that they fit firmly. Tying-in is often not essential, although helpful because it imparts extra strength to the graft. Single joints need tying-in. Ultra-adhesive graft sealants, such as those with a high resin content, applied in a hot liquid state, fulfil a similar function.

GRAFTING MACHINES USING KNIFE BLADE CUTTERS

Use of a knife blade for cutting the wood reduces laceration and bruising compared with the saw blade, but it is essential that the edge is kept extremely sharp and is properly set up.

Machines are available in two main types. One has blades mounted on a table which slides above a base plate and cuts the material fed in from the side, very much like a horizontal guillotine. The other is better described as a grafting tool because the blades are fitted into a secateurs-style mount.

GRAFTING TOOLS

Several grafting tools using shaped blades in a secateurs-style mount are available (Figure K.1). The tool is operated using the handles to make the cuts as for secateurs. Introduction of a foot operated system or compressed air power source speeds up cutting time and reduces operator fatigue (Figure K.2). There are various blade patterns but the most usual are notch, omega and chip bud. Under the categories discussed earlier, the first two constitute inlay and the third veneer grafts.

The notch removes a 'V' shaped portion of wood into which is placed a similarly shaped 'V' cut made from the same blade. The notch is produced by two straight-edged blades, which may be removed and individually sharpened to a good cutting edge. Due to the angle of engagement, some cracking and rough laceration of the wood on the emerging side of the cut is unavoidable.

An alternative cutting pattern is the omega, made from a blade shaped to the Greek omega symbol (Ω) (Figure 2.10 in Chapter 2: Machine and Tool-Made Grafts). Advantages are that, when

---

* Davis II (1977).
† Law (1977).

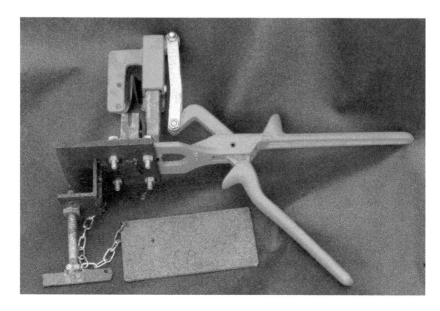

FIGURE K.1   Grafting tool, a hand operated secateurs-style model to produce a 'V' shaped notch or inlay graft. Attached is an 'in-house' fabricated base plate, with in-built screw clamp to enable bolting to a bench top.

assembled, the understock and scion are more positively locked together and tying-in is less necessary. The disadvantage of the omega pattern is that, since sharpening the blade is more difficult, the quality of the cut may suffer with an adverse effect on takes.

Grafting tools offer more flexibility than machines. The material to be grafted can be fed onto the anvil at varying angles and depth. This enables more judgement to be made with regard to the shape and form of the cuts. Skill is required for using these tools in this way. The 'V' notch blade has been used to cut rootstocks to produce a type of long tongue side graft in Magnolia (really a side inlay graft), the scion being cut by hand. The results achieved were comparable with handmade, long tongue side veneer grafts, but there was no advantage in output and only sufficiently large diameter rootstocks could be safely processed.

Some grafting tools could be better described as grafting aids and are useful in some situations. A single blade mounted in a hand-operated guillotine-type tool (Figure K.3) can produce a single cut for a wedge graft into virtually all shapes and sizes of rootstock. Removal of one blade from a 'V' wedge tool would produce a similar result. These tools make such cuts more easily and safely than by hand, especially when using thick wood. The scion wedge graft can also be cut to shape by a tool to produce a 'V' wedge, but because of the potential for damage to the wood mentioned above, hand cutting is normally preferred. For easier species this type of wedge graft linked to effective after-care can produce excellent results. Small operations could achieve a similar result by using double-cut secateurs to produce the same cut.

FIGURE K.2   Mechanised version of the hand-operated, secateurs-style grafting tool, with pneumatic ram fitted to replace hand operation, and speed-up cutting rates (F.P. Matthews Ltd.).

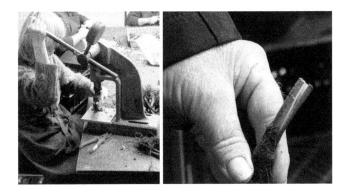

FIGURE K.3   LEFT, hand operated guillotine style tool to produce a wedge cut, normally in the rootstock, in a range of wood thicknesses and strengths. This tool is part of a graft production system, the scion cuts made by hand and the grafts assembled separately from the prepared components. RIGHT, close-up of the cut produced (Verhelste Nursery).

Appendix L: Workstations: Integral, Static and Mobile

INTEGRAL/STATIC

See Figure 7.1 in Chapter 7 for a view of a single person integral/static workstation.

An area 2500mm long by 1800–2200mm wide is required for a single grafter in a standing position; to help reduce summer temperatures, the ceiling needs to ideally be a minimum of 2100mm above floor level. This might contain a grafting bench 1400–1800mm by 800mm and 1000–1100mm high. Within the workstation can be included a free-standing electrical heater, electric fan (for summer cooling) and knife sharpening station (Tormek Super Grind® or similar). Additional convenience is provided by a number of shelves beneath the bench for storing tools, ties, cleaning pads and sterilants, etc. The integral lighting system is suspended 900mm above the bench top. Grafters in a seated position could utilise a wider space (2500mm).

MOBILE

A number of options for mobile workstations are available. The layout below is appropriate for walk-in grafting polytunnels in the open (Figure L.1).

A similar layout, mounted on wheels, would be suitable for use in access paths within multi-bay structures. The model depicted could be moved using a hand- or machine-operated winch. Serviced by appropriate materials handling equipment, this layout offers minimal requirements for additional support staff beyond ensuring a regular supply of rootstocks, and standing-down completed grafts in the grafting polytunnel.

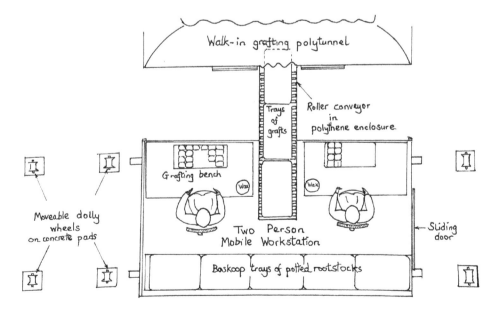

FIGURE L.1   Two-person mobile workstation moved by runners on demountable dolly wheels, or using a more complicated structure having built-in wheels. Rootstocks may be fed into the grafting area via back-door access. Completed grafts fed to the standing down beds in the tunnel via roller conveyor access.

Appendix M: Grafting Bench Design and Construction

A plan view of the grafting bench (Figure M.1) represents a recommended design and layout for grafting materials for a right-handed grafter, which follows suggested grafting sequence detailed in Chapter 4: Sequence of Actions. For a left-handed grafter, disposition of materials and equipment should be reversed.

Side view is of a horizontal bench fitted with an additional low-level bench. The additional bench is favoured by some for preparing splice graft rootstocks. This may be fixed loosely to the bench leg for right- or left-handed grafters as appropriate and may be adjusted for height by using a lock nut.

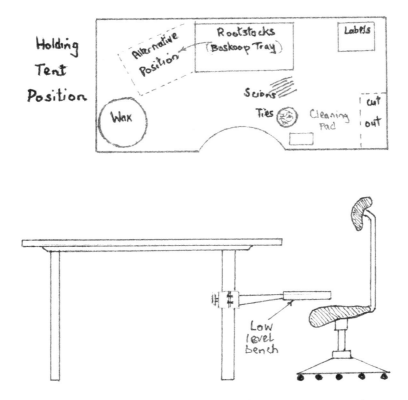

FIGURE M.1   Plan view and side view of grafting bench for right-handed, seated grafter. Note the optional addition of height-adjustable, low, level bench.

Appendix N: Seat Design

To avoid long-term lumbar damage, well-designed seating is essential. Because of the complex movements involved in grafting, it is not possible to designate the 'perfect' seat which meets all requirements.

Guidelines for good design are available: the seat should be between 350mm and 450mm in depth and 500–560mm wide; seat cushioning should be 38–50mm, deeper at the back than the front. Too soft a seat base, though it initially may feel comfortable, eventually causes a rise in skin temperature, increasing pressure under the thighs and leads to lower blood circulation. An excessively hard base produces too much pressure (85–100psi) on the 'seat bones'. A low-level backrest (125–225mm high) supports only the lumbar region, all that is required for grafting, but a medium-level rest (650–675mm high) does provide the opportunity to occasionally lean back. Both backrests should have a curve of 25–50mm between 125–225mm high from the seat of the chair. Height, back rest adjustment and the ability to swivel are basic necessities.

Height of the chair in relation to the height of the bench is crucial to ensure that no long-term damage is done to the back. Examples of chairs set too low, too high and at the correct height are shown in Figure N.1. Changing the angle of the seat, and some fore and aft movement relative to its ground position, linked to a slightly downward sloping bench work-top are ideal.

These facilities result in high costs; a good seat is an expensive item. Modern automobile seats provide most of the desired features. An enterprising approach, likely to save on costs, is to purchase second hand car seats from a demolition yard. With ingenuity these may be attached to a suitable base assembly with height adjustment, slope and swivel facilities (Figure N.2).

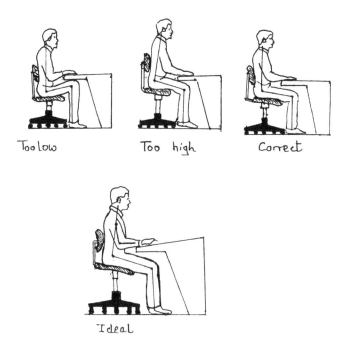

FIGURE N.1   Height of the chair in relation to the bench is important to avoid damage to the spine and provide a comfortable working position. A slightly sloping bench and chair base suits many grafters better than the conventional, 'correct', level chair base.

FIGURE N.2   Modern car seat obtained from car breakers, subsequently attached to an office swivel chair base. The adjustable height seat is set for top-working grafts of 'Patio plants' of *Corylus avellana* Red Majestic. The rootstock, *Corylus colurna* stem (1+1+1P 8–10mm Ø at 80cm high), potted into a 2-litre container, is seen standing on a low bench that also has height adjustment, to provide a convenient level for making the graft. Two scions are placed on the rootstock at 80cm high using the short tongue side veneer method, carried out in late summer.

Appendix O: Lighting Levels for Grafting Benches

Codes for light-level requirements are produced by various organisations. These are expressed as lux or lumen. Lumen refers to the strength of the light measured at source and lux is the light 'density' at the point of use.

The Health and Safety Executive (UK) suggest 500 lux for work requiring fine detail.

Specialist lighting contractors will be able to provide appropriate equipment to achieve the suggested levels.

In practice, experience is that good light level is provided by two 36-watt fluorescent tubes, 1200mm long. These are mounted parallel, and in close proximity, 1–1.2 metres above an 1800mm × 800mm grafting bench. Theoretically, this may provide more than suggested requirements.

Good illumination reduces eye strain and facilitates accurate work for long periods.

TABLE O.1

Suggested Light Level Requirements for Various Activities*

Activity	Lux per Metre Squared
Normal technical drawing work Mechanical workshops	1000
Hospital operating theatre Detailed technical drawing work	1500–2000

*Figures from CIBSE (Chartered Institute of Building Services Engineers).

Appendix P: Effect of Fungicides on Graft Success

In Holland, comparisons were made between chemicals using captan and carbendazim, sprayed and scion-dipped, on a range of Birch cultivars. Captan was unsuccessful but carbendazim spray increased takes by 24% over the untreated control of 54%, giving a 78% take. Dipping scions improved results over spraying by a further 10%, giving 88% success[*]. Carbendazim (Aagrano®) applied to winter and summer grafts of *Acer palmatum* cultivars were also tested. Summer grafts failed to respond but winter grafts were improved, especially *Acer shirasawanum* Aureum, which showed increases up to 50%. The following season (1967) takes of winter grafts of *Acer palmatum* Elegans Purpureum were improved by 22% using carbendazim (Aagrano®)[†].

Other trials of fungicides applied at grafting were carried out at the Boskoop Proefstation over many years. For Birch grafts, carbendazim and thiophanate methyl proved the most successful[‡]. Subsequently, results were improved by applying thiophanate methyl (Topsin M®) as a spray to grafts of *Betula pendula* Trisitis using rootstocks which had been potted in a mix incorporating carbendazim in the compost. Untreated grafts produced takes of 54%, which increased to 70% by treatment[§].

---

[*] van Doesburg & de Vogel (1959).
[†] van Elk (1966) and van Elk (1967).
[‡] Perquin (1974).
[§] Perquin (1975).

Appendix Q: Design and Construction of Free-Standing, Walk-In Grafting Polytunnels

In this type of structure, straight sides for the first 120cm are an advantage for internal movement and access. A 75 × 50mm pressure-treated timber base plate, fixed to a ground level concrete slab or single concrete block, is supplemented by an identical plate at fixed height of 90–120cm above. Both plates are attached to galvanised steel circular or box section tubes, placed vertically to match hoop positions, so that the polytunnel hoops can also be fitted at the same point. The plates are separated and held rigidly in position by vertical timber stanchions (also 75 × 50mm) attached to the steel verticals. Additional vertical timber supports for heating pipes, to hold insulation material, etc., are spaced between the main supports. These can be lighter grade (50 × 25mm). This construction provides an extremely rigid, durable and strong support structure (Figure Q.1).

Polythene cladding is best fixed in two sections. The upper, covering the hoops, is fixed to the upper plate and comprises an inflated double skin, ideally including a 'bubble polythene' lining. This can either be placed between the polythene double skins or be supported by horizontal galvanised wires running the length of the tunnel. The advantage of the latter is that one or more layers of fleece can also be placed in position tucked under the wires, easily done if the double skin is deflated during fixing. This will virtually eliminate any drips when the house is used for grafts on open beds rather than in poly-tents.

The lower section is a single skin of polythene (clear or black) or suitable cladding material such as exterior grade plywood. Some grafters consider extra light at grafting bed level provided by clear polythene is advantageous. Personal experience indicates it is not necessary, and more durable material such as thick gauge black polythene or plywood can be used for this area. Good insulation along the sides between basal and upper plates is important. Double layers of 'bubble polythene' when using clear polythene, or glass fibre insulation sheets with black polythene or plywood, should be fixed to reduce condensation and heat loss.

Dependent upon the length of the tunnel and whether fans can be fitted in end walls, side ventilation may be required, but this is best avoided if possible. If required, wooden frames clad with polythene or plywood can be made to fit snugly between alternate pairs of vertical stanchions. The interface between the ventilator and tunnel side must be as tight as possible to prevent water vapour loss from the structure when high humidity is required.

The ends of the tunnel may also be clad with clear polythene sheet, but plywood provides additional strength and convenience for mounting end doors and fixing the cladding (see Chapter 14, Figure 14.2).

FIGURE Q.1   View of free-standing grafting polytunnel. The side cladding is removed and shows construction details. In this example, side mounted PVC heating pipes are shown in position. Vertical 75 × 50mm timbers are seen at the end, and tunnel hoop positions with intermediate verticals are visible between. Part of the plywood gable end is just visible.

Appendix R: Hot-Pipe Solid Cover Heating Chamber Dimensions

Suggested dimensions for a heating chamber containing an electrical heating cable or hot water pipe are shown in the diagram below (Figure R.1). On the right are shown larger, stronger, side edging components required to house hot water pipe heating.

Length of standard sheets of extruded polystyrene (2400mm) may be extended by using a glued butt joint further strengthened by use of 8mm Ø wooden dowels glued into place. It is not recommended to build lengths longer than 3200mm as handling becomes more difficult and damage to the section more likely. A completed extruded polystyrene heating chamber is shown in Chapter 16, Figure 16.6.

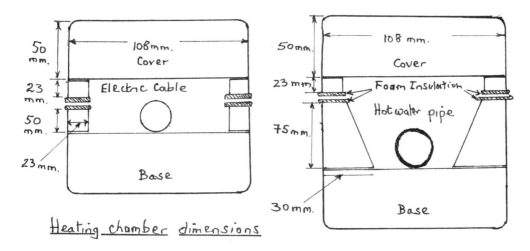

FIGURE R.1 Polystyrene hot-pipe heating chambers for electrical heating cable (left) and hot water pipe (right). Suggested dimensions are shown in cross-section diagrams. Not to scale.

Appendix S: 'Economy' Hot-Pipe Design

Suggested construction and dimensions are shown in the cross-section diagram (Figure S.1).

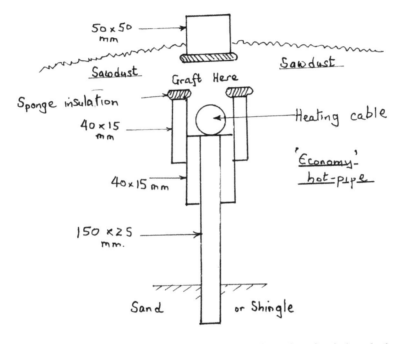

FIGURE S.1 'Economy' hot-pipe. Suggested construction and dimensions for timber planks are shown in the cross-section diagram. Not to scale. (See also Figure 16.13.)

Bibliography

WORKS CITED

Adlam, J. (2007, May). *Rhododendron Growing the German Way*. East Malling: Horticultural Development Council.

Agromillora California. (2017, December). Clonal Walnuts. Retrieved from Agromillora Nursery: http://www.agromillora-ca.com/products/nut-trees/walnuts/.

Ahrens, J. F. (1979). Cuttings from Herbicide-Treated Nursery Stock - What Can We Expect? *The International Plant Propagators' Society Combined Proceedings*, 29, pp. 348–357.

Aldrich, R. A., & Bartok, J. W. (1994). *Greenhouse Engineering* (33rd ed.). Ithaca, NY: Northeast Regional Agricultural Engineering Service.

Alexander III, J. H. (1998). A Summary of Graft Compatibility from the Records of the Arnold Arboretum. *The International Plant Propagators' Society Combined Proceedings*, 48, pp. 371–372.

Alexander III, J. H. (2001). Modified Nurse Seed Grafting of Aesculus. *The International Plant Propagators' Society Combined Proceedings*, 51, pp. 417–420.

Alkemade, J. P., & van Elk, B. C. (1989). Het Enten Van Boomkwekerijgewassen (Voorlichting Boomteelt, Brochure nr. 12). Boskoop: Proefstation voor de Boomkwekerij (PB).

Alley, C. J. (1965). Materials and Equipment Used in Grafting and Budding. *The International Plant Propagators' Society Combined Proceedings*, 15, pp. 275–277.

Amory, M. H. (2009). *The Oaks of Chevithorne Barton*. London, UK: Adelphi.

Anderson, D. (1976). Hardy Ornamental Stock Beds. *The International Plant Propagators' Society Combined Proceedings*, 26, pp. 136–139.

Aradhya, M., Potter, K. D., Gao, F., & Simon, C. J. (2007). Molecular Phylogeny of Juglans (Juglandaceae) a Biographic Perspective. *Tree Genetics and Genomes*, 3, pp. 363–378.

Ashburner, K., & McAllister, H. A. (2013). *The Genus Betula A Taxonomic Revision of Birches*. Richmond, UK: The Royal Botanic Gardens Kew.

Atwood, J. G. (1999). *Hardy Nursery Stock Project 78*. Kenilworth, Warwickshire, UK: Horticultural Development Council.

Auders, A. G., & Spicer, D. P. (2012). *Royal Horticultural Society Encyclopedia of Conifers A Comprehensive Guide to Cultivars and Species*. Nicosia: Kingfisher Publishing Limited.

Auvil, T. (2016, October 10). *Apple Rootstocks* (W. Jones, Editor). Retrieved February 2018, from WSU Tree Fruit Comprehensive Tree Fruit Site: http://treefruit.wsu.edu/web-article/apple-rootstocks-2/.

Bailey, L. H. (1920). *The Nursery Manual*. New York: The MacMillan Co.

Baker, R. L. (1976, July). Some Observations of Japanese Horticulture. *The Plant Propagator*, 22(3), p. 11.

Bakker, D. (1992). Grafting of Junipers. *The International Plant Propagators' Society Combined Proceedings*, 42, pp. 439–441.

Bakker, D. (1995). Grafting on Bare-Root Stock of Small Standard Trees. *The International Plant Propagators' Society Combined Proceedings*, 45, pp. 464–465.

Baldwin, H. (2017, March). Mexican Oaks in Cultivation. *The Plantsman*, 16(1), pp. 10–17.

Barnes, H. W. (2005). The Grafting of *Cedrus libani* pendula onto *Picea abies*. *The International Plant Propagators' Society Combined Proceedings*, 55, pp. 441–442.

Barnes, H. W. (2006). Grafting White Barked Birches onto *Betula nigra*: Practice and Possibilities. *The International Plant Propagators' Society Combined Proceedings*, 56, pp. 407–410.

Barnes, H. W. (2008). Grafting of *Cedrus atlantica* onto *Pinus strobus*. *The International Plant Propagators' Society Combined Proceedings*, 58, pp. 334–335.

Barnett, J. R., & Miller, H. (1994, January). The Effect of Applied Heat on Graft Union Formation in Dormant *Picea sitchensis* (Bong.). *Journal of Experimental Botany*, 45(1), pp. 135–143.

Bazzani, L. (1990). Practical Root Grafting. *The International Plant Propagators' Society Combined Proceedings*, 40, pp. 38–40.

Bean, W. J. (1976). *Trees and Shrubs Hardy in the British Isles* (8th Revised ed., *Vol.* 1 and 3). (S. Taylor, Ed.). London, UK: John Murray Ltd.

Beeson, R. C., & Proebsting, W. M. (1998). Carbon Metabolism in Scions of Colorado Blue Spruce. 1. Needle Starch. *Journal of the American Society of Horticultural Science*, 113, pp. 796–799.

Begum, S., Nakaba, S., Oribe, Y., Kubo, T., & Funada, R. (2007, September). Induction of Cambial Reactivation by Localized Heating in a Deciduous Hardwood Hybrid Poplar (*Populus sieboldii* × *P. grandidentata*). *Annals of Botany*, 100(3), pp. 439–447.

Bennison, J. (2010, March). Grafting Tree Peonies. *The Plantsman (New Series)*, 9(Part 1), 28–31.

Benson, D. M., & Jones, R. K. (2015, July 30). Phytophthora Root Rot and Its Control on Established Woody Ornamentals OD13. Retrieved April 2017, from North Carolina State University Plant Pathology Extension: https://www.ces.ncsu.edu/depts/pp/notes/oldnotes/odin13/od13.htm.

Berg, S. (1999). Sticking the Knife in: Grafted Nursery Stock Production at Yorkshire Plants. *The International Plant Propagators' Society Combined Proceedings*, 49, pp. 540–541.

Biggs, A. R. (1992). Anatomical and Physiological Responses of Bark Tissues to Mechanical Injury. In R. A. Blanchette, & A. R. Biggs (Eds.), *Defense Mechanisms of Woody Plants Against Fungi*, pp. (13–40). Berlin: Springer-Verlag.

Bir, R. E., & Ranney, T. G. (2000). Engineering Heat and Flood Tolerant Birches Through Grafting. *Combined Proceedings, International Plant Propagators' Society*, 50, p. 374.

Boomteeltproeftuin Midden-Nederland (Lienden). (1985). *Stimuleren van de groei van schoksgewijze groeinde gewassen zoals Quercus robur Fastigiata*. Boskoop: Jaarboek, Proefstation voor de Boomteelt en het Stedelijk Groen.

Bowes, B. G. (1999). *A Colour Atlas of Plant Propagation and Conservation*. London, UK: Manson Publishing Ltd.

Bradley, P. (1982). Bench Grafting Ornamentals and Fruits. *The International Plant Propagators' Society Combined Proceedings*, 32, pp. 220–223.

Briggs, B. (1977a). Questionbox re. Pinus sp. *The International Plant Propagators' Society Combined Proceedings*, 27, p. 499.

Briggs, B. A. (1977b). Manipulation of Herbicides and Effect of Herbicides on Rooting. *The International Plant Propagators' Society Combined Proceedings*, 27, pp. 463–466.

Brotzman, T. (2000). How Grafting Technique on Aesculus Can Influence Healing. *The International Plant Propagators' Society Combined Proceedings*, 50, p. 392.

Brotzman, T. C. (2012). Budding and Grafting – If You Think You Know It All ... Think Again! *International Plant Propagators' Society Combined Proceedings*, pp. 255–261.

Brown, H. C. (1963, June). Handling Juniper Grafts. *The Plant Propagator*, 9(2), p. 8.

Buchholz, T. (2014a, March 28). Greenhouse Twelve. Retrieved from Flora Wonder Blog: http://florawonder.blogspot.com/2014/03/greenhouse-twelve.htm.

Buchholz, T. (2014b, March 7). Conifer Grafting. Retrieved from Flora Wonder Blog: http://florawonder.blogspot.co.uk/2014/03/conifer-grafting.html.

Buchholz, T. (2015, December 11). Weepers. Retrieved April 2017, from Flora Wonder Blog: http://florawonder.blogspot.co.uk/2015/12/weepers.html.

Buchholz, T. (2016a, March 4). Making Babies. Retrieved from Flora Wonder Blog: http://florawonder.blogspot.com/2016/03/making-babies.html.

Buchholz, T. (2016b, May 13). Grandfather's Garden. Retrieved from Flora Wonder Blog: http://florawonder.blogspot.com/2016/05/grandfathers-garden.html.

Buchholz, T. (2016c, May 21). Conifers in GH28. Retrieved from Flora Wonder Blog: http://florawonder.blogspot.com/2016/05/conifers-in-gh28.html.

Buchholz, T. (2016d, September 9). Maples on My Miind. Retrieved from Flora Wonder Blog: http://florawonder.blogspot.com/2016/09/maples-on-my-mind.html.

Buchholz, T. (2017a, February 10). Winter Propagation. Retrieved from Flora Wonder Blog: http://florawonder.blogspot.com/2017/02/winter-propagation.html.

Buchholz, T. (2017b, July 7). Grafting This on That. Retrieved from Flora Wonder Blog: http://florawonder.blogspot.com/2017/07/grafting-this-on-that.html.

Buchholz, T. (2017c, September 22). Arrivals and Departures. Retrieved from Flora Wonder Blog: http://florawonder.blogspot.com/2017/09/arrivals-and-departures.html.

Burchell, T. W. (2002). The Use and Propagation of Wingnut (Pterocarya spp.) as a Phytophthora-Resistant Rootstock for Walnut (Juglans spp.). *The International Plant Propagators' Society Combined Proceedings*, 52, pp. 599–603.

Burchell Nursery. (n.d.). A General Guide to Rootstocks. Retrieved December 2017, from Burchell Nursery: http://www.burchellnursery.com/rootstock-options.php.

Bush, R. F. (2003). The Pitfalls of Grafting. *The International Plant Propagators' Society Combined Proceedings*, 45, pp. 296–297.

Byrne, C., & Byrne, G. (2004). Ten Years' Experience of Grafting Using the Hot-Pipe Callusing Technique. *The International Plant Propagators' Society Combined Proceedings*, 54, pp. 217–219.

Camelbeke, K., & Aeillo, A. S. (2016). Tree of the Year *Corylus fargesii*. In *International Dendrology Society Yearbook* (pp. 12–29).

Caron, J. E. (1959). Proeven Met Seringe-Onderstammen. Boskoop: Jaarboek, Proefstation voor de Boomkwekerij.

Carter, A. R. (1979). Discussion Group Report Daphne Propagation. *The International Plant Propagators' Society Combined Proceedings*, 29, pp. 248–251.

Carville, L. L. (1970). Environmental Control for Grafting. *The International Plant Propagators' Society Combined Proceedings*, 20, pp. 232–238.

Cave, P. B. (1982). Grafting Unusual Betula Cultivars. *The International Plant Propagators' Society Combined Proceedings*, 32, pp. 368–369.

Challice, J. S., & Westwood, M. N. (1973, September). Numerical Taxonomic Studies of the Genus *Pyrus* Using Both Chemical and Botanical Characters. *Botanical Journal of the Linnean Society*, 67(2), pp. 121–148.

Chong, C. (1981). Simultaneous Grafting and Rooting of Upright Junipers. *The International Plant Propagators' Society Combined Proceedings*, 31, pp. 504–505.

Clark, D. N. (1979). Planning Propagation Facilities for the 1980's. *The International Plant Propagators' Society Combined Proceedings*, 29, pp. 229–233.

Clark, R. (1981). Grafting Walnut Trees in a Cold Climate. *The International Plant Propagators' Society Combined Proceedings*, 31, pp. 337–339.

Clark, S. (1976). Cutting-Grafts of Camellia reticulata. *The International Plant Propagators' Society Combined Proceedings*, 26, pp. 415–416.

Clement, W. L., & Donoghue, M. J. (2012, May 30). Barcoding Success as a Function of Phylogenetic Relatedness in Viburnum, a Clade of Woody Angiosperms. *BMC Evolutionary Biology*, 12, 73.

Coggeshall, M. V. (1996). Oak Grafting Techniques. *The International Plant Propagators' Society Combined Proceedings*, 46, pp. 481–486.

Coombes, A. J. (2015). Tree of the Year: *Quercus rysophylla*. In C. Boisset (Ed.), *International Dendrology Society Yearbook*, (pp. 22–34).

Cooper, A. J. (1973). Root Temperature and Plant Growth – A Review. *Commonwealth Bureau of Horticulture and Plantation Crops, Commonwealth Agricultural Bureau, Research Review No. 4*.

Copes, D. (1969a). External Detection of Incompatible Douglas Fir Grafts. *The International Plant Propagators' Society Combined Proceedings*, 19, pp. 97–102.

Copes, D. (1969b). Graft Union Formation in Douglas Fir. *American Journal of Botany*, 56(3), pp. 285–289.

Copini, P., den Ouden, J., Decuyper, M., Mohren, G. M., Loomans, A. J., & Sass-Klaassen, U. (2014, January 1). Early Wound Reactions of Japanese Maple During Winter Dormancy: The Effect of Two Contrasting Temperature Regimes. *AoB Plants*, 6.

Cornell Human Factors and Ergonomics Research Group (CHFERG), Department of Design and Environmental Analysis at Cornell University. (2015, April). Sitting and Chair Design. Retrieved April 2015, from Cornell University Ergonomics Web (CU Ergo): http://ergo.human.cornell.edu/DEA3250Flipbook/DEA3250notes/sitting.html.

Cox, P. A. (1979, September). Breaking Growth and Saving Fuel in Rhododendron Propagation. *The Plant Propagator*, 25(3), pp. 12–13.

Crane, M. B., & Lawrence, W. J. (1952). *The Genetics of Garden Plants* (4th ed.). New York: Macmillan.

Creech, J. L. (1954). A Review of Vegetative Propagation of Beech and Linden. *Plant Propagators Society*.

Cross, D. (1995). Lagerstedt Hot Callusing Pipe. *The International Plant Propagators' Society Combined Proceedings*, 45, pp. 541–452.

Curtis, B. (1978). Propagation and Production of *Acer Palmatum* "Dissectum" Cultivars. *The International Plant Propagators' Society Combined Proceedings*, 28, pp. 351–353.

Curtis, W. J. (1962). The Grafting of Koster Spruce, Cedrus atlantica Glauca, Copper Beech, Pink and Variegated Dogwoods. *The International Plant Propagators' Society Combined Proceedings*, 62, pp. 249–253.

Davies, C. C., Fritsch, P. W., Li, J., & Donoghue, M. J. (2002). Phylogeny and Biogeography of Cercis (Fabaceae): Evidence from Nuclear Ribosomal ITS and Chloroplast ndhF Sequence Data. *Systematic Botany*, 27(2), pp. 289–302.

Davis II, B. (1976). Bench Grafting Plum and Apricot as Compared to T-Budding. *The International Plant Propagators' Society Combined Proceedings*, 26, pp. 253–255.

Davis II, B. (1977). Addendum to Law "Manual Grafting Versus Machine Grafting". *The International Plant Propagators' Society Combined Proceedings*, 27, p. 369.

Davis II, B. (1982). Grafting of *Pyrus calleryana* Cultivars. *The International Plant Propagators' Society Combined Proceedings*, 32, pp. 509–511.

Davis, P. (2015). Lighting: The Review. Retrieved 2017, from AHDB Horticulture CP125: https://horticulture.ahdb.org.uk/sites/default/files/u3089/Lighting.

Decker, B. M. (1998). Creating Separate Environments to Improve the Grafting Success of Specific Evergreen Species. *The International Plant Propagators' Society Combined Proceedings*, 48, pp. 342–344.

Decker, B. M. (2013). Grafting Rhododendron on pH Neutral Understocks. *The International Plant Propagators' Society Combined Proceedings*, 63, pp. 227–229.

Deering, T. (1979, September). Bench Grafting Betula Species. *The Plant Propagator*, 25(3), p. 8.

Deering, T. D. (1991). Walnut Propagation Using Bench Grafting. *The International Plant Propagators' Society Combined Proceedings*, 41, pp. 64–67.

Denk, T., Grimm, G. W., Manos, P. S., Deng, M., & Hipp, A. L. (2017). An Updated Infrageneric Classification of the Oaks: Review of Previous Taxonomic Schemes and Synthesis of Evolutionary Patterns. Retrieved 2018, from *Cold Spring Harbour Laboratory BioRxiv*: https://www.biorxiv.org/content/early/2017/07/31/168146.

deWilde, R. C. (1964). Production and Breeding of Lilacs. *The International Plant Propagators' Society Combined Proceedings*, 14, pp. 107–111.

Dirr, M. A., & Heuser Jr, C. A. (2006). *The Reference Manual of Woody Plant Propagation From Seed to Tissue Culture* (2nd ed.). Portland, OR: Timber Press.

Dixon, G. R. (2016, September). Roots and Root Health in Gardens. (M. Grant Ed.) *The Plantsman*, 15(Part 3), pp. 182–187.

Dorsman, C. (1966). Grafting of Woody Plants in the Glasshouse. *Proceedings 17th International Horticultural Congress*, 1, p. 366.

Driver, C. A. (1988). *Utility of Pyrus Species and Cultivars 6.4*. Boskoop: Jaarboek, Proefstation voor de Boomkwekerij.

Duncan, R., Connell, J., & Edstrom, J. (2010, November 30). Rootstocks for California Almond Orchards. Retrieved 2017, from The Almond Doctor: http://thealmonddoctor.com/2010/11/30/rootstocks-for-california-almond-orchards/.

Dunn, N. D. (1995). The Use of Hot Pipe Callusing for Bench Grafting. *Combined Proceedings, International Plant Propagators' Society*, 45, pp. 139–141.

Dupee, S. A., & Clemens, J. (1981). Propagation of Ornamental Grevillea. *The International Plant Propagators' Society Combined Proceedings*, 31, pp. 198–208.

Eichelser, J. (1967). Simultaneous Grafting and Rooting Techniques as Applied to Rhododendrons. *The International Plant Propagators' Society Combined Proceedings*, 17, p. 112.

Elkins, R. (2012). *Evaluation of Potential New Size Controlling Rootstocks for European Pears*. Research Project NC 140, University of California Coop Extension Lake and Mendocino Counties.

Esau, K. (1960). *Anatomy of Seed Plants*. New York: John Wiley & Sons, Inc.

Fan, C. (2001, June). Phylogenetic Relationships Within Cornus (Cornaceae) Based on 26S rDNA Sequences. *American Journal of Botany*, 88(6), pp. 1131–1138.

Farmer Jr, R. E., & Hatmaker, J. F. (1970, September). Graft Failure of Black Cherry Scions on Four Prunus Rootstock Species. *The Plant Propagator*, 16(3), pp. 8–9.

Farrell, D. (2011). Research Chemist Trials Glue for Grafting. *IPPS Review* (Autumn/Winter), 4.

Figlar, R. B. (1985, September). Grafting Magnolia on Liriodendron. *The Plant Propagator*, 31(3), pp. 9–11.

Fillmore, R. H (1951). A General Review of Woody Plant Propagation. *Plant Propagators Society*, 1, p. 40.

Flemer III, W. (1962). The Vegetative Propagation of Oaks. *Plant Propagators' Society Combined Proceedings*, 12, pp. 168–173.

Flemer III, W. (1980). Linden Propagation – A Review. *The International Plant Propagators' Society Combined Proceedings*, 30, pp. 333–336.

Flemer III, W. (1986). New Advances in Bench Grafting. *The International Plant Propagators' Society Combined Proceedings*, 36, pp. 545–549.

Foster, S. (1992). Propagation of Japanese Maples by Softwood Cutting and Grafting. *The International Plant Propagators' Society Combined Proceedings*, 42, pp. 373–374.

Fowler Nurseries Inc. (n.d.). Rootstocks. Retrieved 2017, from Fowler Nurseries Inc.: http://www.fowlernurseries.com/New%20Pages/Rootstocks.htm.

Frank, P., & Matthews Ltd. (2018). Fruit Tree Rootstocks. Retrieved February 2018, from Frank P Matthews Trees for Life: https://www.frankpmatthews.com/advice/fruitrootstocks.

Gaggini, J. B. (1971). A Time Measurement Study: Bench Grafting of Woody Plants Under Glass. *The International Plant Propagators' Society Combined Proceedings*, 21, pp. 275–292.

Gardiner, J. (2000). *Magnolias A Gardeners Guide* (revised ed.). Portland, Oregon: Timber Press, Inc.

Garner, R. J. (2013). *The Grafter's Handbook – 6th edition revised and updated by S. Bradley*. London: Octopus Publishing Group.

Gernandt, D. S., López, G. G., García, S. O., & Liston, A. (2005). Phylogeny and Classification of Pines. *Taxon*, 54(1), pp. 29–42.

Gilbert, A. (2005). 'Long John' Grafts. *The International Plant Propagators' Society Combined Proceedings*, 55, pp. 117–119.

Gleeson, M. (2007). Grafting Red-Flowering Gum:Trying a New Method. *The International Plant Propagators' Society Combined Proceedings*, 57, pp. 718–720.

Goldammer, T. (2018). *Grape for Growers Handbook. A Guide to Viticulture for Wine Production* (3rd ed.). USA: Apex Publishers.

Gregg, P. (2013). Conifer Grafting at Iseli Nursery: Fundamentals to Creating a Great Product. *The International Plant Propagators' Society Combined Proceedings*, 63, pp. 151–153.

Halda, J. J. (1998). Some Taxonomic Problems in the Genus Daphne. Retrieved from *Acta Mus.Richnov*: http://www.moh.cz/pdf/amr/8.

Hall, P. M. (1977). Techniques of Juniper Grafting. *The International Plant Propagators' Society Combined Proceedings*, 27, pp. 246–251.

Harber, P. J. (2005). *Applicability of No-Fines Concrete as a Road Pavement. Dissertation for Course ENG4111/4112*. Retrieved from University of Southern Queensland Faculty of Engineering and Surveying: https://eprints.usq.edu.au/472/1/PaulHARBER-2005.pdf.

Harris, A., Frawley, E., & Wen, J. (2017, June 7). *The Utility of Single-Copy Nuclear Genes for Phylogenetic Resolution of Acer and Dipteronia (Aceraceae, Sapindaceae)*. (F. Z. Board, Producer) Retrieved 2018, from Bio One – Research Evolved: http://www.bioone.org/doi/full/10.5735/085.054.0603.

Harris, D. C. (1976). Propagation of Japanese Maples by Grafting. *The International Plant Propagators' Society Combined Proceedings*, 26, pp. 169–170.

Harrison, V. (1978). Some Observations on the Influence of Temperature on Walnut Propagation. *The International Plant Propagators' Society Combined Proceedings*, 28, pp. 280–282.

Harrison-Murray, R. S. (2002). *Examination of Techniques to Raise Humidity in Mist Houses. Hardy Nursery Stock (HNS) Project 76*. Horticultural Development Council.

Hartmann, H. T., & Kester, D. E. (1959). *Plant Propagation Princples and Practices*. Englewood Cliffs, N.J.: Prentice Hall, Inc.

Hartmann, H. T., Kester, D. E., Davies Jr, F. T., & Geneve, R. L. (2014). *Hartmann & Kester's Plant Propagation: Principles and Practices* (8th ed.). Harlow, UK: Pearson Education Limiited.

Hasey, J. (2016, July 14). Selecting the Right Clonal Rootstock for Managing Soil and Pest Problems. Retrieved 2017, from Sacramento Valley Orchard Source: http://www.sacvalleyorchards.com/walnuts/.

Hatch, D. (1982). Grafting of Pinus, Picea, and Abies. *The International Plant Propagators' Society Combined Proceedings*, 32, pp. 215–217.

Haworth-Booth, M. (1963). *The Moutan or Tree Peony*. Letchworth, Hertfordshire, UK: The Garden Book Club.

Heimstra, J., & van der Sluis, B. (2005). The Development of Verticillium-Resistant Acer Rootstocks. *The International Plant Propagators' Society Combined Proceedings*, 55, pp. 421–422.

Hellriegel, F. C. (1982). The Nature of Callus and Its Importance to the Plant Propagator. *The International Plant Propagators' Society Combined Proceedings*, 32, pp. 65–66.

Hendricks, B. (2013). Grafting for Quick Turn- Bigger Can Be Better. *Combined Proceedings, International Plant Propagators' Society*, 63, pp. 303–305.

Hess, C. E. (1953). Magnolia Propagation by Grafting. *The International Plant Propagators' Society*, 3, p. 113.

Hesselein, R. (2005). Chip Budding Hard-to-Root Magnolias. *The International Plant Propagators' Society Combined Proceedings*, 55, pp. 384–386.

Hewson, A. (2012). *Hardy Nursery Stock Propagation Guide*. Horticultural Development Company (HDC) Division of Agriculture and Horticulture Development Board (AHDB).

Holden, V. (2001). Grafting Tips. *The International Plant Propagators' Society Combined Proceedings*, 51, p. 274.

Holland, B. T., Warren, S. L., Ranney, T. G., & Eaker, T. A. (2001). Rootstock Selection and Graft Compatibility of Chamaecyparis Species. *The International Plant Propagators' Society Combined Proceedings*, 51, pp. 461–465.

Hoogendoorn, D. P. (1981). Grafting Upright Junipers. *The International Plant Propagators' Society Combined Proceedings*, 31, pp. 501–505.

Hoogendoorn, D. P. (1988). Grafting *Viburnum carlesii* Compactum. *The International Plant Propagators' Society Combined Proceedings*, 38, pp. 563–566.

Hooper, V. (1990). Selecting and Using Magnolia Clonal Rootstocks. *The International Plant Propagators' Society Combined Proceedings*, 40, pp. 343–346.

Hooper, V. (2010). Twenty Years Watching Rootstocks. *The International Plant Propagators' Society Combined Proceedings*, 60, pp. 151–160.

Hottovy, S. A. (1986). Japanese Maple Propagation at Monrovia Nursery. *The International Plant Propagators' Society Combined Proceedings*, 36, pp. 89–91.

Howard, B. (1990). Improving Propagation Techniques: Getting The Environment Right. *The Garden* (December), pp. 646–649.

Howard, B. H. (1977). A Personal View of the Role of a Propagation Research Worker. *The International Plant Propagators' Society Combined Proceedings*, 27, pp. 78–83.

Howard, B. H. (1995). Opportunities for Developing Clonal Rootstocks from Natural Seedlings of Tilia spp. *The Journal of Horticultural Science & Biotechnology*, 70(5), pp. 775–786.

Howard, B. H., & Oakley, W. (1996). *Factors Determining Bud-Take in Ornamental Trees. Hardy Nursery Stock (HNS) Project 45*. East Malling Research Station, Bradbourne: Horticultural Development Council.

Howard, B. H., & Oakley, W. (1997a). Bud-Grafting Difficult Field-Grown Trees. *The International Plant Propagators' Society Combined Proceedings*, 47, pp. 328–333.

Howard, B. H., & Oakley, W. (1997b). Bud-Grafting in Acer Crimson King Related to Root Growth. *Journal of Horticultural Science*, 72(5), pp. 697–704.

Howard, B. H., Harrison-Murray, R. S., & Mackenzie, K. A. (1984). *Rooting Responses to Wounding Winter Cuttings of M.26 Apple Rootstock*. Retrieved 2017, from Taylor and Francis On Line – *Journal of Horticultural Science*: https://www.tandfonline.com/doi/abs/10.1080/00221589.1984.11515179.

Howe, C. F. (1976). The Production of Container-Grown Trees by Bench Grafting – Some Criteria for Success. *The International Plant Propagators' Society Combined Proceedings*, 26, pp. 145–149.

Hummel, R. L. (1999). Discussion Group: Budding and Grafting Made Better. *The International Plant Propagators' Society Combined Proceedings*, 49, pp. 567–569.

Humphrey, B. (1978). Propagation by Grafting Under Glass at Hilliers Nursery. *The International Plant Propagators' Society Combined Proceedings*, 28, pp. 482–488.

Humphrey, B. E. (1979). Propagation of Hamamelis and Related Plants. *The International Plant Propagators' Society Combined Proceedings*, 29, pp. 256–260.

Humphrey, B. E., & Dummer, P. (1967). Stock/Scion Relationships. *The International Plant Propagators' Society Combined Proceedings*, 17, pp. 389–391.

Hunt, P. B. (1971, May). The Compatibility Response of Lilac Hybrids on Rootstocks of Other Members of Oleaceae. *The Plant Propagator*, 17(2), pp. 9–15.

Hyde, D. (1996). IPPS Workshop: Summer Grafting. *The International Plant Propagator's Society Newsletter*, pp. 23–25.

Intven, W. J., & Intven, T. J. (1989). Apical Grafting of *Acer palmatum* and Other Deciduous Plants. *Combined Proceedings, International Plant Propagators' Society*, pp. 409–412.

Jackson, B. (2018). *All Things Fibre*. Growing Media Association (GMA). Kenilworth: ADHB Growing Media Review.

Jaynes, R. A. (2003). Chip Budding Aralia. *Combined Proceedings, International Plant Propagators' Society*, 53, p. 459.

Johnson, R. S., & DeJong, T. (2009–2012). Peach Rootstock Trial. Retrieved 2017, from University of California Fruit Report: http://ucanr.edu/sites/fruitreport/rootstocks/nc-140_rootstock_trials/2009_peach_rootstock_trial/.

Joustra, M. K., Verhoeven, P. A., & Ruesink, J. B. (1985). *The Use of Auxins with Grafting Robinia Pseudoacacia Frisia, 1.13* (pp 58–59). Boskoop: Jaarboek, Proefstation voor de Boomteelt en het Stedelijk Groen.

Joustra, M. K. (1985, December). Grafting of Amelanchier. *The Plant Propagator*, 31(4), p. 4.

Joustra, M. K., & Ruesink, J. B. (1985). *Pinus-cultivars, groeistoffen bij het enten.1.14.1* (pp. 60–62). Boskoop: Jaarboek, Proefstation voor de Boomteelt en het Stedelijk Groen.

Joustra, M. K., Ruesink, J. B., & Verhoeven, P. A. (1989). *Enten van Pinus*. Boskoop: Jaarboek, Proefstation voor de Boomkwekerij – Boomtelt Praktijkonderzoek.

Joustra, M. K., Ruesink, J. B., & Verhoeven, P. A. (1990). *Enten van Pinus*. Boskoop: Jaarboek, Proefstation voor de Boomteelt – Boomteelt Prakijkonderzoek.

Kaeiser, M., Jones, J. H., & Funk, D. T. (1975, January). Interspecific Walnut Grafting in the Greenhouse. *The Plant Propagator*, 20(4), pp. 2–7.

Karlsson, I., & Carson, D. (1985). Survival and Growth of Abies amabilis Scions Grafted on Four Species of Understock. *The Plant Propagator*, 31(2).

Kawarada, K. (2008). The Propagation of Momijii Garden Plants. In *The Maple Society Symposium 2008 in Japan* (pp. 29–31). Saitama, Japan: Nurserymen's Association.

Kayange, C. W., & Scarborough, I. P. (1976, July). Chip Budding Unrooted Rootstocks of Tea (*Camellia sinensis*). *The Plant Propagator*, 22(3), p. 14.

Kern, C. (1956). The Use of Grafts to Obtain Own-Rooted Lilacs. *Plant Propagators' Society Proceedings*, 6, pp. 88–92.

Kester, D. E. (1965). The Physiology of Grafting. *Combined Proceedings, International Plant Propagators' Society*, 15, pp. 261–273.

Kim, S., Chong-Wook, P., Yong-Dong, K., & Youngbae, S. (2001). Phylogenic Relationships in Family Magnoliaceae Inferred from ndhF Sequences. *American Journal of Botany*, 88(4), pp. 717–728.

Klapis Jr, A. J. (1964). Grafting Junipers. *Combined Proceedings, International Plant Propagators' Society*, 14, pp. 101–103.

Kloosterhuis, W. E. (1975). *Proeven op de Boomteeltproeftuin te Horst*. Boskoop: Jaarboek, Proefstation voor de Boomkwekerij.

Kloosterhuis, W. E. (1977). *Proeven op de Boomteeltproeftuin de Horst (Table 57)*. Boskoop: Jaarboek, Proefstation voor de Boomwekerij.

Knuckey, D. (1969). Bud-Grafting Magnolias. *The International Plant Propagators' Society Combined Proceedings*, 19, pp. 221–222.

Kroin, J. (1992). Advances Using Indole-3-Butyric Acid (IBA) Dissolved in Water for – Rooting Cuttings, Transplanting, and Grafting. *The International Plant Propagators' Society Combined Proceedings*, 42, pp. 489–492.

Krüssmann, G. (1964). *Die Baumschule*. Berlin: Verlag Paul Parey.

Lagerstedt, H. B. (1969). Grafting: A Review of Some Old and Some New Techniques. *The International Plant Propagators' Society Combined Proceedings*, 19, pp. 91–96.

Lagerstedt, H. B. (1981). A Device for Hot Callusing Graft Unions of Fruit and Nut Trees. *The International Plant Propagators' Society Combined Proceedings*, 31, pp. 151–159.

Lane, C. (1976). Production of Norway Maple Cultivars by Bench Grafting. *The International Plant Propagators' Society Combined Proceedings*, 26, pp. 150–153.

Lane, C. (2005). *Witch Hazels*. Cambridge, UK: Timber Press.

Lane, C. (2012). Personal Experiences of Grafting High Value Plants. *The International Plant Propagators' Society Combined Proceedings*, 62, pp. 187–188.

Lane, C. G. (1982). Bench Grafting Under Heated Glass. *The International Plant Propagators' Society Combined Proceedings*, 32, pp. 217–219.

Lane, C. G. (1993a). Magnolia Propagation. *The International Plant Propagators' Society Combined Proceedings*, 43, pp. 163–165.

Lane, C. G. (1993b). Propagation of the Genus Betula. *Betula: Proceedings of the IDS Betula Symposium 2–4 October 1992* (pp. 51–60). International Dendrology Society.

Lantos, A. (1995). Winter Bench Grafting of Walnut Varieties. *The International Plant Propagators' Society Combined Proceedings*, 45, pp. 146–148.

Larson, R. A. (2006). Grafting: A Review of Basics as well as Special Problems Associated with Conifer Grafting. *The International Plant Propagators' Society Combined Proceedings*, 56, pp. 318–322.

La Rue, J. H., & Johnson, R. S. (1989). *Peaches Plums & Nectarines – Growing and Handling for Fresh Market*. Davis, California, USA: UC Davis.

Law, J. (1977). Manual Grafting Versus Machine Grafting. *The International Plant Propagators' Society Combined Proceedings*, 27, pp. 368–369.

Leach, D. G. (1960). Outside Green Grafting of Rhododendrons under Polythene. *Plant Propagators' Society Combined Proceedings*, 10, pp. 90–93.

Leach, D. G. (1961). *Rhododendrons of the World*. New York, USA: Charles Scribner's Sons.

Légaré, M. (2007). The Future of Grafting. *The International Plant Propagators' Society Combined Proceedings*, 57, pp. 380–384.

Leiss, J. (1966). Trials with Three Juniperus Understocks. *The International Plant Propagators' Society Combined Proceedings*, 16, pp. 215–217.

Leiss, J. (1969). Hamamelis Propagation. *The International Plant Propagators' Society Combined Proceedings*, 19, pp. 349–352.

Leiss, J. (1971). Grafting Tree Peonies. *The International Plant Propagators' Society Combined Proceedings*, 21, pp. 387–388.

Leiss, J. (1977). Propagation of *Aralia elata* Variegata. *The International Plant Propagators' Society Combined Proceedings*, 27, pp. 461–463.

Leiss, J. (1984). Root Grafting of Oaks. *The International Plant Propagators' Society Combined Proceedings*, 34, pp. 526–528.

Leiss, J. (1986). Modified Side Graft for Deciduous Trees. *The International Plant Propagators' Society Combined Proceedings*, 36, pp. 543–544.

Leiss, J. (1988). Piece Root Grafting of Oaks: An Update. *The International Plant Propagators' Society Combined Proceedings*, 38, pp. 531–532.

Leiss, J. (1995). Grafting on Roots. *The International Plant Propagators' Society Combined Proceedings*, 45, pp. 469–471.

Lentz, P. A. (1999). Grapevine Propagation with an Emphasis on Grafting. *The International Plant Propagators' Society Combined Proceedings*, 49, pp. 633–636.

Leon, C. (1994). *Hazard Assessment of Poisonous Plants in the UK Horticultural Trade*. London: The National Poisons Unit, Guy's Hospital; Kew: The Royal Botanic Gardens; The Royal Horticultural Society.

Leslie, C., Hackett, W., & Robinson, R. (2012). Clonal Propagation of Walnut Rootstock Genotypes for Genetic Improvement. Retrieved 2017, from UC Davis Eruit & Nut Research & Information Center: http://ucanr.edu/sites/cawalnut/category/RootProp/.

Lienden Trial Group. (1985). *Lienden Tree Trial*. Yearbook. Boskoop: Lienden Tree Trial Station.

Loach, K. (1981). Propagation Under Mist and Polythene: History, Principles, and Development. *Askham Bryan Horticultural Technical Course*, 21, pp. 23–31.

Long, L. E., Brewer, L. J., & Kaiser, C. (2014). *Cherry Roostocks for the Modern Orchard. 57th Annual I.F.T.A. Conference*. Kelowna, British Columbia, Canada: Department of Horticulture Oregon State University.

Lu, H., Jiang, W., Ghiassi, M., Lee, S., & Nitin, M. (2012, January). Classification of Camellia (Theaceae) species Using Leaf Architecture Variations and Pattern Recognition Techniques. *PLoS ONE*, 7(1), p. e29704.

Luby, J. J. (2003). Taxonomic Classification and Brief History. In D. C. Ferree, & I. J. Warrington (Eds.), *Apples: Botany Production and Uses* (pp. 9–12). Wallingford, UK: CABI.

Macdonald, B. (1974). Camellia Propagation Discussion Group Reports. *The International Plant Propagators' Society Combined Proceedings*, 24, pp. 152–153.

Macdonald, B. (1979). Propagation of Picea Discussion Group Report. *The International Plant Propagators' Society Combined Proceedings*, 29, pp. 260–265.

Macdonald, B. (1986). *Practical Woody Plant Propagation for Nusery Growers: Volume 1*. Portland, Oregon: Timber Press.

Macdonald, B. (1988). Bench Top-Working of Ornamental Trees, Shrubs, and Conifers. *The International Plant Propagators' Society Combined Proceedings*, 38, p. 92.

Macdonald, B. (1989). Bench Grafting Colorado Blue Spruce - Criteria for Success. *The International Plant Propagators' Society Combined Proceedings*, 39, pp. 131–134.

MacDonald, P. T. (2014). *The Manual of Plant Grafting*. Portland, OR: Timber Press.

Mahlstede, C. (1962). A New Technique in Grafting Blue Spruce. *The International Plant Propagators' Society Combined Proceedings*, 12, pp. 125–126.

Mahlstede, J. P. (1958). Graft Failures in Apple Scions. *The International Plant Propagators' Society Combined Proceedings*, 8, pp. 153–158.

Mathers, H. (1999). Discussion Group: Rootstock of Choice. *The International Plant Propagators' Society Combined Proceedings*, 49, pp. 626–628.

Mathews, B. (1982). The Economics of Grafting. *The International Plant Propagators' Society Combined Proceedings*, 32, pp. 230–232.

Mathies, J. (1985, March). Selection of Understock, and Chip-Budding vs. T-Budding in the Propagation of Pyrus Cultivars, 'Bradford' 'Chanticleer' and 'Redspire'. *The Plant Propagator (The International Plant Propagators' Society)*, 31(1), p. 9.

McCulley, M. E. (1983). Structural Aspects of Graft Development. In R. Moore (Ed.), *Vegetative Compatibility Responses in Plants* (pp. 71–88). Waco, TX: Baylor University Press.

McGranahan, G., Hackett, W. P., Lampinen, B. D., Leslie, C., Manterola, N., Gong, Y., et al. (2007–2012). Clonal Propagation of Walnut Rootstock Genotypes for Genetic Improvement. Retrieved 2017, from UC Davis Fruit & Nut Research & Information Center: http://ucanr.edu/sites/cawalnut/category/RootProp/.

McGuire, J. J., Johnson, W. M., & Dawson, C. (1987). Leaf-Bud or Side-Graft Nurse Grafts for Difficult-to-Root Rhododendron Cultivars. *The International Plant Propagators' Society Combined Proceedings*, 37, pp. 447–449.

McMillan Browse, P. (1971, May). Propagation of Aesculus. *The Plant Propagator*, 17(2), pp. 5–6.

McMillan Browse, P. (1979a). *Selection of Rootstocks*. Brooksby Agricultural College East Midlands UK: I.P.P.S Gb & Ireland Region.

McMillan Browse, P. (1979b, March). *Notes on Selection of Rootstocks*. UK: International Plant Propagators' Society Papers from regional meetings, p. 6.

McMillan Browse, P. D. (1970). Notes on the Propagation of Viburnums. *The International Plant Propagators' Society Combined Proceedings*, 70, pp. 378–382.

McPhee, G. (2007). Getting Grafts to Take ... or Why Grafts Don't Take ... or 1000 Excuses for a Poor Result. *The International Plant Propagators' Society Combined Proceedings*, 57, pp. 608–610.

Meacham, G. (2003). Timing for Top Grafting Cercis and Cercidiphyllum Cultivars. *The International Plant Propagators' Society Combined Proceedings*, 53, p. 388.

Meacham, G. E. (1995). Bench Grafting, When Is the Best. *Time*? *The International Plant Propagators' Society Combined Proceedings*, 45, pp. 301–304.

Meacham, G. E. (2008). Summer Veneer Bench Grafting of *Acer palmatum* Cultivars. *The International Plant Propagators' Society Combined Proceedings*, 58, pp. 310–311.

Megre, D., Kondratovičs, U., & Dokāne, K. (2007). Simultaneous Graft Union and Adventitious Root Formation During Vegetative Propagation in Elepidote Rhododendrons. Vol 723 Biology. *Acta Universitatis Latviensis*, 723, pp. 155–162.

Meyer Jr, M. M. (1977, September). Grafting and Budding as an Educational Exercise Producing Novelty Ornamental Plants. *The Plant Propagator*, pp. 12–13.

Mezitt, R. W. (1973). Grafting to Obtain Unusual Shapes and Forms. *The International Plant Propagators' Society Combined Proceedings*, 23, pp. 335–338.

Micke, W. C. (1996). *Almond Production Manual*. UCANR Publications.

Millais, D. (1996). Propagation of Rhododendrons at Millais Nurseries. *The International Plant Propagators' Society Combined Proceedings*, 46, pp. 190–193.

Miller, W. P. (1979). The Production of Rhododendrons by Grafting. *The International Plant Propagators' Society Combined Proceedings*, 29, pp. 158–161.

Moore, J. C. (1963). Propagation of Chestnuts and Camellia by Nurse Seed Grafts. *The International Plant Propagators' Society Combined Proceedings*, 13, pp. 141–143.

Mudge, K. W. (2013). Anatomy and Physiology of Grafting Union Formation. Retrieved from The *How, When & Why of Grafting* (Online Course from Department of Horticulture, Cornell University): http://www.hort.cornell.edu/grafting/anatomy.physiology/BiolGB.html.

Mudge, K., Janick, J., Scofield, S., & Goldschmidt, E. E. (2009, October 7). A History of Grafting (J. Janick, Ed.). *Horticultural Reviews*, 35, pp. 449–470.

Mundey, G. (1952, December). A Method of Propagating Magnolias. *Journal of the Royal Horticultural Society*, pp. 449–450.

Murphy, N. J. (2005). Propagation of *Cercis canadensis* Forest Pansy. *The International Plant Propagators' Society Combined Proceedings*, 55, pp. 273–276.

Mustard, M. J., & Lynch, S. J. (1977). Nursery Propagation and the Anatomical Union of Cleft Grafted Gardenias. *Proceedings of the Florida State Horticultural Society*, 90, pp. 347–349.

National Mechanical Insulation Committee (NMIC). (2016, June). Mechanical Insulation Design Guide – Materials and Systems. Retrieved from Whole Building Design Guide (WBDG) – a program of the National Institute of Building Sciences: https://www.wbdg.org/design/midg_materials.php.

Nelson, S. H. (1968). Incompatibility Survey Among Horticultural Plants. *The International Plant Propagators' Society Combined Proceedings*, 18, pp. 343–350.

Nesbitt, M. L., Goff, W. D., & Stein, L. A. (2002, April–June). Effect of Scionwood Packing Moisture and Cut-End Sealing on Pecan. *HortTechnology*, 12(2), pp. 257–260.

Neubauer, H. (1998). Field Propagation of Cercis and Hamamelis. *The International Plant Propagators' Society Combined Proceedings*, 48, pp. 409–411.

Ohse, B., Hammerbacher, A., Seele, C., Meldau, S., Reichelt, M., Ortmann, S., & Wirth, C. (2017, February). Salivary Cues: Simulated Roe Deer Browsing Induces Systemic Changes in Phytohormones and Defence Chemistry in Wild-Grown Maple and Beech Saplings. *Functional Ecology*, 31(2), pp. 340–349.

Okken, G. (1998). Review of Scion/Understock Compatibilities Results at Okken Nurseries. *The International Plant Propagators' Society Combined Proceedings*, 48, pp. 368–370.

Omar, W. (1962). The Grafting of *Acer palmatum*. *The International Plant Propagators' Society Combined Proceedings*, 12, pp. 256–257.

Orange Pippin Ltd. (2018). Rootstocks for Apple Trees. Retrieved February 2018, from Orange Pippin Trees: https://www.orangepippintrees.co.uk/articles/rootstocks-for-apple-trees.

Orel, G., & Curry, T. (2017). New Camellia Species from Vietnam. *The Plantsman*, 16(1), pp. 37–43.

Parks, C. (2017). *Camellia species*. Retrieved April 2017, from American Camellia Society: https://www. americancamellias.com/care-culture-resources/the-camellia-family/camellia-specie.

Patrick, B. (1992). Budding and Grafting of Fruit and Nut Trees at Stark Bro's. *The International Plant Propagators' Society Combined Proceedings*, 42, pp. 354–356.

Peek, R. (1986). Propagating Rhododendron Yakushimanum by Cutting-Grafts. *The International Plant Propagators' Society Combined Proceedings*, 36, pp. 330–332.

Pellekooren, A. (1974). Grafting of *Picea*. Boskoop: Jaarboek, Proefstation voor de Boomkwkerij.

Perquin, F. W. (1974). Enten Van *Betula*. Boskoop: Jaarboek, Proefstation voor de Boomkwekerij.

Perquin, F. W. (1975). Enten Van *Betula*. Boskoop: Jaarboek, Proefstation voor de Boomkwekerij.

Pigott, D. (2012). *Lime-Trees and Basswods. A Biological Monograph of the Genus Tilia*. Cambridge, UK: Cambridge University Press.

Pontikis, C. A., Papalexandris, C. X., & Aristeridou, M. (1985, March). The Effect of BA and GA3 on Patch Budding Success of Persian Walnut Seedlings. *The Plant Propagator*, 31(1), pp. 13–14.

Proefstation voor de Boomkwekerij. (1982). *Onderstammen voor Tilia*. Boskoop: Jaarboek, Proefstation voor de Boomteelt en het Stedlijk Groen.

Proefstation voor de Boomkwekerij. (1984). *Vroeg oculeren van sierkersen met gekoeld hout*. Boskoop: Jaarboek, Proefstation voor de Boomteelt en het Stedelijk Groen.

Ramsbottom, A., & Toogood, A. (1999, November). Some Like It Hot. *The Garden*, pp. 850–851.

Ravesloot, M. B. (1995). 7.5 Het enten van laanbomen. Boskoop: Jaarboek, Proefstation voor de boomteelt en het stedelijk groen.

Reath, D. (1982). Propagation of Tree Peonies. *The International Plant Propagators' Society Combined Proceedings*, 32, pp. 512–515.

Rehder, A. (1940). *Manual of Cultivated Trees and Shrubs Hardy in North America* (2nd ed.). MacMillan Company.

Richards, M. (1972, June). Bare Root Grafting of Cedrus. *The Plant Propagator*, 18(2), p. 8.

Roller, J. B. (1973). Program for Growing Crabapples. *The International Plant Propagators' Society Combined Proceedings*, 23, pp. 333–334.

Ruesink, J. B., & van Kuik, A. J. (1991). *9.2. Gebruik van phytophthora resistente Chamaecyparis als onderstam voor Chamaecyparis lawsoniana en Chamaecyparis obtusa Cultivars*. Boomteelt Praktijkonderzoek Jaaverslag Boskoop. Boskoop: Boomteelt Prakijkonderzoek.

Ryan, F. G. (1966). Grafting *Eucalyptus ficifolia*. *The Plant Propagator*, 12(2), pp. 4–6.

Sacramento Valley Orchard Source. (2016). Selecting the Right Clonal Rootstock for Managing Soil and Pest Problems. Retrieved December 2017, from Sacramento Valley Orchard: http://www.sacvalleyorchards. com/walnuts/.

Santamour Jr, F. S. (1988a). Graft Compatibility in Woody Plants: An Expanded Perspective. *Journal of Environmental Horticulture*, 6(1), pp. 27–32.

Santamour Jr, F. S. (1988b). Cambial Peroxidase Enzymes Related to Graft Compatibility in Red Oak. *Journal of Environmental Horticulture*, 6(3), pp. 87–93.

Santamour Jr, F. S. (1996). Potential Causes of Graft Incompatibility. *The International Plant Propagators' Society Combined Proceedings*, 46, pp. 339–342.

Savella, L. (1977). Poly Tent Versus Open Bench Grafting. *The International Plant Propagators' Society Combined Proceedings*, 27, pp. 369–371.

Schepker, H. (2016). The Inkaro Story. In P. Hayward (Ed.), *Rhododendrons Camellias and Magnolias Group Yearbook* (pp. 186–196).

Schupp, J., & Crassweller, R. (2018). Apple Rootstocks: Capabilities and Limitations. Retrieved February 2018, from PennState Extension: https://extension.psu.edu/apple-rootstocks-capabilities-and-limitations.

Scott, M. (1979). Taking Stock – Management of Stock Blocks. *The International Plant Propagators' Society Combined Proceedings*, 29, pp. 233–241.

Scott, M. (2003). New Rootstocks for the Nursery Industry. *The International Plant Propagators' Society Combined Proceedings*, 53, pp. 501–504.

Serr, E. F. (1964). Walnut Rootstock. *The International Plant Propagators' Society Combined Proceedings*, 14, pp. 327–329.

Serr, E. F. (1968, February). Dwarfing Interstocks for Persian Walnuts. *The International Plant Propagators' Society*, 14(1), pp. 10–13.

Sexton, D. (1965). Grafting of Selected Ornamentals. *The International Plant Propagators' Society Combined Proceedings*, 15, pp. 278–279.

Sheat, W. G. (1965). *Propagation of Trees Shrubs and Conifers*. London: Macmillan and Co. Ltd.

Shippy, W. B. (1930, April). Influence of Environment on Callousing of Apple Trees and Grafts. *American Journal of Botany*, 17(4), pp. 290–327.

Sierra Gold Nurseries. (2017a). Cherries – Clonal Rootstocks. Retrieved 2017, from Sierra Gold Nurseries: https://www.sierragoldtrees.com/cherry-rootstocks.

Sierra Gold Nurseries. (2017b). Walnuts Clonal Rootstocks. Retrieved December 2017, from Sierra Gold Nurseries: https://www.sierragoldtrees.com/walnut-rootstocks.

Sierra Gold Nurseries. (2017c, November). Almond Rootstocks. Retrieved 2017, from Sierra Gold Nurseries: https://www.sierragoldtrees.com/almond-rootstocks.

Standbrook, N. (2005). Experiences in Propagation of *Cornus mas* and *Cornus florida* Cultivars. *The International Plant Propagators' Society Combined Proceedings*, 55, pp. 268–269.

Starrett, M. C. (2008). Chapter 12 – Techniques of Grafting. Retrieved from University of Vermont – Commercial Plant Propagation PSS 138 Spring 2008 Syllabus: http://www.uvm.edu/~mstarret/plant-prop/chapter12.pps.

Stebbins, R. (1995). OSU Extension Catalogue - Choosing Pear Rootstocks for the Pacific Northwest. Retrieved 2017, from OSU Outreach and Engagement Extension Service: https://catalog.extension.oregonstate.edu/pnw341.

Stegemann, S., & Bock, R. (2009, May). Exchange of Genetic Material Between Cells in Plant Tissue Grafts. *Science*, 324(5927), pp. 649–651.

Stoner, H. E. (1974). Grafting, from Scion to Plant. *The International Plant Propagators' Society Combined Proceedings*, 24, pp. 401–405.

Sweet, J. B., Goodall, R. A., & Campbell, A. I. (1978). Improvement of Hardy Nursery Stocks. *The International Plant Propagators' Society Combined Proceedings*, 28, pp. 220–223.

Sziklai, O. (1967). Grafting Techniques in Forestry. *The International Plant Propagators' Society Combined Proceedings*, 17, pp. 124–129.

Teese, A. (1979). Grafting Maples. *The International Plant Propagators' Society Combined Proceedings*, 29, pp. 580–582.

Tereschenko, A. P., & Duarte, J. (1998). *Patent No. US5832662 – Method for Grafting Rootstock Using Improved Grafting Film*. USA.

The Garden. (2017, February). Newton's apple trees to link UK scientists. *The Garden*, p. 11.

The Plantsman. (2016, December). Plant Focus. Wentworth elm rediscovered at Holyrood. *The Plantsman (New Series)*, 15(Part 4), p. 209.

Thompson, B. (2005). Understock- Keys to Success. *The International Plant Propagators' Society Combined Proceedings*, 55, p. 380.

Thompson, W. D. (1998). Summer Grafting. *The International Plant Propagators' Society Combined Proceedings*, 48, pp. 345–346.

Thomsen, A. (1978). Propagation of Conifers by Cuttings and Grafting. *The International Plant Propagators' Society Combined Proceedings*, 28, pp. 215–220.

Thorburn, G. C. (1973). Certain Aspects of Propagation in Holland and Germany. *The International Plant Propagators' Society Combined Proceedings*, 23, pp. 167–170.

Trehane, J. (1998). *Camellias the Complete Guide to their Cultivation and Use*. London: B.T. Batsford Ltd.

Tubesing, C. E. (1988). Strange Grafts I Have Known. *The International Plant Propagators' Society Combined Proceedings*, 38, p. 186.

Twombly, K. (1996). Trials and Tribulations of Producing *Acer pensylvanicum* Erythrocladum. *The International Plant Propagators' Society Combined Proceedings*, 46, pp. 532–533.

Upchurch, B. (2008). The Mysteries of Grafting and Some Forgotten Basics. *The International Plant Propagators' Society Combined Proceedings*, 58, pp. 203–207.

Upchurch, B. (2009). The Basics of Grafting. *The International Plant Propagators' Society Combined Proceedings*, 59, pp. 577–582.

Upchurch, B. L. (2005). Grafting Deciduous Plants: Before and Aftercare. *The International Plant Propagators' Society Combined Proceedings*, 55, pp. 546–549.

Valavanis, W. N. (1982). Propagation of Plant Cultivars with "Yatsubusa" Characteristics. *The International Plant Propagators' Society Combined Proceedings*, 32, pp. 503–506.

Vanderbilt, R. (1960). The Establishment and Maintenance of a Stock Block of Hardy Hybrid Rhododendrons. *The International Plant Propagators' Society Combined Proceedings*, 10, pp. 99–100.

van Doesburg, J. (1960). *Het Enten van Picea breweriana*. Boskoop: Jaarboek, Proefstation voor de Boomkwekerij.

van Doesburg, J. (1961). *Het Enten Van Berken*. Boskoop: Jaarboek, Proefstation voor de Boomkwekerij.

van Doesburg, J. (1962). *Het Enten Van Berken*. Boskoop: Jaarboek, Proefstation voor de Boomkwekerij.

van Doesburg, J., & de Vogel, P. (1959). *Het Enten van Berken in Kas en Bak*. Boskoop: Jaarboek, Proefstation voor de Boomkwekerij.

van Doesburg, J., & Pellekooren, A. (1962). Het Enten en Oculeren van Aralia. Boskoop: Jaarboek, Proefstation voor de Boomkwekerij.

van Doesburg, J., & van Elk, B. C. (1963). Het Enten van Eiken. Boskoop: Jaarboek, Proefstation voor de Boomkwekerij.

van Doesburg, H. H., Schalk, G., & Stam, J. (1977). *Experiments at the Research Station on Avenue Trees*. Boskoop: Jaarboek, Proefstation voor de Boomteelt en het Stedelijk Groen.

van Elk, B. C. (1965a). *Enten van Picea*. Boiskoop: Jaarboek, Proefstation voor de Boomkwekerij.

van Elk, B. C. (1965b). *Het Enten van Cornus*. Boskoop: Jaarboek, Proefstation voor de Boomkwekerij.

van Elk, B. C. (1966). *Het Enten van Betula*. Boskoop: Jaarboek, Proefstation voor de Boomkwekerij.

van Elk, B. C. (1967). *Enten*. Boskoop: Jaarboek, Proefstation voor de Boomkwekerij.

van Elk, B. C. (1968a). *Enten Acer*. Boskoop: Jaarboek, Proefstation voor de Boomkwekerij.

van Elk, B. C. (1968b). *Enten Picea Cultivars*. Boskoop: Jaarboek, Proefstation voor de Boomkwekerij.

van Elk, B. C. (1978). *Enten*. Boskoop: Jaarboek, Proefstation voor de Boomkwekerij.

van Gelderen, C. (2001, Summer). Hidden Frontiers in Acers (P. Gregory, Ed.). *The Maple Society Newsletter*, 11(2), pp. 12–17.

van Gelderen, D. M., de Jong, P. C., & Oterdoom, H. J. (1994). *Maples of the World*. Portland, Oregon, USA: Timber Press.

Vasudeva, R., & Reddy, B. M. (2015). An Informal Network of Grafting Experts to Help the Communities to Conserve and Utilize Wild-Aromatic Pickle-Mango (*Mangifera indica*) Diversity in the Central Western Ghats, India. *Acta Horticulturae*, 1101, pp. 63–68. Retrieved May 2018, from ActaHortic: https://www.actahort.org/books/1101/1101_10.htm.

Verhoeven, P. A. (1985a). *Onderstammen voor Chamaecyparis nootkatensis Pendula*. Boskoop: Jaarboek, Proefstation voor Boomteelt en het Stedelijk Groen.

Verhoeven, P. A. (1985b). Onderstammen *voor* Hamamelis – Cultivars. Boskoop: Jaarboek, Proefstation voor de Boomteelt en Het Stedelijk Groen.

Vermeulen, J. P. (1983). Side Veneer Grafting. *Combined Proceedings, International Plant Propagators' Society*, 33, pp. 422–425.

Vertrees, J. D. (1975). Observations on *Acer circinatum* Propagation. *The Plant Propagator*, 21(4), p. 11.

Vertrees, J. D. (1978). Notes on Propagation of Certain Acers. *The International Plant Propagators' Society Combined Proceedings*, 28, pp. 95–96.

Vertrees, J. D. (1991). Understocks for Rare Acer Species. *The International Plant Propagators' Society Combined Proceedings*, 41, pp. 272–274.

Villanueva, G. G., & Maniaci, S. (1996). Machine Grafting of Grapevines Using the Spinks Grafting Machine. *The International Plant Propagators' Society Combined Proceedings*, 46, pp. 309–310.

Villis, N. (1975). Propagation of Aesculus X neglecta Erythroblastos. *The Plant Propagator*, 21(2), p. 5.

Wagner, G. (1967). Speeding Production of Hard-to-Root Conifers. *The International Plant Propagators' Society Combined Proceedings*, 17, pp. 113–114.

Waite, H., Grainaje, D., Whitelaw-Weckert, M., Torlay, P., & Hardie, W. J. (2013). Soaking Grapevines Cuttings in Water: A Potential Source of Cross Contamination by Micro-Organisms. *Phytopathologia Mediterranea*, 52(2), pp. 359–368.

Waite, H., Whitelaw-Weckert, M., & Torley, P. (2015). Grapevine Propagation: Principles and Methods for the Propagation of High-Quality Grapevine Planting Material. *New Zealand Journal of Crop and Horticultural Science*, 43(2), pp. 144–161.

Warren, P. (1973, September). Propagation of Cercis Cultivars by Summer Budding. *The Plant Propagator*, 19(3), pp. 16–17.

Washington State University. (2017a). Rootstocks – Sweet Cherry. Retrieved 2017, from Washington State University: http://treefruit.wsu.edu/varieties-breeding/rootstocks/.

Washington State University. (2017b). Rootstocks – Pear. Retrieved 2017, from WSU Tree Fruit Comprehensive Tree Fruit Site: http://treefruit.wsu.edu/varieties-breeding/rootstocks/.

Washington State University. (2017c). Rootstocks for Apple. Retrieved February 2018, from WSU Tree Fruit Comprehensive Tree Fruit Site: http://treefruit.wsu.edu/varieties-breeding/rootstocks/.

Watanabe, S. (1999). The Japanese Tradition of Grafting. *The International Plant Propagators' Society Combined Proceedings*, 49, pp. 657–658.

Wedge, D. (1956). How We Propagate French Lilacs at the Wedge Nursery. *Plant Propagators' Society Proceedings*, 6, pp. 75–78.

Wedge, D. (1977). Propagation of Hybrid Lilacs. *The International Plant Propagators' Society Combined Proceedings*, 27, pp. 4332–4434.

Weguelin, D. (1972). Bench Grafting Discussion Group. *The International Plant Propagators' Society Combined Proceedings*, 22, p. 270.

Weinstock, D. (2016, September 15). The Big Push for Dwarfing Rootstock. *The Good Fruit Grower*, p. 1.

Wells, J. S. (1955). *Plant Propagation Practices*. New York: MacMillan Company.

Wells, R. (1986). Camellia Grafting at Monrovia Nursery. *The International Plant Propagators' Society Combined Proceedings*, 36, pp. 99–101.

Wells, R. H. (1980). Conifer and Magnolia Grafting. *The International Plant Propagators' Society Combined Proceedings*, 30, pp. 89–92.

Widmoyer, F. B. (1962). Anatomical Aspects of Budding and Grafting. *Plant Propagators Society*, 12, pp. 132–135.

Wiegrefe, S. J. (2003). Graft Compatibilities of Hornbeam (Carpinus) Species and Hybrids. *The International Plant Propagators' Society Combined Proceedings*, 53, pp. 549–554.

Willard, F. (1968). Notes on the Grafting of Picea pungens Kosteriana. *The International Plant Propagators' Society Combined Proceedings*, 18, pp. 84–87.

Wilson, K. (2000, March). Apple Rootstocks Agdex 211/36. Retrieved February 2018, from Ontario Ministry of Agriculture,Food and Rural Affairs: http://www.omafra.gov.on.ca/english/crops/facts/00-007.htm.

Wolff, R. P. (1973). Successes and Failures in Grafting Japanese Maples. *The International Plant Propagators' Society Combined Proceedings*, 23, pp. 339–345.

Wood, T. (1996). The Right Rootstock for a Good Graft Stick. *The International Plant Propagators' Society Combined Proceedings*, 46, pp. 167–168.

Xiang, Q.-Y., Thomas, D. T., Zhang, W., Manchester, S. R., & Murrell, Z. (2006, February). Species Level Phylogeny of the Genus Cornus (Cornaceae) Based on Molecular and Morphological Evidence. *Taxon – International Association for Plant Taxonomy*, 55(1), pp. 9–30.

Yoshikawa, F. T., Ramming, D. W., & LaRue, J. H. (1989). *Peaches, Plums and Nectarines, Growing and Handling for Fresh Market* (J. H. LaRue, & R. S. Johnson, Eds.). University of California.

Zaczek, J. J., Heuser Jr, C. W., & Steiner, K. C. (1999). Low Irradiance During Rooting Improves Propagation of Oak and Maple Taxa. *Journal of Environmental Horticulture*, 17(3), pp. 130–133.

Zenginbal, H., Ozcan, M., & Demir, T. (2006). An Investigation on the Propagation of Kiwifruit (Actinidia deliciosa) by Grafting Under Turkey Ecological Conditions. I. *International Journal of Agricultural Research*, 1(6), pp. 597–602.

Zhang, Y.-L., Li, X.-P., Guo, W.-Z., & Fen, S.-C. (2016). Cluster-Flowering Camellias in Shanghai Botanical Garden. *Rhododendrons Camellias and Magnolias Group*. Royal Horticultural Society, pp. 215–220.

Ziraldo, D. J. (1974). Grafting of Grape Vines. *The International Plant Propagators' Society Combined Proceedings*, 24, pp. 367–369.

Plant Index

Subject Index